This book provides a comprehensive overview of the fundamental principles and applications of semiconductor diode laser arrays. All of the major types of arrays are discussed in detail, including coherent, incoherent, edge- and surface-emitting, horizontal- and vertical-cavity, individually addressed, lattice-matched or strained-layer systems.

The first five chapters deal with various aspects of coherent arrays, covering such topics as lasers, amplifiers, external-cavity control, theoretical modelling and operational dynamics. Spatially incoherent arrays are then discussed, with particular emphasis on their high-power capability and reliability, and on practical packaging and pumping schemes. After dealing with vertical-cavity surface-emitter arrays, which have potential applications in parallel optical-signal processing, the book concludes with a description of individually addressable arrays of use in multi-channel optical recording.

Diode laser arrays have a host of actual and potential applications in a variety of fields, and this detailed review of their properties and uses will be of great value to engineers and scientists studying applications of such arrays, as well as to graduate students and established researchers in physics, chemistry and electrical engineering.

CAMBRIDGE STUDIES IN MODERN OPTICS: 14

Series Editors

P. L. KNIGHT

Optics Section, Imperial College of Science, Technology and Medicine

A. MILLER

Department of Physics and Astronomy, University of St Andrews

DIODE LASER ARRAYS

TITLES IN PRINT IN THIS SERIES

Interferometry (second edition)
W. H. Steel

Optical Holography – Principles, Techniques and Applications
P. Hariharan

Fabry–Perot Interferometers
G. Hernandez

Holographic and Speckle Interferometry (second edition)
R. Jones and C. Wykes

Laser Chemical Processing for Microelectronics
edited by K. G. Ibbs and R. M. Osgood

The Elements of Nonlinear Optics
P. N. Butcher and D. Cotter

Optical Solitons – Theory and Experiment
edited by J. R. Taylor

Particle Field Holography
C. S. Vikram

Ultrafast Fiber Switching Devices and Systems
M. N. Islam

Optical Effects of Ion Implantation
P. D. Townsend, P. J. Chandler and L. Zhang

Diode Laser Arrays
edited by D. Botez and D. R. Scifres

DIODE LASER ARRAYS

Edited by

DAN BOTEZ

Philip Dunham Reed Professor,
Department of Electrical and Computer Engineering
University of Wisconsin, Madison

and

DON R. SCIFRES

President, SDL Inc.

CAMBRIDGE UNIVERSITY PRESS
Cambridge, New York, Melbourne, Madrid, Cape Town, Singapore, São Paulo

Cambridge University Press
The Edinburgh Building, Cambridge CB2 2RU, UK

Published in the United States of America by Cambridge University Press, New York

www.cambridge.org
Information on this title: www.cambridge.org/9780521419758

First published 1994
This digitally printed first paperback version 2005

A catalogue record for this publication is available from the British Library

Library of Congress Cataloguing in Publication data
Diode laser arrays / edited by Dan Botez and Don R. Scifres.
p. cm. – (Cambridge studies in modern optics ; 14)
Includes bibliographical references and index.
ISBN 0 521 41975 1
1. Semiconductor lasers. I. Botez, Dan. II. Scifres, Don R.
III. Series.
TA1700.D56 1994
621.36′6 – dc20 93-41258 CIP

ISBN-13 978-0-521-41975-8 hardback
ISBN-10 0-521-41975-1 hardback

ISBN-13 978-0-521-02255-2 paperback
ISBN-10 0-521-02255-X paperback

Contents

List of contributors — *page* xi
Preface — xiii

1 Monolithic phase-locked semiconductor laser arrays — 1
DAN BOTEZ
1.1 Introduction — 1
1.2 Array modes — 5
1.3 Arrays of positive-index guides — 7
1.4 Arrays of antiguides — 22
1.5 Future prospects for all-monolithic antiguided arrays — 62
References — 67

2 High-power coherent, semiconductor laser, master oscillator power amplifiers and amplifier arrays — 72
DAVID F. WELCH AND DAVID G. MEHUYS
2.1 Introduction — 72
2.2 Monolithically integrated MOPAs and amplifier arrays — 73
2.3 Monolithic active-grating master oscillator power amplifiers — 85
2.4 Discrete-element MOPA performance — 96
2.5 Integrated flared-amplifier MOPAs — 109
2.6 Conclusions — 119
References — 120

3 Microoptical components applied to incoherent and coherent laser arrays — 123
JAMES R. LEGER
3.1 Requirements of incoherent and coherent systems — 124
3.2 Microoptics — 128
3.3 Coherent techniques — 139
3.4 Incoherent techniques — 162
3.5 Conclusion — 175
References — 176

4 Modeling of diode laser arrays 180
G. RONALD HADLEY
4.1. Introduction 180
4.2 Edge-emitting arrays 182
4.3 Vertical-cavity surface-emitting arrays 212
4.4 Conclusion 221
Appendix 222
References 223

5 Dynamics of coherent semiconductor laser arrays 226
HERBERT G. WINFUL AND RICHARD K. DEFREEZ
5.1 Introduction 226
5.2 Coupled lasers 226
5.3 Coupled-mode theory for laser array dynamics 228
5.4 A propagation model for laser array dynamics 234
5.5 Experimental observations 241
5.6 Conclusions 252
References 253

6 High-average-power semiconductor laser arrays and laser array packaging with an emphasis on pumping solid state lasers 255
RICHARD SOLARZ, RAY BEACH, BILL BENETT, BARRY FREITAS, MARK EMANUEL, GEORG ALBRECHT, BRIAN COMASKEY, STEVE SUTTON AND WILLIAM KRUPKE
6.1 Introduction 255
6.2 High-power semiconductor laser array requirements 256
6.3 Approaches to packaging high-average-power two-dimensional semiconductor laser arrays 263
6.4 Performance comparison of devices fabricated to date 268
6.5 Laser diode array performance 276
6.6 Fundamental limits to high-average-power operation and potential improvements 287
References 291

7 High-power diode laser arrays and their reliability 294
D. R. SCIFRES AND H. H. KUNG
7.1 Introduction 294
7.2 Failure mechanisms of high-power laser arrays 308
7.3 Lifetests of high-power diode arrays 316
7.4 Environmental tests of packaged diode laser arrays 326
7.5 Future directions for high-power diode laser array research 329
References 330

8 Strained layer quantum well heterostructure laser arrays 336
JAMES J. COLEMAN
8.1 Introduction 336
8.2 Strained layer metallurgy and critical thickness 338
8.3 The effects of strain on emission wavelength 341
8.4 Threshold current density in strained layer structures 344

8.5 Antiguiding in strained layer lasers 351
8.6 Reliability 355
8.7 Conventional edge emitter and surface emitter laser arrays 356
8.8 Leaky mode and antiguided strained layer laser arrays 359
8.9 Other strained layer laser materials systems 360
References 363

9 Vertical cavity surface-emitting laser arrays 368
CONNIE J. CHANG-HASNAIN
9.1. Introduction 368
9.2 Vertical cavity surface-emitting laser design 370
9.3 VCSEL array fabrication 373
9.4 VCSEL characteristics 377
9.5 Two-dimensional multiple wavelength VCSEL array 387
9.6 Phase-locked VCSEL arrays 400
9.7 Addressable VCSEL array 402
9.8 Future prospects 410
References 411

10 Individually addressed arrays of diode lasers 414
DONALD B. CARLIN
10.1 Introduction 414
10.2 Multichannel optical recording 415
10.3 Individually addressable arrays 416
10.4 Interelement isolation and array packaging 424
10.5 Representative array structures and reported results 430
10.6 Surface emitting arrays 435
10.7 Summary 439
References 440

Index 444

Contributors

Chapter 1
Monolithic phase-locked semiconductor laser arrays

DAN BOTEZ
Department of Electrical and Computer Engineering, University of Wisconsin at Madison, 1415 Johnson Drive, Madison, WI 53706-1691, USA

Chapter 2
High-power coherent, semiconductor laser, master oscillator power amplifiers and amplifier arrays

DAVID F. WELCH AND DAVID G. MEHUYS
SDL Inc., 80 Rose Orchard Way, San Jose, CA 95134, USA

Chapter 3
Microoptical components applied to incoherent and coherent laser arrays

JAMES R. LEGER
Department of Electrical Engineering, University of Minnesota, Minneapolis, MN 55455, USA

Chapter 4
Modeling of diode laser arrays

G. RONALD HADLEY
Sandia National Laboratories, Albuquerque, NM 87185, USA

Chapter 5
Dynamics of coherent semiconductor laser arrays

HERBERT G. WINFUL
Department of Electrical Engineering and Computer Science, University of Michigan, Ann Arbor, MI 48109, USA
and

RICHARD K. DEFREEZ
Department of Applied Physics and Electrical Engineering, Oregon Graduate Institute of Science and Technology, Beaverton, OR 97006, USA

Chapter 6
High-average-power semiconductor laser arrays and laser array packaging with an emphasis on pumping solid state lasers

RICHARD SOLARZ, RAY BEACH, BILL BENETT, BARRY FREITAS, MARK EMANUEL, GEORG ALBRECHT, BRIAN COMASKEY, STEVE SUTTON AND WILLIAM KRUPKE
Lawrence Livermore National Laboratory, Livermore, CA 94550, USA

Chapter 7
High-power diode laser arrays and their reliability

D. R. SCIFRES AND H. H. KUNG
SDL Inc., 80 Rose Orchard Way, San Jose, CA 95134, USA

Chapter 8
Strained layer quantum well heterostructure laser arrays

JAMES J. COLEMAN
Microelectronics Laboratory, University of Illinois, Urbana, IL 61801, USA

Chapter 9
Vertical cavity surface-emitting laser arrays

CONNIE J. CHANG-HASNAIN
Department of Electrical Engineering, Stanford University, Stanford, CA 94305, USA

Chapter 10
Individually addressed arrays of diode lasers

DONALD B. CARLIN
David Sarnoff Research Center, 201 Washington Road, Princeton, NJ 08540, USA

Preface

Diode laser arrays have been studied for the last 15 years. Initially the interest was in creating phase-locked arrays of high coherent powers delivered in narrow, diffraction-limited beams for applications such as free-space optical communications. Around 1983 interest arose in spatially incoherent arrays as efficient pumps for solid-state lasers. This remains nowadays the most widespread use of diode laser arrays. In recent years there has been a demand for individually addressable one-dimensional and two-dimensional arrays to be used in parallel optical-signal processing, optical interconnects, and multichannel optical recording. We have attempted in this book to cover the development and features of all types of arrays demonstrated to date. The first five chapters treat various aspects of coherent arrays: lasers, amplifiers, external-cavity control, modeling and operational dynamics. Chapters 6–8 are dedicated primarily to spatially incoherent arrays. High-power capability, reliability, packaging and pumping schemes are discussed as they relate to the major application: solid-state-laser pumping. Individually addressable arrays of both the surface-emitting type (i.e. vertical-cavity surface emitters) and edge-emitting type are treated in Chapters 9 and 10, respectively.

Coherent arrays have proved a particularly challenging task. For the first ten years of research (1978–1988) the best that could be achieved was 50–100 mW in a diffraction-limited beam; that is, pretty much the same power as from a single-element device. The breakthrough occurred with the development of resonant leaky-wave-coupled arrays (also called ROW arrays) which allowed for the first time for global coupling (i.e. each element equally coupled to all others) in structures of strong optical-mode confinement. As a result, by 1993 watt-range coherent powers have been demonstrated in both cw and pulsed operation. An alternate approach to achieving high coherent power has been the recent development of master-oscillator power-amplifier (MOPA)-type devices. Grating-outcoupled surface-emitting devices have demonstrated close to 1 W of diffraction-limited peak pulsed power, while edge-emitting MOPAs of the fanout type have demonstrated 3 W cw coherent power (at 5 °C). In parallel to the development of high-power monolithic coherent sources research has been carried out in the area of external-cavity controlled arrays. The best result to date is 0.9 W of diffraction-limited power from arrays of 20 elements coherently coupled in Talbot-type external cavities.

Extensive modelling: exact theory, coupled-mode formalism, Bloch-function method, has been performed for phase-locked arrays. The analysis has been successful in explaining the behavior of both gain- and index-guided arrays. A notable recent success has been the application of the Bloch-function method to antiguided arrays.

Array dynamics has also been an area of intense activity. It has been generally found that evanescent-wave-coupled and Y-junction-coupled phase-locked arrays are basically unstable, while antiguided arrays are fundamentally stable as long as saturable absorption is avoided in the interelement regions.

For the user of high power laser diode arrays in systems products, Chapters 6 and 7 provide insight into both the state-of-the-art performance and the practical characteristics and limitations of these devices in field operation. The performance and tradeoffs between various types of laser array structures such as quasi-cw bars, cw bars, stacked arrays, and two-dimensional surface emitting arrays, among others, are discussed. Limitations on output power and laser reliability are presented as are methods for improving reliability of products in the field under a wide range of environmental conditions.

Strained-layer lasers seem particularly suited to the fabrication of laser diode arrays owing to the increased range of wavelengths available from such devices relative to their fully lattice-matched counterparts. In addition, the reduced threshold currents of such devices allow high power operation with reduced interelement interaction. The physics, capabilities and reliability of strained-layer laser arrays is treated in Chapter 8.

Although studied since 1979 vertical-cavity surface emitters (VCSE) did not really 'take off' until 1988. Ever since, feverish activity at many laboratories around the world has led to low-threshold, low-voltage, efficient devices, which can be configured in two-dimensional (2-D) surface-emitting array geometries. By using a 4 × 4 array and frequency-division multiplexing aggregate bandwidths as high as 80 Gb/s have been achieved. The development of 2-D VCSE array technology is crucial for the practical implementation of parallel optical-signal processing, optical interconnects and optical computing.

Separately addressed single-transverse-mode arrays have found utility in optical data transmission, data storage and data processing. An overview of such arrays used primarily for optical data storage is presented in Chapter 10.

The progress to date for all diode laser array types is thus presented and discussed. The future looks bright for both research and development of semiconductor laser arrays.

1

Monolithic phase-locked semiconductor laser arrays

DAN BOTEZ

1.1 Introduction

Phase-locked arrays of diode lasers have been studied extensively over the last 15 years. Such devices have been pursued in the quest to achieve high coherent powers (>100 mW diffraction limited) for applications such as space communications, blue-light generation via frequency doubling, optical interconnects, parallel optical-signal processing, high-speed, high-resolution laser printing and end-pumping solid-state lasers. Conventional, narrow-stripe (3–4 μm wide), single-mode lasers provide, at most, 100 mW reliably, as limited by the optical power density at the laser facet. For reliable operation at watt-range power levels, large-aperture (≥ 100 μm) sources are necessary. Thus, the challenge has been to obtain single-spatial-mode operation from large-aperture devices, and maintain stable, diffraction-limited-beam behavior to high power levels (0.5–1.0 W).

By comparison with other types of high-power coherent sources (master oscillator power amplifier (MOPA), unstable resonators), phase-locked arrays have some unique advantages: graceful degradation; no need for internal or external isolators; no need for external optics to compensate for phasefront aberrations due to thermal- and/or carrier-induced variations in the dielectric constant; and, foremost, beam stability with drive level due to a strong, built-in, real-index profile. The consequence is that, in the long run, phase-locked arrays are bound to be more reliable than either MOPAs or unstable resonators.

The four basic types of phase-locked arrays are shown schematically in Figure 1.1: leaky-wave-coupled, evanescent-wave-coupled, Y-junction-coupled and diffraction-coupled. Leaky-wave-coupled devices are arrays for which the lasing modes have the major field-intensity peaks in the low-index array regions, so called leaky-array modes[1,2]. The first phase-locked array to be published[3] was a gain-guided array, which by its nature

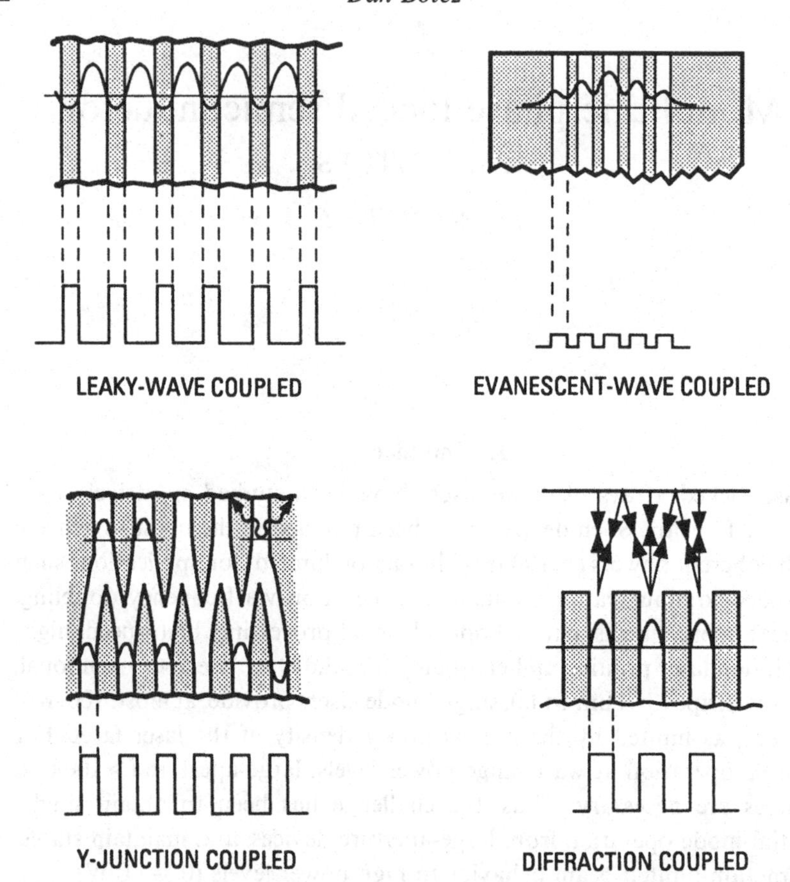

Figure 1.1. Schematic representation of basic types of phase-locked linear arrays of diode lasers.The bottom traces correspond to refractive-index profiles.

is a leaky-wave-coupled device. Evanescent-wave-coupled devices[4] use array modes whose field-intensity peaks reside in the high-index array regions. They have been the subject of intensive research over the 1983–8 time period. Since operation of evanescent-type arrays was difficult to achieve in the in-phase mode (i.e., single central lobe), Y-junction-coupled[5] devices and diffraction-coupled[6] devices were proposed to select in-phase operation via wave interference and diffraction respectively.

Up to 1988 the results were not at all encouraging: maximum diffraction-limited single-lobe powers of ≈50 mW or coherent powers (i.e., fraction of the emitted power contained within the theoretically defined diffraction-limited-beam pattern) never exceeding 100 mW. Thus, the very purpose of fabricating arrays, to surpass the reliable power level of single-element

Figure 1.2. Types of overall interelement coupling in phase-locked arrays: (*a*) series coupling (nearest-neighbor coupling, coupled-mode theory); (*b*) parallel coupling. (After Ref. 91.)

devices, was not achieved. The real problem was that researchers had taken for granted the fact that strong nearest-neighbor coupling implies strong overall coupling. In reality, as shown in Figure 1.2, nearest-neighbor coupling is 'series coupling', a scheme plagued by weak overall coherence and poor intermodal discrimination[7]. Strong overall interelement coupling occurs only when each element couples equally to all others, so-called 'parallel coupling'[7]. In turn, intermodal discrimination is maximized and full coherence becomes a system characteristic. Furthermore, parallel-coupled systems have uniform near-field intensity profiles, and are thus immune to the onset of high-order-mode oscillation at high drive levels above threshold.

Parallel coupling can be obtained in evanescent-wave-coupled devices, but only by weakening the optical-mode confinement, and thus making the devices vulnerable to thermal- and/or injected-carrier-induced variations in the dielectric constant. For both full coherence and stability it is necessary to achieve parallel coupling in structures of strong optical-mode confinement (e.g., built-in index steps ≥ 0.01). As shown below, only strongly guided, leaky-wave-coupled devices meet both conditions. A stable, parallel-coupled source is highly desirable for systems applications since it has graceful degradation; that is, the failure or obscuration of some of its components does not affect the emitted beam pattern.

Table 1.1 gives a brief history of phase-locked array research and development. In 1978 Scifres *et al.*[3] reported the first phase-locked array: a five-element gain-guided device. It was a further eight years before Hadley[8] showed that the modes of gain-guided devices are of the leaky type. Gain-guided arrays have generally operated in leaky, out-of-phase (i.e., two-lobed) patterns with beamwidths many times the diffraction limit due to poor intermodal discrimination. Furthermore, gain-guided devices, being generated simply by the injected-carrier profile, are very weakly

Table 1.1. *Brief history of phase-locked diode laser array research and development*

Time period	Array type	Preferred array mode	Overall interelement coupling	Single-mode selectivity	Max. diffraction-limited power
1978–	Gain-guided	Leaky in- or out-of-phase	Series	Poor	–
1981	Antiguided	Leaky in- or out-of-phase	Series	Poor	–
1983–8	Positive-index-guided	Evanescent out-of-phase	Series	Moderate	0.2 W
1988	Antiguided	Leaky in- or out-of-phase	Series	Moderate	0.2 W
1989–	In-phase resonant antiguided	Leaky in-phase	Parallel	Excellent	2.0 W

Note: The dashes in the final column imply insignificant value.

guided and thus vulnerable to thermal gradients and gain spatial hole burning. The first real-index-guided, leaky-wave-coupled array (i.e., so-called antiguided array) was realized in 1981 by Ackley and Engelmann[9]. While the beam patterns were stable, the lobes were several times the diffraction limit, with in-phase and out-of-phase modes operating simultaneously due to the lack of a mode-selection mechanism. Positive-index-guided arrays came next in array research (1983–8). In-phase, diffraction-limited-beam operation could never be obtained beyond 50 mW output power. Some degree of stability could be achieved in the out-of-phase operating condition such that, by 1988, two groups[10,11] reported diffraction-limited powers as high as 200 mW.

In 1988 antiguided arrays were resurrected and, from the first attempt, researchers obtained 200 mW diffraction-limited in-phase operation[12]. Hope for achieving high coherent powers from phase-locked arrays was rekindled, although there was no clear notion as to how to obtain single-lobe operation. The breakthrough occurred in 1989 with the discovery of resonant leaky-wave coupling[13], which, as shown in Table 1.1 allowed parallel coupling among array elements for the first time, and thus the means of achieving high-power, single-lobe, diffraction-limited operation. The experimental coherent powers quickly escalated such that, to date, up to 2 W has been achieved in a diffraction-limited beam[14].

Looking at Table 1.1 it is apparent that positive-index-guided devices were just one stage in array development, and that what has finally made phase-locked arrays a success has been the discovery of a mechanism for selecting in-phase leaky modes of antiguided structures.

In this chapter we first discuss the modal content of periodic arrays of real-index variations. Then positive-index-guided arrays and coupled-mode theory are treated briefly, mostly for historical reasons. The bulk of the treatment is dedicated to antiguided arrays: Bloch-function analysis, resonant leaky-wave coupling, mode discrimination mechanisms, and relevant results. Finally, the extension of resonant leaky-wave coupling in two dimensions and its implications are presented and discussed.

1.2 Array modes

A monolithic array of phase-coupled diode lasers can be described simply as a periodic variation of the real part of the refractive index. Two classes of mode characterize such a system: evanescent-type array modes, for which the fields are peaked in the high-index array regions; and leaky-type array modes, for which the fields are peaked in the low-index array regions (see Figure 1.3). Another distinction is that while evanescent-wave modes have propagation-constant values between the low and high refractive-index values, leaky modes have propagation-constant values below the low refractive-index value[1,15]. For both classes of mode the locking condition is said to be 'in-phase' when the fields in each element are cophasal, and 'out-of-phase' when fields in adjacent elements are a π phase-shift apart.

Historically, evanescent-type modes were the first to be analyzed, simply because it could be assumed that they arose from the superposition of individual-element wavefunctions, and thus could readily be studied via coupled-mode formalism[16,17]. While coupled-mode theory has proved quite useful in understanding early work on phase-locked arrays it has severe limitations: it does not apply to strongly coupled systems, and it does not cover leaky-type array modes. Ironically, control of leaky modes turns out to be the key for high-power, phase-locked operation. Thus, after many years of use, coupled-mode theory is suddenly obsolete as far as the design and analysis of high-power coherent devices is concerned.

Not being solutions of the popular coupled-mode theory is one major reason why leaky array modes have been overlooked for so long. The other reason is that for few-element (up to five) arrays, leaky modes play a minor role since they are very lossy[15]. However, for high-power devices

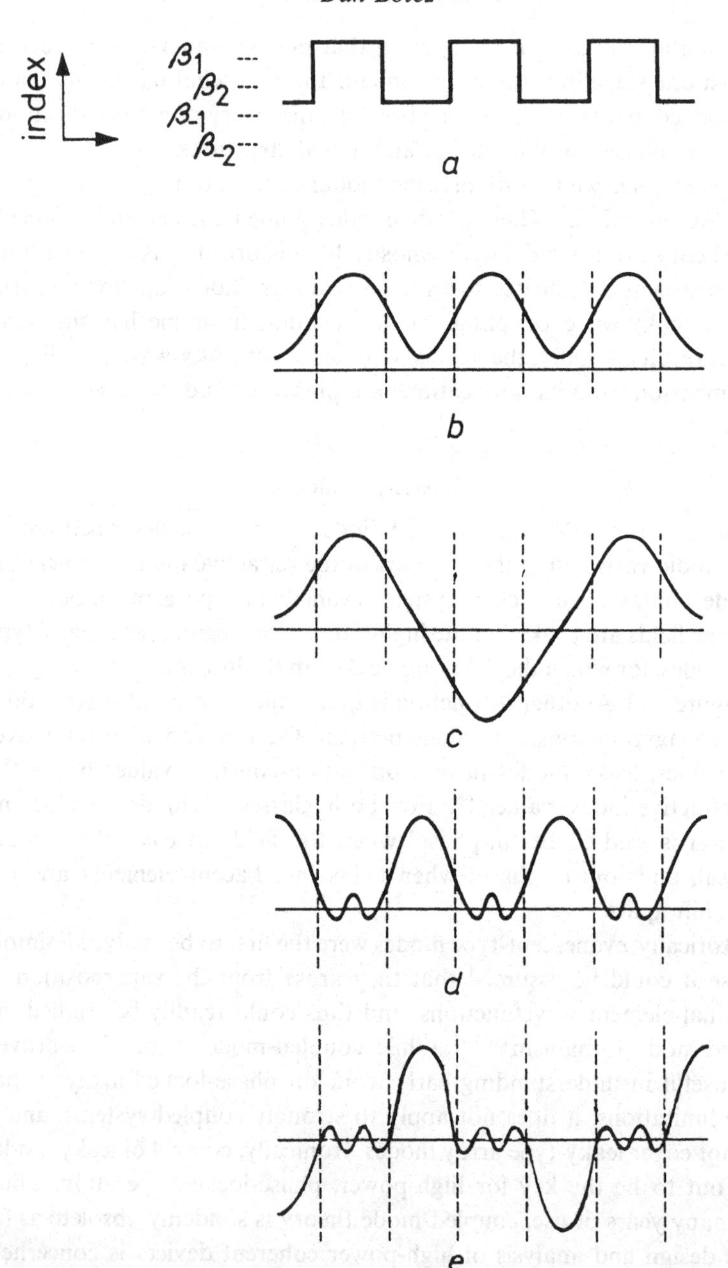

Figure 1.3. Modes of array of periodic real-index variations: (*a*) index profile; (*b*) in-phase evanescent-wave type; (*c*) out-of-phase evanescent-wave type; (*d*) in-phase leaky-wave type; (*e*) out-of-phase leaky-wave type. Respective propagation constants are β_1, β_2, β_{-1} and β_{-2}. (After Ref. 91.)

(ten or more elements), leaky-mode operation is what ensures stable diffraction-limited-beam operation to high drive levels.

Evanescent-type array modes are of importance only when the modal gain in the high-index regions is higher than the modal gain in the low-index regions. Starting with Streifer *et al.*[18], many workers[19–21] have shown that excess gain in the high-index array regions generally favors oscillation in the evanescent out-of-phase mode, in close agreement with experimental results. There is, however, one major limitation: the built-in index step, Δn, has to be below the cutoff for high-order (element) modes[18,22]. For typical devices, $\Delta n \leq 5 \times 10^{-3}$. In turn, the devices are sensitive to gain spatial hole burning[23] and thermal gradients.

Leaky-type modes are favored to lase when gain is preferentially placed in the low-index regions[1,24,25]. Unlike evanescent-type modes there is no limitation on Δn; that is, no matter how high Δn is, the modes favored to lase comprise fundamental element modes coupled in-phase *or* out-of-phase[24]. This fact has two important consequences. First, one can fabricate single-lobe-emitting structures of high-index steps (0.05–0.02), that is, stable against thermal- and/or carrier-induced index variations. Second, it becomes clear why predictions of coupled-mode theory that excess gain in the low-index array regions[18,22] favor an in-phase evanescent-type mode have failed. The simple reason is that leaky modes could not be taken into account in coupled-mode analysis. In fact, the very array structures proposed by Streifer *et al.* for in-phase-mode operation[18], when analyzed using exact theory by Fujii *et al.*[26], were found to primarily favor operation in an out-of-phase leaky mode. To analyze both array-mode types for a given structure one has to use either exact theory[1,26] or the Bloch-function method[24,27,28].

We classify phase-locked arrays into two types: positive-index-guided arrays, for which evanescent-mode operation dominates; and negative-index-guided (antiguided) arrays, for which leaky-mode operation dominates.

1.3 Arrays of positive-index guides

1.3.1 History and devices types

Around 1983 it became apparent that gain-guided arrays, while relatively easy to fabricate, consistently provided unstable beam patterns with increasing drive or change in drive conditions. These problems were caused by the extremely weak optical-mode confinement inherent to gain-guided devices. In order to fabricate stable devices, researchers concentrated their work on arrays with 'built-in' refractive-index profiles. The simplest approach was to preferentially etch a series of ridges in planar double-heterojunction (DH) material (Figure 1.4(*a*)), thus providing mode confinement via variations in the effective index as the cladding-layer thickness varied[29–32]. The next level of sophistication was to regrow material over such patterns as to create buried-ridge waveguide arrays[33,34] (Figure 1.4(*b*)). Two other index-guided array types depended on the crystal growth properties of metal–organic vapor-phase epitaxy (MOVPE) and liquid-phase epitaxy (LPE) over channeled substrates (Figure 1.4(*d*) and (*c*), respectively). When using MOVPE, the substrate channels are replicated identically and a constant-thickness active layer with periodic bends is obtained[35]. The bends in the active layer create lateral mode control[36]. When using LPE, channels readily fill up such that a nonuniform-thickness cladding layer is achieved[37]. Waveguides are created in the vicinity of the channel by the proximity of the substrate on both sides of the channel[37–40]. Yet another means of achieving arrays of positive-index guides was preferential impurity diffusion (Figure 1.4(*e*,*f*)): Zn for DH-type material[4] and Si for quantum-well material[41]. In either case, dopant diffusion lowers the index locally, thus indirectly creating arrays of positive-index guides.

All devices were created with the idea of causing coupling via evanescent-wave overlap, that is, to excite only the evanescent-type array modes (see Figure 1.3(*b*, *c*)). By-and-large, this happened since the very fabrication of these early arrays created lossy low-index array regions, thus giving no chance for leaky-type array modes to be excited.

With very few exceptions[33,34,39], the devices generally operated in the out-of-phase condition, simply because the out-of-phase evanescent mode, having field nulls in the lossy interelement regions, has better field overlap with the 'gainy' element regions than does the in-phase mode[18]. However, this was only part of the problem. The other part had to do with the stability of the beam quality; that is, starting with near-diffraction-limited beams at threshold the devices' beam patterns rapidly broadened above 1.5 × threshold, such that the coherent powers never exceeded ≈50 mW. Two quests ensued: the quest to obtain in-phase-mode operation controllably, and the quest to achieve beam stability to high drive levels above threshold.

The goal of achieving single-lobe operation was pursued in two ways: varying the gain and/or index (lateral) profile in longitudinally uniform devices to favor in-phase evanescent-mode operation; and creating longitudinally nonuniform devices for which the interelement coupling

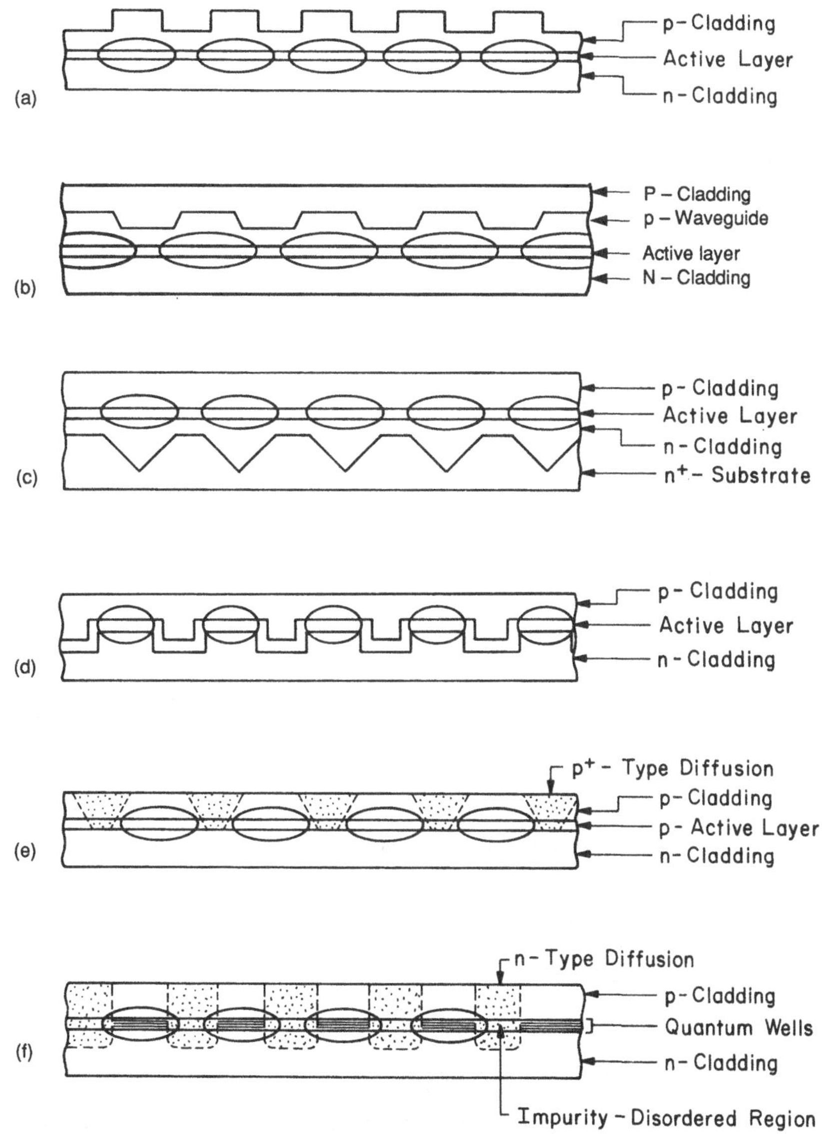

Figure 1.4. Schematic representation of positive-index-guided phase-locked arrays fabricated by: (*a*) chemical etching of ridge-type waveguides; (*b*) etch and regrowth over ridges; (*c*) liquid-phase epitaxial (LPE) growth over channeled substrates; (*d*) metal–organic vapor-phase epitaxial (MOVPE) growth over channeled substrates; (*e*) preferential p-type-dopant diffusion; (*f*) quantum-well-structure disordering induced by preferential n-type-dopant diffusion. The structure can be either of the AlGaAs/GaAs type (i.e., GaAs active layer and AlGaAs cladding layers) or of the InGaAsP/InP type (i.e., InGaAsP active layer and InP cladding layers). (After Ref. 56, © 1986 IEEE.)

mechanism (*not* evanescent wave) forces in-phase-mode operation (see bottom of Figure 1.1). The in-phase-mode quest is treated in Subsections 1.3.3 and 1.3.4.

Chen and Wang[23], followed by Whiteaway *et al.*[42], showed that, due to gain spatial hole burning, in-phase evanescent-mode arrays are fundamentally unstable, explaining the ready beam broadening observed experimentally above $\approx 1.5 \times$ threshold. In 1987, Thompson *et al.*[43] showed that the out-of-phase evanescent-mode is stable against gain spatial hole burning. A flurry of activity ensued[10,44–46] to fabricate stable out-of-phase-operating arrays; single-lobe operation then remained to be obtained by using π phase-shifter coatings[44,47] or plates[48,49]. The best result thus achieved was 200 mW cw in a diffraction-limited beam pattern[11].

1.3.2 The coupled-mode formalism

The first analysis of phase-locked arrays was published by Scifres *et al.*[50], who interpreted their experimental data by considering the diffraction pattern from a uniformly illuminated grating with equally spaced slits corresponding to individual laser-array elements. This is the so-called simple diffraction theory, which, while proving useful in interpreting some experimental results[50–52], provides no method for describing the allowed oscillating modes of an array of coupled emitters. The first application of the coupled-mode theory to phase-locked arrays was performed by Otsuka[53]. However, Otsuka analyzed an array with an infinite number of emitters and infinite aperture and found only two modes of operation: 0°-phase-shift and 180° phase-shift, which is no different from what had been claimed by proponents of simple diffraction theory. Butler *et al.*[16] have published the first coupled-mode analysis for an array of N coupled, identical elements. They found that an array of N emitters has N normal modes or eigenmodes, which they chose to call *array modes.*

The main findings of coupled-mode analysis are depicted schematically in Figure 1.5 for an array of nine elements. For each array mode, the near-field element amplitudes vary across the array. There is a succession from the 'fundamental' array mode, $L = 1$, the in-phase mode, to the last high-order array mode, $L = N$, the out-of-phase mode. For the case shown in Figure 1.5, $N = 9$. Shown by dashed lines are the envelope functions of the array-mode near-field amplitude profiles, which correspond to the modes of an infinite potential well of width $(N + 1)S$.

By solving an eigenvalue equation, and assuming only nearest-neighbor

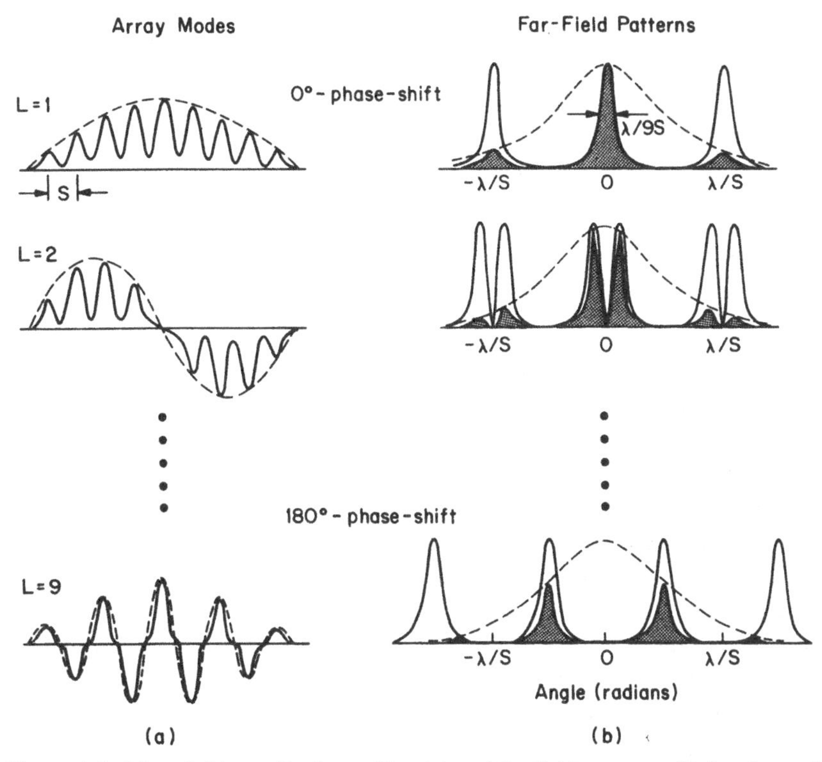

Figure 1.5. Near-field amplitude profiles (*a*) and far-field patterns (*b*) for three of the nine evanescent-type modes of a nine-element array. L is the array-mode number and λ is the free-space wavelength. The dashed curves in (*a*) are the envelope functions of the near-field amplitude profiles; and the dashed curves in (*b*) correspond to the individual-element far-field pattern. (After Ref. 56, © 1986 IEEE.)

coupling, the modes' eigenvalues are given by

$$\delta\beta_L = -\frac{k_0^2 c}{\beta}\cos\left[\frac{L\pi}{N+1}\right], \tag{1.1}$$

where β is the two-dimensional propagation constant, c is the coupling coefficient and L is the array-mode-number.

The far-field beam patterns corresponding to each array mode are shown in Figure 1.5(*b*). For an array of N coupled, identical elements, the far-field intensity distribution is given by[16,50]

$$F(\Theta) = |E(\Theta)|^2 I(\Theta), \tag{1.2}$$

where Θ is the angle with respect to the normal to the facet, $E(\Theta)$ is the far-field amplitude distribution for one of the laser-array elements and $I(\Theta)$ is a function characterizing the array and representing the interference effect of the coupled emitters in the array, the so-called grating function; $I(\Theta)$ is given by[54]

$$I(\Theta) = I_L(u) \propto \frac{\sin^2\left[\frac{(N+1)u}{2} + \frac{L\pi}{2}\right]}{\left[\sin^2\left(\frac{u}{2}\right) - \sin^2\left(\frac{L\pi}{2(N+1)}\right)\right]^2} \qquad L = 1, 2, \ldots, N, \qquad (1.3)$$

with $u = k_0 s \sin\Theta$, where k_0 is the free-space wavenumber ($k_0 = 2\pi/\lambda$), and s is the spacing between two array elements. The function $I(\Theta)$ is shown in Figure 1.5(*b*) by solid curves and represents the array pattern for the case where the array elements are point sources. In reality, the array elements have a finite size that provides an individual far-field intensity distribution $|E(\Theta)|^2$, and thus the array pattern is the product of $|E(\Theta)|^2$ and $I(\Theta)$. From a practical point of view, the desired beam pattern is that for which the energy is primarily contained within a single, forward-looking beam, that is, the $L = 1$ pattern with most of the light concentrated in the central lobe. Since the fundamental mode of positive-index guides can be well approximated by a Gaussian[55] it is possible to find analytically a condition for which at least 80% of the emitted energy resides in the main lobe. The condition is[55]

$$s\sqrt{d}(\Delta n)^{3/4} < \lambda^{3/2}/[4n^{3/4}], \qquad (1.4)$$

where s is the interelement spacing, d is the element width, Δn is the lateral index step, λ is the vacuum wavelength and n is the average index of refraction. Design curves for single-lobe operation at $\lambda = 0.80$ μm are shown in Figure 1.6. For instance, for a typical Δn of 0.007, when the element width is 3 μm the interelement spacing should be below 1.6 μm in order that $\geq 80\%$ of the energy is in the main lobe.

The diffraction-limited beamwidth (full width at half-power) is loosely defined to be the ratio of the wavelength over the emitting aperture, i.e., λ/NS. A close look shows that the diffraction-limited beamwidth is a function of the element number, N[54]. For arrays of few elements (two to four), the lobe beamwidth is virtually the same as λ/NS. Arrays of 10–40 diodes have beamwidth values in the 1.09–$1.16\lambda/NS$ range. In the limit $N \to \infty$ for a finite aperture, the beamwidth tends to $1.19\lambda/NS$, as expected for a source with a raised-cosine amplitude profile.

Coupled-mode formalism proved an instant success since it explained

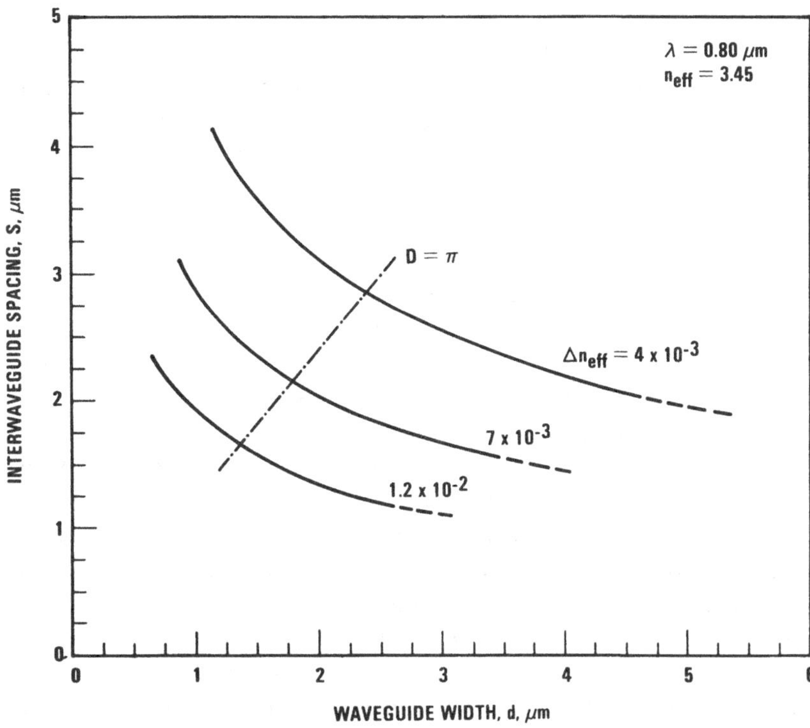

Figure 1.6. Condition that 80% of light is emitted in the main lobe for three different (lateral) index differentials and fixed wavelength ($\lambda = 0.8$ μm) and n_{eff}. For a given Δn_{eff}, the area below the curve corresponds to >80% light in the main lobe. The dot-and-dash curve corresponds to the first-order-mode cutoff condition. (After Ref. 55, © 1988 IEEE.)

most experimental data from positive-index-guided arrays[56]. It provided a simple design tool, but it also contained the grains of its eventual demise. The fact that frequency splitting between adjacent evanescent-type modes is inversely proportional to the number of elements[16] suggested some difficulty in achieving diffraction-limited-beam operation from arrays of large-element number ($N > 10$). Even more ominous was the fact that it applies only to weakly coupled systems, which, as time has shown, are not suitable for high-power coherent operation. Yet, when the object of coupling oscillators is not high power, but, rather, just an electro-optic function such as controlled beam steering, coupled-mode formalism is quite adequate for describing array behavior.

1.3.3 The quest for in-phase-mode operation: longitudinally uniform devices

The first positive-index-guided arrays produced[4,37] invariably operated in the out-of-phase condition. By using coupled-mode theory[18,20] it was demonstrated that in structures with excess gain in the high-index array regions, the out-of-phase evanescent mode overlaps better with the gain medium than does the in-phase mode, and thus it is favored to lase. The most straightforward apparent solution for obtaining in-phase-mode operation was to create high (modal) gain in the low-index array regions[18–20,22,39,57]. Twu *et al.*[57] preferentially pumped the low-index regions of a ridge-guide array and obtained an in-phase-like beam pattern, although it was much broader than the diffraction limit. Preferential interelement pumping was also used for two-element devices[39,58], with a best result of 150 mW being obtained in the in-phase mode. Mukai *et al.*[33], Kapon *et al.*[34] and Kaneno *et al.*[59] produced buried-rib-guide (five-to-eight)-element arrays for which the low-index regions had higher modal gain (i.e., higher transverse optical-mode confinement factor Γ_0) than the high-index regions, and indeed obtained in-phase, narrow-beam operation, although to modest powers (40–70 mW).

These initial, hopeful results occurred only in structures having no interelement loss and relatively few elements (up to eight). The reason for such behavior is clear from Fujii *et al.*'s analysis[26] (Figure 1.7). The diagram shows the modal gain of in-phase and out-of-phase evanescent-type modes, and the out-of-phase leaky mode as a function of the relative gain between the interelement and element regions. When gain is dominant in the high-index regions the out-of-phase evanescent mode is favored, as expected. At the other extreme (i.e., dominant gain in the low-index regions) the favored mode (of five- and seven-element devices) is the leaky out-of-phase, in spite of its edge radiation losses. The in-phase evanescent mode is favored over a limited range only for arrays of up to seven elements when the gain in the low-index regions is slightly higher than that in the high-index regions (e.g., by creating higher Γ_0 in the low-index regions)[22]. The few in-phase-beam experimental results[33,34,59] can thus be explained. However, the main conclusion is that for many-element (ten or more) arrays, as required for high-power operation, placing excess gain in the low-index regions will not favor in-phase evanescent-mode operation.

Another means of obtaining in-phase-mode operation is to create array structures for which the in-phase and out-of-phase modes have highly dissimilar lateral field-intensity profiles[60–62], and then preferentially pump the region where the in-phase-mode profile peaks. These so-called 'chirped'

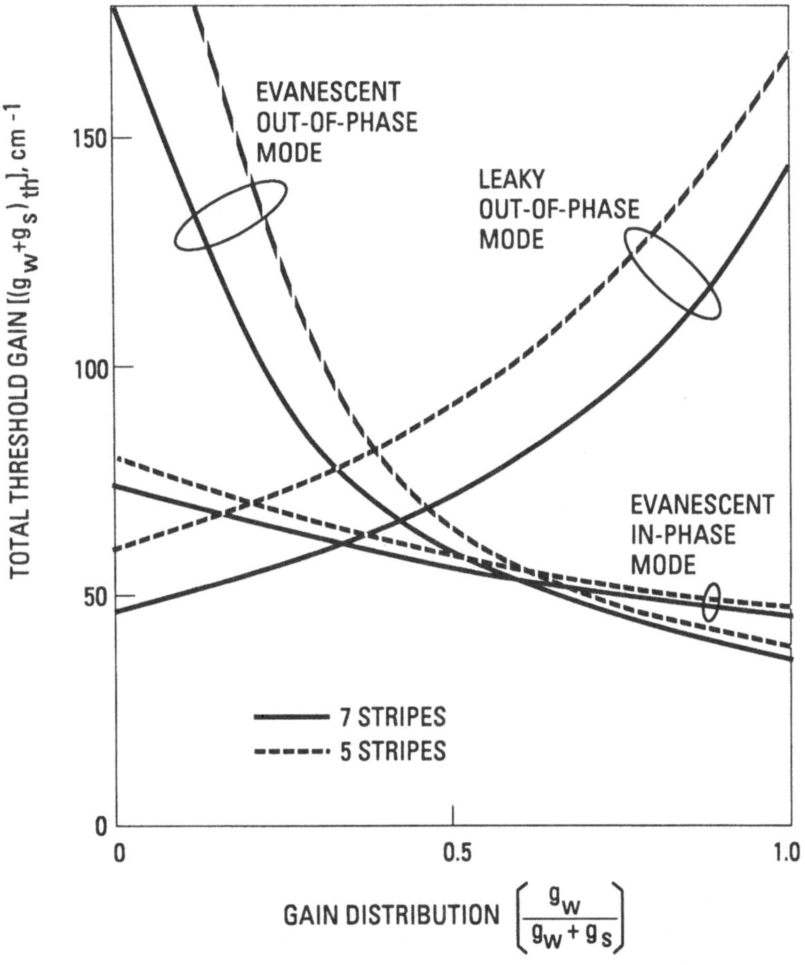

Figure 1.7. Modal threshold gain vs ratio of waveguide gain, g_w, to total local gain; g_s is the interelement (local) gain. (Ref. 26.)

arrays are depicted schematically in Figure 1.8. By symmetrically or asymmetrically varying the elements' width across the array, the in-phase mode intensity profile is forced to peak in the array center or at one of its edges. The best result was 100 mW single-lobe diffraction-limited-beam operation from a three-element device[63]. The main drawback of the approach is that the emitting area is relatively small compared to the array[60], thus defeating the purpose of fabricating arrays: the achievement of a large emitting area. Furthermore, when the effect of injected carriers

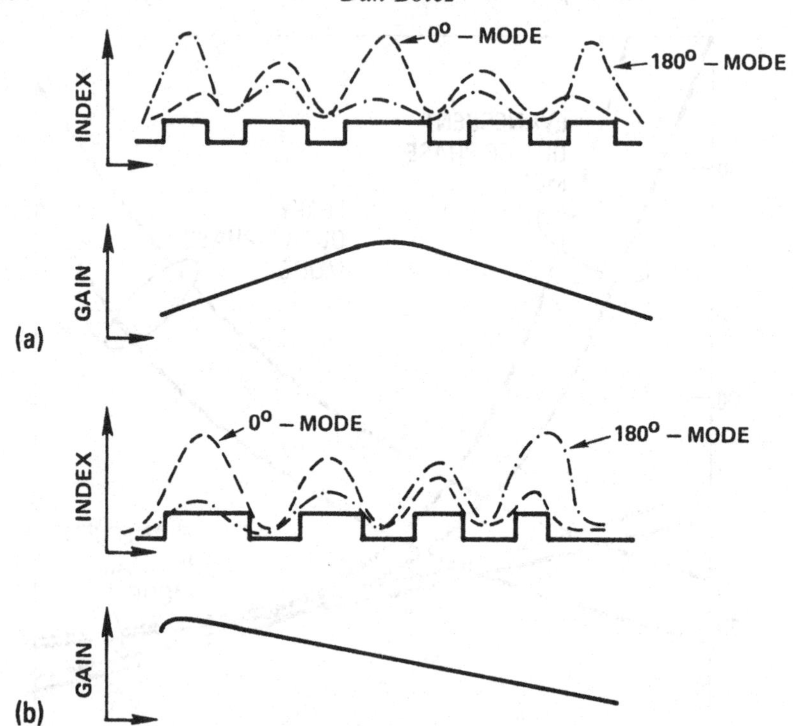

Figure 1.8. Schematic representation of chirped-type array types: (*a*) symmetric; (*b*) asymmetric.

on the active-layer dielectric constant is taken into account[62], the preferred array mode may not even be the in-phase one.

Finally, allowing for the natural tendency of positive-index-guided arrays to operate in the out-of-phase condition, one can place π phase-shifters in front[47–49] of alternate array elements. Figure 1.9 shows such a structure[47]: a variable-thickness facet coating ensuring a π phase-shift between adjacent elements, as well as the same optical-mode reflectivity. Matsumoto *et al.*[44] succeeded in demonstrating such a device.

1.3.4 The quest for in-phase-mode operation: longitudinally nonuniform devices

As a consequence of the difficulties encountered in obtaining stable, single-lobe operation from evanescently coupled arrays, alternative coupling schemes were devised for ensuring in-phase array operation. The two

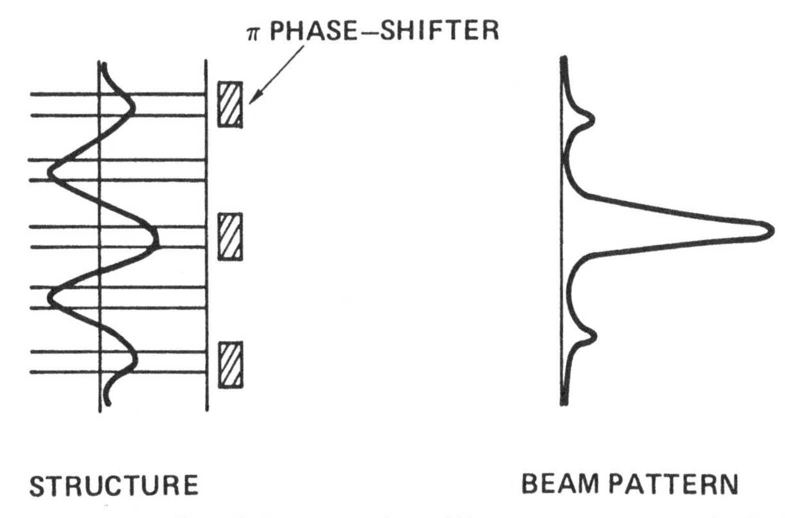

Figure 1.9. The effect of alternate π phase-shifters on an array operating in the out-of-phase condition.

basic approaches pursued were: interferometric devices[5,40,64,65] and diffraction-coupled devices[6,66–68].

Interferometric devices generally used Y-shaped single-mode waveguides. At each Y-shaped junction (Figure 1.10) fields from adjacent waveguides couple efficiently to a single waveguide only if they are in phase with each other. The combination of out-of-phase fields is simply lost as a radiation mode (i.e., destructive interference). Thus Y-junction devices[64,65,69,70] severely suppress out-of-phase mode operation. The concept of Y-junction arrays was originally put forward by Chen and Wang[5] and then refined by Streifer *et al.*[69]. Initial results were encouraging[64]: 400 mW peak pulsed and 200 mW cw power in stable in-phase-like beam patterns. However, the beams were rather broad[64,70]: approximately four times the diffraction limit, mainly due to poor discrimination against adjacent modes[71,72]. This weak intermodal discrimination reflects the fact that interelement coupling in Y-junction arrays is of the nearest-neighbor type (i.e. series coupling). Recently, an interferometric array composed of Y-shaped and X-shaped branches has been proposed[72] (Figure 1.11), such that a certain degree of parallel coupling can be achieved; whereby, as expected, intermodal discrimination increases dramatically. Another interesting proposal for achieving parallel coupling is the 'tree-array' device[73]: a fan-out of light via Y branches starting from a single waveguide and ending with a large linear array of emitters. However, the practical

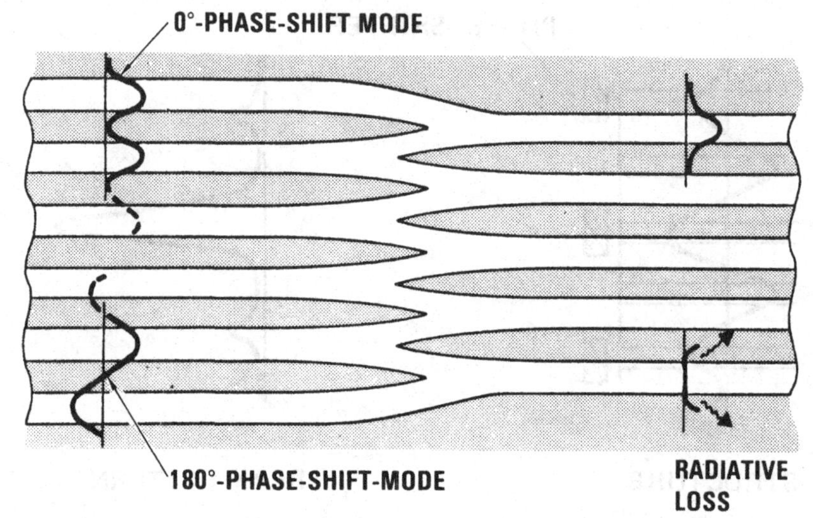

Figure 1.10. Schematic representation of Y-junction array operation.

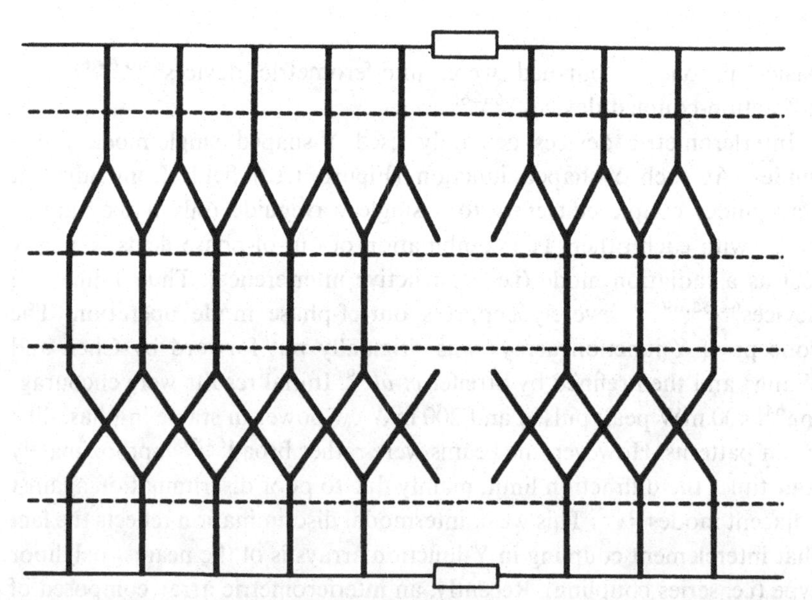

Figure 1.11. Schematic representation of X–Y-junction-type array. (Ref. 72.)

implementation of such devices is nontrivial because of the difficulty of fabricating low-loss, symmetric Y junctions.

After all the intensive work on Y-junction-coupled devices, the best results were obtained only from single-Y-branch devices[65,74]: 60–90 mW cw in a stable diffraction-limited beam. That is understable since in single-Y-branch devices there are only two modes: in-phase and out-of-phase, of which the latter is effectively suppressed.

Diffraction-coupled devices were based on the concept that fields from adjacent waveguides can be made to oscillate in-phase by placing a feedback mirror at an appropriate distance from the waveguides' apertures. The most complete analysis of such devices was performed by Mehuys *et al.*[68]. However, since, just as for Y-junction devices, the interelement coupling is of the nearest-neighbor type, the best results were single-lobe patterns of beamwidth several times the diffraction limit[68] (100 mW in a beam 2.5 × diffraction limit).

1.3.5 The quest for array-mode stability

Even though diffraction-limited in-phase-mode operation could be achieved, it occurred only near threshold. At ≈1.5 × threshold the beam of evanescent-wave-coupled devices would invariably broaden with increasing drive level. The reason for this behavior was twofold: (1) self-focusing of the in-phase mode due to gain spatial hole burning[23,42]; and (2) excitation of higher-order modes taking advantage of the gain unused by the in-phase mode[23]. The effect is shown schematically in Figure 1.12. The near-field profile of the in-phase mode of a uniform evanescent-wave-coupled array has a raised-cosine-shaped envelope. Thus, with increasing drive above threshold, gain saturation due to local photon density is uneven. The gain is saturated mostly in the central element which, in turn, due to a decrease in carrier density, creates a local increase in the refractive index. The net effect on the in-phase mode is self-focusing, which further accentuates the mode-profile nonuniformity. Thus a positive feedback

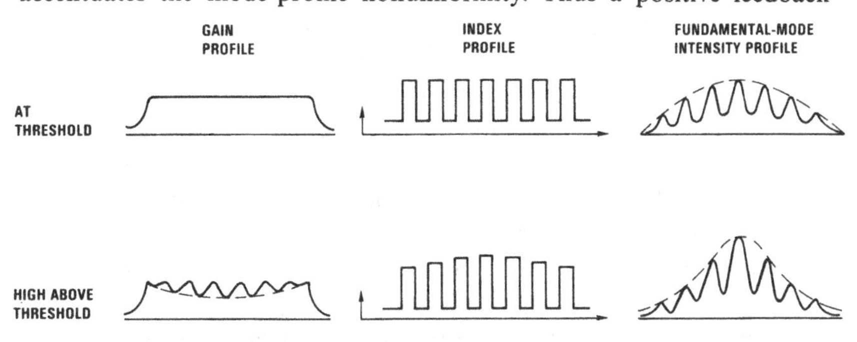

Figure 1.12. The effect of gain spatial hole burning on the fundamental (evanescent-wave) array mode.

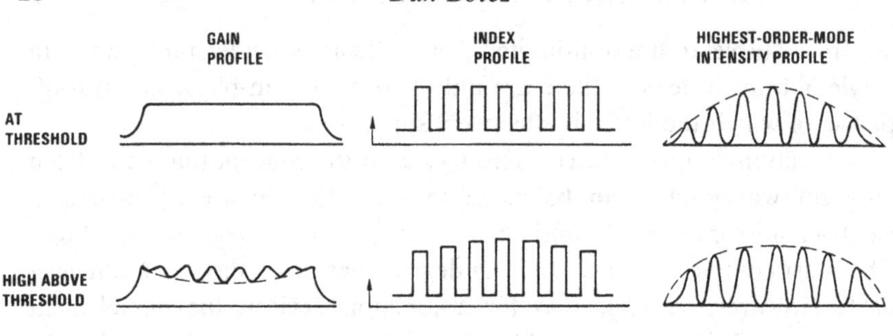

DEFOCUSING

Figure 1.13. The effect of gain spatial hole burning on the out-of-phase (evanescent-wave) array mode.

mechanism is created which, with increasing drive, narrows the near-field and broadens the far-field beamwidth. At the same time, since the area of the array is uniformly pumped, more and more gain is available for use by the high-order modes, which in turn reach threshold and cause further broadening of the emitted beam. Thus, as concluded by Chen and Wang[23] and Thompson *et al.*[43], the in-phase mode of evanescently coupled arrays is fundamentally unstable with increasing drive level.

Thompson *et al.*[43] found, however, that the out-of-phase mode is stable against spatial hole burning (see Figure 1.13). The out-of-phase mode of a uniform array also has a cosine-shaped near-field intensity envelope but behaves like a high-order mode of a guide as large as the array; that is, the refractive-index peaking due to spatial hole burning, while focusing the fundamental mode, causes defocusing of the out-of-phase mode. Thus, with increasing drive the out-of-phase mode profile shape changes from raised-cosine to uniform. This is a negative feedback effect which makes the out-of-phase mode fundamentally stable with increasing drive.

Experiments have readily confirmed[10,11,45,46,75,76] that high-order array modes are stable with increasing drive level. In fact, a structure was designed to favor out-of-phase-mode oscillation: the X-junction array (Figure 1.14). Pairs of single-mode waveguides are linked via Y junctions with a two-moded waveguide section. The Y-junction angle ($\approx 6°$) was designed to introduce large radiation losses for the fundamental mode of the connecting region, but negligible losses for the first-order mode of the connecting region. Thus the out-of-phase array mode is strongly favored to lase. Diffraction-limited-beam operation was achieved to 200 mW at $2.8 \times$ threshold (Figure 1.15(top)). The near-field intensity profiles close to and high above threshold (Figure 1.15(bottom)) confirm the defocusing

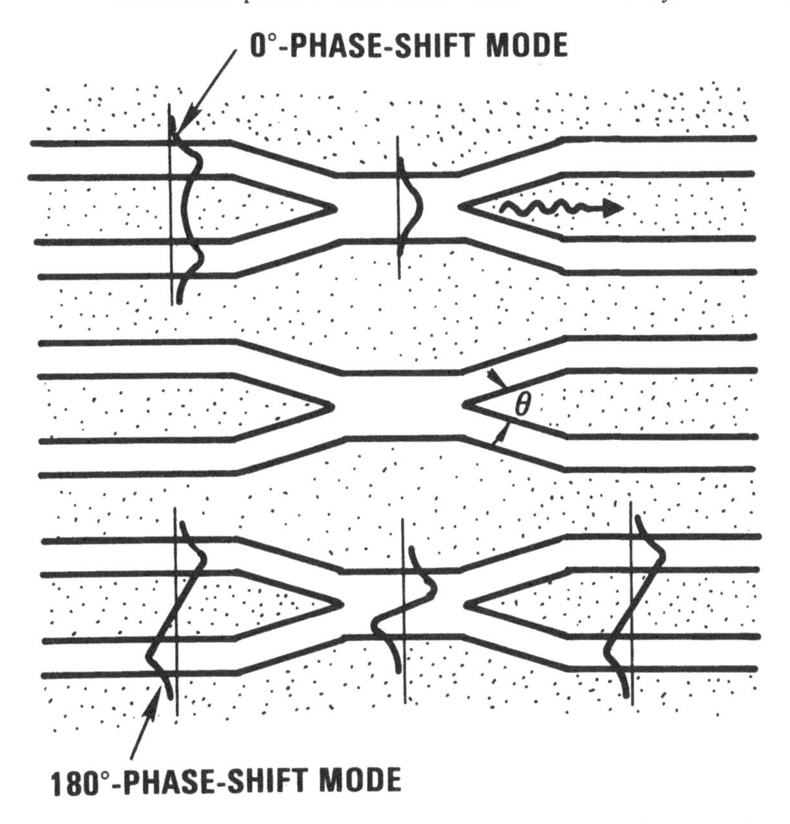

Figure 1.14. Top view of X-junction-type array. Typical branching angle θ is 6°. (After Ref. 10.)

effect postulated by Thompson *et al.*[43]. To obtain a single-lobed pattern, one can use the π phase-shifter coatings or plates described in Subsection 1.3.3.

Another proposal for stabilizing the array mode with increasing drive to was create a device whose in-phase-mode intensity profile is uniform to start with[77,78]. However, Thompson *et al.*[43] claimed that rendering the in-phase-mode near field uniform only has the effect of partially reducing its fundamental instability. Also, there has been no experimental support that the proposed designs[77] could indeed provide a stable array mode.

1.3.6 Conclusion

Positive-index-guide arrays were studied intensively over a five-year period and the results are summarized in Table 1.2. The promise of

Table 1.2. *Summary of results from positive-index-guide phase-locked arrays*

Coupling mechanism	Overall interelement coupling	Optical-mode confinement	Maximum power	
			*DL**	Narrow beam
Evanescent-wave				
• in-phase	Series	Moderate	0.05 W**	0.2 W; 3 × *DL*
• out-of-phase	Series	Moderate	0.2 W	0.3 W; 1.5 × *DL*
Y-junction				
• 1–2 elements	Parallel	Moderate	0.09 W	
• 10–11 elements	Series	Strong	–	0.4 W; 5 × *DL*
Diffraction	Series	Strong	–	0.1 W; 3 × *DL*

Notes:
* Diffraction-limited.
** From arrays of five to ten elements. Three-element devices were shown[63] to provide 0.1 W cw in a diffraction-limited beam.

high-power (>0.1 W) in a single, diffraction-limited beam never materialized. It is conceivable that with further effort (e.g., optimization of out-of-phase evanescent-coupled devices or X–Y-junction devices) coherent powers in the 0.2–0.5 W range could be achieved. However, the recent success of antiguided arrays does not justify such efforts.

There are, however, several useful lessons to be learned: (1) series coupling prevents many-element (ten or more) arrays from operating in a single spatial mode; (2) gain spatial hole burning in evanescent-coupled devices prevents stable operation of the fundamental (i.e., in-phase) mode; (3) parallel coupling, if achievable in strongly index-guided devices, would ensure stable, single-mode operation to high powers.

Finally, it should be said that although not successful for high-power coherent applications, positive-index-guided arrays may still be of use. One clear application would be controlled beam steering. For instance, one could create a phase ramp across the aperture of an evanescent-wave-coupled array by simply tailoring the injected-carrier profile. By varying currents through separate contact pads, the beam pattern could then be made to shift controllably.

1.4 Arrays of antiguides

1.4.1 History and array types

The basic properties of a single real-index antiguide are shown schematically in Figure 1.16. The antiguide core has an index n_0 lower than the

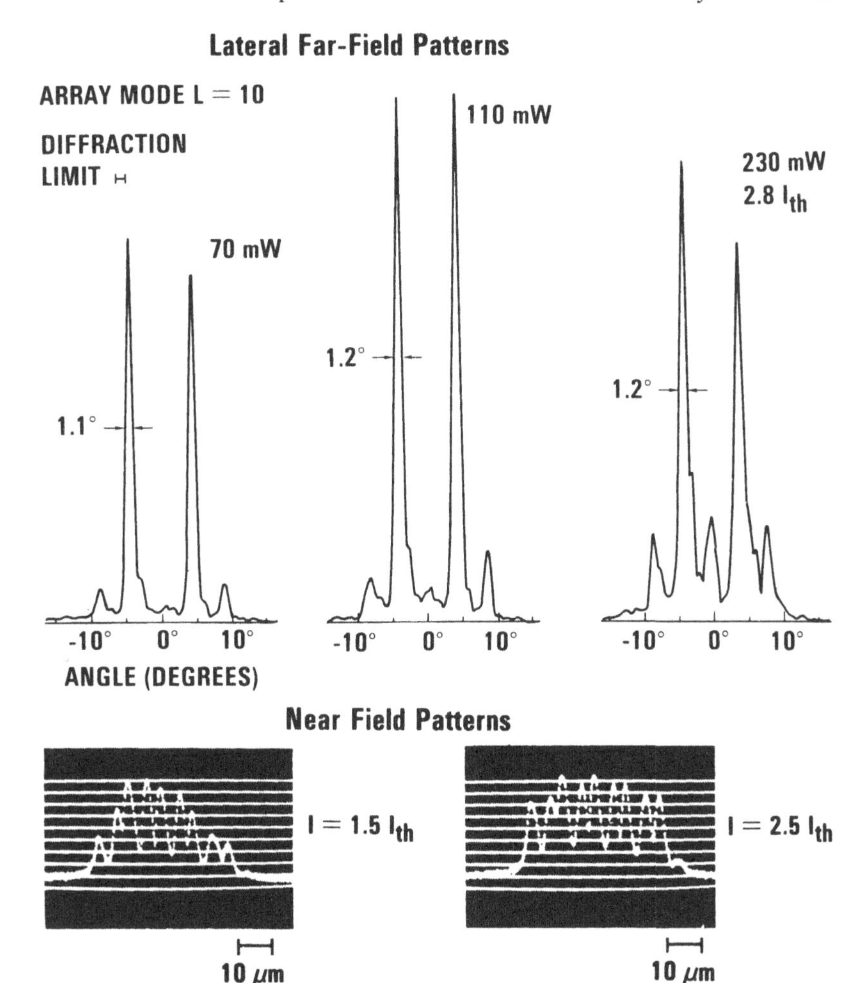

Figure 1.15. Typical electro-optical characteristics of X-junction array at different power levels. The threshold current, I_{th}, is $\approx$300 mA. (After Ref. 10.)

index of the cladding n_1. The index depression, Δn, is $(1\text{–}2) \times 10^{-3}$ for gain-guided lasers and $(2\text{–}5) \times 10^{-2}$ for strongly index-guided lasers. The propagation constant of the fundamental mode β_0 is below the index of the core. The quantum-mechanical equivalent is thus a quasibound state above a potential barrier. By contrast, the quantum-mechanical equivalent of the fundamental mode in a positive-index guide is a bound state in a potential well. Whereas in a positive-index guide, radiation is trapped via total internal reflection, in an antiguide, radiation is only partially

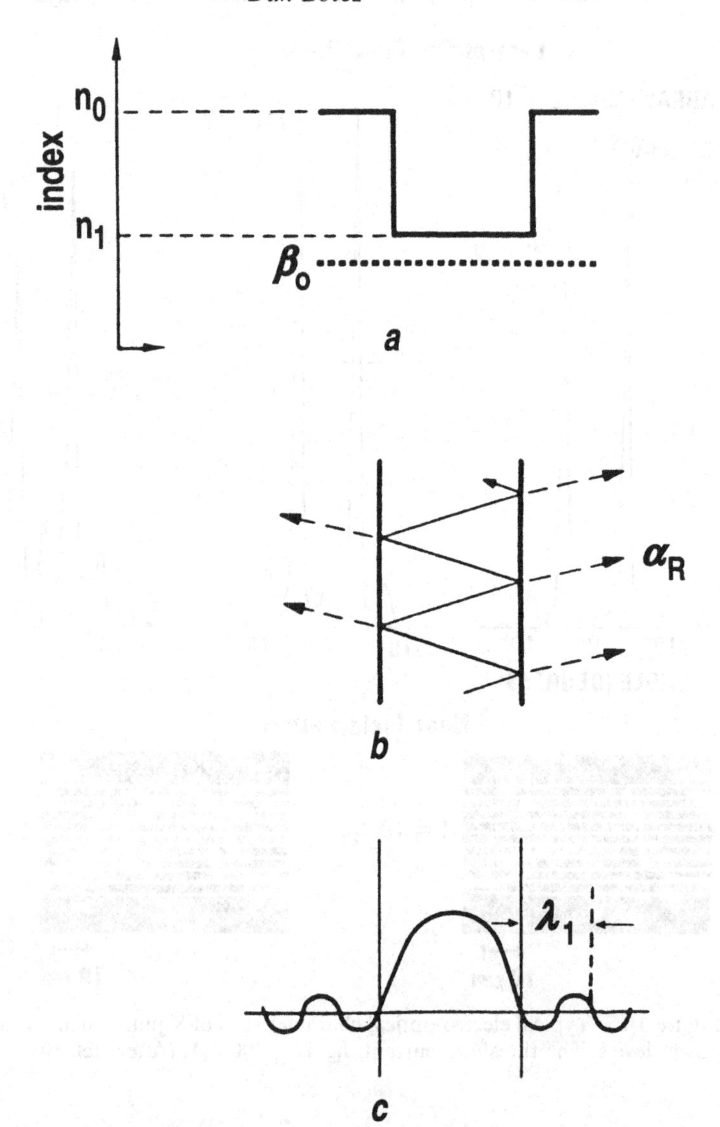

Figure 1.16. Schematic representation of real-index antiguide: (*a*) index profile; (*b*) ray-optics picture (α_R is edge radiation loss coefficient); (*c*) near-field amplitude profile (λ_1 is the leaky-wave periodicity in the lateral direction). (After Ref. 91.)

reflected at the antiguide-core boundaries (Figure 1.16(*b*)). Light refracted into the cladding layers is radiation leaking outwardly with a lateral wavelength λ_1 (Figure 1.16(*c*))[79]:

$$\lambda_1 \cong \lambda/\sqrt{[2n\,\Delta n + (\lambda/2d)^2]}, \tag{1.5a}$$

and can be thought of as a radiation loss:

$$\alpha_R = (l + 1)^2\lambda^2/d^3 n\sqrt{(2n\Delta n)}, \qquad (1.5b)$$

where d is the antiguide-core width, Δn is the lateral refractive-index step, n is the average index value, λ is the vacuum wavelength, and l is the (lateral) mode number. For typical structures ($d = 3$ μm, $\Delta n = 2$–3×10^{-2}) at $\lambda = 0.85$ μm, typical λ_1 and α_R values are 2 μm and 100 cm^{-1} respectively. Since $\alpha_R \propto (l + 1)^2$, the antiguide acts as a lateral-mode discriminator. For a proper mode to exist, α_R has to be compensated for by gain in the antiguide core[80]. Single antiguides have already been used for quite some time in CO_2 'waveguide' lasers.

Historically, the first arrays of antiguided lasers were gain-guided arrays[3] (Figure 1.17(*a*)) since an array of current-injecting stripe contacts provides an array of (carrier-induced) index depressions for which the gain is highest in the depressed-index regions. While for a single antiguide the radiation losses can be quite high[80], closely spacing antiguides in linear arrays reduces the device losses[12,81] significantly since radiation leakage

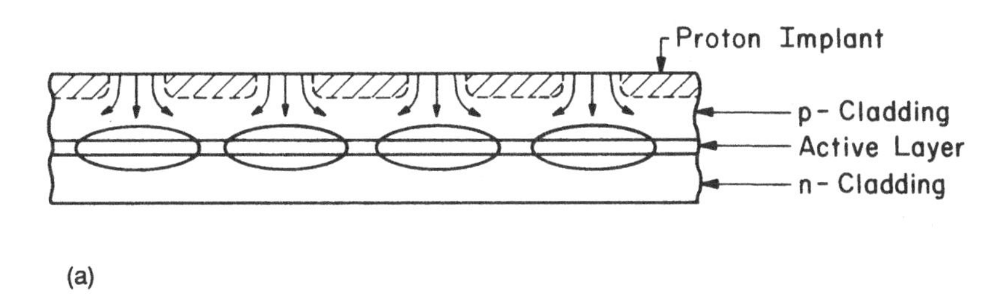

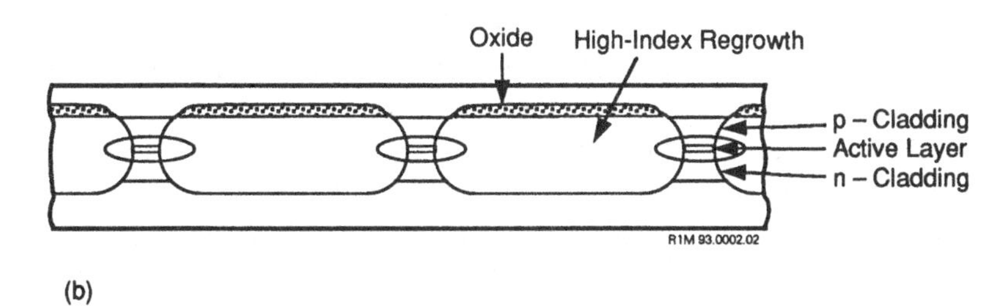

Figure 1.17. Schematic representation of early types of negative-index-guided (antiguided) phase-locked arrays: (*a*) gain-guided arrays; (*b*) buried-heterostructure (BH) arrays.

from individual elements mainly serves the purpose of coupling the array elements.

The first real-index antiguided array was realized by Ackley and Engelmann[9]: an array of buried heterostructure (BH) lasers designed such that the interelement regions had higher refractive indices than the effective refractive indices in the buried active mesas (Figure 1.17(*b*)). Since the high-index interelement regions had no gain, only leaky array modes could lase. The device showed definite evidence of phase locking (in-phase and out-of-phase) but had relatively high threshold-current densities (5–7 kA/cm^2) since the elements were spaced far apart (13–15 μm), thus not allowing for effective leaky-wave coupling.

For practical devices the high-index regions have to be relatively narrow (1–3 μm), which is virtually impossible to achieve using BH-type fabrication techniques. Instead, one can fabricate narrow, high-effective-index regions by periodically placing high-index waveguides in close proximity (0.1–0.2 μm) to the active region[1,12,15] (see Figure 1.18(*a*)). In the newly created regions the fundamental transverse mode is primarily confined to the passive guide layer; that is, between the antiguided-array elements the modal gain is low. (The effect of the presence of a first-order transverse mode as well is treated in Subsection 1.4.6.) To further suppress oscillation of evanescent-wave modes an optically absorbing material can be placed between elements (Figure 1.18(*a*).)

The first closely spaced, real-index antiguided array was realized by liquid-phase epitaxy (LPE) over a patterned substrate[1,12]. Initially, a passive-waveguide structure ($Al_{0.3}Ga_{0.7}As/Al_{0.1}Ga_{0.9}As$) is grown on top of a GaAs substrate. Then channels are etched and a planar DH-type laser structure is regrown taking advantage of the LPE-growth characteristics over patterned substrates[37]. Each buried interchannel mesa had both higher effective refractive index than the channel regions[1] and lower modal gain than the channel regions[12]. Devices made in recent years are fabricated by metal–organic chemical vapor deposition (MOCVD) and can be classified into two types: the complimentary-self-aligned (CSA) stripe array[13,15] (Figure 1.18(*b*)); and the self-aligned-stripe (SAS) array[14,82,83] (Figure 1.18(*c*)). In CSA-type arrays preferential chemical etching and MOCVD regrowth occurs in the interelement regions. For SAS-type arrays the interelement regions are built-in during the initial growth, and then etching and MOCVD regrowth occur in the element regions. Note that for SAS-type devices the passive guides and loss regions between elements can be incorporated in just one layer[82]. Furthermore, some SAS-type arrays[83] appear to have no interelement loss.

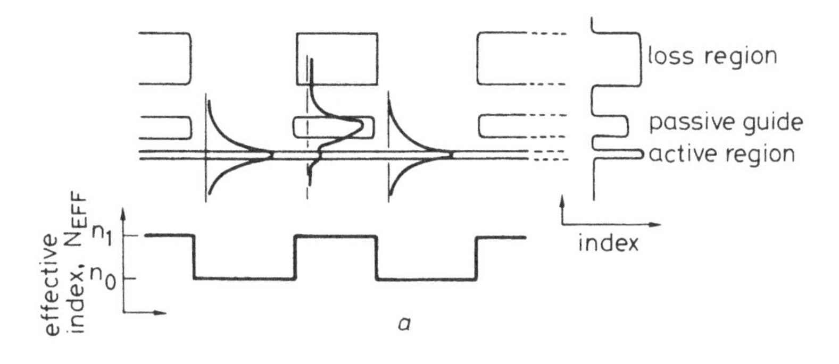

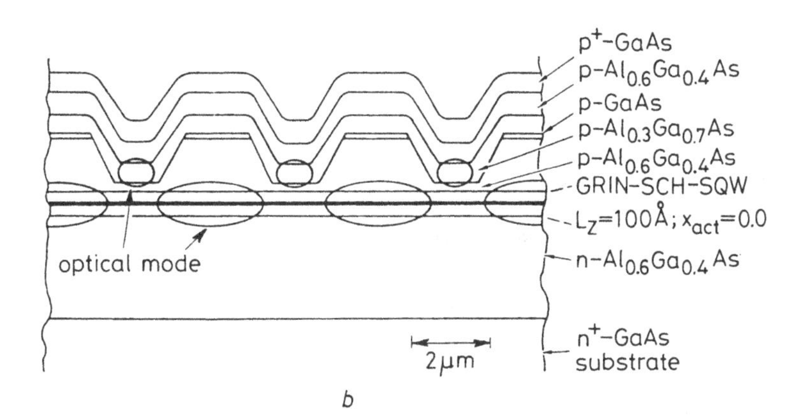

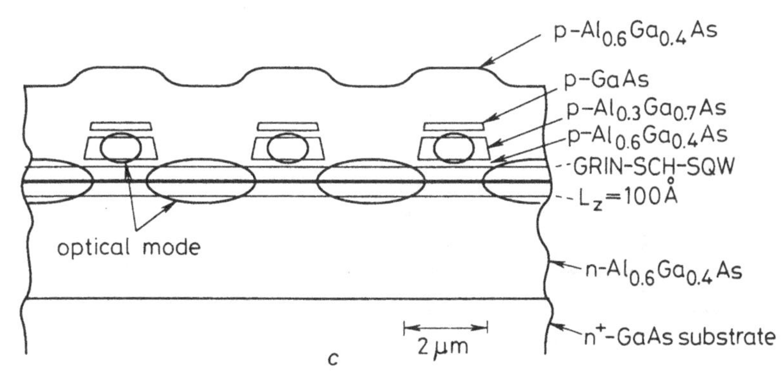

Figure 1.18. Schematic representation of modern types of arrays of (closely spaced) antiguides: (*a*) practical way of fabricating arrays of closely spaced antiguides; (*b*) complimentary-self-aligned (CSA) array type (Refs. 13 and 15); (*c*) self-aligned-stripe (SAS) array type (Refs. 14, 82, and 83). (After Ref. 91.)

1.4.2 Bloch-function analysis

Antiguided arrays have been studied using exact analysis[1,2,15,26] (see Chapter 4). However, a simple formalism that provides a ready design tool is the Bloch-function method[24,27,28]. Eliseev, Nabiev and Popov[24] were the first to apply the method to a periodic variation in the real part of the dielectric constant, and solve for both evanescent-type array modes and leaky-type array modes. Botez and Holcomb[27] completed the analysis by taking into account the points of resonant leaky-wave coupling.

A periodic variation of the real part of the dielectric constant is considered (lower part of Figure 1.19): period Λ, antiguide-core width d, interelement-region width s, and index step $\Delta n = n_1 - n_0$. A simple notation is used to identify the leaky array modes:

$$(l, m); \quad l = 0, 1, \ldots; \quad m = 0, 1, \ldots,$$

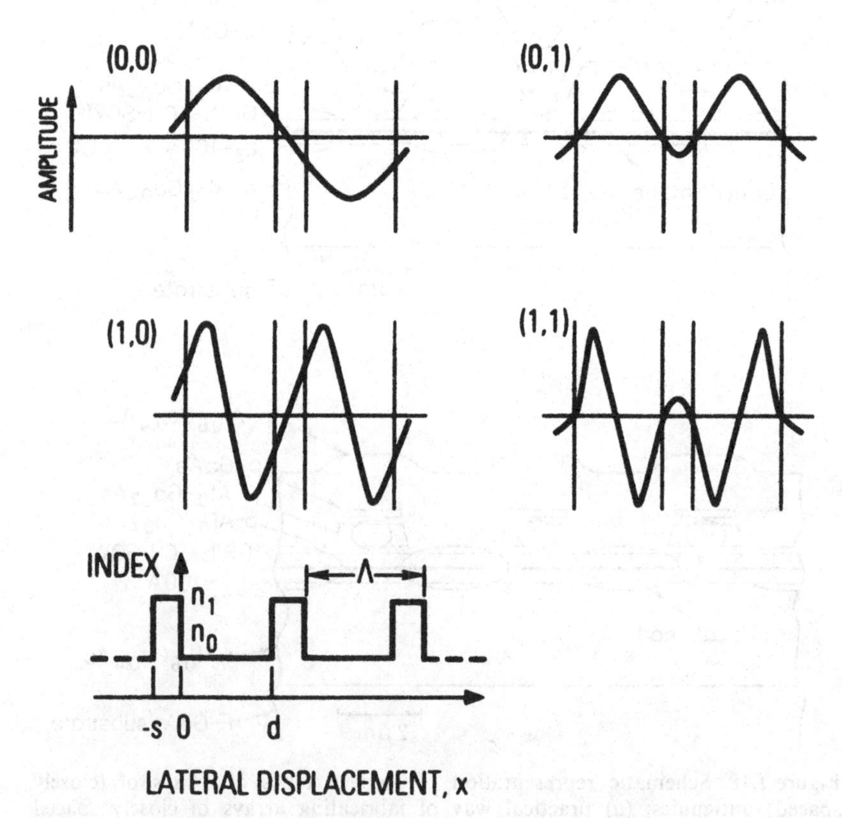

Figure 1.19. Schematic representation of leaky-type array modes for a periodic variation of the real part of the dielectric constant. (After Ref. 27.)

where l is the (lateral) mode order in the antiguide-core region[15], and m is the number of field-intensity peaks in the interelement region[15]. The basic modes of this infinite-extent array are depicted schematically in Figure 1.19. When l and m are of opposite parity, the modes are 'in-phase' type. When l and m have the same parity, the modes are 'out-of-phase' type. The 'fundamental' array mode is (0, 0), a mode comprising out-of-phase-coupled fundamental (element) modes[15]. The most desirable mode is (0, 1), a mode comprising in-phase-coupled fundamental (element) modes[15].

For optical waves the periodic index distribution of Figure 1.19 is analogous to the Kronig–Penney potential for the electron-wavefunction in a periodic potential[84]. We consider a wave propagating in the z direction (with a propagation constant β) normal to the transverse direction, x: $\Psi(x)\exp(-i\beta z)$. The Bloch solutions of the wave equation for a periodic potential are of the form $\Psi(x) = \exp(iqx)\phi(x)$, where $\phi(x)$ is a periodic function of x, and q, the Bloch wave number, varies from $-\pi/\Lambda$ to π/Λ. When the appropriate solution forms are used, the following eigenvalue equation is obtained for leaky-type TE modes[24]:

$$\cos(q\Lambda) = \cos(h_1 s)\cos(h_0 d) - (1/2)(h_1/h_0 + h_0/h_1)\sin(h_1 s)\sin(h_0 d), \quad (1.6)$$

where

$$h_1 = \sqrt{(\varepsilon_1 k^2 - \beta^2)} \quad \text{and} \quad h_0 = \sqrt{(\varepsilon_0 k^2 - \beta^2)}$$

are the transverse propagation constants in the interelement and antiguide-core regions respectively[15]; ε_1 and ε_0 are the complex dielectric constants in the interelement and antiguide-core regions. The wavefunction $\Psi(x)$ is then[24]

$$\Psi(x) = \exp(iqx)\{E_1 \exp[ih_1(x - n\Lambda)] + E_2 \times \exp[-ih_1(x - n\Lambda)]\}\exp[-iq(x - n\Lambda)] \quad (1.7a)$$

for all interelement regions (i.e., $n\Lambda - s < x < n\Lambda$), and

$$\Psi(x) = \exp(iqx)\{F_1 \exp[ih_0(x - n\Lambda)] + F_2 \times \exp[-ih_0(x - n\Lambda)]\}\exp[-iq(x - n\Lambda)] \quad (1.7b)$$

for all antiguide-core regions (i.e., $n\Lambda < x < n\Lambda + d$), where

$$E_j = -h_1\sin(h_0 d) - (-)^j ih_0\exp\{i[q\Lambda - (-)^j h_1 s]\} + (-)^j ih_0\cos(h_0 d), \quad j = 1, 2, \ldots$$

$$F_j = -h_0\sin(h_1 s)\exp(iq\Lambda) + (-)^j ih_1\exp[(-)^j ih_0 d] - (-)^j ih_1\cos(h_1 s)\exp(iq\Lambda).$$

The above equations define the near-field amplitude profiles of the leaky array modes, and can be used to compute far fields[24,28]. The only solutions of practical relevance are for $q = 0$ and π/Λ (i.e., in-phase and out-of-phase element coupling). There are no Bloch-function solutions at points of leaky-wave resonant coupling[15] (e.g., $h_1 = \pi/s$, $h_0 = \pi/d$ for mode (0, 1)). The resonance points correspond to (lateral) standing waves, sines or cosines, equivalent to Bragg-resonance solutions for distributed feedback (DFB) lasers[85]. In fact, resonant arrays operating in mode (0, 1) can be thought of as second-order lateral DFB structures[85].

When small changes in dielectric constant occur in the antiguide cores and claddings, Eq. (1.6) becomes

$$-\Lambda \sin(q\Lambda)\delta q = (A_1/2h_1^2)k^2\delta\varepsilon_1 + (A_0/2h_0^2)k^2\delta\varepsilon_0 - (A_1\beta/h_1^2 + A_0\beta/h_0^2)\delta\beta, \quad (1.8)$$

where

$$A_j = a[\pm\cos(a) - \cos(b)]/\sin(a) - 0.5(-)^j(h_0/h_1 - h_1/h_0)\sin(a)\sin(b),$$

with $a = h_0 d$, $b = h_1 s$ for $j = 0$; $a = h_1 s$, $b = h_0 d$ for $j = 1$; and $\pm$ signs correspond to in-phase and out-of-phase conditions respectively. The left-hand side of Eq. (1.8) is zero since the practical Bloch wave numbers are $q = 0$, π/Λ. Then by defining β_i and ε_{ji} as the imaginary parts of the propagation constant β and the dielectric constants ε_0 and ε_1, it follows that

$$\delta\beta_i = [(A_1/2h_1^2\beta)k^2\delta\varepsilon_{1i} + (A_0/2h_0^2\beta)k^2\beta\varepsilon_{0i}]/(A_1/h_1^2 + A_0/h_0^2), \quad (1.9)$$

and the differential gain, when the antiguide-core gain varies, is

$$\delta\beta_i/k\delta\varepsilon_{0i} = A_0 k/[2h_0^2\beta(A_1/h_1^2 + A_0/h_0^2)] \cong (1/2n_0)[\delta(\Gamma g)/\delta(\Gamma_0 g)], \quad (1.10)$$

where Γ is the two-dimensional array-mode (optical) confinement factor[86,87], Γ_0 is the transverse confinement factor in the antiguide core, and g is the gain in the active region; $\delta\beta_i/k\delta\varepsilon_{0i}$ is $1/(2n_0)$ times the rate of increase in (array) modal gain, $G_a = \Gamma g$, when the antiguide-core local gain $G_0 = \Gamma_0 g$ increases (e.g., by raising Γ_0 or by preferentially injecting carriers in the core). In Figure 1.20 we show $\delta\beta_i/k\delta\varepsilon_{0i}$ plots for leaky modes of an array with $d = 3$ μm and $s = 1$ μm. It has been shown[87] that Γ can be expressed as a linear function of differential gain; that is, the mode of highest differential gain has the highest modal gain and will thus be the one favored to lase. For the case shown in Figure 1.20, the in-phase leaky mode [i.e., (0, 1)] is the one favored to lase.

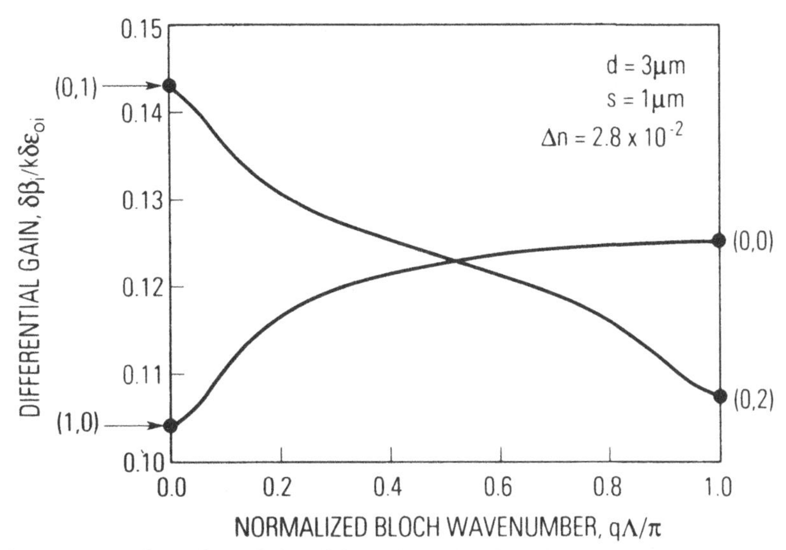

Figure 1.20. Dispersion of the differential gain, $\delta\beta_i/k\delta\varepsilon_{0i}$ (Eq. (1.10)), for leaky modes of an antiguided-array structure. The 0.0 axis corresponds to in-phase modes, while the 1.0 axis corresponds to out-of-phase modes. (After Ref. 27.)

Eliseev *et al.*[24] have plotted the dispersion diagrams for the differential gains of both leaky-type and evanescent-type array modes. One such diagram is shown in Figure 1.21. Excess gain is placed in the low-index array regions. For *evanescent* array modes we use the notation:

$$[n, p]; \quad n = 0, 1, \ldots; p = 0, 1, \ldots,$$

where n is the (lateral) mode number in the waveguide core (i.e., the high-index regions), and p is the number of field-intensity nulls in the low-index regions. The in-phase evanescent array mode comprising coupled fundamental (element) modes (so-called fundamental mode in coupled-mode theory) is then mode $[0, 0]$.

Figure 1.21 makes it clear that in arrays with excess gain in the low-index regions, the leaky-type modes are much more favored to lase than the evanescent-type modes. Ironically, evanescent mode $[0, 0]$, claimed as the winner by coupled-mode theory[18], is one of the least favored modes to lase. Eliseev *et al.*[24] have also plotted the differential-gain diagrams for the case where gain is preferentially placed in the high-index regions. Then, in good agreement with coupled-mode analyses[18–20], the out-of-phase evanescent mode (i.e., $[0, 1]$) is the one favored to lase.

For both cases shown here (Figures 1.20 and 1.21) the in-phase leaky mode is the one favored to lase. However, as shown by Eliseev *et al.*[24],

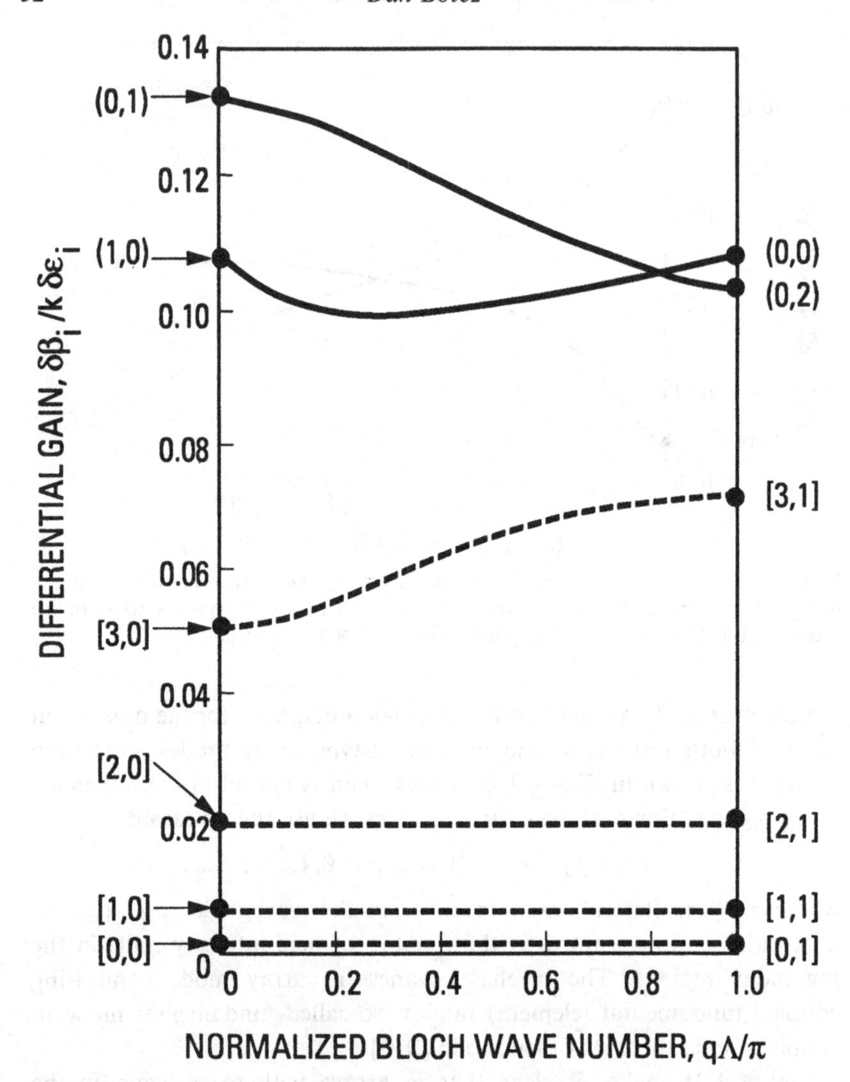

Figure 1.21. Dispersion of the differential gain, $\delta\beta_i/k\delta\varepsilon_i$, for leaky modes (solid curves) and evanescent modes (dashed curves) for an array structure with $\Delta n = 0.03$; $d = s = 3$ μm. The mode nomenclature is given in the text. The 0 axis corresponds to in-phase modes, while the 1.0 axis corresponds to out-of-phase modes. (Data taken from Ref. 28.)

and then clarified by Botez and Holcomb[27] and Botez and Mawst[87], this is not always the case. For arrays of antiguides the leaky array mode favored to lase can be either in-phase *or* out-of-phase, depending on the specific structure under consideration[87].

1.4.3 Resonant leaky-wave coupling

Owing to lateral radiation, a single antiguide can be thought of as a generator of laterally propagating travelling waves of wavelength λ_1 (Figure 1.16(*a*)). Then, in an array of antiguides, elements will resonantly couple in-phase or out-of-phase when the interelement spacings correspond to an odd or even integral number of (lateral) half-wavelengths ($\lambda_1/2$) respectively (see Figure 1.22). The resonance condition is

$$s = m\lambda_1/2 \qquad \left.\begin{array}{ll} m = \text{odd} & \text{resonant in-phase mode} \\ m = \text{even} & \text{resonant out-of-phase mode,} \end{array}\right\} \qquad (1.11)$$

where s is the interelement spacing. Typical s values are 1 μm. Then, for in-phase mode resonance, $\lambda_1 = 2$ μm.

When the resonance condition is met, the interelement spacings become Fabry–Perot resonators in the resonance condition. The situation is the opposite of that for antiresonant reflecting optical waveguide(ARROW)-type structures[88,89]. In the ARROW device antiresonance is sought such that $s = (2m + 1)\lambda_1/4$. Arrays of ARROW passive waveguides[90] have, as a salient feature, a lack of optical crosstalk between elements. By contrast, in resonant arrays of antiguides we have full communication between elements which, in effect, achieves parallel coupling. To emphasize complementarity to ARROW devices, resonant arrays of antiguides have been named resonant optical waveguide (ROW) arrays. To illustrate parallel coupling, in Figure 1.23 we show light propagation in a ROW array. Light launched in the central antiguide is partially reflected at core–cladding interfaces and totally transmitted through the interelement (cladding)

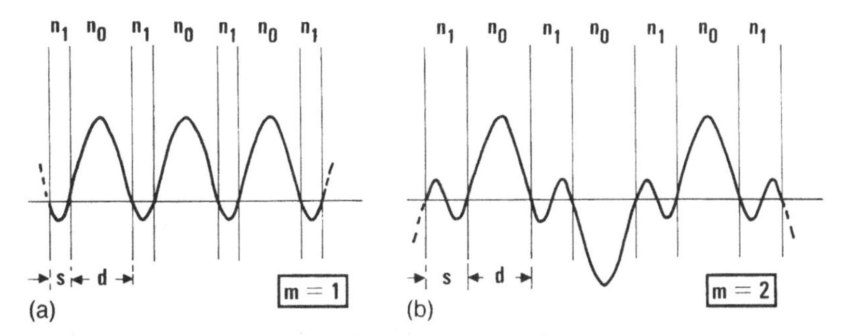

Figure 1.22. Near-field amplitude profiles in resonantly coupled arrays: (*a*) in-phase resonant mode; (*b*) out-of-phase resonant mode; *m* is number of interelement near-field intensity peaks. (After Ref. 13.)

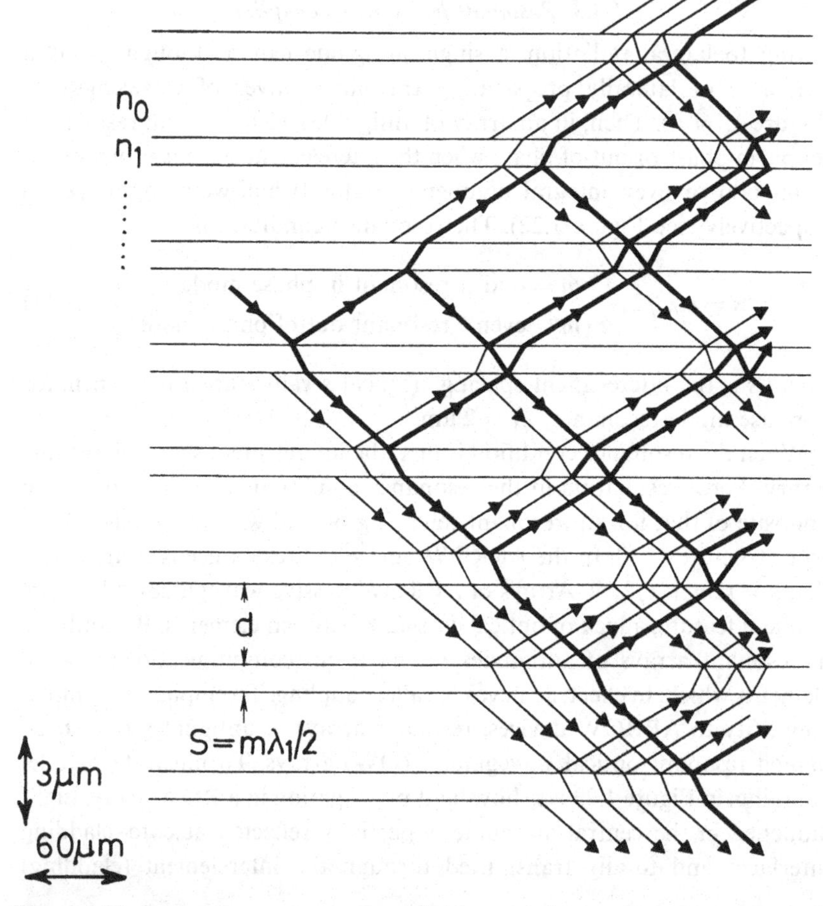

Figure 1.23. Ray optics representation of leaky-wave fan-out in resonant array of antiguides. (After Ref. 91.)

regions. Upon entering adjacent antiguides it couples in-phase or out-of-phase with rays there. A fan-out occurs, highlighted by heavier lines, reminescent of tree-array types made from Y junctions[73]. It can easily be seen that after several bounces, rays originating in the central antiguide couple to rays in all antiguides (i.e., long-range coupling). Also, it is apparent that rays originating in other antiguides generate their respective 'tree-array' networks; that is, a ROW array can be thought of as interconnected 'tree-array' networks of rays, a fitting description for parallel coupling. As expected, at and near resonance, ROW arrays have uniform near-field intensity profiles[15], and maximum intermodal

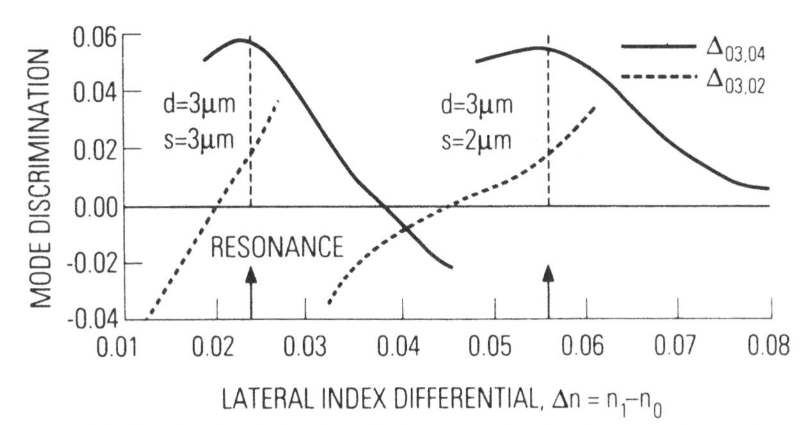

Figure 1.24. Mode discrimination Δ between the in-phase mode (0, 3) and out-of-phase modes (0, 2) and (0, 4) for two different array structures. Arrows indicate resonances (see Ref. 15) $\Delta n_{l,m} = (\lambda^2/8n_0)([m^2/s^2] - [l+1]^2/d^2)$, of mode (0, 3) for each structure ($\lambda = 0.86$ μm, $n_0 = 3.40$). The dashed curves end at the (0, 2)- and (0, 4)-mode cutoffs respectively. (After Ref. 27.)

discrimination[15,26,86]. The lateral resonance effect in antiguided arrays is quite similar to Bragg resonances in DFB-type structures[85], as detailed in the following subsection.

Interesting effects occur when the differential gain (Eq. (1.10)) is studied near resonance. The amount of discrimination between array-mode types is defined as

$$\Delta_{lm,l'm'} = (\delta\beta_i/k\delta\varepsilon_{0i})_{(l,m)} - (\delta\beta_i/k\delta\varepsilon_{0i})_{(l'm')},$$

simply the difference in differential gain between modes (l, m) and $(l'm')$. In Figure 1.24 we show the Δs between mode (0, 3) and out-of-phase modes, around Δn values corresponding to resonance[15] for arrays with $d = 3$ μm and $s = 3$ and 2 μm respectively. In both cases, mode (0, 3) is favored to lase around its respective resonances[15], with maximum discrimination occurring on the high-Δn side of resonance. The reason for this behavior is evident by looking at Figure 1.25. At and near its resonance the resonant mode has negligible field between elements. The out-of-phase modes (i.e., (0, 2) and (0, 4)), being nonresonant, have a large interelement field. Mode (0, 4) always has a larger percentage of interelement field than mode (0, 2) and is thus more discriminated against than mode (0, 2). Thus, maximum intermodal discrimination occurs when the Δ curves corresponding to mode (0, 2) (dashed lines in Figure 1.24) stop (due to the mode cutoff)[15] on the high-Δn side of resonance. This is fortunate since for finite-size arrays intermodal discrimination based on different edge

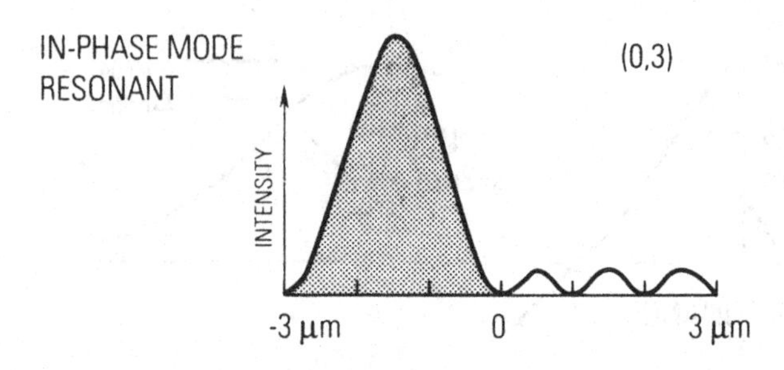

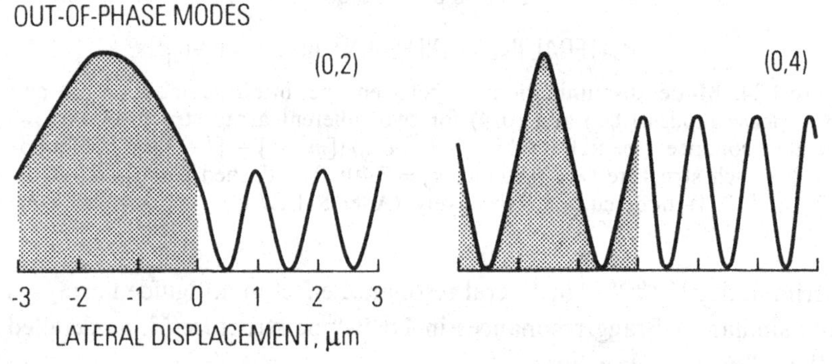

Figure 1.25. Near-field intensity profiles of modes (0, 2), (0, 3) and (0, 4) at the resonance of mode (0, 3) ($\Delta n = 2.4 \times 10^{-2}$ in Figure 1.24). Shaded areas correspond to antiguide cores. (After Ref. 27.)

radiation losses between resonant modes and adjacent modes also reaches a maximum near resonance on the high-Δn side[15].

It can easily be shown by using previous analysis[15] that the high-Δn side of resonance on a mode-loss vs Δn plot corresponds to the high-s side of resonance on a mode-loss vs s plot. Then when plotting modal gain vs s, one expects maximum discrimination (between in-phase and out-of-phase modes) on the high-s side of resonance. Indeed, that is what Hadley[25] and the present author[91] have found by using exact analyses, and what has more recently been shown by employing the Bloch-function method[87].

Figure 1.24 shows intermodal discrimination around points of resonant in-phase mode coupling. Similar behavior occurs around points of resonant out-of-phase mode coupling[87]; that is, in arrays of antiguides the array mode favored to lase is the *resonant* leaky mode: in-phase or out-of-phase.

1.4.4 Resonant modes and distributed-feedback analogy

By considering the eigenvalue value for leaky array modes (Eq. (1.5)), one can solve for the allowed values of β. Since q must be real, the values of β in a photonic lattice are limited to a series of bands similar to energy bands in a crystal lattice. Figure 1.26 shows the allowed values of β/k_0 as a function of s when $d = 3$ μm, wavelength $\lambda = 0.85$ μm, $n_0 = 3.402\,95$, and $n_1 = 3.426\,46$. The shaded areas indicate the allowed bands, and the solid lines are the band edges when $q\Lambda = N\pi$, where N is an integer. For a given value of β, the band edges are periodic with a period $\lambda_1/2$. It is useful to also define $\lambda_0 = 2\pi/h_0$ as the lateral wavelength in the element region, and then a β vs d plot for fixed s has $\lambda_0/2$ periodicity. The periodic structure continues for arbitrarily large s or d. By contrast, for the evanescent-wave-coupled modes of infinite arrays[92], as s increases, the allowed values of β simply converge to the discrete β values for a single element. Such behavior is expected, since evanescently coupled elements eventually decouple at large s values. By contrast, leaky-wave coupling can occur over large distances (i.e., when $s \gg d$)[93].

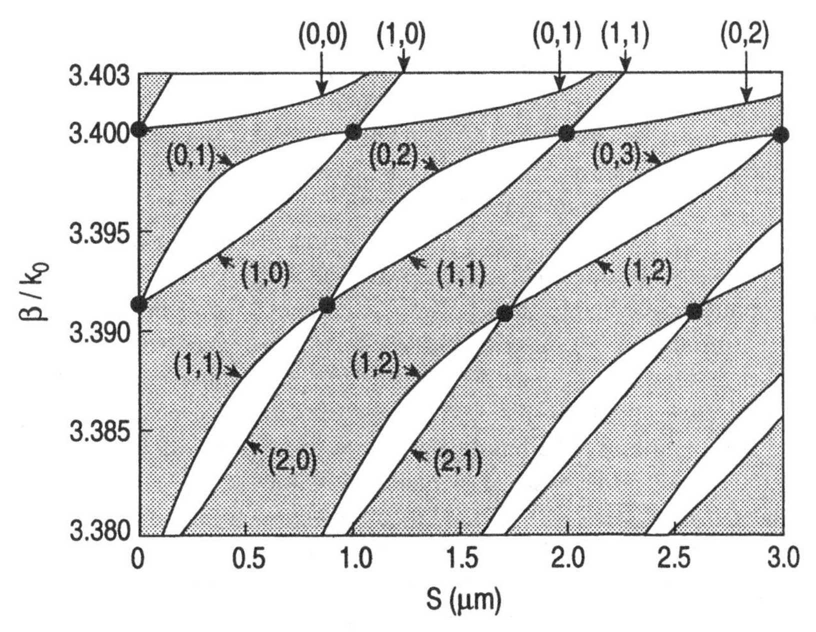

Figure 1.26. Allowed values of the normalized propagation constant β/k_0 (shaded regions) as a function of interelement spacing, s, for the case $d = 3$ μm, $n_0 = 3.402\,95$ and $n_1 = 3.426\,46$. Solid lines at the band edges correspond to (l, m) leaky array modes (Ref. 27), and solid dots indicate resonance points. (After Ref. 85.)

The only modes of the infinite array that have practical relevance are band-edge modes. The band-edge modes are identified in Figure 1.26 by the (l, m) notation. The crossing points of the band edges, where the gap vanishes, are of particular interest. It can be shown that the (lateral) Bragg condition (i.e., $h_1 s + h_0 d = N\pi$) is satisfied in the middle of the gap. Thus, the only points that can simultaneously satisfy the Bragg condition and have an allowed solution are the points where the band edges cross. These points also correspond to geometries for which resonant, leaky-wave coupling occurs. Note that coupled fundamental modes $(0, m)$ are defined from cutoff $(\beta/k_0 = n_0)$ to the resonance points of the coupled first-order modes $(l, m-1)$. There, the $(0, m)$ modes transform into coupled second-order modes $(2, m-2)$.

A formulation which treats the Bloch solutions at the band edges as standing waves[94] can be used to compute the (l, m) modes at the resonance points. The pair of band-edge solutions that border a gap, Ψ_{I} and Ψ_{II}, are associated with the Nth Bragg order $(q\Lambda = N\pi)$. The solutions are shown in Table 1.3. The coefficients A, A', B, and B' can easily be determined by matching the boundary conditions. If N is even, the period for $\Psi(x)$ is Λ (i.e., in-phase array modes), and the solutions Ψ_{I} and Ψ_{II} can be labeled symmetric and antisymmetric[95] respectively. When N is odd, the period is 2Λ (i.e., out-of-phase array modes). For example, the $(0, 1)$ and $(1, 0)$ modes constitute the pair of band-edge solutions when $N = 2$. Since at resonance the stopband is zero, modes $(0, 1)$ and $(1, 0)$ become degenerate, and can be constructed simply from lateral half-cosines and half-sines (see Figure 2(*b*) in Ref. 85).

Figure 1.27 shows the solutions (β) as a function of s for modes 8–60 in a ten-element antiguided array[15], with the same parameters used for the infinite case. The modes are numbered sequentially from left to right and correspond to the mode numbers for a ten-element array[15]. The band-edge infinite-array solutions are indicated by dotted lines. All the allowed values of β for the finite array lie between the infinite-array band-edge solutions. The band-edge modes of finite arrays have similar characteristics to the band-edge modes of infinite arrays. For example, the near field of mode 18[15] is similar to the near field of the $(0, 1)$ mode, since it corresponds to the $(0, 1)$ band edge (Figure 1.26). Consequently, the resonance point for mode 18 is identical to the resonance point for mode $(0, 1)$. The upper set of resonance points in Figure 1.27 corresponds to resonances of in-phase-coupled (modes 18 and 36) and out-of-phase-coupled (mode 27) fundamental modes as s equals odd and even numbers of $\lambda_1/2$ respectively. The lower set of resonance points corresponds to

Table 1.3. *Band-edge solutions Ψ_I and Ψ_{II} for infinite-extent ROW arrays. At resonance, $h_0 = N_0\pi/d$, $h_1 = N_1\pi/s$, where N_0 and N_1 are integers such that $N_0 + N_1 = N$*

N	Ψ_I for $\lvert x\rvert < d/2$	Ψ_I for $d/2 < x < d/2 + s$	Ψ_{II} for $\lvert x\rvert < d/2$	Ψ_{II} for $d/2 < x < d/2 + s$
Even	$\cos(h_0x)$	$A\cos[h_1(x - \Lambda/2)]$	$\sin(h_0x)$	$B\sin[h_1(x - \Lambda/2)]$
Odd	$\cos(h_0x)$	$A'\sin[h_1(x - \Lambda/2)]$	$\sin(h_0x)$	$B'\cos[h_1(x - \Lambda/2)]$

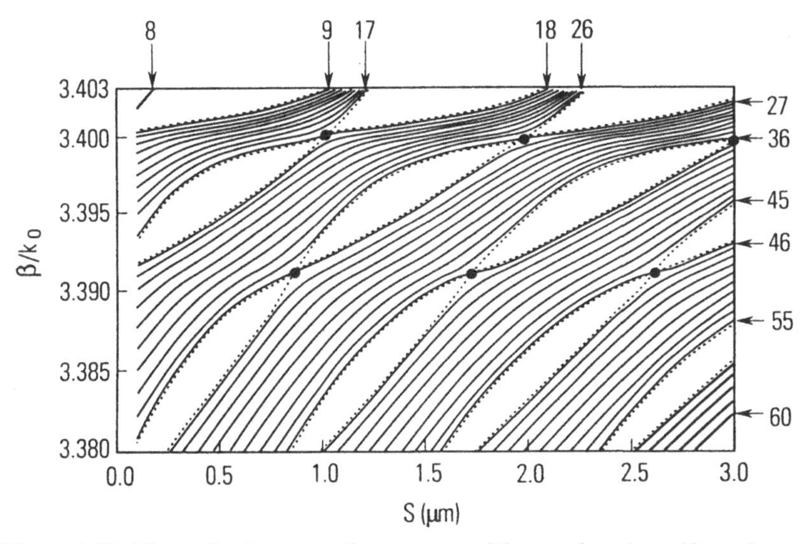

Figure 1.27. Normalized propagation constant β/k_0 as a function of interelement spacing, s, for leaky modes 8–60 of a ten-element antiguided array for the case $d = 3$ μm, $n_0 = 3.40295$ and $n_1 = 3.42646$. The solid dots indicate resonance points, and the dotted lines indicate the band-edge solutions for the infinite-extent array. (Ref. 85.)

resonances of in-phase and out-of-phase-coupled first-order (element) modes[15] (i.e., modes 28, 37, and 46).

The desired (0, 1) mode is resonant when the second-order Bragg condition is exactly satisfied; that is, when $h_1s + h_0d = 2\pi$ ($\beta/k_0 = 3.4$, $s = 1$ μm in this case). Therefore, a ROW array can be thought of as a second-order DFB. Figure 1.28 is a schematic of a ROW array as a second-order DFB structure. The lateral wave vectors $\pm h_0$ and $\pm h_1$ are analogous to the forward and backward waves in a second-order DFB

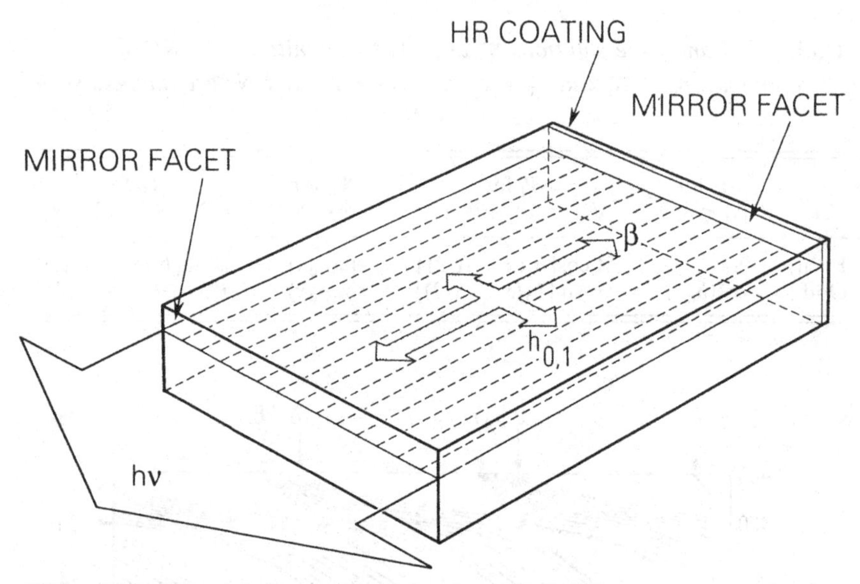

Figure 1.28. Schematic illustrating the analogy of a ROW array to a second-order DFB laser. The lateral wave vectors $\pm h_0$ and $\pm h_1$ are analogous to the forward and backward waves of a second-order DFB laser, while β is analogous to the (first-order) radiation component of a second-order DFB structure. Feedback is provided by mirror facets on the radiative component. (Ref. 85.)

laser, while the longitudinal component β is analogous to the (first-order) radiation component (surface emission) of a second-order DFB laser. For second-order DFB lasers, only the symmetric mode radiates[95], and similarly for finite arrays of antiguides at the (0, 1) resonance, only the symmetric mode (i.e., mode 18 in Figure 1.27) exists.

1.4.5 Intermodal discrimination

There are three intrinsic intermodal discrimination mechanisms in arrays of antiguides:

(1) modal overlap with the gain region[25,91], so-called Γ-effect[87];
(2) edge radiation losses[12–15];
(3) interelement loss[1,15].

There is also an 'extrinsic' modal discrimination mechanism: diffraction in Talbot-type spatial filters[96]. To understand them we plot the near-field intensity profiles of the in-phase mode in Figure 1.29(*a*), the out-of-phase mode in Figure 1.29(*b*) and the next highest mode to the in-phase mode,

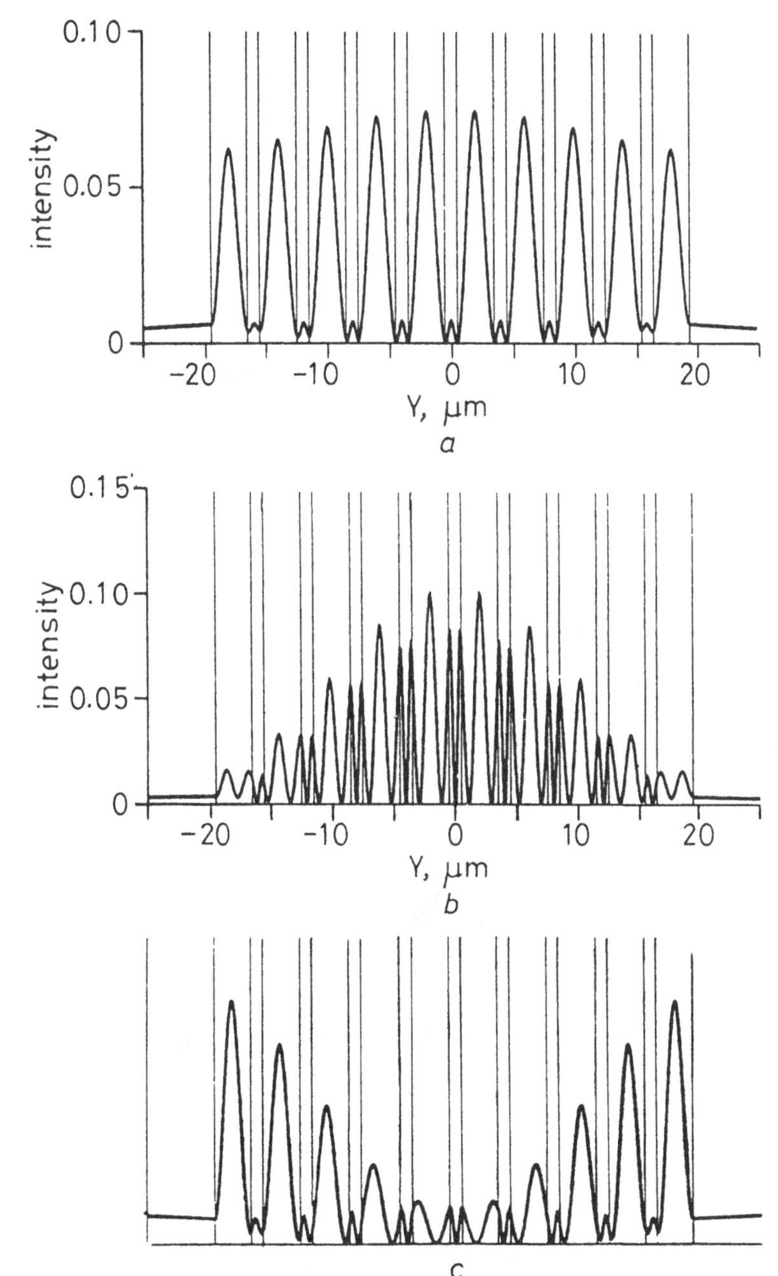

Figure 1.29. Near-field intensity profiles for selected array modes of a ten-element ROW array; $d = 3$ μm, $s = 1$ μm, $\lambda = 0.86$ μm, $\Delta n = 2.8 \times 10^{-2}$: (*a*) in-phase mode ($L = 18$); (*b*) out-of-phase mode ($L = 27$); (*c*) adjacent mode ($L = 19$). (Ref. 91.)

the so-called upper adjacent mode, in Figure 1.29(*c*) for a ten-element array designed to operate close to the in-phase mode resonance[15].

1.4.5.1 Γ effect

The low-index array regions will have higher modal gain than the high-index array regions by simply providing higher (transverse) field overlap with the active region than in the high-index regions[12,25,87]. We define Γ_{x0} and Γ_{x1} as transverse optical-mode confinement factors in the element and interelement regions respectively. For arrays of ten or more elements, leaky-wave modes are favored to oscillate over evanescent-wave modes as long as[24,25] $\Gamma_{x0} > \Gamma_{x1}$. For an array mode we are concerned with field overlap with the active region, both transversely as well as laterally; that is, the parameter of interest is the *two-dimensional* optical-confinement factor Γ. For an infinite-extent array Γ is equal to the two-dimensional (2-D) optical confinement factor *within one period*. If, for a given array mode (l, m), the percentage of energy in the antiguide core within one period is $\Gamma_{y,lm}$, it follows that

$$\Gamma_{lm} = \Gamma_{x0}\Gamma_{y,lm} + \Gamma_{x1}(1 - \Gamma_{y,lm}), \tag{1.12}$$

where Γ_{lm} is the 2-D optical-mode confinement factor for array mode (l, m). For example, Figure 1.30 shows the percentage of energy in the antiguide core for the in-phase mode (0, 1): $\Gamma_{y,01}$.

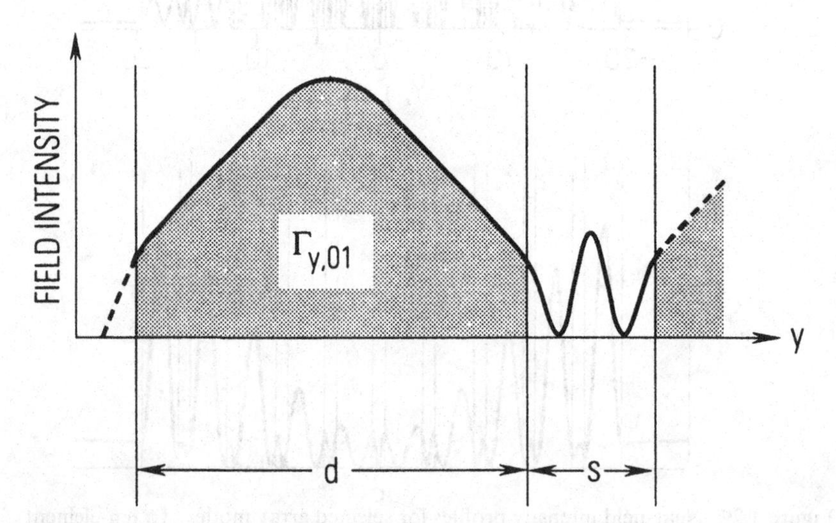

Figure 1.30. Near-field intensity profile for array mode (0, 1) over one array period; $\Gamma_{y,01}$ is the percentage of energy in the antiguide core *within* one period. (After Ref. 87.)

Going back to Eq. (1.9) for the differential gain, and using standard definitions[97] for β_i and ε_{0i} it follows that:

$$\frac{\delta\beta_i}{k\delta\varepsilon_{0i}} = \frac{1}{2n_0}(\delta G_{lm}/\delta G_0), \tag{1.13}$$

where G_{lm} is the array modal gain and G_0 is the local modal gain in the antiguide core ($G_0 = \Gamma_{x0}g_0$, where g_0 is the material gain in the active region). For uniform pumping $G_{lm} = \Gamma_{lm}g_0$, and then from Eqs. (1.10), (1.12) and (1.13) we obtain

$$\Gamma_{y,lm} = \frac{n_0 k}{\beta_{lm}} \cdot \frac{1}{1 + (A_{1,lm}/A_{0,lm})(h_{0,lm}/h_{1,lm})^2}. \tag{1.14}$$

While in positive-index waveguides[98] n_0 can be quite different from $n_{\text{eff}} = \beta/k$, in antiguides and antiguided arrays[15] $n_0 \approx n_{\text{eff}}$. Then a good approximation for Γ_y is

$$\Gamma_{y,lm} \approx [1 + (A_{1,lm}/A_{0,lm})(h_{0,lm}/h_{1,lm})^2]^{-1}. \tag{1.15}$$

By using Eq. (1.15) in Eq. (1.12), it follows that

$$\Gamma_{lm} \approx \Gamma_{x0}/[1 + (A_{1,lm}/A_{0,lm})(h_{0,lm}/h_{1,lm})^2] + \Gamma_{x1}/[1 + (A_{0,lm}/A_{1,lm})(h_{1,lm}/h_{0,lm})^2]. \tag{1.16}$$

At resonance points, where the wavefunctions are exact sines or cosines[85], Γ_{lm} can be shown to be[87]

$$\Gamma_{lm,\text{res}} = \Gamma_{x0}/\{1 + (s/d)^3[(1+1)/m]^2\} + \Gamma_{x1}/\{1 + (d/s)^3[m/(1+1)]^2\}. \tag{1.17}$$

In Figure 1.31 we plot Γs for the case where $\Gamma_{x0} = 0.06$, $\Gamma_{x1} = 0.01$, $d = 3$ μm, $\lambda = 0.86$ μm, $n_0 = 3.40$, and $n_1 = 3.424$ as s varies between 0 and 4 μm. Resonances for modes (0, 1), (0, 2), (0, 3) and (0, 4) occur for[15] $s = m\lambda_1/2$ (1, 2, 3 and 4 μm in Figure 1.31) and are indicated with vertical arrows. Two features stand out: (1) as s varies, the mode favored to lase (i.e., of highest Γ) is alternately in-phase or out-of-phase; and (2) a given mode is always favored to lase at and near its resonance. Both features are due to the fact that resonant modes have least field in the low-Γ_x inter-element regions[15,27]. The asymmetry in modal discrimination with respect to resonance has the same cause as the asymmetry in modal discrimination when Δn varies[87] (Figure 1.24): significantly different amounts of inter-element field for the two closest nonresonant modes (see Figure 1.25).

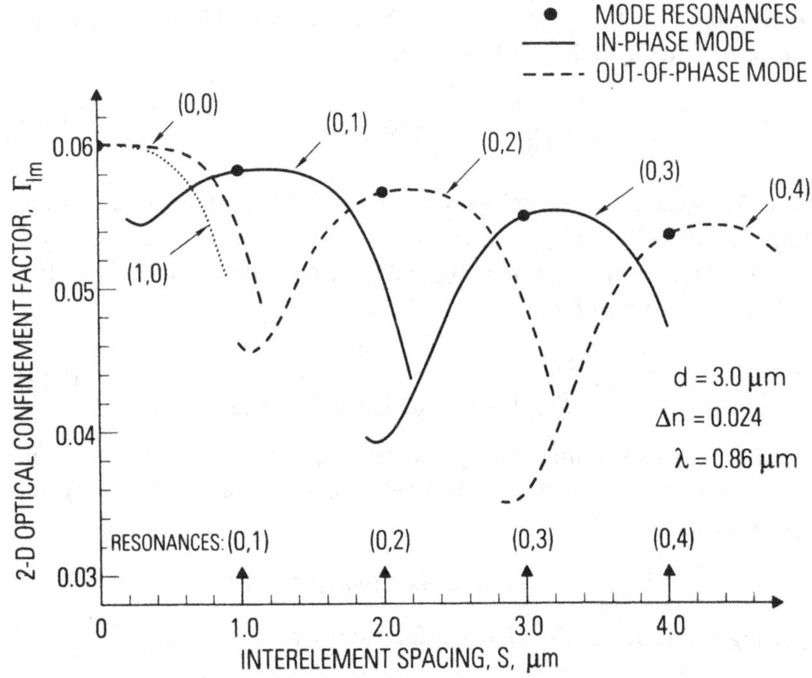

Figure 1.31. 2-D optical-mode confinement factor as a function of interelement spacing for various modes of an infinite-extent array (Ref. 27). The dotted curve corresponds to coupled first-order element modes: array mode (1, 0) (Ref. 27). (After Ref. 87.)

Note that for small s values (≤ 0.8 μm) the out-of-phase mode (0, 0) is favored to lase. Since lowering s is equivalent to lowering Δn, it follows that for small Δn values mode (0, 0) is the one favored to lase. This explains Fujii *et al.*'s results[26] for weakly index-guided arrays (Figure 1.7), as well as Mehuys and Yariv's[99] results of Bloch-function analysis of gain-guided AlGaAs/GaAs arrays. In both cases the array mode favored to lase is a (0, 0)-like leaky mode of a finite array.

In Figure 1.32 we also plot the Γ curve for array mode (1, 0), that is, coupled first-order (element) modes[15]. Such modes are not affected by Talbot-type filters that suppress out-of-phase modes[15]. For Talbot-type devices modes (0, 0) and (0, 2) in Figure 1.31 become irrelevant, while the in-phase mode (0, 1) operates over a wide s range except for $s < 0.64$ μm, when mode (1, 0) takes over. We note that this is quite similar to the situation for low-Δn Talbot-type devices[27].

For Figure 1.31 we employ the results of the Bloch-function analysis using the effective-index method. However, for actual structures, one has

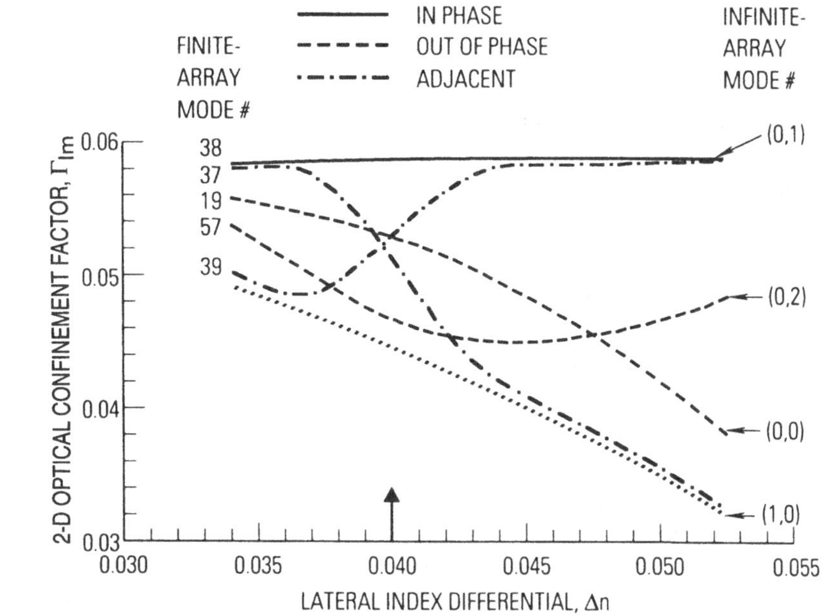

Figure 1.32. 2-D optical-mode confinement factor, Γ, as a function of lateral index step for relevant modes of a 20-element array of antiguides (see Ref. 14). On the right-hand side we indicate the *equivalent* infinite-array modes (Ref. 27) for in-phase mode 38 and out-of-phase modes 19 and 57. Mode (1, 0) (i.e., dotted curve) is the asymptote for adjacent modes 37 and 39 above and below the in-phase-mode resonance respectively. (After Ref. 87.)

to use a 2-D analysis[25,91]. Results of such a rigorous treatment are shown in Figure 1.32 for a 20-element array[14] with $d = 3$ μm, $s = 1$ μm and $\lambda = 0.85$ μm. The behavior of the Γs for in-phase and out-of-phase modes is quite similar to that given in Figure 1.31 and Ref. 27; that is, the resonant in-phase mode (mode 38) is favored at resonance ($\Delta n = 0.04$) and over a wide Δn range on the high-Δn side of resonance. The most interesting aspect is the behavior of the adjacent array modes[15]: 37 and 39. Below and above resonance, modes 37 and 39, respectively, have similar Γs to the in-phase mode 38, as expected, due to their negligible interelement field. As known from exact analysis[15], at resonance the adjacent array modes transform into coupled first-order-element modes. Figure 1.32 confirms this in that mode 39, below resonance, and mode 37, above resonance, approach the Γ curve for mode (1, 0). The practical implication is that on the high-Δn-side of resonance, where high-power coherent arrays are operated[14,82,83,86], one cannot rely on the Γ effect alone to suppress the upper adjacent mode (e.g., mode 39). If there are no means to suppress the adjacent mode, the beam patterns will be $\approx 2 \times$ diffraction limit[82,83,86]; that is, the Γ effect does not ensure pure diffraction-limited-beam operation. In order to achieve diffraction-limited beams to high powers the upper adjacent mode has to be suppressed via edge radiation losses in the array[15] and/or edge diffraction losses in Talbot-type filters[86,100].

1.4.5.2 Edge radiation losses

The edge radiation loss for a single antiguide is given by Eq. (1.5b). In an array, radiation that is normally lost laterally for a single antiguide is used to couple all elements. Only the edge elements leak outwardly giving a loss at resonance expressed by[15,131]

$$\alpha_{RR} \approx \alpha_R / N, \tag{1.18}$$

where N is the number of array elements. For a 20-element array, $\alpha_{RR} \approx 5\ \text{cm}^{-1}$, a value that hardly affects the threshold. The radiation loss vs. index-step curve is well approximated near resonance by a parabola[131]. Adjacent array modes (Figure 1.29(*c*)) have relatively large field at the array edges and, in turn, larger edge losses. The difference $\Delta\alpha$ is shown in Figure 1.33. For a ten-element array[15], $\Delta\alpha = 10\text{–}15\ \text{cm}^{-1}$, which strongly suppressed oscillation of the adjacent mode and thus allows sole in-phase mode operation (i.e., diffraction-limited-beam operation) to high powers. In general $\Delta\alpha = \alpha_{RR}$ with less than 20% error[132].

Figure 1.33 also shows the edge loss for an out-of-phase mode. Being nonresonant, mode 27 has only nearest-neighbor-type coupling and subsequently a cosine-shaped envelope of negligible field at the array edges. However, owing to the Γ-effect and interelement loss[15], oscillation of the out-of-phase mode is suppressed in spite of its lower edge loss compared to the in-phase mode.

1.4.5.3 Interelement loss

Interelement loss can be introduced by placing highly absorbing material in the interelement regions (see Figure 1.18(*a*)). Similarly to the Γ effect, nonresonant modes 'see' significantly more loss than do resonant modes. Results of the Bloch-function formalism can be used to model the effect. Assuming uniform pumping and an interelement loss coefficient α_1, the array modal gain is given by

$$G_{lm} = \Gamma_{lm} g_0 - (1 - \Gamma_{y,lm})\alpha_1, \tag{1.19}$$

where $\Gamma_{y,lm}$ is given by Eq. (1.15). The interelement loss can be due to

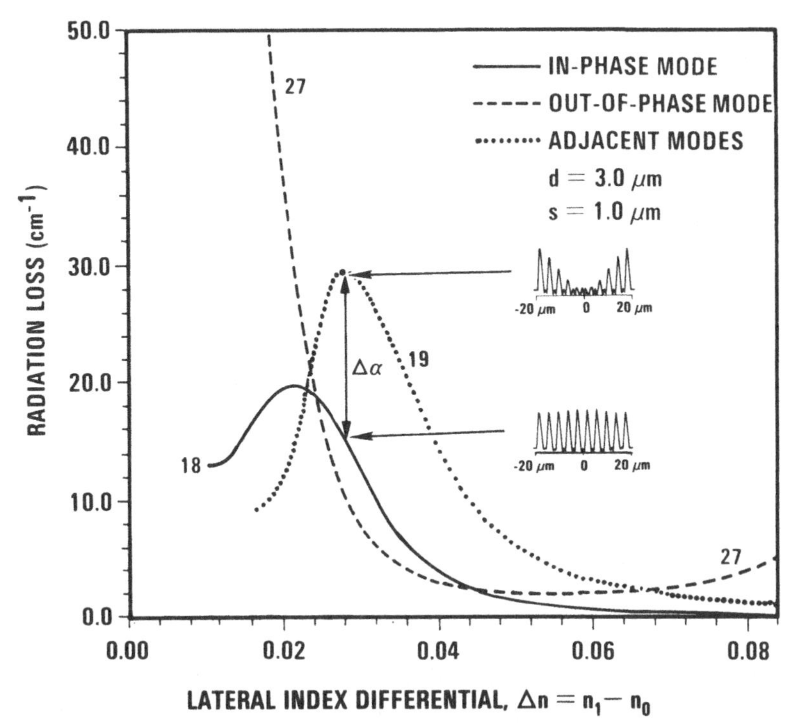

Figure 1.33. Modal radiation loss against lateral index step for modes 18, 19 and 27 of ten-element ROW array; $\lambda = 0.86$ μm (Ref. 15). Insets are near-field intensity profiles of modes 18 and 19 near resonance. (After Ref. 91.)

absorption and/or transverse antiguiding[15]. It has been shown[15] for a ten-element structure with $\alpha_1 = 50$–$90\ \text{cm}^{-1}$ that interelement loss strongly suppresses the out-of-phase nonresonant mode while having a negligible effect on the in-phase resonant mode (Figure 17 of Ref. 15).

1.4.5.4 Diffraction in Talbot-type spatial filters

Talbot-type spatial filters[96–102] employ the imaging properties of freely diffracting radiation from periodically spaced sources[103,104]. The distance at which both in-phase and out-of-phase sources self-image is called the Talbot distance:

$$Z_T = \frac{2\Lambda^2 n_{\text{eff}}}{\lambda}, \tag{1.20}$$

where Λ is the array period, n_{eff} is the effective index, and λ is the vacuum wavelength. At a distance of $Z_T/2$ the in-phase mode image is displaced

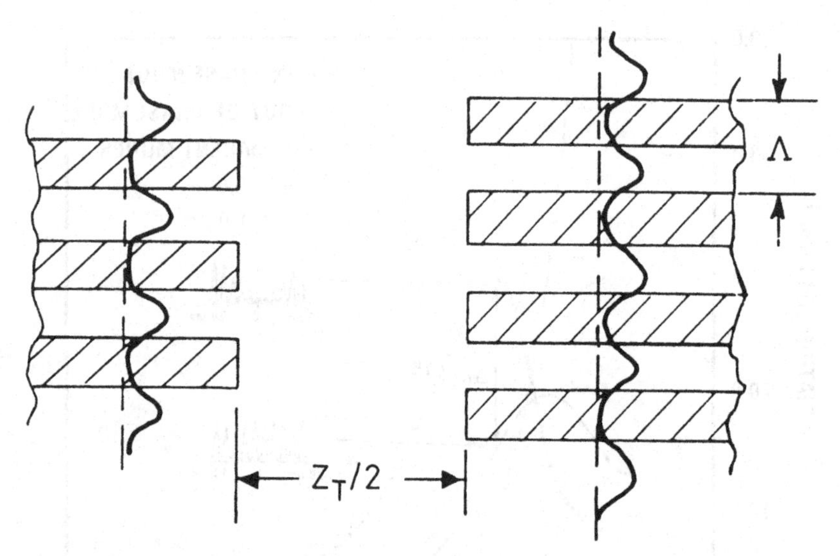

Figure 1.34. Schematic representation of Talbot-type intercavity spatial filter; Z_T is given by Eq. (1.20). (After Ref. 91.)

transversely by a half-period, whereas the out-of-phase mode self-images[103,104]. Thus, a structure made of noncollinear arrays, spaced a distance $Z_T/2$ apart (Figure 1.34), strongly favors in-phase mode operation over out-of-phase operation. In addition, because the adjacent array modes have high field at the array edges, they suffer more edge diffraction losses in the Talbot filter than does the in-phase mode[15,100]. Van Eyck *et al.*[100], in particular, have analyzed how the Talbot filter affects adjacent array modes in 10–11-element arrays.

In general, Talbot filters have the following two effects:

(1) For resonant arrays they provide 'cleaner' far-field patterns at some cost in efficiency[105].
(2) For near-resonant arrays they ensure diffraction-limited-beam operations[86].

Table 1.4 summarizes the effects of the various discrimination mechanisms in arrays of antiguides. At resonance there is no need for Talbot-type spatial filters[106]. Off resonance, whereas the out-of-phase mode may be suppressed by intrinsic array discrimination mechanisms, adjacent modes are not, which in turn will give beamwidths (2–3) × diffraction limit, and, to ensure diffraction-limited-beam operation, Talbot filters are necessary.

Table 1.4. *Mode suppression mechanisms that ensure diffraction-limited-beam in-phase array-mode operation*

Suppression mechanisms	Resonant array	Near-resonant array
	Array modes suppressed	
Γ-effect	out-of-phase	out-of-phase*
Interelement loss	out-of-phase	out-of-phase*
Edge radiation losses	adjacent	–
Talbot-type filter	out-of-phase and adjacent	out-of-phase and adjacent

Note:
* Primarily on the high Δn-side (high s-side) of resonance in the modal loss against $\Delta n(s)$ plots.

The intermodal discrimination mechanisms have a direct effect on the array-mode threshold-current density:

$$J_{\text{th},L} = J_0 + (\beta\Gamma_L)^{-1}[\alpha_{0,L} + (1/2L_0)\ln(1/R_1R_2T^m_{\text{Tf},L})], \quad (1.21)$$

where J_0 is the transparency (area) current density, β is the gain coefficient, Γ_L is the two-dimensional confinement factor for mode L, $\alpha_{0,L}$ is the mode intracavity loss, L_0 is the device length, R_1 and R_2 are the mirror reflectivities, $T_{\text{Tf},L}$ is two-way-averaged transmissivity of a Talbot-type spatial filter for mode L, and m is the number of Talbot filters encountered in a cavity round trip.

For arrays of more than ten elements Γ_L of in-phase and out-of-phase modes is well approximated by Γ_{lm} (Eq. (1.16)). The intracavity loss, $\alpha_{0,L}$, is composed of free-carrier absorption ($\alpha_a \approx 3\ \text{cm}^{-1}$), edge radiation losses, α_e, and interelement loss, α_i. For in-phase and out-of-phase modes at resonance, α_e is given by Eq. (1.18); α_i for mode L is provided by

$$\alpha_{iL} = (1 - \Gamma_{y.lm})\alpha_1, \quad (1.22)$$

as can be seen from Eq. (1.19).

The Talbot-filter transmissivity is asymmetrical since one filter end has at least one more element than the other filter end (see Figure 1.34). The two-way-averaged transmissivity is then defined as

$$T_{\text{Tf}} = \sqrt{(T_+T_-)}, \quad (1.23)$$

where T_+ is the mode transmissivity for light propagating from the filter's narrow end to its wide end, and T_- is the transmissivity for the opposite case.

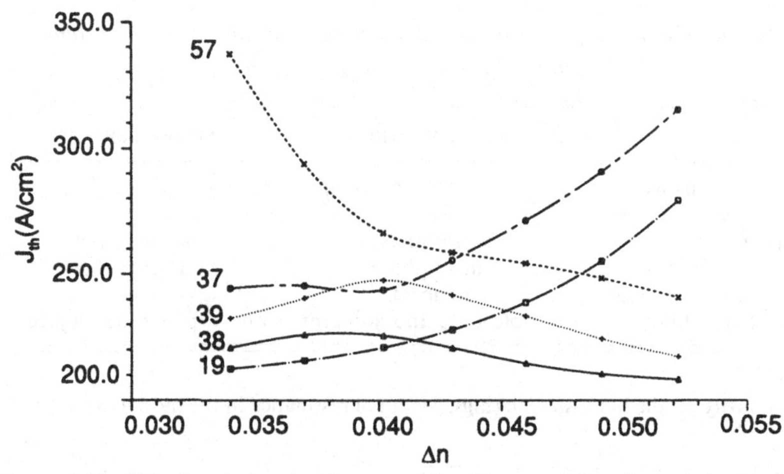

Figure 1.35. Calculated threshold-current densities as a function of Δn for in-phase mode (38), adjacent modes (37, 39), and out-of-phase modes (57, 19). The in-phase mode has lowest threshold on the high-Δn side of its resonance ($\Delta n_{res} = 0.04$). (After Ref. 14.)

Calculated J_{th} curves vs Δn for an actual 20-element structure[14] are shown in Figure 1.35. The calculations take into account the Γ-effect, edge radiation losses and interelement loss. The in-phase mode (number 38) has lowest threshold on the high-Δn side of resonance ($\Delta n_{res} = 0.04$) over the range 0.042 to 0.052, which corresponds to a 3% variation in the Al content of the passive-guide layer. If Talbot filters are considered, the discrimination is much stronger since out-of-phase modes 19 and 57 are totally suppressed, and the threshold of adjacent mode 39 increases significantly at and above resonance.

1.4.6 Rigorous modelling

So far we have discussed antiguided arrays by using the effective-index method (EIM)[15]. Hadley has analyzed antiguided arrays using a finite-difference method[25]. However, the EIM is invalid when more than one transverse mode is supported in the interelement regions[107,108], and the finite-difference approach cannot readily be used for device design.

The basic structure to start with is a multiclad lateral waveguide (Figure 1.36) which supports a single transverse mode in the core region and two transverse modes (an even and an odd one) in the outer regions. EIM is a good approximation only when the overlap integral of the fields of the

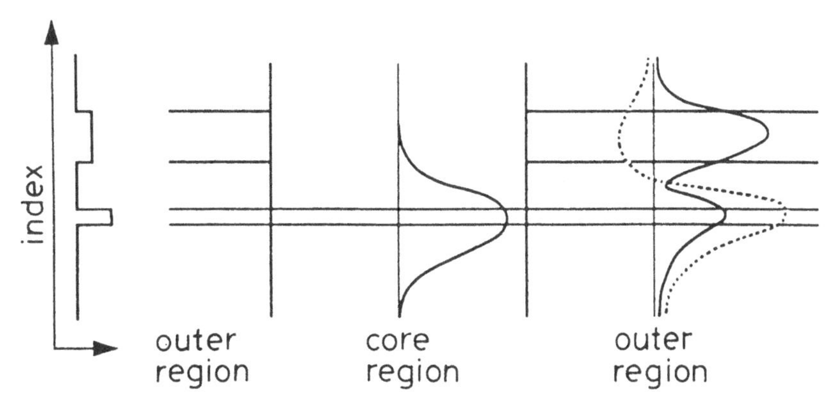

Figure 1.36. Schematic representation of transverse modes in multiclad lateral waveguide. In the outer region(s) the solid and dotted curves correspond to the even and odd transverse modes respectively. (After Ref. 91.)

transverse mode in the core region and one of the modes in the outer region is close to unity. When substantial overlap occurs with the even mode (i.e., $|C_{01}|^2 \approx 1$)[107] the lateral waveguide is close to a pure antiguide, a structure also known as LCSP[109]. When substantial overlap occurs with the odd mode (i.e., $|C_{02}|^2 \approx 1$) the lateral waveguide is a positive-index guide, the familiar CSP[110]. For intermediate cases the only rigorous solution is that pointed out by Amann[108], i.e., match boundary conditions for fields and field derivatives of supported modes and four to ten radiation modes. The number of modes in the core and outer regions must be the same; that is, if N radiation modes are considered in the core, $N - 1$ radiation modes should be considered in the outer regions.

Hadley *et al.*[106] and ourselves[91] have extended Amman's model for mutliclad waveguides to arrays of antiguides. Ten to twenty transverse modes are used for each element and interelement region. The novel 2-D code that we developed is called MODEM[91] and has been applied to 20-element 3 to 1-geometry GRINSCH structures[86,105]. Figure 1.37 shows the main differences between EIM calculations and MODEM calculations. We plot the modal radiation losses of in-phase mode 38 and adjacent modes 37 and 39 as a function of Δn (which is the lateral index step in EIM) and the difference between the (interelement) even-mode propagation constant, and the transverse-mode propagation constant in the core in MODEM. The major difference is that resonance (i.e., the point of maximum radiation loss and uniform near-field intensity profile for in-phase mode 38) occurs at significantly higher Δn values than for EIM. This means that to achieve resonance we have to incorporate

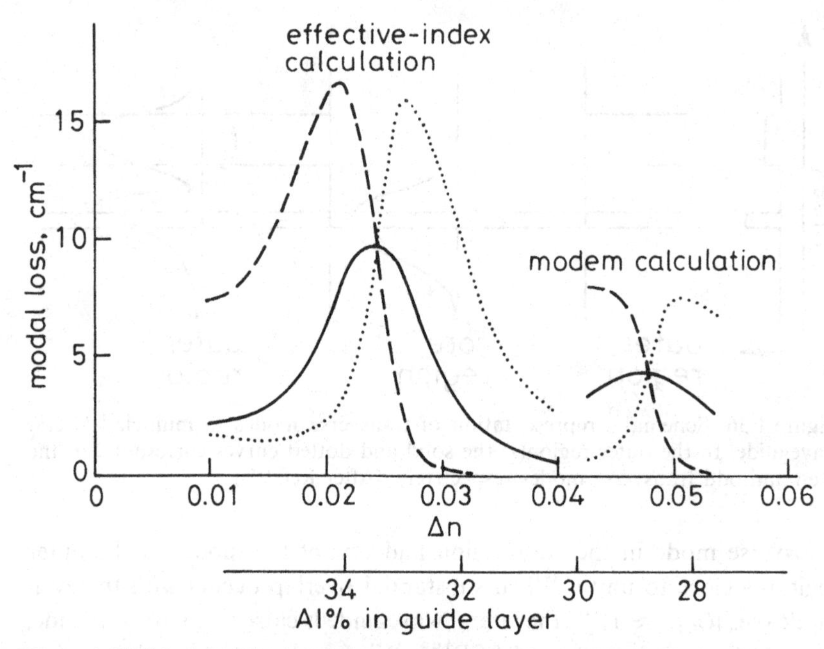

Figure 1.37. Modal radiation loss of modes 38, 37 and 39 of 20-element array (Ref. 105) as function of Δn, lateral index step in the effective-index method calculation, and difference between propagation constants of even (interelement) transverse mode and fundamental (element) transverse mode in the MODEM calculation. The solid, dashed and dotted curves correspond to in-phase mode 38, adjacent mode 37 and adjacent mode 39 respectively. (After Ref. 91.)

$\approx 5\%$ less Al in the passive-guide layers of the interelement regions. We also note that the amount of discrimination between the in-phase mode and the adjacent modes decreases while remaining significant (4–5 cm^{-1}). Although the practical impact is not serious, the conceptual impact is at the very least puzzling: the interelement region at resonance is no longer an odd number of lateral half-wavelengths. The reason is that interelement regions contain both traveling waves (the leaky components), as well as the tails of evanescent waves. Thus, the interelement region cannot be characterized by an (effective) index of refraction.

Recent work on dual-state Fabry–Perot etalons[111] has clarified the origin of the resonant-transmission-peak shift found in Figure 1.37. Total internal reflection within the interelement regions creates a phase shift $\Delta\phi$, which is a function of the overlap integral $|C_{01}|^2$. As shown in Figure 1.38 for a typical structure[85], $\Delta\phi$ varies between 0 and 0.6π as $|C_{01}|^2$ varies between 1 and 0. The structure analyzed in Figure 1.37

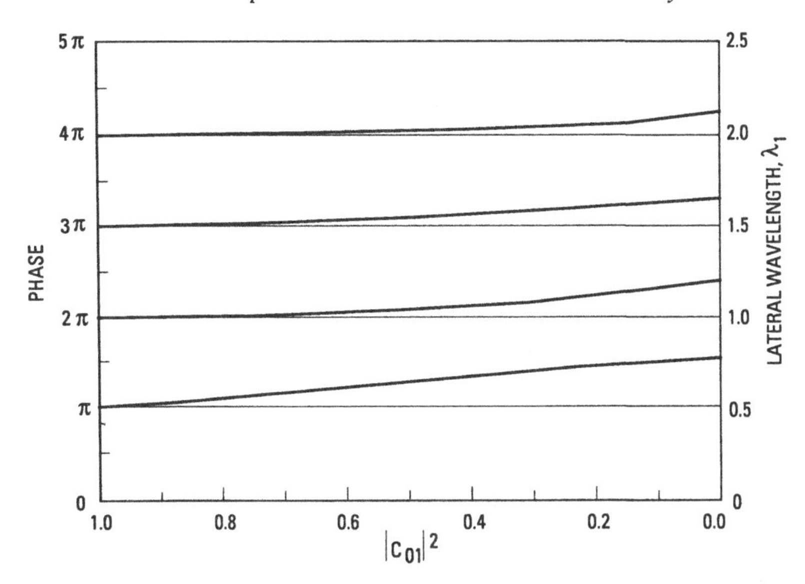

Figure 1.38. Resonant-transmission-peak location as a function of the field-overlap parameter between the transverse mode in the antiguide core and the even mode in the outer region, $|C_{01}|^2$ (Ref. 107). (After Ref. 111.)

has poor field overlap $|C_{01}|^2 = 0.31$ and thus the phaseshift with respect to the half-wave condition should be substantial. While for a purely antiguided device (i.e., resonance at $\Delta n = 2.4 \times 10^{-2}$) the interelement region corresponds to a phase change of π, for the structure with $|C_{01}|^2 = 0.31$ (i.e., resonance at $\Delta n = 4.7 \times 10^{-2}$) the interelement region corresponds to a phase change of 1.385π; that is, as deduced from the exact calculation in Figure 1.37, $\Delta\phi$ is 0.385π. This value agrees, within 5%, with the $\Delta\phi$ value (0.405π) obtained from Figure 1.38. As can be seen from Figure 1.38, for structures with $|C_{01}|^2 \geq 0.65$, EIM provides a good approximation as to where resonance occurs, as observed in prior work[15,106].

1.4.7 Relevant electro-optical characteristics

In pulsed operation, 20-element devices have demonstrated diffraction-limited beams to 1.5 W and 10.7 × threshold[112], and to 2.1 W and 13.5 × threshold[14] (Figure 1.39). At the 2.1 W power level only 1.6 W is coherent, uniphase power and the main lobe contains 1.15 W, so-called 'power in the bucket'. The fact that diffraction-limited beams can be maintained to very high drive levels is a direct consequence of the

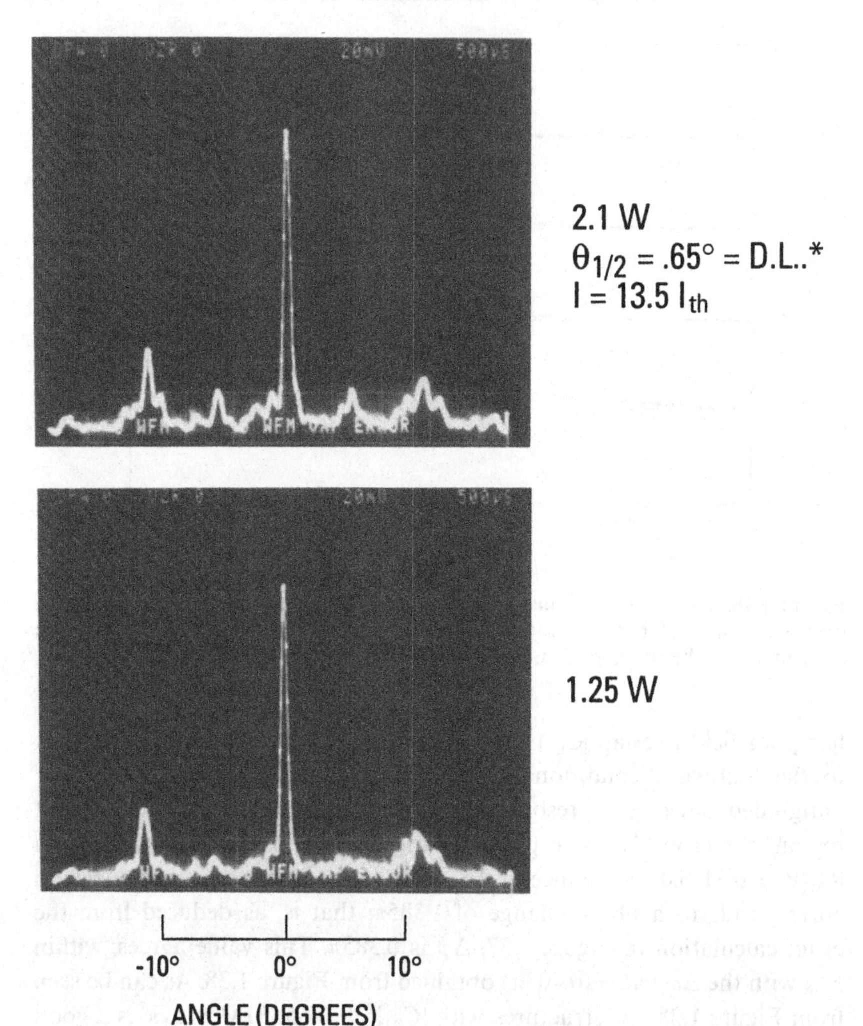

Figure 1.39. Lateral far-field beam patterns at several power levels for a resonant array with a monolithic Talbot filter. A diffraction-limited beam is maintained above the 2 W output power level. (After Ref. 14.)
* D.L. = diffraction limited.

inherent self-stabilization of the in-phase resonant mode with increasing drive level[113].

When carrier diffusion is taken into account[114], it has been shown that for devices with $\Delta n = 0.025$, the adjacent mode reaches threshold (i.e., multimode operation starts) at five to seven times the threshold of the in-phase mode. Furthermore, if the index step is increased to ≈ 0.16, the

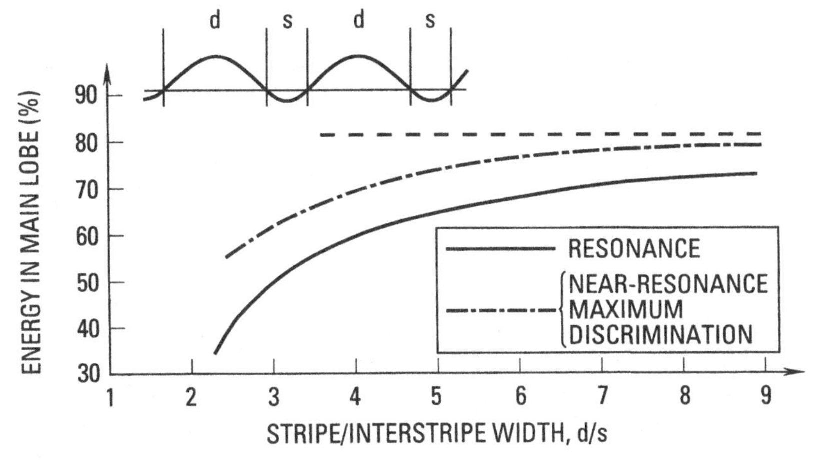

Figure 1.40. Percentage of power in the main far-field lobe as a function of element/interelement width ratio. (After Ref. 118.)

effect of gain spatial hole burning is negligible, such that one can achieve single-(in-phase)-mode operation to (20–25) × threshold[114].

For devices with a 3 to 1 element to interelement width ratio the percentage of coherent power in the main lobe is typically 60%. Figure 1.40 shows that with increasing element/interelement width ratio the main-lobe power percentage will increase to the theoretical limit of 81%[115]. In any event, as long as the emission is coherent, one can use aperture-filling optical techniques[116] to gather close to 100% of the energy in the main lobe. The far-field dependence with varying heatsink temperature has also been studied (Figure 1.41). As can be seen from the diagram, the beam pattern is stable at 0.2 W over a 90 °C range, and at 0.5 W over a 65 °C range. These values are in good agreement with theoretically expected values of ≈100 °C, corresponding to the region over which the device is resonant as the wavelength peak varies with temperature (≈3 Å/°C).

In cw operation, one is limited by thermal effects. The best result in a diffraction-limited beam is 0.5 W cw (Figure 1.42). The result has been achieved by optimizing the device overall efficiency. For 1000 μm-long devices with HR and AR facet coatings (98 and 4% reflectivity), slope efficiencies of 48–50% are observed and the power conversion efficiency reaches values in the 20–25% range at 0.5 W output. Preliminary lifetests show room-temperature extrapolated lifetimes in excess of 5000 h[117].

Recently, by using large-aperture (120 μm), 20-element structures of

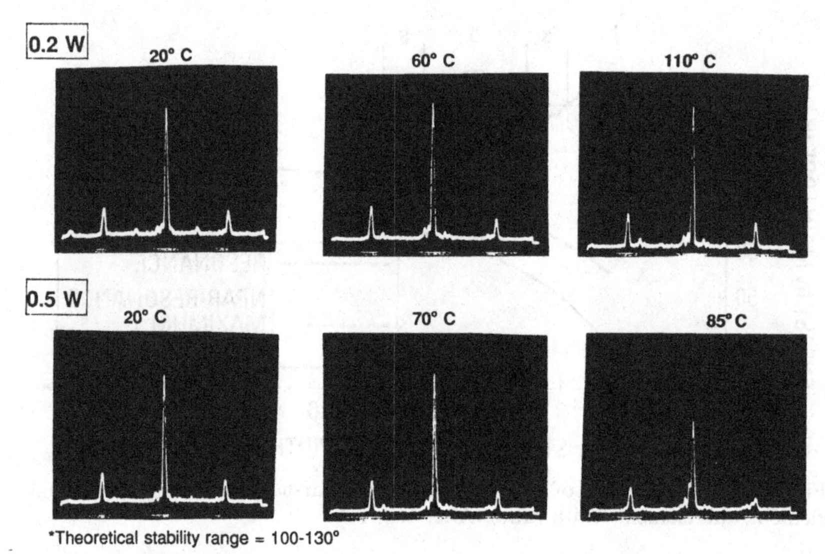

Figure 1.41. The far-field pattern of a resonant array as a function of heat-sink temperature at two different power levels.

5 to 1 element/interelement width ratio, we were able to obtain[118] 1 W cw operation in a beam 1.7 × diffraction limit (Figure 1.43). Close to threshold (0.1 W) the beam is diffraction limited, and 75% of the energy resides in the main lobe (as expected from Figure 1.40). The 1 W value represents the highest cw coherent power achieved to date from any type of fully monolithic diode emitters (i.e. without phase-correcting optics).

The results presented so far are for AlGaAa/GaAs structures ($\lambda = 0.84$–0.86 μm). Antiguided arrays have also been made from strained-layer quantum-well material ($\lambda = 0.92$–0.98 μm) in both the CSA[119] and the SAS-type array configurations[82,83]. Resonant devices were achieved only with the CSA configuration, and provided 1 W pulsed diffraction-limited-beam operation[119]. From SAS-type devices Major *et al.*[83] have demonstrated stable and efficient cw operation to 0.5 W in a beam 1.5 × diffraction limit.

Nonresonant devices typically have beams (2–3) × diffraction limit. First of all, the in-phase mode has a raised-cosine-shaped envelope and thus gain spatial hole burning is nonuniform across the device causing mode self-focusing[113,114], similar to evanescent-wave coupled devices[23]. Then adjacent modes can readily reach threshold and the beamwidth increases to (2–3) × diffraction limit. The combination of the in-phase mode and one or two adjacent modes uses the available gain efficiently, thus not

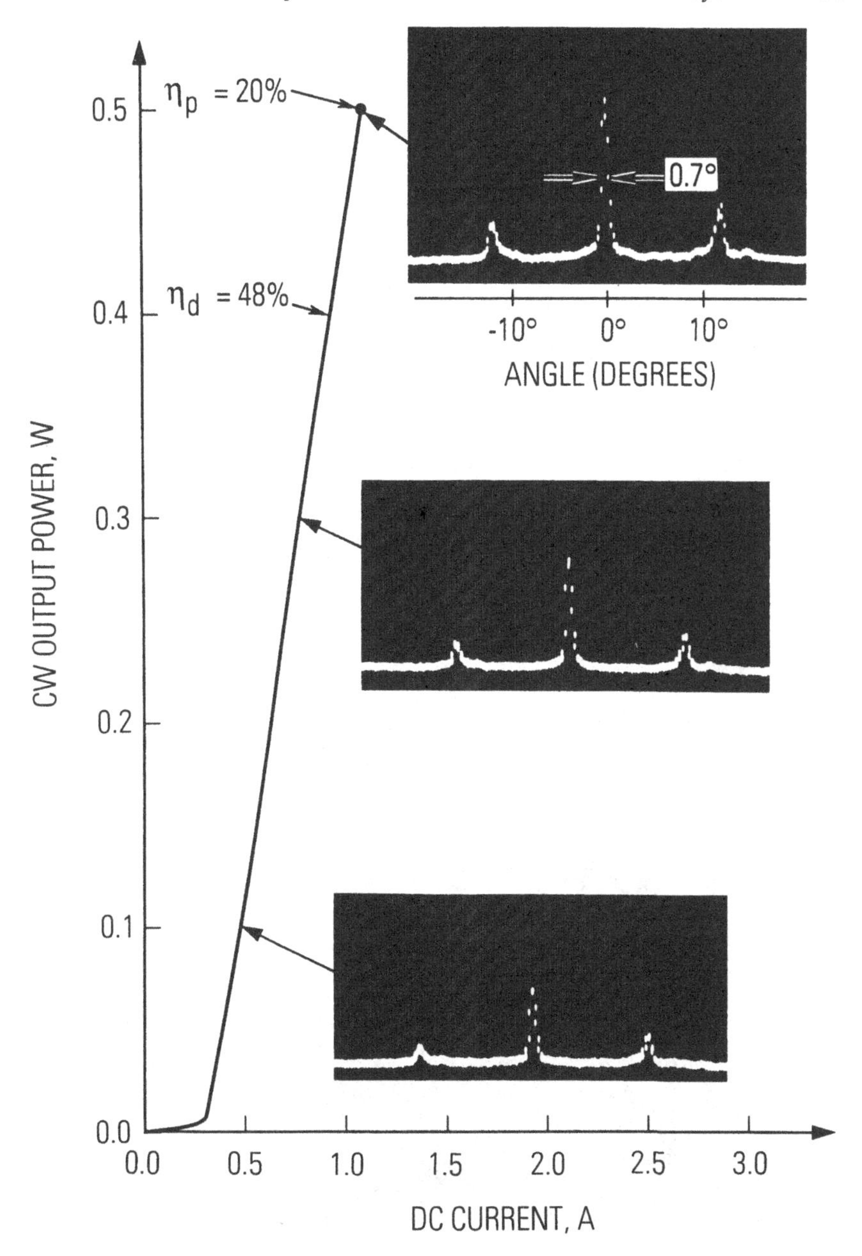

Figure 1.42. CW operation of optimized 20-element ROW arrays: light-current characteristic and lateral far-field patterns at various power levels. Beam pattern is diffraction-limited up to 0.5 W. (After Ref. 91.)

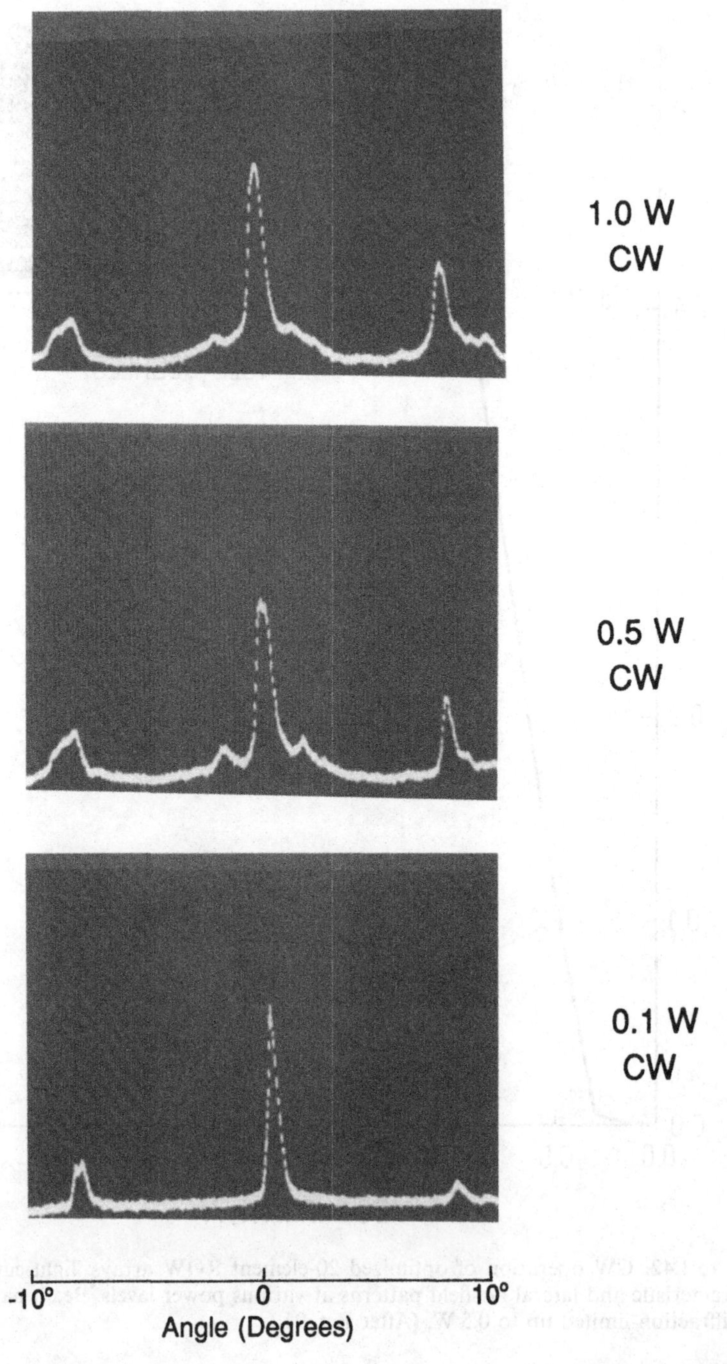

(*see caption opposite*)

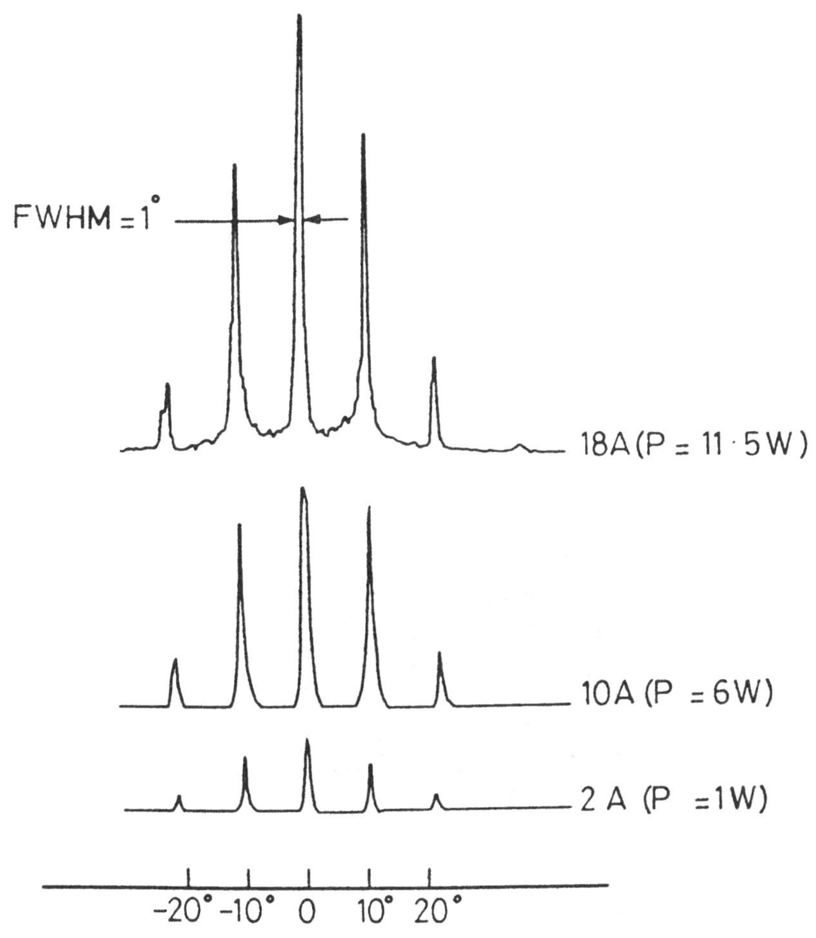

Figure 1.44. Pulsed far-field patterns of 40-element antiguide array with large index step ($\Delta n = 0.19$). The beam is 2.5 × diffraction limit up to 36 × threshold (11.5 W). (After Ref. 120.)

allowing other modes to come in. This explain why stable beams can be maintained to very high drive levels and peak powers: 5 W to 45 × threshold from 80 μm-wide aperture devices[86], and 11.5 W from 185 μm-wide aperture devices[120] (Figure 1.44). For the latter, up to 32 W was achieved in a similar beam[133]. It should also be mentioned that the insertion of Talbot-type filters in near-resonant devices provides virtually

---

Figure 1.43. CW far-field patterns of 20-element ROW array with 5 μm-wide elements and 1 μm-wide interelements at various power levels. (After Ref. 118.)

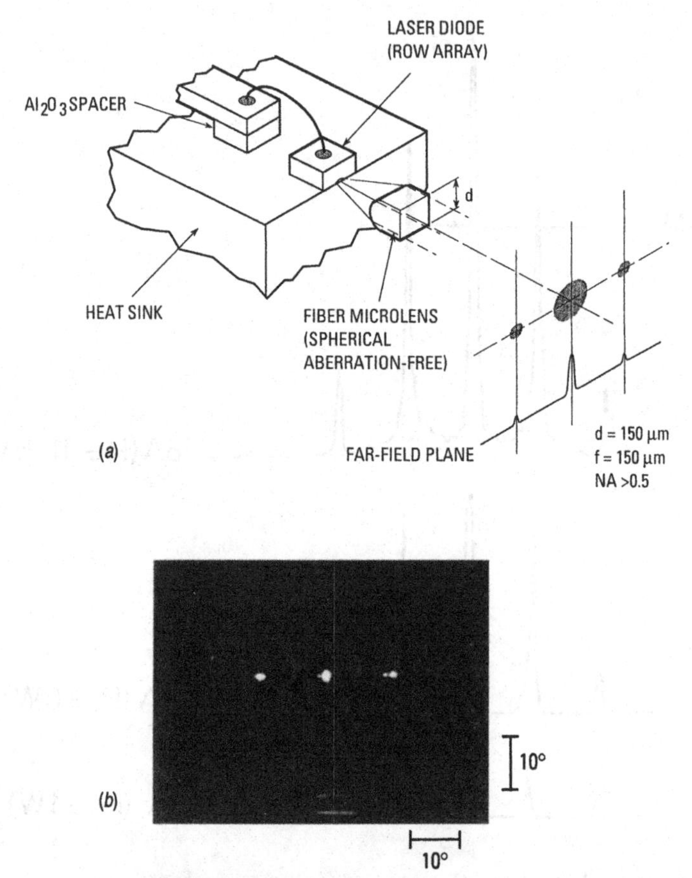

Figure 1.45. Collimation of ROW-array beam pattern: (*a*) schematic diagram of fiber microlens setup; (*b*) far-field pattern at 0.25 W cw.

diffraction-limited beam patterns[86], since the adjacent modes are suppressed via edge diffraction losses in the spatial filter(s).

The concern so far has been to obtain a narrow ($\leq 1°$) diffraction-limited beam in the plane of the junction. However, perpendicular to the junction the beamwidth is quite large ($\theta_{1/2} \approx 40°$), which gives a highly elliptical beam. By using aberration-free fiber microlenses developed by J. Snyder[121] (see Figure 1.45(*a*)), we were able to obtain virtually circular diffraction-limited beams ($0.7° \times 0.6°$, as shown in Figure 1.45(*b*)) up to 0.25 W cw collimated output power.

There has been concern that, owing to their large size, arrays will have relatively slow pulse response. In fact, as shown in Figure 1.46,

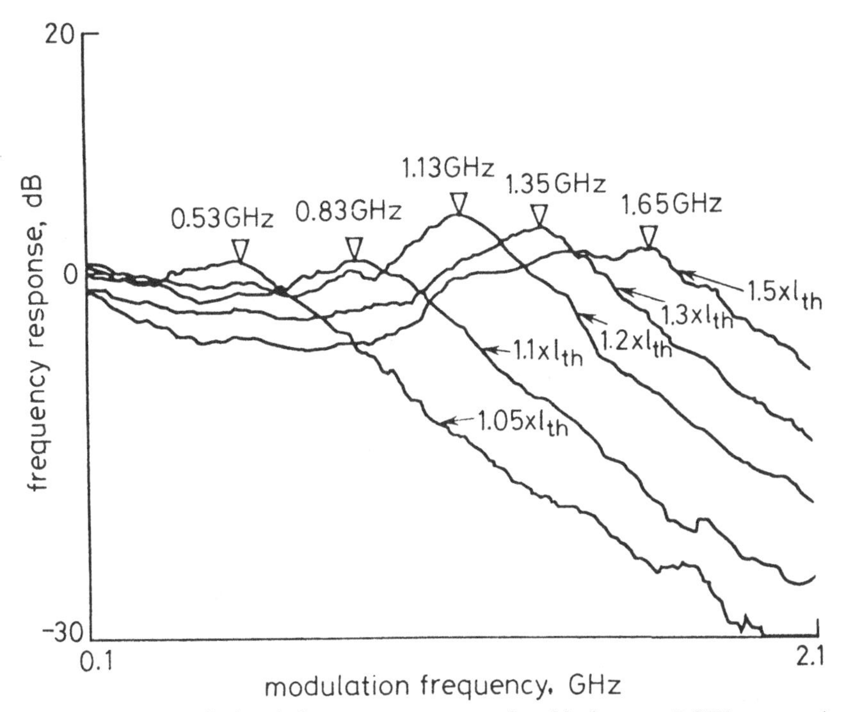

Figure 1.46. Small-signal frequency response for 20-element ROW array. At 1.5 × threshold the 3 dB modulation bandwidth is ≈1.85 GHz.

3 dB modulation bandwidths of 1.8 GHz can be achieved at only 1.5 × threshold[122]. Theoretically these meet expectations when considering that single-element devices are theoretically capable of 30–35 GHz bandwidths. Preliminary tests show pulse response times of <150 ps. AM large-signal modulation may prove quite a challenge, in which case FM modulation may be an attractive alternative for frequency-stabilized devices.

ROW arrays can be made single-frequency by injection locking[123] or by the use of an external diffraction grating[124]. Only a narrow (5 μm-wide) injection spot on the device back facet is needed for full array injection, a consequence of parallel coupling in the device. The master-oscillator wavelength can be changed over a ≥30 Å interval, before locking is no longer possible. The most interesting result is that while tuning the wavelength the emitted beam is stable[123], unlike the beam steering experienced in injection-locked broad-area oscillators.

1.4.8 Conclusions

Gain-guided devices were the first antiguided-type array structures. Some notable results are: 0.8 W peak pulsed power in an out-of-phase pattern with beamwidths 10–20 × diffraction limit[125], and 0.22 W cw near-diffraction-limited, single-lobed operation from a chirped, interferometric-type device[126]. However, due to the inherently weak optical-mode confinement of gain-guided devices, the beam pattern changes with injected current, and from pulsed to cw drive conditions.

Parallel coupling in strong index-guided structures has allowed ROW arrays to reach stable diffraction-limited operation to powers 20–30 times higher than for other array types. This finally fulfils the phase-locked arrays promise of vastly improved coherent power by comparison with single-element devices. Table 1.5 summarizes the best results to date.

1.5 Future prospects for all-monolithic antiguided arrays

For edge-emitting coherent arrays the eventual limitation may just be thermal; that is, just as for incoherent devices, maximum emitting apertures would be 400–500 μm wide. Then the maximum projected DL powers would be in the 3–5 W range.

Beyond 5 W, one will have to resort to 2-D surface-emitting arrays. The 2-D array unit cells will be ROW arrays resonantly coupled via radiation leakage (Figure 1.47). Since ROW arrays leak radiation laterally at predetermined angles (8–10° for an index step of $(3–5) \times 10^{-2}$)[79,93], one can build a diamond-shaped configuration for which four ROW arrays are mutually coupled (Figure 1.47). Radiation is outcoupled via 45° micromachined turning mirrors (Figure 1.48(*b*, *c*)) or it could be grating outcoupled. Interunit electrodes ensure, via carrier-induced changes in the dielectric constant, that adjacent units are resonantly coupled

Table 1.5. *Performance to date of strong index-antiguided arrays*

	Maximum power	
Drive condition	DL*	Narrow beam
cw	0.5 W	1 W; 1.7 DL
Pulsed	2.1 W	32 W; 2.3 DL

Notes:
* DL = diffraction limited.

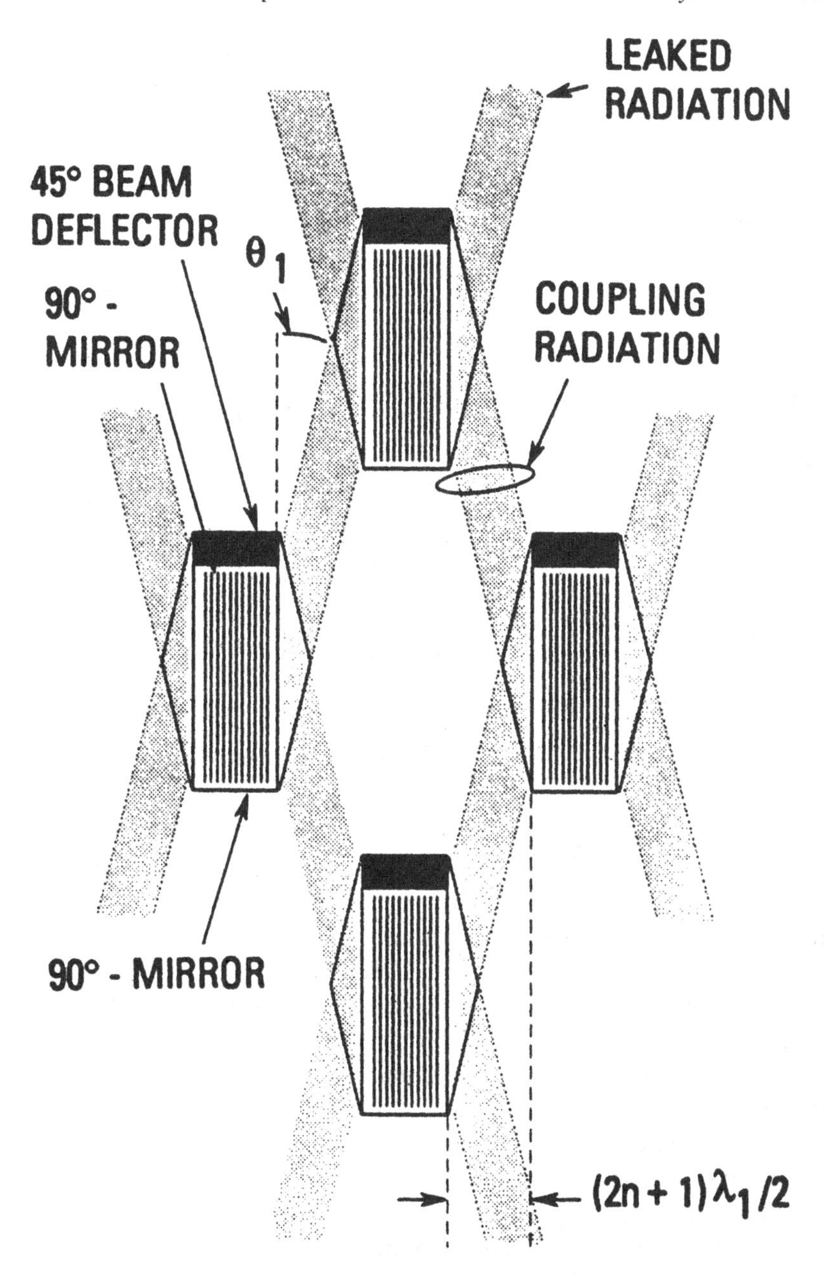

Figure 1.47. Schematic diagram of the four-array, two-dimensional interarray-coupling geometry. Radiation leaked at a fixed angle (θ_1) injection locks neighboring arrays, creating a spatially coherent ensemble. (After Ref. 127, © 1993 IEEE.)

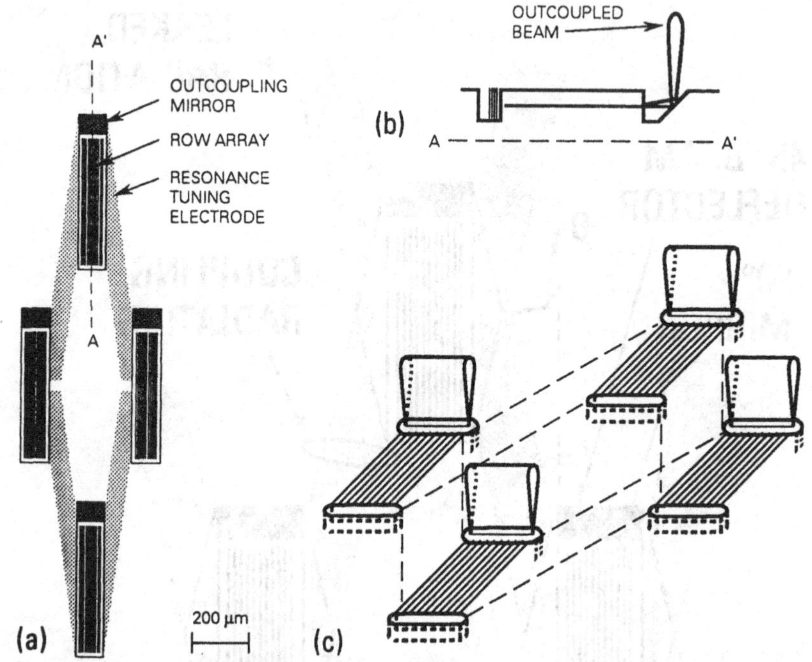

Figure 1.48. Schematic representation of: (*a*) top view of four-array unit cell; (*b*) cross-section of one laser cavity, illustrating the surface emission via 45° ion-milled deflecting mirrors; and (*c*) three-dimensional view of four-array unit cell. (After Ref. 127, © 1993 IEEE.)

(Figure 1.48(*a*)). Leaky-wave coupling of ROW arrays over large distances (90–176 μm) has already been demonstrated in both linear configurations[93] as well as 2-D configurations[127]. In the 2-D diamond-shaped configuration, 2-D beam patterns in excellent agreement with theory (Figure 1.49) have been achieved up to 1.2 W peak power. Preliminary fringe-visibility data are only 35%, but with further work and the possible insertion of diffraction gratings for frequency stabilization, visibility values close to 100% are expected.

Starting from the four-unit diamond configuration one can build large 2-D arrays (Figure 1.50) all locked via resonant leaky-wave coupling. The fact that each array unit is coupled to four nearest neighbors is reminiscent of the parallel-coupling mechanism in ROW arrays. As opposed to previous 2-D array schemes[128] the 2-D ROW array has three significant advantages: (1) it ensures phase locking, *not* frequency locking[128]; (2) it represents the first parallel-coupled 2-D monolithic array; (3) the interunit coupling is independent of the feedback and/or

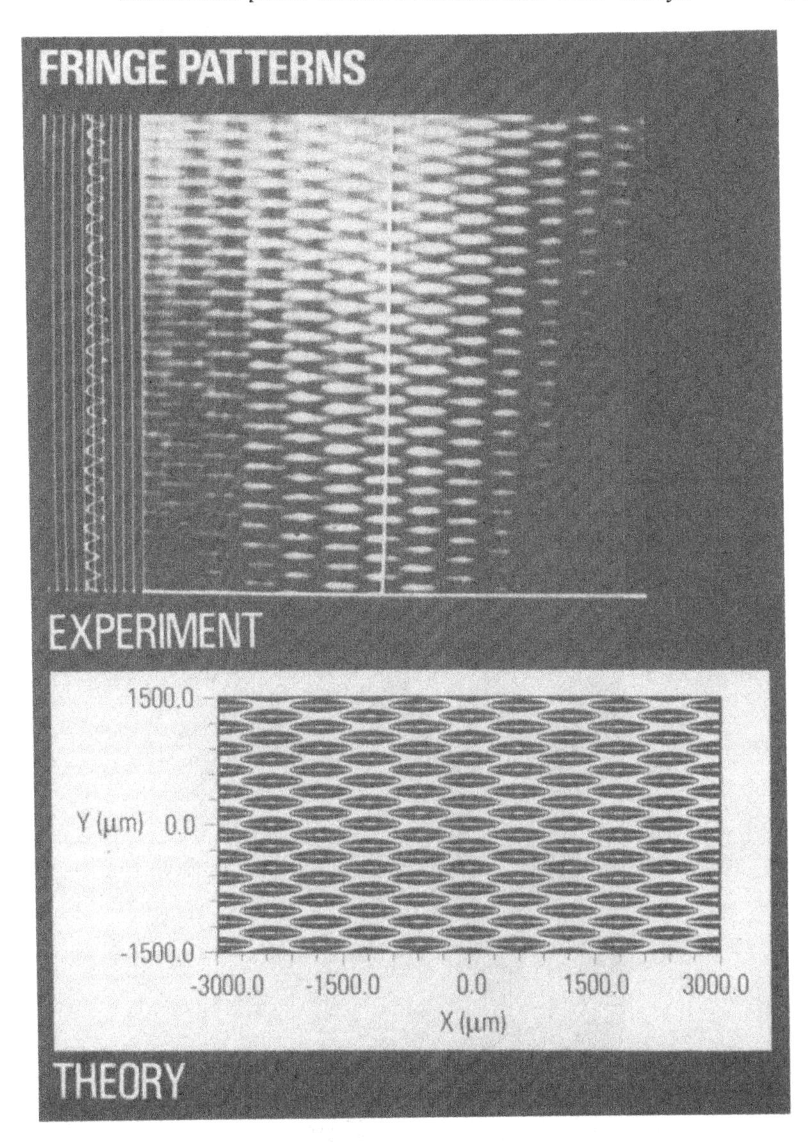

Figure 1.49. Experimental and theoretical far-field patterns for the four-array unit cell operating in the phase-locked condition. The saucer-shaped fringe pattern confirms spatial coherence between all four array sections. A fringe visibility of 35% was measured.

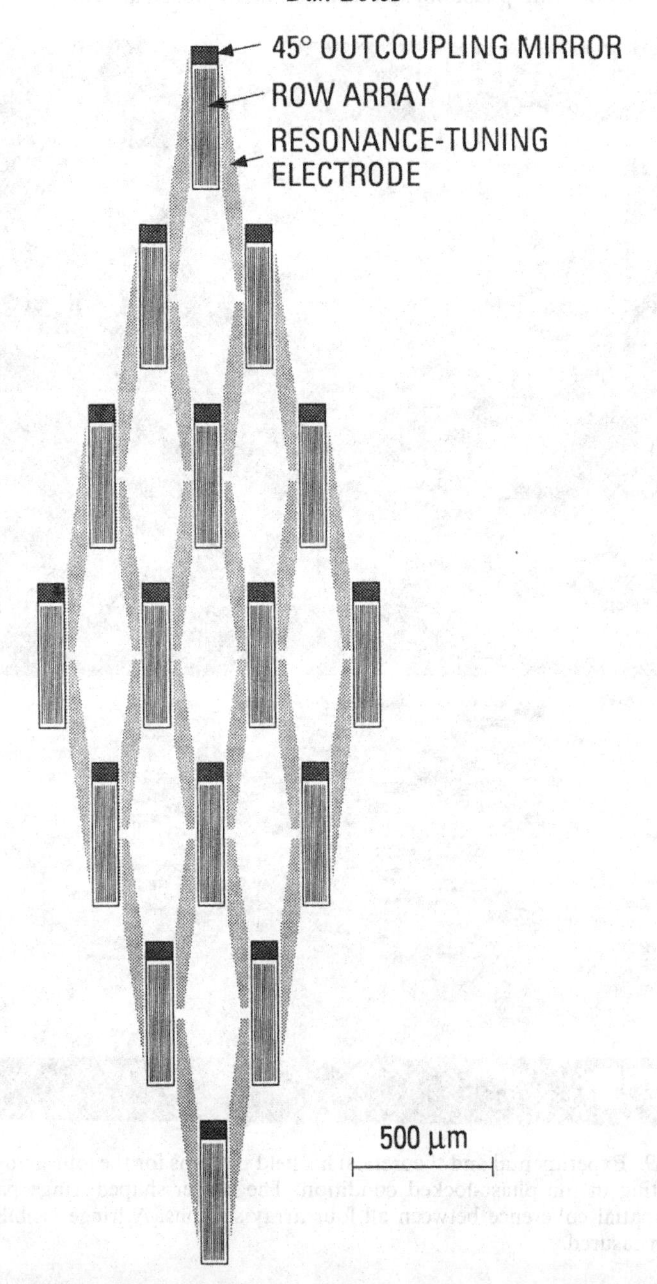

Figure 1.50. Top view schematic of a 16-array 2-D structure. Parallel coupling is possible since each array section has multiple coupling paths via leaked radiation. (After Ref. 127, © 1993 IEEE.)

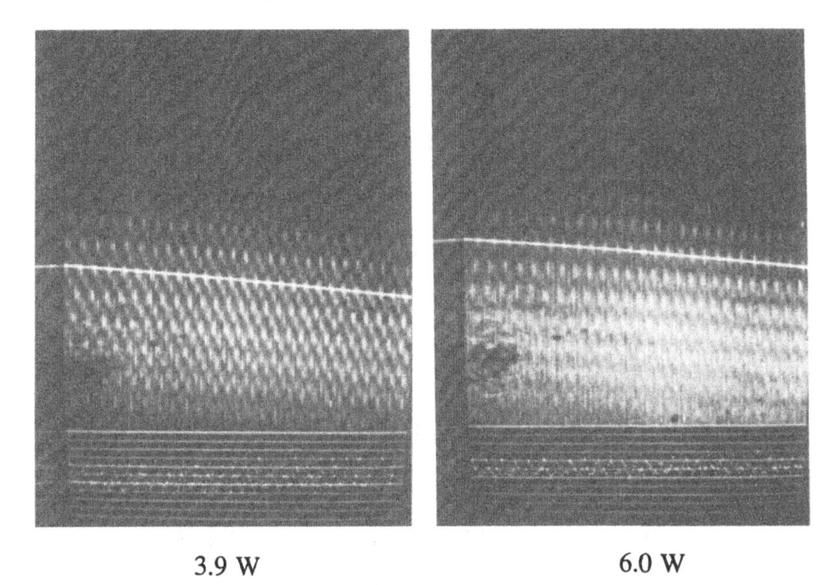

Figure 1.51. Far-field fringe patterns from nine-array surface emitter operating at 3.9 W, and 6.0 W output power levels. Spatial coherence is verified by the strong degree of fringe visibility observed (45% at 3.9 W). (After Ref. 127, © 1993 IEEE.)

beam-outcoupling mechanisms. Results from nine-unit arrays are quite encouraging: 3.9 W with 45% visibility and 6 W with 33% visibility (Figure 1.51). For a very large number of units (≥ 100) there is always concern that the device may operate in several mutually coherent regions and thus, to control hundreds of emitters, a master oscillator (i.e., injection locking) may be required[129,130]. All in all we foresee that 2-D ROW arrays are capable of coherent powers in excess of 50 W.

References

1. D. Botez and G. Peterson, *Electron. Lett.*, **24**, 1042 (1988).
2. G. R. Hadley, *Opt. Lett.*, **14**, 308 (1989).
3. D. R. Scifres, R. D. Burnham and W. Streifer, *Appl. Phys. Lett.*, **33**, 1015 (1978).
4. D. E. Ackley, *Appl. Phys. Lett.*, **42**, 152–4 (1983).
5. K. L. Chen and S. Wang, *Electron. Lett.*, **21**, (8), 347–9 (1985).
6. J. Katz, S. Margalit and A. Yariv, *Appl. Phys. Lett.*, **42**, 554 (1983).
7. W. J. Fader and G. E. Palma, *Opt. Lett.*, **10**, 381 (1985).
8. G. R. Hadley, J. P. Hohimer and A. Owyoung, *Appl. Phys. Lett.*, **49**, 684 (1986).
9. D. E. Ackley and W. H. Engelmann, *Appl. Phys. Lett.*, **39**, 27 (1981).
10. D. Botez, P. Hayashida, L. J. Mawst and T. J. Roth, *Appl. Phys. Lett.*, **53**, 1366 (1988).
11. H. Hosoba, M. Matsumoto, S. Matsui, S. Yano and T. Hijikata, *The Review of Laser Engineering*, **17**, 32 (1989) (in Japanese).

12. D. Botez, L. J. Mawst, P. Hayashida, G. Peterson and T. J. Roth, *Appl. Phys. Lett.*, **53**, 464 (1988).
13. D. Botez, L. J. Mawst, G. Peterson and T. J. Roth, *Appl. Phys. Lett.*, **54**, 2183 (1989).
14. L. J. Mawst, D. Botez, C. A. Zmudzinski, M. Jansen, C. Tu, T. J. Roth and J. Yun, *Appl. Phys. Lett.*, **60**, 668 (1992).
15. D. Botez, L. J. Mawst, G. Peterson and T. J. Roth, *IEEE J. Quantum Electron.*, **QE-26**, 482 (1990).
16. J. K. Butler, D. E. Ackley and D. Botez, *Appl. Phys. Lett.*, **44**, 293 (1984).
17. E. Kapon, J. Katz and A. Yariv, *Opt. Lett.*, **9**, 125 (1984).
18. W. Streifer, A. Hardy, R. D. Burnham and D. R. Scifres, *Electron. Lett.*, **21**, 118 (1985).
19. S. R. Chinn and R. J. Spiers, *IEEE J. Quantum Electron.*, **QE-20**, 358 (1984).
20. J. K. Butler, D. E. Ackley and M. Ettenberg, *IEEE J. Quantum Electron.*, **QE-21**, 458 (1985).
21. I. Suemune, H. Fujii and M. Yamanishi, *J. Lightwave Techn.*, **LT-4**, 730 (1986).
22. W. Streifer, A. Hardy, R. D. Burnham, R. L. Thornton and D. R. Scifres, *Electron. Lett.*, **21**, 505 (1985).
23. Kuo-Liang Chen and Shyh Wang, *Appl. Phys. Lett.*, **47**, 555 (1985).
24. P. G. Eliseev, R. F. Nabiev and Yu. M. Popov, *J. Sov. Las. Res.*, **10**, 6, 449 (1989).
25. G. R. Hadley, *Opt. Lett.*, **14**, 16, 859 (1989).
26. H. Fujii, I. Suemune and M. Yamanishi, *Electron. Lett.*, **21**, 713 (1985).
27. D. Botez and T. Holcomb, *Appl. Phys. Lett.*, **60**, 539 (1992).
28. R. F. Nabiev and A. I. Onishchenko, *IEEE J. Quantum Electron.*, **QE-28**, 2024 (1992).
29. V. I. Malakhova, Y. A. Tambiev and A. D. Yakubovich, *Sov. J. Quantum Electron.*, **11**, 1351 (1981).
30. Y. Twu, A. Dienes, S. Wang and J. R. Whinnery, *Appl. Phys. Lett.*, **45**, (7), 709–11 (1984).
31. H. Temkin, R. A. Logan, J. P. van der Ziel, C. L. Reynolds, Jr and S. M. Tharaldsen, *Appl. Phys. Lett.*, **46**, (5), 465–7 (1985).
32. N. K. Dutta, L. A. Koszi, S. G. Napholtz, B. P. Segner and T. Cella, *Tech. Digest of CLEO Conference*, Paper TuF5, Baltimore, MD, May 21–24 (1985).
33. S. Mukai, C. Lindsey, J. Katz, E. Kapon, Z. Rav-Noy, S. Margalit and A. Yariv, *Appl. Phys. Lett.*, **45**, 834 (1984).
34. E. Kapon, L. T. Lu, Z. Rav-Noy, M. Yi, S. Margalit and A. Yariv, *Appl. Phys. Lett.*, **46**, 136 (1985).
35. D. R. Scifres, R. D. Burnham and W. Streifer, *Electron. Lett.*, **18**, 549 (1982).
36. J. J. Yang, R. D. Dupuis and P. D. Dapkus, *J. Appl. Phys.*, **53**, 7218 (1982).
37. D. Botez and J. C. Connolly, *Appl. Phys. Lett.*, **43**, 1096 (1983).
38. D. E. Ackley, *Electron. Lett.*, **20**, 695–7 (1984).
39. C. B. Morrison, L. M. Zinkiewicz, A. Burghard and L. Figueroa, *Electron. Lett.*, **21**, 337 (1985).
40. D. Botez, T. Pham and D. Tran, *Electron. Lett.*, **23**, 416 (1987).
41. P. Gavrilovic, K. Meehan, J. E. Epler, N. Holonyak, Jr, R. D. Burnham, R. L. Thornton and W. Streifer, *Appl. Phys. Lett.*, **46**, 857 (1985).
42. J. E. A. Whiteaway, G. H. B. Thompson and A. R. Goodwin, *Electron. Lett.*, **21**, 1194 (1985).
43. G. H. B. Thompson and J. E. A. Whiteaway, *Electron. Lett.*, **23**, 444 (1987).
44. M. Matsumoto, M. Taneya, S. Matsui, S. Yano and T. Hijikata, *Appl. Phys. Lett.*, **50**, 1541 (1987).

45. M. Sagawa and T. Kajimura, *Appl. Phys. Lett.*, **55**, 1376 (1989).
46. M. Sagawa and T. Kajimura, *Appl. Phys. Lett.*, **56**, 1837 (1990).
47. D. Ackley, D. Botez and B. Bogner, *RCA Rev.*, **44**, 625 (1983).
48. J. R. Heidel, R. R. Rice and H. R. Appelman, *IEEE J. Quantum Electron.*, **QE-22**, 749 (1986).
49. S. Thaniyavarn and W. Dougherty, *Electron. Lett.*, **23**, 5 (1987).
50. D. R. Scifres, W. Streifer and R. D. Burnham, *IEEE J. Quantum Electron.*, **QE-15**, 917 (1979).
51. J. P. van der Ziel, R. M. Mikulyak, H. Temkin, R. A. Logan and R. D. Dupuis, *IEEE J. Quantum Electron.*, **QE-20**, 1259 (1984).
52. D. E. Ackley and R. W. H. Engelmann, *Appl. Phys. Lett.*, **39**, 27 (1981).
53. K. Otsuka, *Electron. Lett.*, **19**, 723 (1983).
54. D. Botez, *IEEE J. Quantum Electron.*, **QE-21**, 1752 (1985).
55. D. Botez, *IEEE J. Quantum Electron.*, **QE-24**, 2034 (1988).
56. D. Botez and D. E. Ackley, *IEEE Circuits and Devices Magazine*, **2**, 8 (1986).
57. Y. Twu, A. Dienes, S. Wang and J. R. Whinnery, *Appl. Phys. Lett.*, **45**, (7), 709–11 (1984).
58. C. B. Morrison, D. Botez and L. M. Zinkiewicz, *Electron. Lett.*, **23**, 1025 (1987).
59. N. Kaneno, T. Kadowaki, J. Ohsawa, T. Aoyagi, S. Hinata, K. Ikeda and W. Susaki, *Electron. Lett.*, **21**, 780 (1985).
60. E. Kapon, C. Lindsey, J. Katz, S. Margalit and A. Yariv, *Appl. Phys. Lett.*, **45**, 200 (1984).
61. C. P. Lindsey, E. Kapon, J. Katz, S. Margalit and A. Yariv, *Appl. Phys. Lett.*, **45**, 722 (1984).
62. J. E. A. Whiteaway, G. H. B. Thompson and A. R. Goodwin, *Electron. Lett.*, **21**, 1194 (1985).
63. J. Ohsawa, S. Hinata, T. Aoyagi, T. Kadowaki, N. Kaneno, K. Ikeda and W. Susaki, *Electron. Lett.*, **21**, 779 (1985).
64. D. F. Welch, W. Streifer, P. S. Cross and D. Scifres, *IEEE J. Quantum Electron.*, **QE-23**, 752 (1987).
65. M. Taneya, M. Matsumoto, S. Matsui, S. Yano and T. Hijikata, *Appl. Phys. Lett.*, **47**, 341 (1985).
66. J. J. Yang, M. Sergant, M. Jansen, S. S. Ou, L. Eaton and W. W. Simmons, *Appl. Phys. Lett.*, **49**, 1139 (1986).
67. T. R. Chen, K. L. Yu, B. Chang, A. Husson, S. Margalit and A. Yariv, *Appl. Phys. Lett.*, **43**, 136 (1983).
68. D. Mehuys, K. Mitsunaga, L. Eng, W. K. Marshall and A. Yariv, *Appl. Phys. Lett.*, **53**, 1165 (1988).
69. W. Streifer, *Appl. Phys. Lett.*, **49**, 59 (1986).
70. A. E. Bazarov, I. S. Goldobin, P. G. Eliseev, O. A. Kobilzhanov, G. T. Pak, T. V. Petrakova, T. N. Pushkim and A. T. Semenov, *Sov. J. Quantum Electron.*, **17**, 551 (1987).
71. C. J. Reinhoudt and C. J. van der Poel, *IEEE J. Quantum Electron.*, **QE-25**, 1553 (1989).
72. W. Streifer, A. Hardy, D. F. Welch, D. R. Scifres and P. S. Cross, *Electron. Lett.*, **26**, 1730 (1990).
73. J. E. A. Whiteaway, D. J. Moule and S. J. Clements, *Electron. Lett.*, **25**, 779 (1989).
74. C. J. van der Poel, P. Opschoor, C. J. Reinhoudt and R. R. Drenten, *IEEE J. Quantum Electron.*, **QE-26**, 1855 (1990).
75. D. Botez, L. Mawst, P. Hayashida, T. J. Roth and E. Anderson, *Appl. Phys. Lett.*, **52**, 266 (1988).

76. L. J. Mawst, D. Botez and T. J. Roth, *Appl. Phys. Lett.*, **53**, 1236 (1988).
77. W. Streifer, M. Osinski, D. R. Scifres, D. F. Welch and P. S. Cross, *Appl. Phys. Lett.*, **49**, 58 (1986).
78. J. Buus, *IEEE J. Quantum Electron.*, **QE-24**, 22 (1988).
79. D. Botez, L. J. Mawst and G. Peterson, *Electron. Lett.*, **24**, 1328 (1988).
80. R. W. H. Engelmann and D. Kerps, *IEE Proc. I, Solid State and Electron. Devices*, **127**, 330 (1980).
81. F. Kappeler, H. Westermeier, R. Gessner and M. Druminski, *Ninth IEEE International Semiconductor Laser Conference*, Paper G-3, Rio de Janeiro, Brazil, 7–10 Aug. (1984).
82. T. H. Shiau, S. Sun, C. F. Schaus and K. Zheng, *IEEE Photon. Tech. Lett.*, **2**, 534 (1990).
83. J. S. Major, D. Mehuys, D. F. Welch and D. R. Scifres, *Appl. Phys. Lett.*, **59**, 2210 (1991).
84. J. P. McKelvey, *Solid-State and Semiconductor Physics* (Harper & Row, New York, 1966).
85. C. A. Zmudzinski, D. Botez and L. J. Mawst, *Appl. Phys. Lett.*, **60**, 1049 (1992).
86. D. Botez, M. Jansen, L. J. Mawst, G. Peterson and T. J. Roth, *Appl. Phys. Lett.*, **58**, 2070 (1991).
87. D. Botez and L. J. Mawst, *Appl. Phys. Lett.*, **60**, 3096 (1992).
88. M. A. Duguay, Y. Kokubun, T. L. Koch and L. Pfeiffer, *Appl. Phys. Lett.*, **49**, 13 (1986).
89. T. L. Koch, E. G. Burkhardt, F. G. Storz, T. L. Bridges and T. Sizer, *IEEE J. Quantum Electron.*, **QE-23**, 889 (1987).
90. T. Baba, Y. Kokubun, T. Sasaki and K. Iga, *J. Lightwave Technol.*, **6**, 1440 (1989).
91. D. Botez, *IEE Proc. J., Optoelectronics*, **139**, 1, 14 (1992).
92. P. Yeh, *Optical Waves in Layered Media* (Wiley, New York, 1988), pp. 326–7.
93. L. J. Mawst, D. Botez, M. Jansen, M. Sergant, G. Peterson and T. J. Roth, *Appl. Phys. Lett.*, **59**, 1655 (1991).
94. R. L. Liboff, *Introductory Quantum Mechanics* (Addison-Wesley, New York, 1980), pp. 284–7.
95. R. J. Noll and S. H. Macomber, *IEEE J. Quantum Electron.*, **QE-26**, 456 (1990).
96. L. J. Mawst, D. Botez, T. J. Roth, W. W. Simmons, G. Peterson, M. Jansen, J. Z. Wilcox and J. J. Yang, *Electron. Lett.*, **25**, 365 (1989).
97. G. P. Agrawal, *IEEE J. Lightwave Technol.*, **LT-2**, 537 (1984).
98. S. Asada, *IEEE J. Quantum Electron.*, **27**, 884 (1991).
99. D. Mehuys and A. Yariv, *Opt. Lett.*, **13**, 571 (1988).
100. P. D. Van Eijk, M. Reglat, G. Vasilieff, G. J. M. Krijnen, A. Driessen and A. J Mouthaan, *IEEE J. Lightwave Technol.*, **9**, 629 (1991).
101. G. R. Hadley, *Opt. Lett.*, **15**, 1215 (1990).
102. J. Z. Wilcox, W. W. Simmons, D. Botez, M. Jansen, L. J. Mawst, G. Peterson and J. J. Yang, *Appl. Phys. Lett.*, **54**, 1848 (1989).
103. A. A. Golubentsev, V. V. Likhanskii and A. P. Napartovich, *Sov. Phys. JETP*, **66**, 676 (1987).
104. J. R. Leger, M. L. Scott and W. B. Veldkamp, *Appl. Phys. Lett.*, **52**, 1771 (1988).
105. L. J. Mawst, D. Botez, T. J. Roth, G. Peterson and J. Rozenbergs, *Appl. Phys. Lett.*, **58**, 22 (1991).

106. G. R. Hadley, D. Botez and L. J. Mawst, *IEEE J. Quantum Electron.*, **QE-27**, 921 (1991).
107. R. Chinn and R. J. Spiers, *IEEE J. Quantum Electron.*, **QE-18**, 984 (1982).
108. M. C. Amann, *IEEE J. Quantum Electron.*, **QE-22**, 1992 (1986).
109. S. J. Lee, L. Figueroa and R. V. Ramaswamy, *IEEE J. Quantum Electron.*, **QE-25**, 1632 (1989).
110. K. Aiki, M. Nakamura, T. Kuroda, J. Umeda, R. Ito, N. Chinone and M. Maeda, *IEEE J. Quantum Electron.*, **QE-14**, 89 (1989).
111. P. Yeh, C. Gu and D. Botez, *Opt. Lett.*, **17**, 1818 (1992).
112. L. J. Mawst, D. Botez, M. Jansen, T. J. Roth and J. Rozenbergs, *Electron. Lett.*, **27**, 369 (1991).
113. R. F. Nabiev, P. Yeh and D. Botez, *Appl. Phys. Lett.*, **62**, 916–18 (1993).
114. R. F. Nabiev, P. Yeh and D. Botez, *SPIE Proceedings OE-LASE '93*, Vol. 1850, pp. 23–36, Los Angeles, CA (1993).
115. D. Botez and L. Frantz, unpublished.
116. J. R. Leger, G. J. Swanson and M. Holtz, *Appl. Phys. Lett.*, **50**, 1044 (1987).
117. L. J. Mawst, D. Botez, M. Jansen, T. J. Roth, C. Zmudzinski, C. Tu and J. Yun, *SPIE Proceedings* **1634**, 2 (1992).
118. C. Zmudzinski, D. Botez and L. J. Mawst, *Appl. Phys. Lett.*, **62**, 2914–16 (1993).
119. C. Zmudzinski, L. J. Mawst, D. Botez, C. Tu and C. A. Wang, *Electron. Lett.*, **28**, 1543–5, 1992.
120. J. S. Major, D. Mehuys and D. F. Welch, *Electron. Lett.*, **28**, 1101 (1992).
121. J. Snyder, *Appl. Opt.*, **30**, 2743 (1991).
122. E. R. Anderson, M. Jansen, D. Botez, L. J. Mawst, T. J. Roth and J. J. Yang, *SPIE Proceedings*, Vol. 1417, Jan., Los Angeles, CA (1991).
123. M. Jansen, D. Botez, L. J. Mawst and T. J. Roth, *Appl. Phys. Lett.*, **60**, 26 (1992).
124. J. Sasian, private communication.
125. D. R. Scifres, C. Lindstrom, R. D. Burnham, W. Streifer and T. L. Paoli, *Electron. Lett.*, **19**, 169 (1983).
126. D. Welch, D. Scifres, P. Cross, H. Kung, W. Streifer, R. D. Burnham, J. Yaeli and T. Paoli, *Appl. Phys. Lett.*, **47**, 1134 (1985).
127. L. J. Mawst, D. Botez, M. Jansen, C. Zmudzinski, S. S. Ou, M. Sergant, T. J. Roth, C. Tu, G. Peterson and J. J. Yang, *IEEE J. Quantum Electron.*, **29**, 1906 (1993).
128. G. A. Evans, D. P. Bour, N. W. Carlson, R. Amantea, J. M. Hammer, H. Lee, M. Lurie, R. Lai, P. Pelka, R. Farkas, J. B. Kirk, S. K. Liew, W. F. Reichert, C. A. Wang, H. K. Choi, J. N. Walpole, J. K. Butler, W. E. Ferguson, R. K. Defreez and M. Felisky, *IEEE J. Quantum Electron.*, **QE-27**, 1594 (1991).
129. A. A. Golubentsev, V. Y. Likhanskii and A. P. Napartovich, Laser-diode technology and applications II, *SPIE Proceedings*, **1219**, 220 (1990).
130. A. A. Golubentsev and V. V. Likhanskii, *Sov. J. Quantum Electron.*, **20**, 522 (1990).
131. D. Botez and A. P. Napartovich, *IEEE J. Quantum Electron*, **30**, April 1994.
132. D. Botez, A. Napartovich and C. Zmudzinski, submitted to *IEEE J. Quantum. Electron.*, 1994.
133. D. Mehuys, J. Major, Jr and D. F. Welch, *SPIE Proc. OE-LASE '93*, Vol. 1850, pp. 2–12, Los Angeles, CA (1993).

2

High-power coherent, semiconductor laser, master oscillator power amplifiers and amplifier arrays

DAVID F. WELCH AND DAVID G. MEHUYS

2.1 Introduction

Semiconductor optical sources exhibit several distinct advantages relative to other solid-state and gas laser systems including compact size, high efficiency, high reliability, robustness and manufacturability. As a result, many of the rapidly growing commercial and consumer markets, including telecommunications, printing and optical data storage, have only been realized through the introduction of semiconductor lasers. The next generation of high-performance electronics will require higher-power semiconductor lasers; but as recently as 1992, commercially available high-power (greater than 200 mW cw) semiconductor lasers did not operate in a single spatial mode and therefore did not radiate in a single diffraction-limited lobe. However, the latest advances in the design and fabrication of semiconductor lasers have resulted in the development of high-power diffraction-limited laser sources.

The architectures that have been studied for high-power diffraction-limited semiconductor lasers can be divided into three categories: oscillators, injection-locked oscillators, and master oscillator power amplifiers (MOPAs). The highest-power diffraction-limited operation from oscillators has been demonstrated from antiguided laser arrays[1–3]. These devices have demonstrated up to 0.5 W cw[4,5] and 1.5[6] to 2.0 W[1] pulsed in a diffraction-limited radiation pattern. Disadvantages of antiguide oscillators are their multi-longitudinal-mode spectra and multi-lobed far-field patterns. Another promising oscillator configuration is the broad-area ring oscillator, which has demonstrated single-frequency operation to greater than 1 W pulsed, and diffraction-limited single-lobed operation to approximately 0.5 W pulsed[7]. Comparable output powers have been demonstrated from injection-locked oscillators[8,9], where approximately 0.5 W cw in a single diffraction-limited beam has been demonstrated. A disadvantage of injection-locked oscillators is the narrow bandwidth determined by the slave-laser cavity resonances[10], within which the master-laser wavelength must be tuned and maintained over time in order to achieve single-frequency operation. Recently, high-power diffraction-limited operation has been demonstrated from hybrid, single-element MOPAs to output powers of 4.5 W cw and 21 W pulsed[11–13]. In this configuration the master oscillator and power amplifier are discrete devices and are coupled together via bulk optics. The MOPA geometry offers the advantages of single-frequency operation, broadband wavelength tunability and diffraction-limited performance. More recently, and the primary focus of this work, is the development of a monolithically integrated MOPA (M-MOPA). In this chapter, both monolithically integrated grating-coupled surface emitters[14–17] and monolithic integrated edge-emitting flared amplifiers[18] will be discussed. From the latter, greater than 3 W cw (at 5 °C) and 7.5 W pulsed in a single-longitudinal-mode, single diffraction-limited beam has been demonstrated.

In addition, monolithic arrays of amplifiers have been fabricated with as many as 400 elements with temporally and spatially coherent output to greater than 19 W pulsed[19], indicating the scalability of the amplifier configurations to even higher coherent output powers.

In this chapter we will discuss the topic of high-power semiconductor master oscillator power amplifiers and will conduct our discussions in two frameworks: (i) the development of monolithically integrated master oscillator power amplifier arrays consisting of cascaded amplifier arrays, parallel amplifier arrays and 2-D amplifier arrays; and (ii) the increase in diffraction-limited output power from a single monolithically integrated master oscillator power amplifier element, that includes discussions of monolithically integrated active-grating master oscillator power amplifiers, discrete broad-area power amplifiers, and monolithically integrated flared-amplifier MOPAs. In this fashion both the topic of single-element performance and scalability to arrays will be addressed. The discussion of these topic areas demonstrates that greater than 3 W cw can be achieved from a single monolithic source, while the extension of the single-element source to higher powers can be achieved by monolithically integrating arrays of such emitters as demonstrated for 400-element amplifier arrays in an optically parallel configuration.

2.2 Monolithically integrated MOPAs and amplifier arrays

The monolithic integration of the master oscillator with a power amplifier was developed almost simultaneously as the result of research on two different topic areas. Work performed at AT&T Bell Laboratories[20,21] and BNR Europe[22], motivated by the need for integrated laser sources in photonic integrated circuits, demonstrated the integration of a single-mode master oscillator coupled to an amplifier in order to overcome the losses associated with photonic integrated circuits. Simultaneously, work on grating-coupled surface-emitting laser arrays for high-power diffraction-limited operations resulted in the development of monolithic integration of master oscillator power amplifiers[14–17]. In order to meet the goal of a multi-watt coherent master oscillator power amplifier, the first implemented configuration consisted of a master oscillator and a cascaded array of amplifiers and surface-emitting grating output couplers[14,15]. The following discussion presents the design and characteristics of such M-MOPAs with cascaded amplifier arrays. Following the discussion of cascaded amplifier arrays it will be shown that parallel arrays of amplifiers are also viable for achieving high-power coherent operation; and finally the coupling of these two technologies demonstrates their utility in 2-D arrays of amplifiers. Subsequent sections will concentrate on increasing the diffraction-limited output power from any single amplifier element.

Figure 2.1 illustrates the M-MOPA configuration initially used for high-power coherent operation, consisting of an array of cascaded power amplifiers. The discussions of this particular design hold true in most characteristics for similar designs performed by other laboratories[15]. The structure consists of a master oscillator which is optically coupled to a cascaded array of amplifiers/output coupler pairs. The master oscillator is

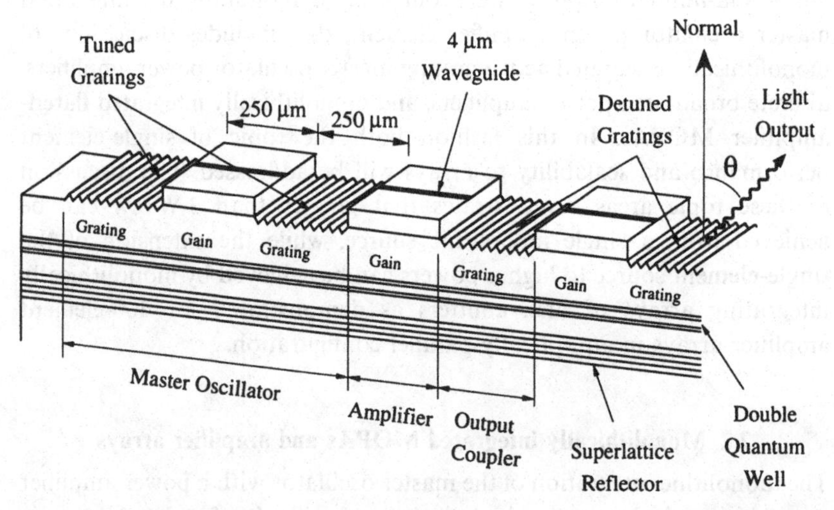

Figure 2.1. Schematic diagram of the M-MOPA configuration. (After Ref. 26, © 1991 IEEE.)

typically a second-order distributed Bragg reflector (DBR) single-mode laser; however it has been shown that first-order DBR lasers can also be used[15]. The DBR laser radiates in a single spectral and spatial mode to an output power in excess of 50 mW. The gratings of the DBR laser are defined such that its output is injected into an amplifying element. The signal is amplified in the gain element and then a portion of the signal is coupled out of the structure with a detuned second-order grating output coupler[14]. The pitch of the output coupling grating is such that the signal is diffracted forward of the normal (towards the direction of propagation). (Alternative designs have also used grating pitches that diffract the traveling-wave light backward; however, these designs may be subject to self-oscillation due to insufficient gain saturation in the amplifier.) As a result of the grating pitch being detuned from the injection signal, the reflectivity of the output coupler to the amplifier gain element is significantly reduced[23]. A portion of the signal is transmitted through the output coupling grating and is injected into the next amplifier stage. The process of amplification and output coupling is repeated for each subsequent amplifier stage. The multi-stage amplifier output coupler increases the total coherent output power generated from a single master oscillator and, as is discussed below, the output power per amplifier stage is approximately 100 mW. As the following discussions show, the cumulative output of the amplifier array is a multi-watt, temporally coherent source. The phase relationship, however, of each emitting aperture is not controlled and will thus require separate individual phase control.

Also incorporated in the M-MOPA structure is a superlattice reflector[24], epitaxially grown below the active region. The superlattice reflector reflects the light that is diffracted towards the substrate back towards the surface, resulting in a higher external differential efficiency. The operation of the superlattice reflector is described elsewhere[23,24].

Plotted in Figure 2.2 is the theoretical diffraction efficiency of the detuned grating as a function of the detuning of the injected signal from the Bragg condition of the grating output coupler. As is shown, at resonance for a 250 μm-long grating the reflectivity of the grating is approximately 50% while the surface-emitting power is 20%. The remainder of the incident power is either transmitted through the grating or is absorbed. When the injected wavelength is detuned from the Bragg condition by 1 μm^{-1}, or approximately 30 nm, the reflected power drops to less than 10^{-4} while the surface-radiated power can exceed 75%. These calculations are performed assuming that a superlattice reflector is positioned to maximize the up-coupling efficiency[23].

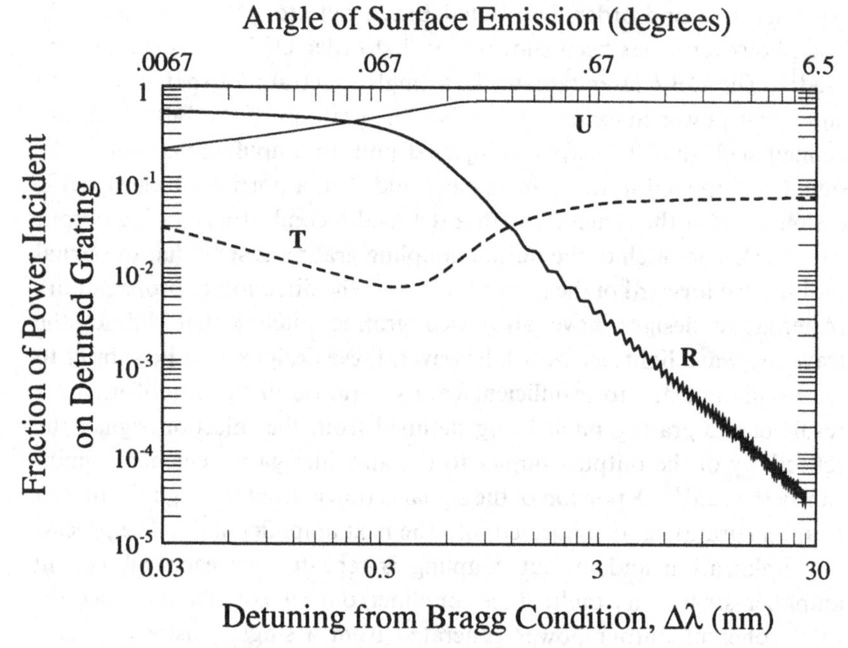

Figure 2.2. Fraction of incident power transmitted (T), reflected (R), and diffracted upward (U) from a 250 μm-long detuned grating, as a function of detuning from the Bragg resonance. (From Ref. 23, © 1991 IEEE.)

The advantage of a detuned grating output coupler is the low residual reflectivity which therefore alleviates the necessity to include an optical isolator between the master oscillator and the power amplifier. One alternative to the detuned grating output coupler is an antireflection (AR) coated facet, which typically has a residual reflectivity of 10^{-3} and is not applicable to amplifier chains.

Optimum performance of the linear chain of amplifier output couplers is obtained when the signal power output from all stages is equal[25,26]. In this condition the radiated power from each output coupler is identical and the Strehl ratio of a phased far-field pattern is maximized. Figure 2.3 plots the theoretical amplified signal and ASE output powers as a function of the input signal power for a 4 μm-wide real refractive index waveguide amplifier. The transverse optical confinement factor of $\Gamma = 0.08$ has been used as appropriate for the double quantum well active region. The amplifier length characterized is 250 μm, and the current density is chosen to be 10 kA/cm^2. At small input powers the signal gain is high while as the input power is increased above the saturation power of the

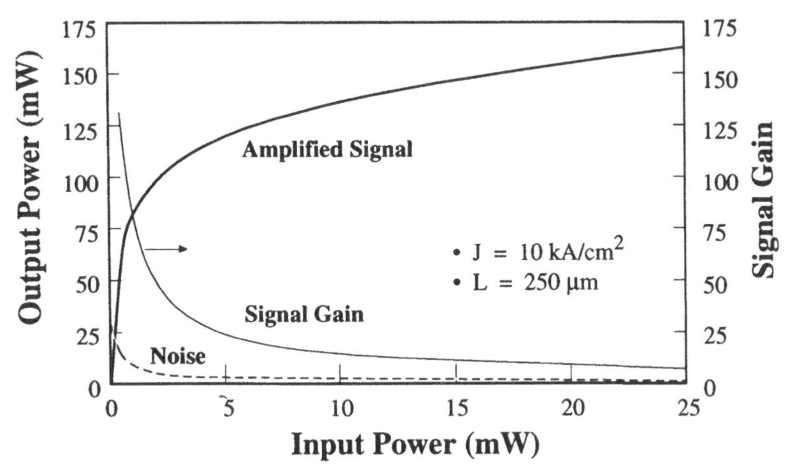

Figure 2.3. Calculated amplified signal output power, spectrally integrated noise output power and signal gain as a function of input power injected into a single-pass waveguide amplifier. The unsaturated (small-signal) gain for this amplifier is 30 dB. (From Ref. 25, © 1991 IEEE.)

amplifier, P_{sat}, in this case approximately 7.5 mW, the gain of the amplifier is reduced.

In order to obtain high-power performance from the amplifier array the injection signal must be sufficient to suppress the noise of the amplifier. The cumulative gain of the amplifier array will tend to either self-oscillate or emit ASE if sufficient noise is present. As is shown in Figure 2.3, the noise level from each individual amplifier stage is less than 2 mW for an injected power of 7.5 mW. To achieve this level of injected power the output coupling grating is designed to transmit greater than 10% of the injected power to the next amplifier stage.

The numerical model of the single-stage amplifier was extended to encompass multiple stages of amplifier/output coupler pairs similar to the array presented in Figure 2.1[25,26]. The signal and noise level for a 12-element amplifier array is presented in Figure 2.4. In this example the amplifier and detuned grating length is 250 μm, the current density is 10 kA/cm^2, the grating transmission is 10% and the input power is 14 mW. The residual reflectivity of each output coupler is assumed to be 10^{-4}. For an input power of 14 mW the gain of the amplifier is 10, which matches the transmission loss of the output coupler. As a result the output of each stage coupled to the grating region is 140 mW. Assuming a surface outcoupling efficiency of 80%, the radiated signal power is 110 mW per output coupler or a total of 1.32 W. Since the injected signal to each

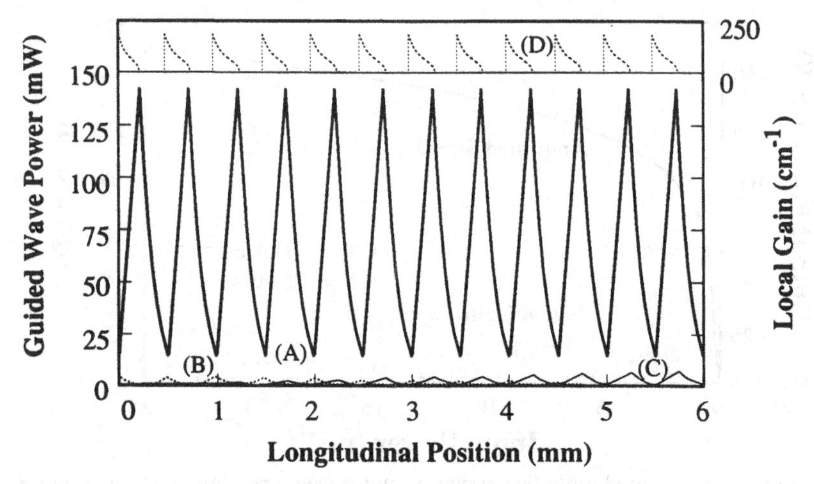

Figure 2.4. Guided-wave power in a 12-stage M-MOPA for the injected power $P_{in} = 14$ mW. Curve (A) is the forward-traveling amplified signal, (B) and (C) are the backward- and forward-traveling, spectrally integrated ASE noise powers and (D) is the self-consistent local modal gain. (From Ref. 25, © 1991 IEEE.)

amplifier exceeds P_{sat}, the power grows linearly in the amplifier regions and decays exponentially in the grating output coupling regions.

The ASE powers are also shown in Figure 2.4. The peak noise level increases linearly with the number of amplifier stages and reaches a maximum of 8 mW in the final stage. Although the gain is equal for the forward- and reverse-traveling waves, the noise is preferentially generated at the front of the amplifier where the gain, and therefore charge density, is greatest. As a result the forward-traveling noise is greater than the backward-traveling noise.

In general the condition of the gain of the amplifier, G, and the transmission of the output coupler, T, will not satisfy $GT = 1$ in all of the amplifier stages. Typically, the injected signal from the master oscillator does not provide the power necessary to equalize the gain of the first amplifier to the transmission of the output coupler. However, after propagation down the amplifier chain the signal will tend to stabilize its gain to satisfy this condition[26].

M-MOPAs were fabricated in a similar design as that presented in Figure 2.1. The master oscillator consists of a gain length of 500 μm and the gain lengths of the amplifier stages consist of 250 μm with 250 μm output coupling regions. The details of fabrication are discussed elsewhere[14]. The epitaxial growth consists of a double quantum well active region and AlGaAs confining and cladding layers. The superlattice reflector is grown below the active region to reflect the substrate-diffracted light towards the top surface, where it reacts with the other diffraction orders of the grating, enhancing the output power. The light is confined throughout the gain and grating regions by a 4 μm-wide real refractive index waveguide. The grating regions are holographically exposed in photoresist and dry-etched into the surface of the epitaxial structure. The pitch of the detuned grating regions is such that the resonant wavelength is 40 nm longer than the resonant wavelength of the oscillator gratings. The M-MOPA is tested under pulsed conditions, 500 ns pulses at a 1 kHz repetition rate.

The light output as a function of the injected current to the amplifiers when operated in parallel is shown in Figure 2.5. With no injected current to the amplifier sections, the output power is approximately 50 mW, which corresponds to the injected signal from the master oscillator when operated at a drive current of 210 mA. As the drive current to the nine-element amplifier array is increased, the gain increases exponentially to an output power of approximately 300 mW, or 33 mW per amplifier, at which point the power grows linearly with injection current to an output power of 1.08 W, or 120 mW per amplifier.

The performance of the M-MOPA agrees well with theory that would

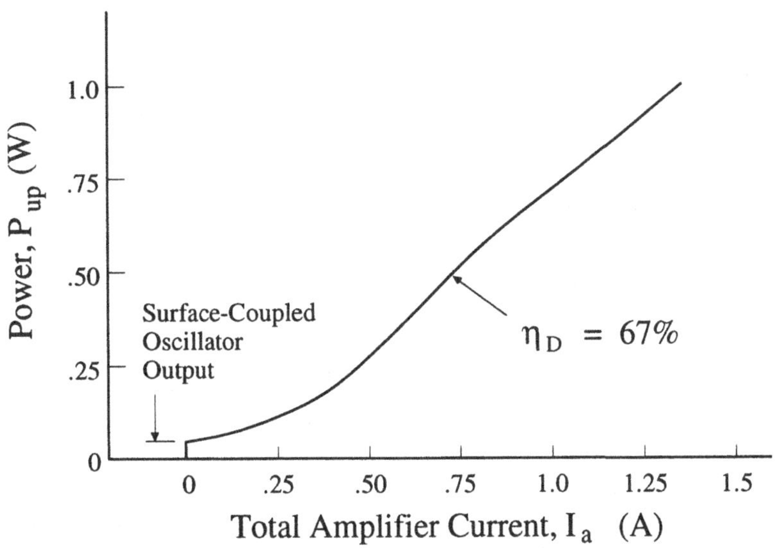

Figure 2.5. Measured light output versus current for a nine-stage M-MOPA. (After Ref. 25, © 1991 IEEE.)

predict a power level of 1.0 W from nine amplifiers at a current density of 10 kA/cm^2. The external differential efficiency of the M-MOPA is 67% once the amplifiers are operated in saturation. The high efficiency indicates that grating-coupled surface emission incorporating a superlattice reflector can match the efficiency of edge-emitting structures. The high differential efficiency also indicates that when amplifiers are operated at a power level above the saturation power of the gain media, the efficiency approaches that of oscillators.

The spectral output of the M-MOPA is a single longitudinal mode which replicates that of the master oscillator. The single-mode spectral characteristic of the M-MOPA designs is a distinct advantage relative to other high-power single-mode laser designs. The spectrum is nominally independent of amplifier drive current and thus high-power operation can be achieved without a significant wavelength chirp.

The far-field radiation pattern of the M-MOPA is governed by the relative phase of the array of output couplers, which is determined by the optical path length of the intervening amplifier section. The optical path length of the amplifier section is determined to first order by the epitaxially grown waveguide structure and to second order by the charge density in the active region. As discussed previously, the charge density in the active region is proportional to the injection current when the local modal power is less than P_{sat}, and is, to first order, independent of injection current when the modal power is greater than P_{sat}. Therefore, the relative phases of the adjacent amplifiers can only be adjusted via current injection to the amplifier in the regime where the modal power is less than the saturation power. In the case of the nine-amplifier array, adequate phase adjustment of the amplifiers can only be accomplished via current injection to the amplifiers at an output power less than 350 mW. Figure 2.6 presents the far-field radiation pattern of the amplifier array when the currents have been tuned to accommodate the relative phases so that all of the output couplers operate in phase with one another. In this condition, the radiation pattern is characteristic of a periodic array of coherent sources and operates in a diffraction-limited radiation pattern. Above these power levels, however, the phases of the amplifiers can no longer be adjusted via current injection into the gain region, and the device no longer operates in a diffraction-limited beam pattern.

The development of amplifier arrays can also be extended to the development of parallel-injected amplifier arrays. As a demonstration of the scalability of amplifier arrays, the structure presented in Figure 2.7 has been fabricated[19]. In this configuration the master oscillator input

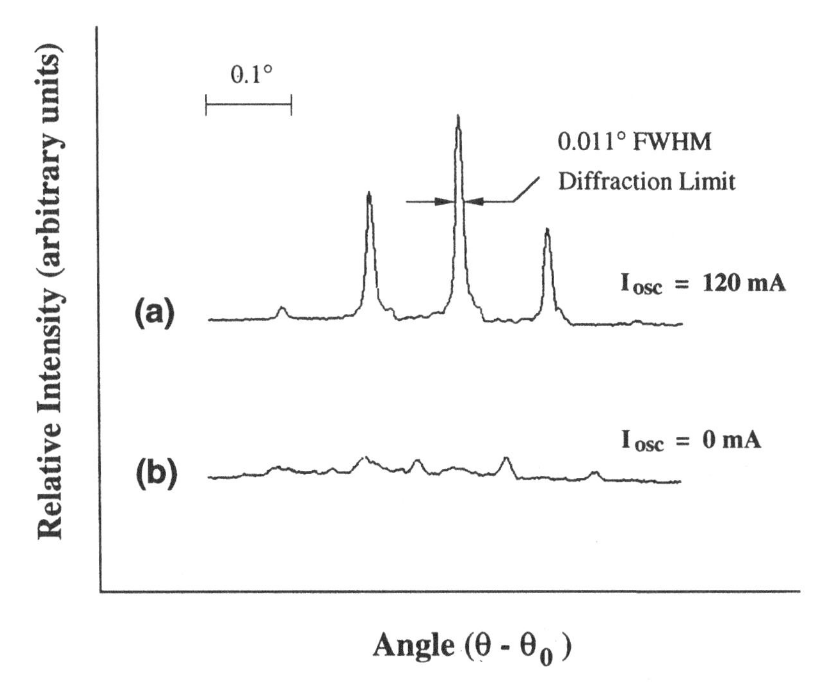

Figure 2.6. The measured radiation pattern from the nine-stage M-MOPA: (*a*) the oscillator is turned on and the amplifier currents are adjusted to phase the outputs (total power 320 mW); (*b*) the oscillator is turned off and the amplifiers are driven as in (*a*). (After Ref. 25, © 1991 IEEE.)

is remotely located and coupled to the amplifier array via optical fiber. The primary branch in the amplifier continues to amplify the signal while simultaneously splitting the signal off to as many as 400 amplifier elements. The amplifier branches are fabricated with dry-etched 45° turning mirrors in order to deflect the light perpendicular to the injected signal axis. Each deflected signal is then branched to accommodate an even larger number of amplifier elements. The light output as a function of input current to the power amplifier region is presented in Figure 2.8. The maximum output power achieved from this configuration is greater than 22 W pulsed (200 μsec, 1% duty factor) where 87% of the output power is temporally coherent; and pairwise fringe measurements on two emitters, separated by 5 mm at the output facet, result in a measured fringe visibility of 0.8. The output coherence has been shown to saturate at an input power of approximately 15 mW from the master oscillator. This work indicates that mutual coherence can be maintained over a large number of parallel-injected amplifiers. However, as in the case of the

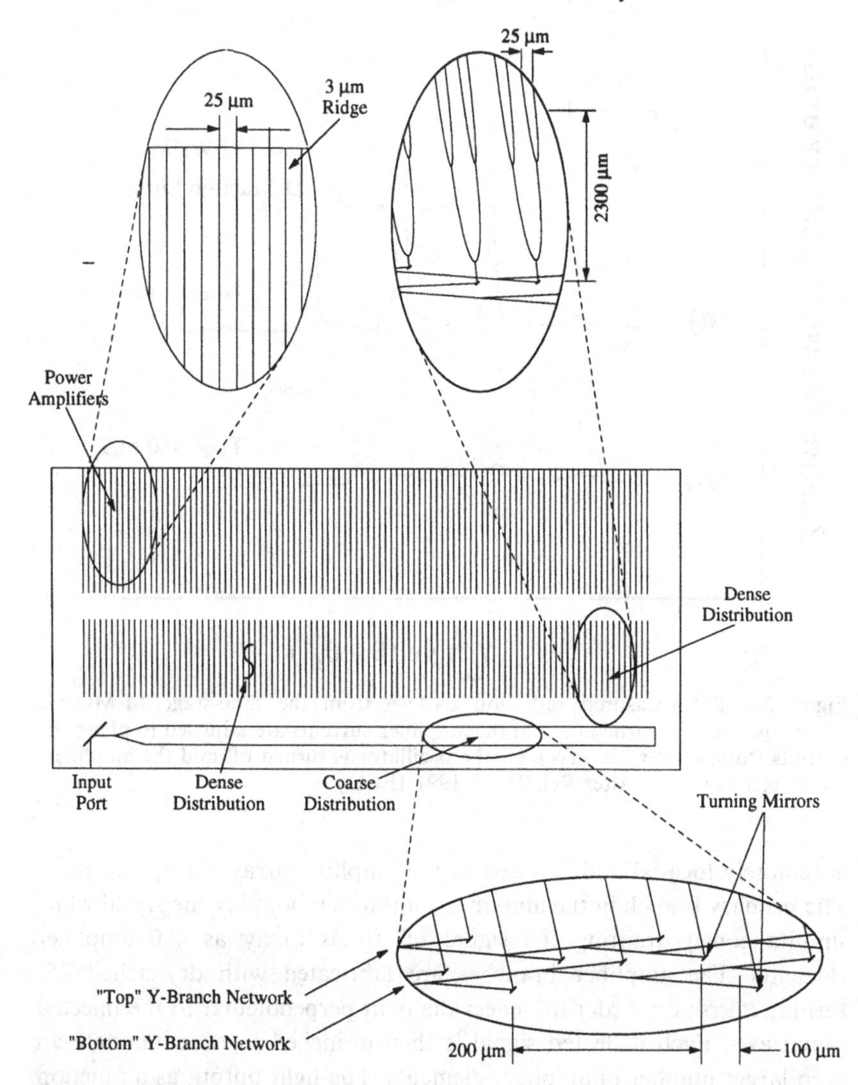

Figure 2.7. Basic layout of the 400-emitter coherent amplifier chip. (From Ref. 19, © 1991 IEEE.)

cascaded arrays, the diffraction-limited operation of the array can only be achieved through independent phase control of each individual emitter.

The concept of an amplifier array can be extended to two dimensions via a single-mode waveguide branching network for parallel injection of amplifier arrays coupled with the cascaded amplifier/output coupler array discussed previously. Presented in Figure 2.9 is the design of such an

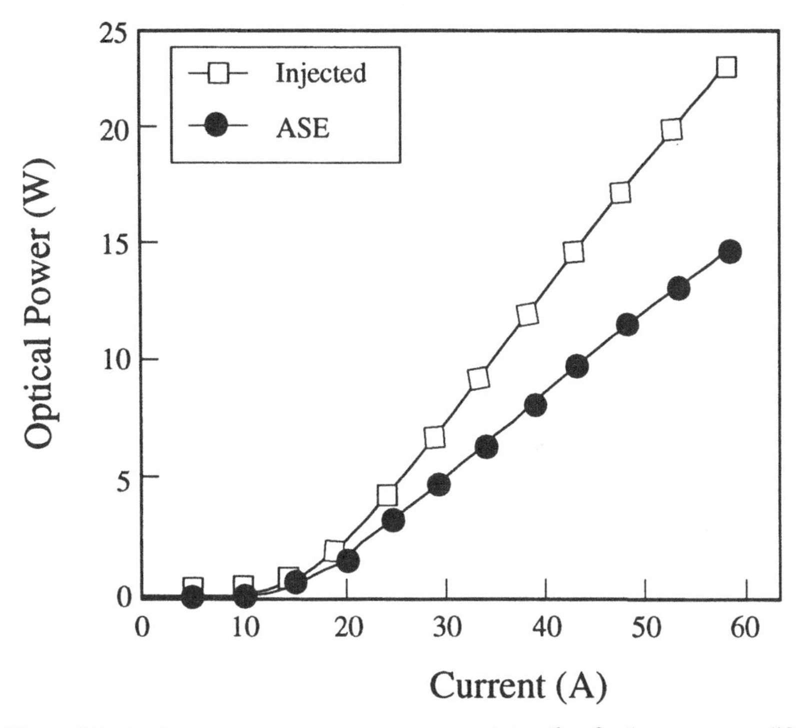

Figure 2.8. Peak output power versus current to the final power amplifier. Currents to the coarse and dense distribution networks were 2 and 12 A respectively. (From Ref. 19, © 1991 IEEE.)

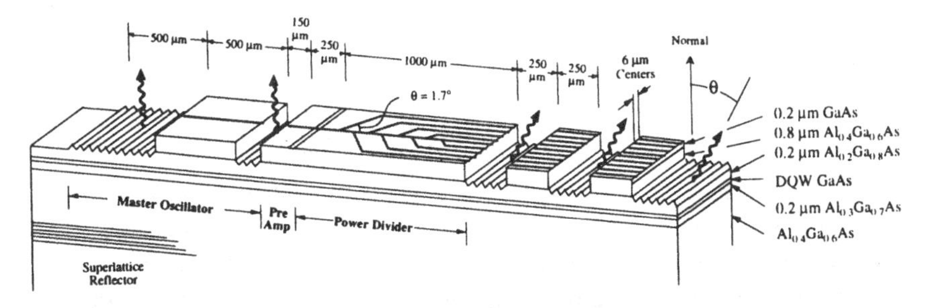

Figure 2.9. Schematic diagram of a 2-D M-MOPA. (From Ref. 27, © 1991 IEEE.)

amplifier array fabricated in a fashion similar to the cascaded amplifier array presented above[27]. The branching network is designed to minimize the noise generation within the power divider section. The output of the power divider is an array of nine parallel waveguides, each of which feed a linear array of seven amplifier-output coupler pairs, resulting in a total of 63 emitting apertures.

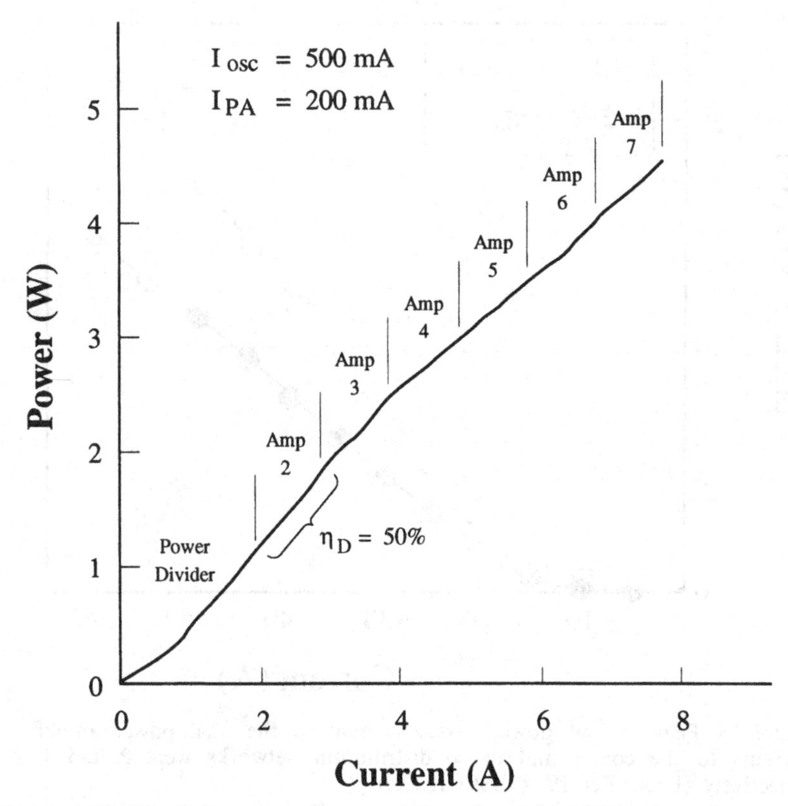

Figure 2.10. The light output characteristics of an M-MOPA where the output of the amplifier is monitored as a function of the input to the amplifier regions only. (From Ref. 27, © 1991 IEEE.)

The light output as a function of the input current is presented in Figure 2.10. The 2-D array of amplifiers emits greater than 4.5 W pulsed, corresponding to an output power greater than 70 mW per emitter. The spectral output of the amplifier array at 4.5 W is similar to the linear amplifier array and is a single longitudinal mode identical to that of the master oscillator. The increased output power indicates the scalability of amplifier technology to high output powers while maintaining a temporally coherent output. The scalability of spatially coherent output from such a structure, however, requires independent phase control of each emitter.

The far-field pattern of the 2-D amplifier array is characteristic of an array of spectrally coherent emitters which are random in phase. As discussed earlier, the phases of the amplifier array can only be adjusted via current injection into the gain region at low output powers and it is thus difficult to phase the elements together at high power levels via injection into the gain region. However, individual phase modulators have shown the ability to phase amplifier outputs. The topic of phased output from M-MOPA designs is presented in the following sections.

2.3 Monolithic active-grating master oscillator power amplifiers

The results presented in the preceding section characterizing the M-MOPA indicate that M-MOPA technology is scalable with the number or length of amplifiers. Output powers in excess of 4.5 W have been demonstrated in a single spectral mode[27]. The relative phase of the cascaded amplifiers, however, is randomly distributed due to the variations in charge density along the length of the individual amplifier regions[25,28]. Where the optical power is below the saturation power of the amplifier, the charge is not clamped and changes in current are proportional to changes in the charge density. Where the charge density is not clamped, the index of refraction, and therefore the relative phases of the amplifier regions, are dependent on the current to the amplifier. As a result, the phases of the individual amplifiers of the cascaded M-MOPA can only be adjusted below the region at which the gain in the individual amplifiers saturate. If, however, the output coupler is integral with the amplifier and the modal power is above the saturated power of the amplifier, phase changes due to current fluctuations along the length of the amplifier are greatly reduced. This is the basis of the development of the monolithically integrated active grating (MAG)-MOPA[16,29].

Figure 2.11 presents a schematic diagram of the MAG-MOPA consisting of (from left to right) a conventional DBR master oscillator, a

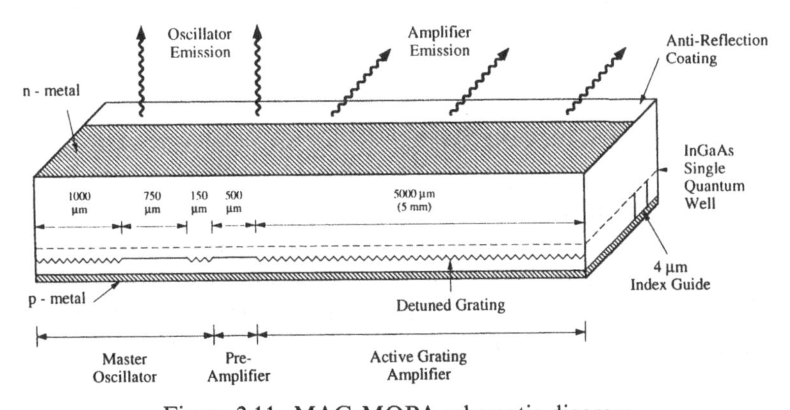

Figure 2.11. MAG-MOPA schematic diagram.

preamplifier, and a long, current-pumped detuned grating output coupler ('active grating'). The preamplifier is used to amplify the master oscillator output and saturate the gain within the active-grating amplifier, thereby suppressing self-oscillation. As will be shown, the guided-wave intensity within the active-grating amplifier region becomes constant along the direction of propagation, since the saturated local gain level is clamped at the level of the guided-mode losses[16,29]. Once this equilibrium is reached, both the modal intensity and the local carrier density remain constant along the remaining length of the aperture.

The active-grating region offers the following advantages:

(1) The guided-wave and radiated-wave near fields are uniform along the length of the aperture, minimizing side lobes in the radiation pattern and maximizing its Strehl ratio.
(2) The carrier density is uniform along the aperture length, alleviating the need for individual phase-control electrodes along its length (as in the case of the M-MOPA) because of the constancy of the modal propagation constant.
(3) There are no interfaces between amplifier and output coupler regions which give rise to residual reflections, thus further suppressing the natural oscillation of the amplifier and reducing coherent feedback to the master oscillator.

As a result of the uniform gain and charge saturation along the length of the power amplifier, the radiation pattern is diffraction-limited. In addition, since there is a single emission aperture, the far-field pattern is primarily a single lobe. Finally, due to the integrated DBR master oscillator, the MAG-MOPA operates to high power in a single longitudinal mode. The following discussion of the MAG-MOPA is divided between theory and experiment.

2.3.1 MAG-MOPA: theory

Figure 2.12 shows a numerical simulation of a current-pumped active grating. The numerical model employed accounts for both forward- and backward-traveling waves and noise generation within the amplifier[25,30]. The amplified signal power is shown as a solid line and the ASE is shown as a dashed line. As Figure 2.12 shows, the guided-wave power grows exponentially near the front end of the amplifier, where the power is less than the saturation power, P_{sat}, of the waveguide amplifier. Above the saturation power, which is typically 5–10 mW for a single-mode waveguide

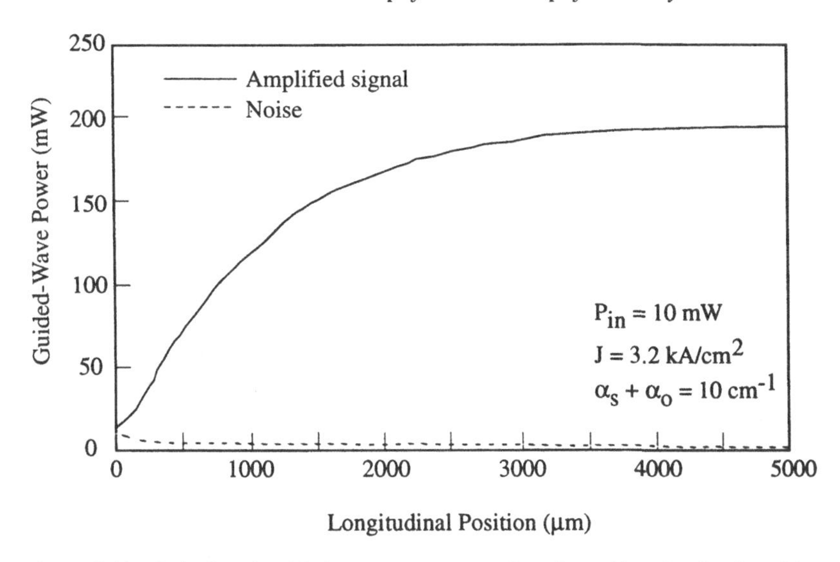

Figure 2.12. Calculated guided-wave power as a function of longitudinal position.

amplifier, the traveling-wave power grows approximately linearly as a function of position before finally saturating. As the traveling-wave power increases along the amplifier, so does the rate of stimulated emission. The corresponding decrease in carrier lifetime decreases the local gain or carrier density in the waveguide. When the local gain decreases to the level of the waveguide losses, whether they be due to absorption, scattering, or output coupling, the traveling wave sees no net gain or loss and, therefore, the guided-wave power saturates and becomes constant as a function of position. In the example shown in Figure 2.12, where the injected power is 10 mW, the current density is 3.2 kA/cm^2 and the grating output coupling loss is 10 cm^{-1}, the guided-wave power saturates at a value of $\approx$200 mW. The distance required to achieve saturation is $\approx$2.5 mm. Obviously, as the grating output coupling level increases, the distance required for saturation decreases, which leads to a lower value for the saturated guided-wave power.

The level at which the guided-wave power saturates can be quantified with the aid of a simple analytical model which treats the active grating as a single-pass, noiseless amplifier including gain saturation. The expression for guided-wave power, P_g, as a function of longitudinal position z within the amplifier is

$$\frac{dP_g}{dz} = \frac{g_0 P_g}{1 + P_g/P_{sat}} - \alpha P_g, \tag{2.1}$$

where g_0 is the unsaturated gain (in cm^{-1}) and α is the total guided-mode loss (in cm^{-1}), which includes absorption, scattering and output coupling.

The first coefficient of P_g on the right-hand side of Equation (2.1) is the saturated gain coefficient. When the saturated gain decreases to the value of the losses α, the guided-mode power saturates. The saturated value of P_g is

$$P_{g,\mathrm{sat}} \approx \frac{g_0}{\alpha} P_{\mathrm{sat}}. \tag{2.2}$$

For example, when the unsaturated gain is 200 cm^{-1} (which is appropriate for a current density of 3.2 $\mathrm{kA/cm^2}$ in a SQW active region) and the saturation power is 10 mW (which is appropriate for a 5 μm-wide waveguide amplifier), the saturated guided-wave power is approximately 200 mW when the output coupling losses are 10 cm^{-1}, as shown in Figure 2.12.

Within the region in which the guided-wave power is linearly increasing, the carrier density is decreasing and therefore the modal propagation constant is also varying. It is, then, optimum to raise the guided-wave power to its saturated value $P_{g,\mathrm{sat}}$ within the preamplifier. Figure 2.13(*a*) shows the schematic of the MAG-MOPA with a preamplifier and Figure 2.13(*b*) and (*c*) show the amplified signal, noise and local gain (carrier density) as functions of position within the preamplifier and active grating. By incorporating a preamplifier, the carrier density within the active grating itself can be made extremely uniform, leading to a constancy of modal propagation constant and a flat phase front for the emitted radiation.

An expression for the radiated power can be derived from Equation (2.1) as follows: assuming that the guided-wave power is above the saturation power (5–10 mW) throughout the active grating, the traveling-wave power $P_g(z)$ which satisfies Equation (2.1) is

$$P_g(z) = P_{\mathrm{in}} \exp(-\alpha z) + P_{\mathrm{sat}}(1 - \exp[-\alpha z]), \tag{2.3}$$

where P_{in} is the power injected from the preamplifier into the active grating. Equation (2.3) can be rewritten in terms of the current I_a into the active grating by using the relations between unsaturated gain and current density. The radiated power can then be derived by integrating the radiated power density along the length of the active grating:

$$P_{\mathrm{rad}} = \alpha_s \int P_g(z)\, \mathrm{d}z, \tag{2.4}$$

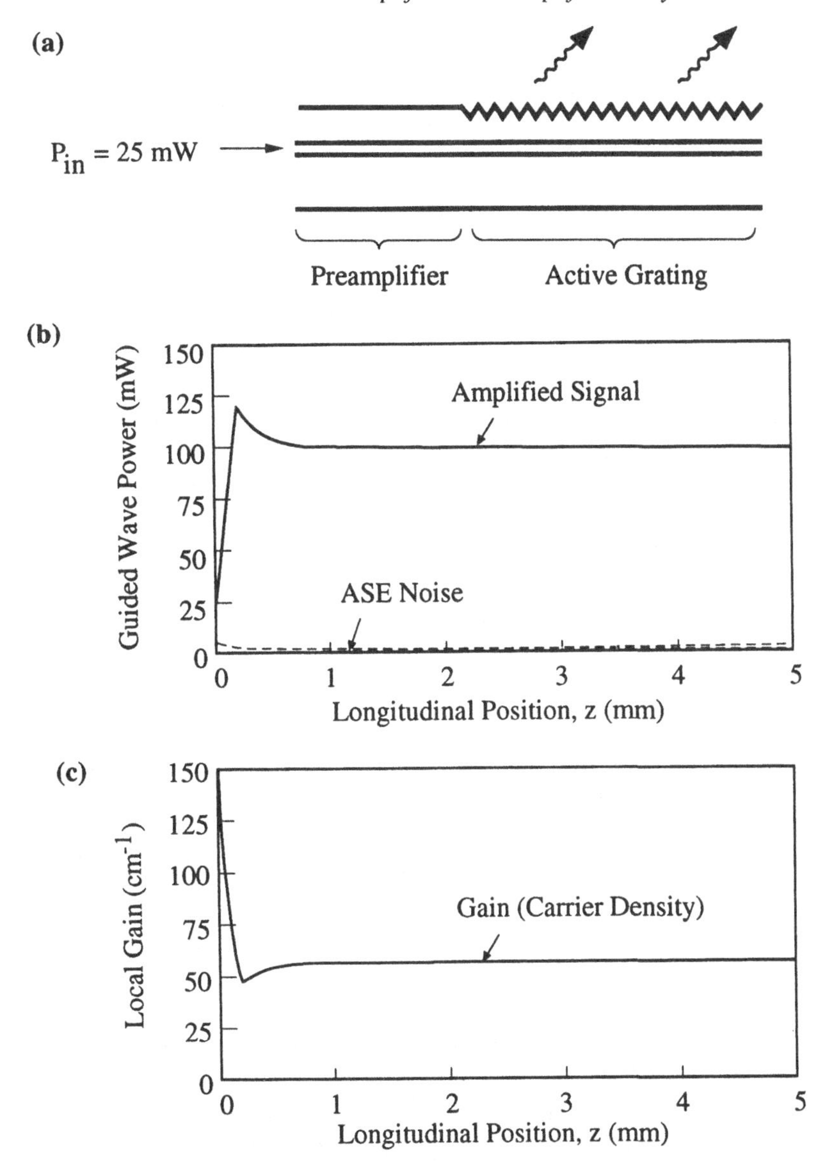

Figure 2.13. Calculations of longitudinal near-field intensity and gain-distribution: (*a*) schematic diagram; (*b*) amplified near field and ASE noise; (*c*) gain distribution.

where α_s is the distributed radiation loss (i.e., the grating output coupling). Note that α_s/α is less than unity, with the remainder being accounted for by absorption and scattering losses. Carrying out the integration in Equation (2.4) leads to the following expression for radiated power:

$$P_{\mathrm{rad}} = \frac{\alpha_{\mathrm{s}}}{\alpha} P_{\mathrm{in}} + \frac{\alpha_{\mathrm{s}}}{\alpha} \eta_{\mathrm{i}} \frac{h\nu}{q} \left(1 - \frac{1}{\alpha L}\right)(I_{\mathrm{a}} - I_{\mathrm{tr}}), \tag{2.5}$$

where I_a is the active grating current, I_{tr} is the active grating transparency current, L is the active grating length, $h\nu$ is the photon energy, q is the electron charge and η_i is the internal quantum efficiency of the amplifier. Note that in deriving Equation (2.5), it was assumed that the guided-wave power becomes saturated within the active grating (i.e., $\alpha L \gg 1$). Equation (2.5) states that without pumping the active grating, a fraction α_s/α of the injected power P_{in} is surface-outcoupled (first term on the right-hand side), which is in agreement with the previous theory of passive grating output couplers (see Ref. 23, and references contained therein). The second term on the right-hand side of Equation (2.5) gives the portion of the output which is amplified and surface-outcoupled. As expected, under the assumption that the guided-wave power exceeds P_{sat}, P_{rad} exhibits a linear dependence on active-grating current.

Figure 2.14 illustrates the numerically generated light–current curve for the device of Figure 2.13, except that the active grating is 2.5 mm long. The device includes a 100 μm-long preamplifier and a 500 μm-long master oscillator. For this simulation, all three gain areas are pumped in parallel. In excess of 1 W of output power is expected at a current density of 10 kA/cm^2. For devices of 1 cm length, up to 4 W of coherent emission would be expected from the active grating.

From Equation (2.5), the external differential efficiency of the MAG-MOPA can be derived. It is the coefficient of the current term; namely,

$$\eta_{\mathrm{D}} = \frac{\alpha_{\mathrm{s}}}{\alpha} \eta_{\mathrm{i}} \left(1 - \frac{1}{\alpha L}\right). \tag{2.6}$$

Equation (2.6) shows that two factors contribute to a high differential efficiency. First, the surface coupling loss, α_s, must comprise a large portion of the total loss α; in other words, the absorption and scattering losses should be insignificant in comparison with the surface coupling loss. Practically speaking, since absorption and scattering losses in a single-mode waveguide with a quantum well active region are typically 5 cm^{-1}, the surface coupling loss should approach 20–25 cm^{-1}. The second factor which contributes to the differential efficiency is the shape of the near field. To maximize carrier extraction from the gain medium via stimulated emission, the near field should be as uniform as possible. In other words, the guided-wave power should be saturated along the full radiating

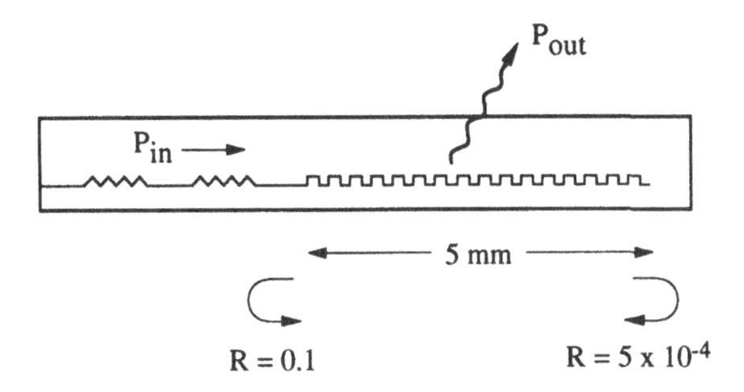

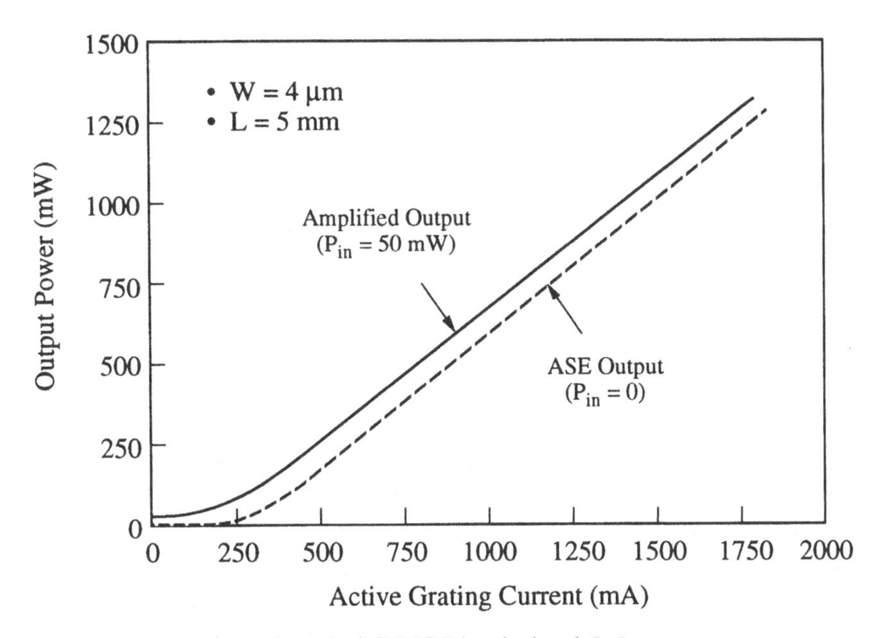

Figure 2.14. MAG-MOPA calculated *L–I* curves.

aperture. This contribution is manifested in the bracketed term on the right-hand side of Equation (2.6). The near field is saturated along the full length of the aperture if $\alpha L \gg 1$.

The effect of grating-surface coupling is illustrated in Figure 2.15, which plots the radiated power density in a 5 mm-long active grating as a function of longitudinal position. The injected power is 5 mW, the current density is 8 kA/cm^2, and the absorption and scattering losses $\alpha - \alpha_s$ are fixed at 3 cm^{-1}. The four curves shown are parametrized by total losses, α, which include a variable amount of surface coupling losses α_s. As these

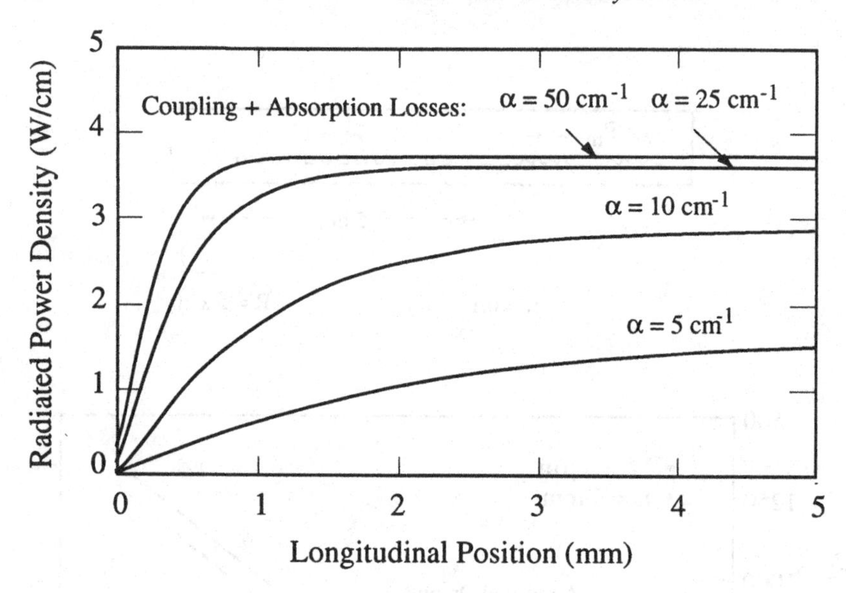

Figure 2.15. Calculated active-grating near field as a function of grating coupling.

curves show, when α_s exceeds 20–25 cm^{-1} the radiating power density becomes uniform along the active-grating length at a maximum level of $\approx$4 W/cm. A more detailed analysis[29] shows that output coupling may decrease again at large α_s due to parasitic gain depletion by ASE noise. If the surface coupling losses are reduced to the level of the scattering and absorption losses, the output-coupled power is low.

Finally, the expected radiation pattern of the MAG-MOPA is easily calculated from the radiating near-field intensity shown in Figure 2.15. Since the period of the active grating is detuned to a longer wavelength than that of the integrated master oscillator, the diffracted beam exits the crystal at an angle θ_0 forward of normal. This angle is defined by the equation

$$\sin \theta_0 = (n_{\text{eff}}\Lambda - \lambda)/\Lambda, \qquad (2.7)$$

where Λ is the grating period, n_{eff} is the modal effective index of refraction, and λ is the emission wavelength in air. If the Bragg wavelength of the detuned grating is $n_{\text{eff}}\Lambda = 50$ nm longer than that of the master oscillator wavelength, λ, then $\theta_0 = 10°$. The radiation pattern of the $\alpha = 25\ \text{cm}^{-1}$ near field in Figure 2.15 was calculated by Fourier transform and is plotted in Figure 2.16. For a 5 mm-long near-field aperture, the diffraction limit in the far field is $\approx$0.01°.

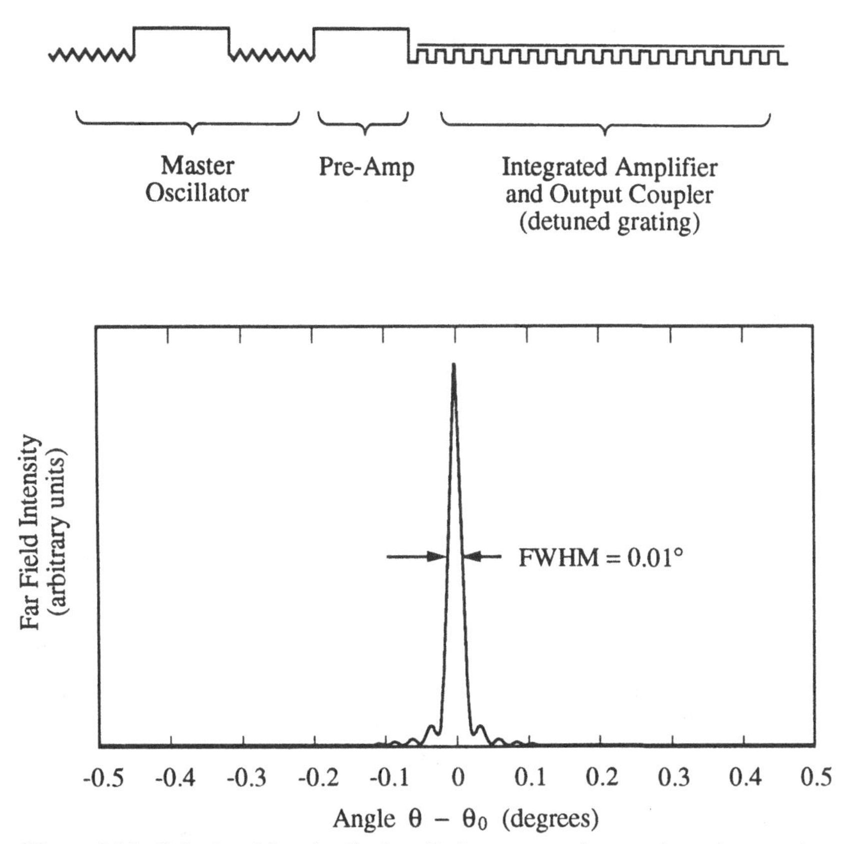

Figure 2.16. Calculated longitudinal radiation pattern from a 5 mm-long active-grating amplifier. For a 50 nm detuning of the active grating from the oscillator grating, the absolute emission angle with respect to the normal $\theta_0 = 10°$.

2.3.2 MAG-MOPA: experiment

Several structures have been studied for implementation of the MAG-MOPA concept[16,17,31,32]; of these only one will be discussed here[31]. The epitaxial structure utilizes a regrown grating region close to the active region to accomplish high grating outcoupling and gain simultaneously.

Figure 2.11 presented the schematic diagram of the MAG-MOPA with a regrown grating region. The epitaxial structure consists of an InGaAs strained quantum well active region surrounded by $Al_{0.3}Ga_{0.7}As$ confining regions and $Al_{0.4}Ga_{0.6}As$ cladding layers. On the p-side the grating is fabricated in $Al_{0.05}Ga_{0.95}As/Al_{0.6}Ga_{0.4}As$ which is within 0.1 μm of the confining layer. The grating is holographically exposed with a HeCd laser

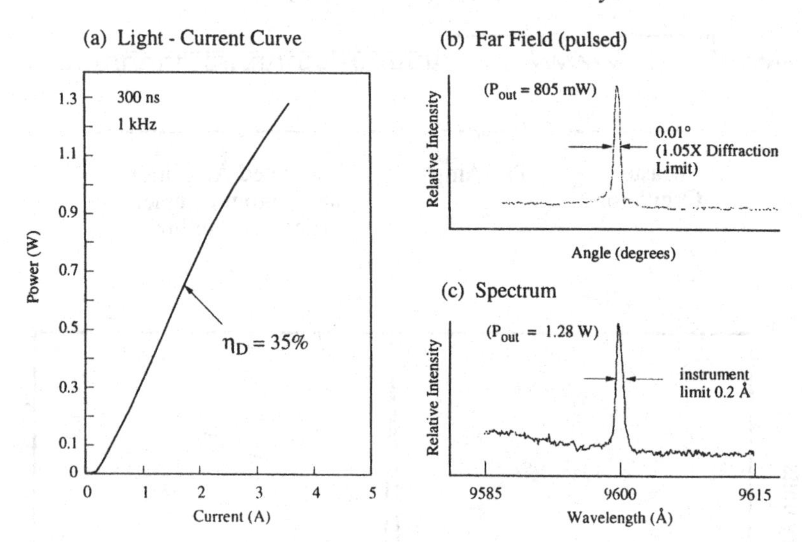

Figure 2.17. MAG-MOPA results under low-duty-cycle pulsed conditions: (*a*) light–current curve; (*b*) far-field pattern; (*c*) spectrum.

and etched into the $Al_{0.05}Ga_{0.95}As$ layer at a depth of 0.08 μm. The DBR grating of the master oscillator and the power amplifier are exposed holographically and etched prior to regrowth on the grating surface, after which a 4 μm-wide real-refractive-index waveguide is fabricated. The master oscillator and preamplifier–power amplifier combination may be contacted separately; however, in some configurations, the contacts are driven in parallel. The light is outcoupled through the substrate and the patterned n-side metallization.

The MAG-MOPA structures were operated pulsed with 500 ns pulses at 1 kHz repetition rate, with separate electrical contacts to the master oscillator and the longitudinal active-grating region. Figure 2.17 presents the pulsed MAG-MOPA characteristics including light–current curve, far-field pattern, and spectrum. As shown in Figure 2.17(*a*), the maximum output power is 1.3 W with a differential efficiency exceeding 35%. With the oscillator operating alone at a current of 100 mA, the spectrum consists of a single longitudinal mode at a wavelength of 960 nm. When the active grating is operated at its maximum output power of 1.3 W, the spectral characteristics of the oscillator emission and the amplifier emission duplicate that of the oscillator within the 0.2 Å resolution of the grating spectrometer, as shown in Figure 2.17(*c*). Figure 2.17(*b*) shows the measured longitudinal radiation pattern of the active-grating emission at an output power of 805 mW. The far field consists of a single diffraction-limited lobe at an angle of 16.5° forward of the normal. This angle corresponds to a 90 nm increase in the pitch of the active grating with respect to the pitch of the resonant DBR reflectors according to Equation (2.7) where $n_{eff} = 3.34$. The FWHM of the amplifier emission is 0.012°, and corresponds to less than 1.1 times the diffraction limit for the 5 mm-long emitting aperture at 960 nm wavelength. Diffraction-limited operation indicates the existence of a near-uniform carrier density along the length of the emitting aperture, in addition to excellent material uniformity. It should be noted that the longitudinal radiation pattern of the MAG-MOPA is imaged through the n-type substrate in the Fourier plane of a long-focal-length ($f = 700$ mm) spherical lens. Due to a spherical curvature induced on the wafer-substrate during polishing, the far-field image plane is displaced slightly from the back focal plane of the imaging lens. The radiation pattern in the transverse direction (perpendicular to the direction of propagation) exhibits a single Gaussian lobe with an FWHM of 9°, diffraction-limited for the 4 μm-wide lateral aperture. Consequently, the 2-D radiation pattern exhibits a single diffraction-limited lobe up to 805 mW peak pulsed output power.

Similar regrown-grating MAG-MOPAs, emitting at 944 nm, were bonded p-side down on a copper heatsink for cw operation. In this configuration, bonding to a patterned submount allows the master oscillator and the preamplifier–power amplifier combination to be driven separately. The resulting *L–I* curve, far-field pattern, and spectrum under cw operation are presented in Figure 2.18. The maximum output power shown in the light–current curve of Figure 2.18(*a*) exceeds 500 mW cw. The spectral characteristic shown in Figure 2.18(*c*) displays a single longitudinal mode at 944 nm wavelength when the DBR master oscillator laser is operated above threshold. The far-field radiation pattern presented in Figure 2.18(*b*) displays a dominant single lobe with FWHM = 0.014°, approximately 1.3 times the diffraction limit, up to 500 mW cw output power. The increase in cw far-field beamwidth over pulsed operation results from non-optimized bonding, which induces non-quadratic curvature in the substrate that cannot be corrected by the spherical imaging lens. However, the stability of the cw far-field beam with drive current indicates that minimal thermal lensing distortion occurs in this design.

The MAG-MOPA device has therefore demonstrated that both single-longitudinal-mode and single-lobed diffraction-limited operation can be achieved simultaneously from a monolithically integrated master oscillator power amplifier. In order to extend MOPA operation beyond the 1 watt power level, however, it is necessary to incorporate a wide-stripe power

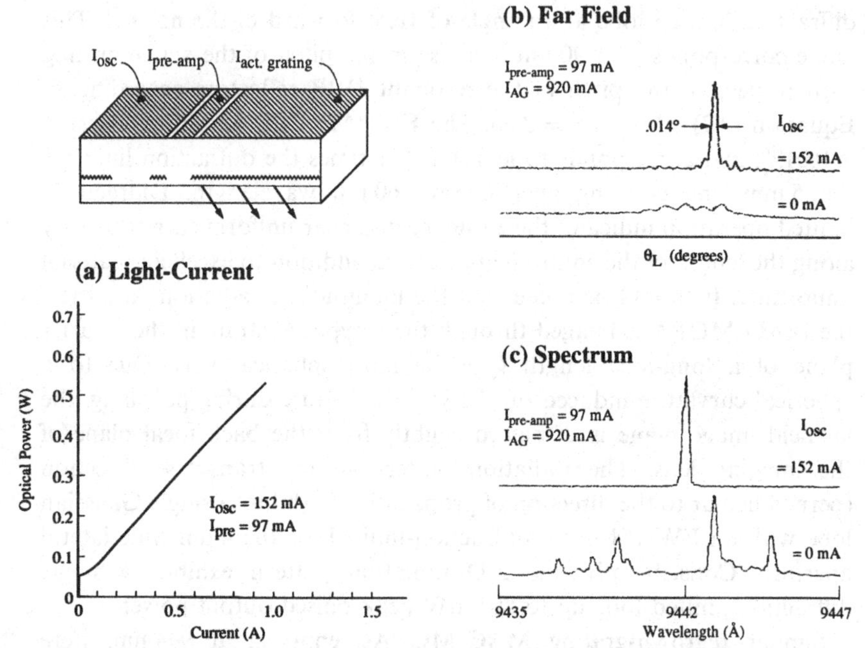

Figure 2.18. MAG-MOPA results under cw conditions: (*a*) light–current curve; (*b*) far-field pattern; (*c*) spectrum.

amplifier. The characteristics of broad-area high-power amplifiers are described in detail in the next section.

2.4 Discrete-element MOPA performance

An alternative design to the MAG-MOPA that has been investigated in order to achieve high-power diffraction-limited operation from a single amplifier is the integration of a broad-area power amplifier with a single-mode master oscillator. As will be discussed in Section 2.5, such an integrated MOPA has resulted in diffraction-limited output powers in excess of 3 W cw. The monolithically integrated M-MOPA designs discussed in Sections 2.3 and 2.5 are directly applicable to similar amplifier arrays discussed in Section 2.2. As a precursor to the discussions of the integrated MOPA, the broad-area amplifier is characterized in discrete form[33,34]. Recently, discrete-element MOPAs have demonstrated multi-watt diffraction-limited output beams[13,35]. Large-area traveling-wave semiconductor amplifiers offer many benefits, including high-power con-

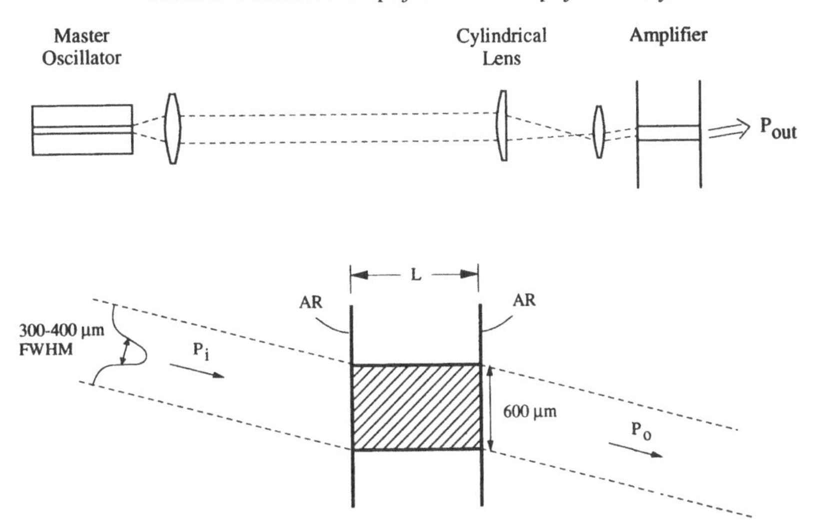

Figure 2.19. Schematic diagram of a single-pass discrete-element MOPA.

version efficiency, tunability, simple fabrication and high-speed modulation capability.

The geometry used for discrete MOPA experiments is shown schematically in Figure 2.19. The master oscillator (MO) is a single-mode laser; both real refractive index-guided semiconductor lasers and Ti:sapphire solid-state lasers have been used to inject GaAs/AlGaAs semiconductor amplifiers. The collimated output of the MO is directed through a cylindrical lens and refocused using a high-numerical-aperture spherical lens on to the input facet of the broad-area semiconductor power amplifier (PA)[13,33]. The PA is typically a separate-confinement-heterostructure single quantum well (SQW) or multiple quantum well (MQW) active-region semiconductor amplifier with a broad-area current-injection contact. Single-pass amplifiers are fabricated by antireflection (AR)-coating both facets, and double-pass amplifiers are fabricated by AR-coating the front facet and depositing a high-reflection (HR) coating on the rear facet.

Referring to Figure 2.19, the combination of cylindrical lens and focusing lens acts as an afocal telescope to inject a broad, Gaussian-shaped beam with a nearly planar wavefront into the active region of the amplifier. The cylindrical lens can be displaced to the left or right of the optical axis in order to vary the incident angle of the injected beam away from normal incidence. The cylindrical lens can also be displaced longitudinally (with respect to the back focal plane of the focusing lens)

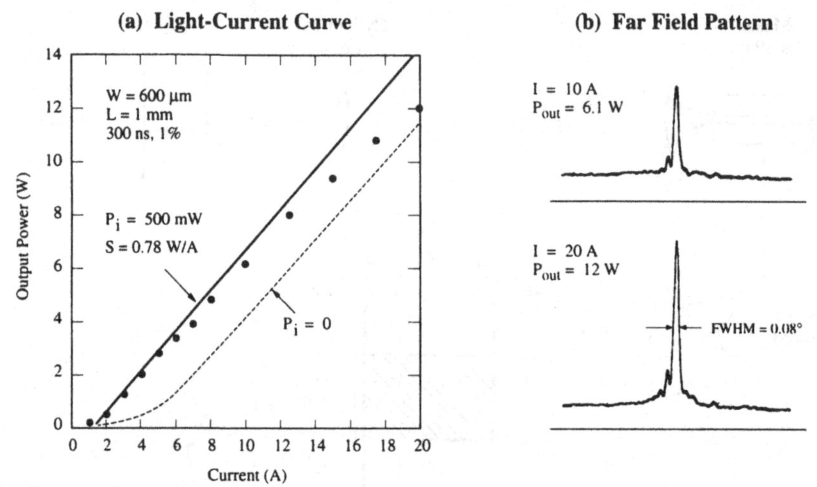

Figure 2.20. Broad-area double-pass amplifier results: (*a*) light–current curve; (*b*) far-field pattern. (After Ref. 35, © 1991 IEEE.)

in order to add or subtract a circular wavefront curvature from the injected beam[33,35,36].

Double-pass and single-pass amplifier performance has been characterized using a Ti:sapphire MO, because of its high power (up to 0.5 W) and broad wavelength tunability, which spans the amplifier gain spectrum. In this section, the generic characteristics of double-pass and single-pass broad-area amplifiers will first be reviewed under Ti:sapphire injection, before discussing amplifier performance under diode laser injection.

Figure 2.20 summarizes the characteristics of double-pass amplifiers 600 μm in width and 1000 μm in length operated under low-duty-cycle pulsed conditions[35]. The injected beam was angled approximately 4° from the optical axis, via a small lateral translation of the cylindrical lens, and served to spatially separate the input and output beams. Figure 2.20(*a*) plots the dependence of the saturated amplified output power on the pulse peak current when the incident power was at its maximum, 500 mW. A maximum peak output power of 12.0 W was measured at a current of 20 A. The measurement of the output power was carried out by tuning the wavelength of the input signal to the gain peak at each current setting, in order to compensate for the carrier density-induced spectral shift of the gain curve. In addition, the longitudinal position of the cylindrical lens was optimized at each current setting to compensate for any carrier-induced or thermally induced quadratic lensing within the amplifier. A maximum slope efficiency of 0.78 W/A was observed, corresponding to a differential quantum efficiency of 52%. This slope efficiency can be compared with the 59% value measured in the same device when operated as a laser before AR coating. The sublinear power dependence of the light–current curve at high current levels is attributed to amplified signal power and amplified spontaneous emission (ASE) noise power not captured by the detector aperture.

Diffraction properties of the amplified beam were determined by directly measuring the far-field diffraction patterns with a linear detector array. The far-field patten of the amplifier of Figure 2.20(*a*) at peak currents of 10 A and 20 A is shown in Figure 2.20(*b*). The measured full-width-at-half-maximum (FWHM) of the center lobe was 0.08°, which is within measurement accuracy of the diffraction limit for a uniformly illuminated 600 μm-wide aperture. The lobe width and position were unchanged for all measured currents. For currents above approximately 15 A, however, an increase in the power contained within the pedestal portion of the far-field pattern was observed.

One factor which was observed to contribute to the deterioration of the far-field pattern emitted by double-pass amplifiers at high output powers was filament formation in the amplifier near field[37]. The near field of the amplified beam was recorded by direct imaging on to a linear detector array. When the injected angle was near the facet normal, $\theta = 0° \pm 0.5°$, the otherwise near-Gaussian near-field distribution broke up into periodically spaced filaments with nearly 100% visibility[37]. Simultaneously, the far-field pattern exhibited large intensity sidelobes. The angular spacing of the sidelobes was consistent with the observed filament period. The formation of filaments also caused the FWHM of the diffraction lobes to broaden, further decreasing the Strehl ratio.

The formation of filaments within the amplified beam is attributed to non-linear interaction of forward- and backward-propagating beams in the reflective amplifier active region[37]. In the presence of gain, the backward-traveling output beam reaches its peak intensity near the output facet where the input beam first enters the amplifier. Therefore, a transverse variation in the output beam intensity can distort the smoothly varying amplitude and phase of the input beam by inducing transverse spatial modulation in both the gain coefficient and the refractive index.

Filamentation within the amplified beam can be largely suppressed by eliminating the strong non-linear interaction which occurs between the forward- and backward-traveling waves – one method is to employ single-pass amplifiers. Single-pass amplifiers are fabricated by depositing antireflection coatings ($R = 0.1\%$) on both cleaved facets and bonding

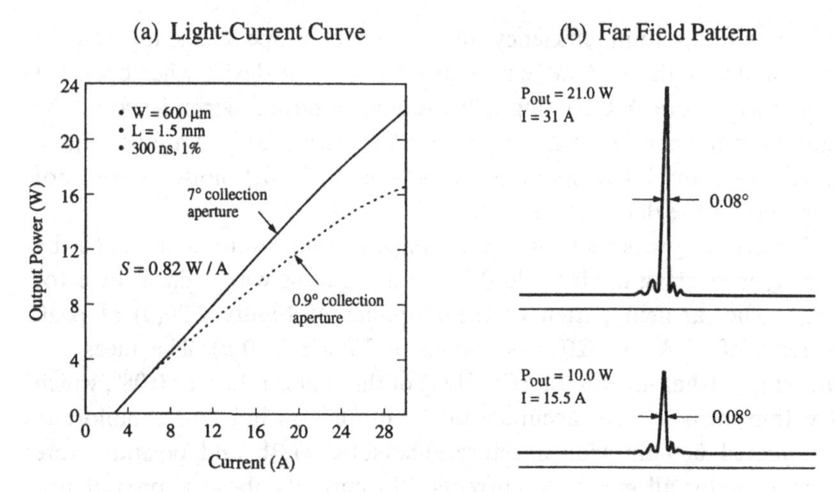

Figure 2.21. Broad area single-pass amplifier results: (*a*) light–current curve; (*b*) far-field pattern.

the amplifier to a matched-length heatsink. Figure 2.21 summarizes the characteristics of a 600 μm-wide by 1500 μm-long single-pass amplifier, operated under low-duty-cycle pulsed conditions (0.3 μs, 1%)[13]. To ensure saturated operation, the amplifier was injected with 0.5 W of Ti:sapphire radiation. Finally, to minimize filamentation at high output power levels, the angle of injection was set at approximately 7°.

Figure 2.21(*a*) plots the amplified power versus peak amplifier current relation. The photodetector used to measure the output power was apertured to collect (i) all power emitted into a 7°-wide pattern centered on the amplified beam, and (ii) only the power emitted into a 0.9° aperture centered on the amplified beam. For the large-angle case, a maximum of 21.0 W of amplified power was collected and a slope efficiency of 0.82 W/A was measured. The smaller angle measurement produced a maximum output of 16.0 W with a corresponding reduction in slope efficiency.

The amplifier far-field pattern was measured directly using a linear detector and is presented in Figure 2.21(*b*) at peak currents of 31 A (output power = 21.0 W) and 15.5 A (output power = 10.0 W). In both cases the FWHM of the dominant central lobe was 0.08°, which, within measurement accuracy, is diffraction-limited for a uniformly illuminated 600 μm-wide aperture.

The single-pass amplifier described here exhibited significantly greater resistance to periodic filamentation than the previously characterized double-pass amplifiers. When the single-pass amplifier was operated at normal incidence, with 500 mW of incident power and 5 μs pulse length, the onset of filamentation occurred at approximately 10 A and 5 W peak output power. This 'threshold' is at least an order of magnitude higher than that observed in double-pass amplifiers.

One other important distinction between single-pass and double-pass amplifiers is their saturation characteristics. Double-pass amplifiers have an inherently higher extraction efficiency than do single-pass amplifiers of the same length because single-pass amplifiers have a longitudinal gain distribution which is unsaturated near the input end[38]. However, the extraction efficiency of single-pass amplifiers can be increased by lengthening the amplifier up to the point where absorption losses cause the power output to saturate[39].

Towards that end, 600 μm-wide by 2200 μm-long single-pass amplifiers have been fabricated for use with a single-mode laser-diode MO[40]. Figure 2.22 summarizes the light–current characteristics and the far-field distribution for such an amplifier operated under low-duty-cycle pulsed conditions (1 μs, 1%). For comparison, Figure 2.22(*a*) plots light–current curves measured using 100 mW diode laser injection, 100 mW Ti:sapphire laser injection, and 400 mW Ti:sapphire laser injection. For Ti:sapphire injection, the incident wavelength was optimized at each current setting to optimize the power output, as described previously. For diode laser injection, the incident wavelength was fixed at 865 nm. Under diode laser injection, a maximum peak output power of 11.6 W was achieved at 25 A

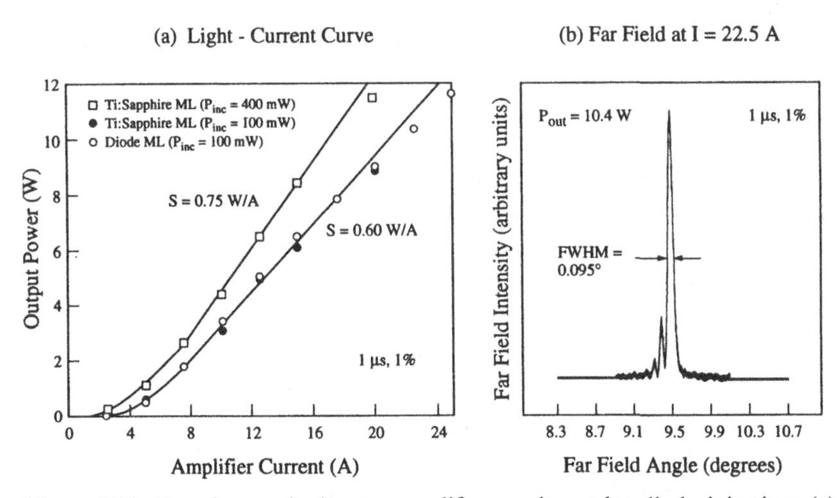

Figure 2.22. Broad-area single-pass amplifier results under diode injection: (*a*) light–current curves comparing diode and Ti:sapphire injection; (*b*) far-field pattern under diode injection at 10.4 W peak output power.

current. The slope efficiency of 0.60 W/A was identical to that measured with 100 mW Ti:sapphire laser injection. The only measured decrease in output power using the laser-diode MO was at low currents, less than 8 A, where the wavelength of peak gain was slightly longer than 865 nm, due to decreased bandfilling. Under higher-power injection with the Ti:sapphire MO, the slope efficiency increased to approximately 0.75 W/A, as expected due to the higher degree of gain saturation near the input end of the amplifier afforded by high-power injection.

The amplifier far-field pattern measured under 100 mW of diode laser injection is shown in Figure 2.22(*b*) at a current level of 22.5 A, corresponding to a peak output power of 10.4 W. The absolute angle of emission was 9.5° from the optical axis, which was required to suppress filamentation and maximize the power emitted into the central lobe of the far-field pattern. The main lobe of the far field was measured to have a FWHM of 0.095°. Although this lobe width was slightly larger than the diffraction limit for a uniformly illuminated 600 μm-wide aperture, within measurement accuracy it was diffraction-limited for the amplified beam near-field width, which was measured to be approximately 500 μm wide. The near-field width at the output facet was necessarily smaller than the aperture width because the off-axis angle of injection caused a lateral shift of approximately 110 μm, based on the effective refractive index of 3.29 measured in these structures[41].

The all-diode-based MOPA of Figure 2.22 represents an important practical demonstration of high-power, diffraction-limited emission from a semiconductor device. A further demonstration of the practicality of broad-area amplifiers as high-power, coherent light generators is diffraction-limited cw operation. In the following paragraphs, cw MOPA operation is described.

For cw operation, a 600 μm-wide by 1.1 mm-long single-pass amplifier was bonded p-side down on to a heatsink which in turn was placed inside a water-cooled fixture[42]. The amplifier output was measured as a function of pump current with a constant incident power of 400 mW corresponding to 270 mW coupled into the amplifier. The light–current curve and amplifier far-field is presented in Figure 2.23. Under cw operation, a maximum of 3.3 W of output power was measured at a current of 7.0 A, as shown in the light–current curve of Figure 2.23(*a*). The output power was collected using a photodetector which captured 1.2° of the far-field emission. Also shown for comparison is the output power collected under the same conditions but with 1 μs low-duty-cycle pulses. A maximum slope efficiency of 0.61 W/A was measured for cw operation and 0.70 W/A

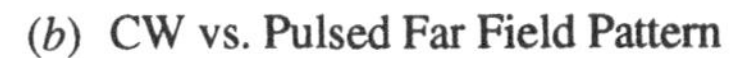

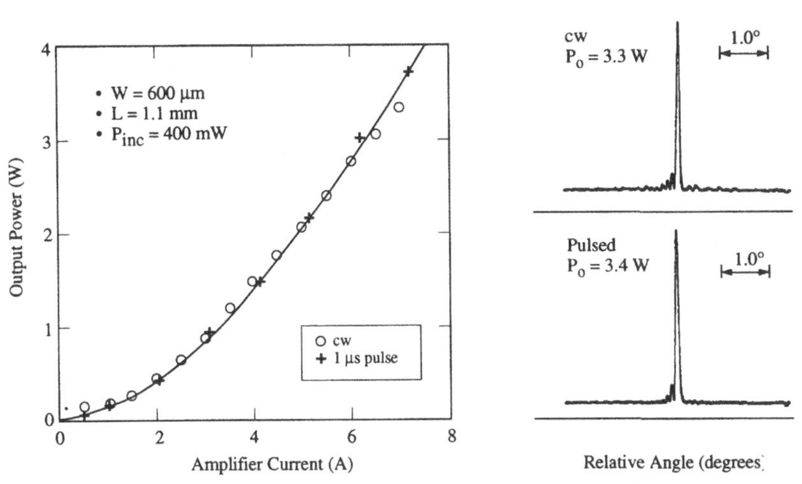

Figure 2.23. CW vs pulsed single-pass amplifier results: (*a*) light–current curve; (*b*) far-field pattern. (After Ref. 42, © 1992 IEEE.)

for pulsed operation. The smaller slope for the cw case corresponded to a slightly higher amount of light scattered outside of the detector aperture, rather than being caused by thermal rollover.

Far-field emission characteristics under pulsed as well as cw operation, shown in Figure 2.23(*b*), consisted of a dominant 0.08°-wide diffraction-limited lobe and some sidelobes. The sidelobes originated from the abrupt clipping of the Gaussian input beam by the amplifier aperture. These diffraction sidelobes were enhanced by the fact that, due to gain saturation, the low-power wings of the Gaussian were preferentially amplified, leading to a near-field distribution at the output facet, which was near-rectangular.

The discrete-element MOPAs described in the preceding paragraphs all consist of single-mode master oscillators and broad-area active region amplifiers. These amplifiers have proven capable of generating multi-watt diffraction-limited beams under pulsed and cw conditions, and offer many benefits including high-power conversion efficiency, tunability, simple fabrication, reliable structure and high-speed modulation capability. Potential drawbacks of the broad-area amplifier structure are: (i) the deterioration of the far-field pattern at high output powers and long pulse lengths, due to filamentation enhanced by interaction of the forward- and backward-traveling waves, and (ii) the high level of injected power required to saturate the output of a broad-stripe amplifier while simultaneously minimizing the ASE noise output. Further improvement of the

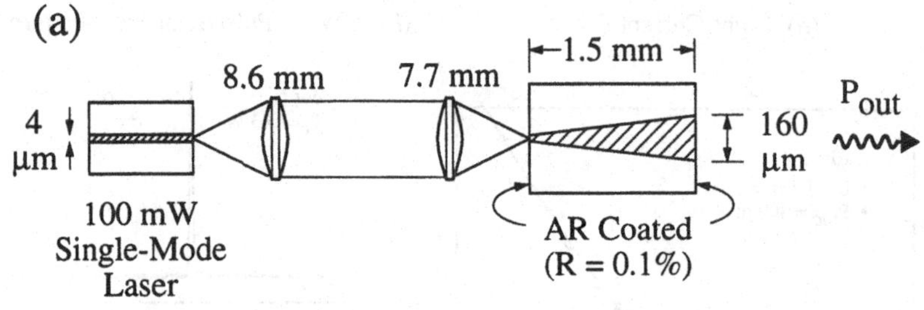

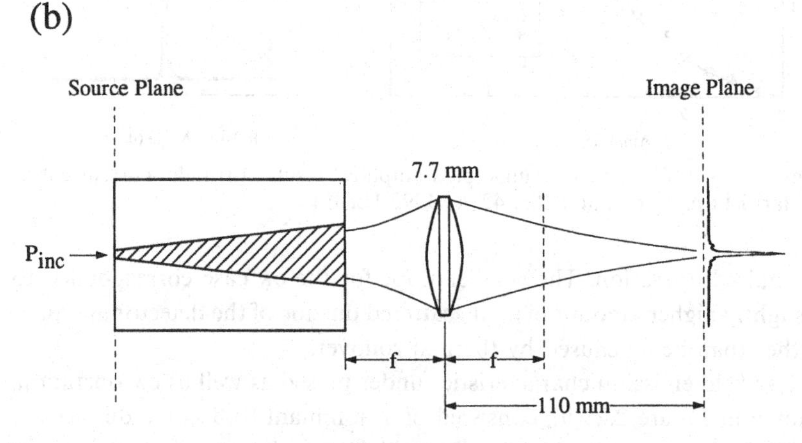

Figure 2.24. Discrete flared amplifier MOPA: (*a*) injection scheme; (*b*) far-field measurement. (After Ref. 44, © 1992 IEEE.)

discrete MOPA performance in these two categories can be gained by amplifying a diverging beam in a tapered-stripe amplifier[43,44], rather than by amplifying a collimated beam in a non-tapered amplifier as described previously.

Figure 2.24(*a*) shows a schematic diagram of a discrete MOPA consisting of a single-mode laser diode MO and a tapered-stripe PA. The MO output is imaged at approximately unit magnification on to the narrow aperture of the tapered single-pass amplifier. The amplifier is 1.5 mm in length, with a gain region which expands up to 160 μm at the output end. The input beam is focused into an approximately 4 μm-wide by 1 μm-high spot at the input aperture, and is allowed to diffract freely in the junction plane as it is amplified, since the gain taper angle of 5.7° exceeds the approximately 3° FWHM of the diverging input beam. The tapered amplifier maximizes its efficiency by expanding the gain volume along the amplifier length, as the amplified power grows, in order to maintain a near-uniform power density and degree of gain saturation[45]. Because the amplifier width is narrow at the input end, the input power required to saturate the output power is decreased in comparison with non-tapered amplifiers having the same output aperture width. Furthermore, since the amplified beam is diverging, the extent of the spatial overlap of the reflected reverse-traveling wave and the input wave is minimized, and is smaller than the filament spacing previously observed in non-tapered amplifiers. Consequently, filamentation is expected to be suppressed to significantly higher optical power densities than the threshold observed in non-tapered amplifiers.

Figure 2.25 presents the tapered amplifier light–current curve and far-field pattern as measured under cw conditions. The light–current curve, shown in Figure 2.25(a), corresponding to an incident MO power of 25 mW, has a slope efficiency of 1.0 W/A, and reaches a maximum of 2.5 W at 3.0 A current. This curve was measured with the MO wavelength detuned to 865 nm, 10 nm longer than the wavelength of peak ASE output. ASE quenching by the injected signal was confirmed by direct observation of the amplifier spectrum with and without MO input. The slope efficiency of 1.0 W/A was nearly constant over the wavelength range 855–865 nm, and corresponded to an external differential quantum efficiency of 70% at 865 nm. This value of differential quantum efficiency is approximately 20% higher than the best value recorded for the equal-length, non-tapered, broad-area amplifiers discussed above, even

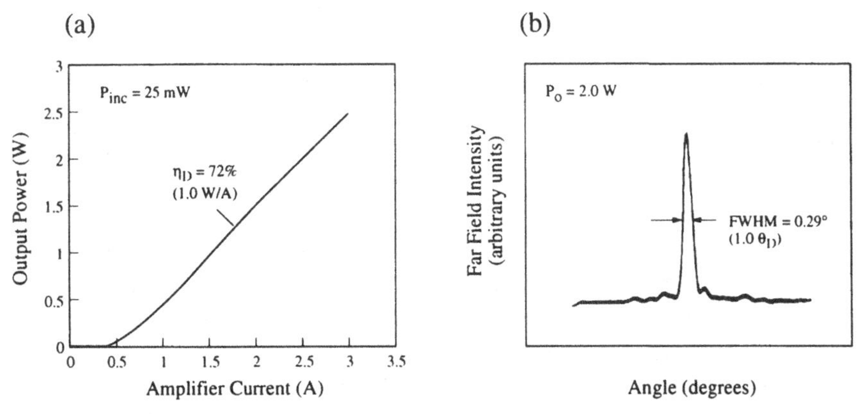

Figure 2.25. Flared amplifier MOPA cw characteristics: (*a*) light–current curve; (*b*) far-field pattern. (After Ref. 44, © 1992 IEEE.)

when injected with as much as 500 mW of Ti:sapphire radiation, and 40% higher than the highest value recorded for non-tapered, broad-area amplifiers injected with 100 mW of laser diode emission.

The radiation pattern directly measured from the tapered amplifier is a broadly diverging beam, similar to the output of an unstable resonator. A more useful measure of the spatial beam quality of the output is to form an image of the virtual source of the diverging beam[43,46]. As shown in Figure 2.24(*b*), a 7.7 mm-focal-length lens, positioned approximately one focal length away from the amplifier output facet, was used to form the image on a linear detector array. The image size so measured, when converted to effective radiation angle, gives the effective far-field pattern with the quadratic diverging component removed[43,44]. The far-field pattern of the tapered amplifier, measured under cw conditions at 2.0 W total output power, is presented in Figure 2.25(*b*). It consists of a dominant Gaussian-shaped lobe with a FWHM of 0.29°, which is diffraction-limited for the 160 μm-wide amplifier output aperture. Near-field observation at the output facet confirmed approximately uniform illumination of the aperture by the amplified beam.

Similar experiments were done by E. S. Kintzer *et al.*[11] at 980 nm by employing strained-layer InGaAs/GaAs/AlGaAs tapered-amplifier devices and a Ti:sapphire MO. A 2-mm-long PA tapering from a 10 μm-wide input aperture to a 215 μm-wide output aperture provided up to 3.5 W cw output power in a diffraction-limited pattern[11], when the incident MO power was 90 mW. The slope efficiency was 0.82 W/A; that is, 65% at the 980 nm emitting wavelength. The central far-field-pattern lobe contains 89% of the total power: 3.1 W; and has a 0.24° lobewidth, which is approximately 1.05 times the diffraction limit for a uniformly illuminated 215 μm-wide aperture.

Due to thermal- and carrier-induced lensing within the tapered amplifier[36,47], the virtual source position is not located precisely at the input facet, but at a position slightly in front of the input facet which depends upon the current level. Based upon experiments performed on rectangular broad-area amplifiers[42,47], carrier-induced lensing is dominant over thermal lensing. At low output powers ($P_0 < 1$ W), where the carrier population is not fully saturated, the virtual source displacement is directly proportional to injection current. At high output powers ($P_0 > 1$ W), where gain saturation effectively clamps the carrier population, the virtual source displacement is minimal. Up to 2.0 W cw output power, however, the far-field width of AlGaAs/GaAs devices remains diffraction-limited, at the drive-dependent image plane.

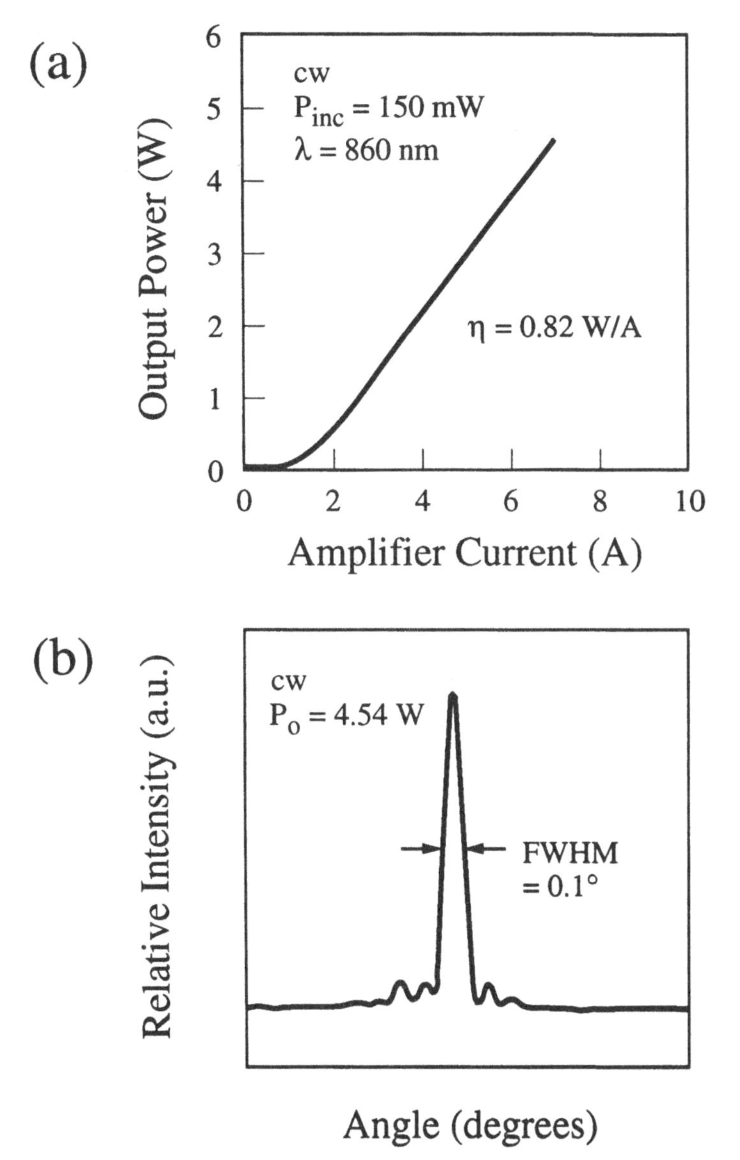

Figure 2.26. CW tapered amplifier characteristics: (*a*) light–current curve; (*b*) far-field pattern at 4.54 W output power. (After Ref. 12, © 1993 IEEE.)

As a result of the high extraction efficiency of the tapered amplifier, the overall electrical-to-optical conversion efficiency of the tapered amplifier MOPA is very high. Accounting for electrical power dissipation of both the MO and PA, the total energy efficiency of the discrete element

MOPA at 2.0 W cw output power is 39.0%, which rivals that of incoherent cw laser arrays.

Even higher cw output powers have been obtained from tapered amplifiers by optimizing the taper geometry. In the above work, the master oscillator output was imaged to a 4 μm-wide spot in the junction plane of the tapered amplifier such that a waist was formed at the input facet. In that case, only 10 mW of incident radiation was required to saturate the amplifier output[44]. In contrast, untapered broad-area amplifiers 600 μm in width require approximately 400 mW of incident Ti:sapphire radiation to saturate the amplifier input. In later work[12], the input aperture of the tapered amplifier has been broadened significantly in order to maximize the amplifier area and take full advantage of the 100–150 mW output power available from commercial single-mode laser diodes. In addition, the taper angle has been increased from 5.7 to 9° to increase the output aperture width for high-power cw operation.

In order to inject the broader-aperture tapered amplifier, an astigmatic injection scheme was used to effectively saturate the amplifier input. The collimated master oscillator input was focused by an $f_c = 300$ mm cylindrical lens and an $f = 7.7$ mm spherical lens. The spherical lens was positioned a distance f from the input facet, and the cylindrical lens was positioned a distance d of a few cm ($d \ll f_c$) to create a virtual waist which was positioned: (i) at the facet perpendicular to the p–n junction, and (ii) approximately 0.2 mm in front of the facet in the junction plane. The collimated input beam had a full-width-at-half-maximum (FWHM) of approximately 4 mm, resulting in a 1.5 μm-wide virtual waist in the junction plane. This waist expanded to approximately 100 μm in width at the input facet by virtue of its 30° divergence in air.

Injection to the tapered amplifier was provided using a cw Ti:sapphire source operated at 150 mW output power. Past experimental results have verified that Ti:sapphire and single-mode laser diode master oscillators can be used interchangeably at these lower powers[40]. The input beam wavelength was fixed at 860 nm wavelength, corresponding to the peak amplifier gain at high output power. The amplifier heatsink was attached to a larger copper block whose temperature was fixed at 10 °C by an underlying thermoelectric cooler. The thermoelectric cooler was backed by a passive, finned heatsink. The cw amplifier light–current curve measured under these conditions is presented in Figure 2.26(*a*). Up to 4.54 W of total output power was measured at a maximum current of 7.0 A, corresponding to a slope efficiency of 0.82 W/A.

The effective far-field pattern of the tapered amplifier emission was determined by imaging the virtual source point with an additional $f = 7.7$ mm-focal-length lens positioned behind the output facet. The far field at the highest power measured, 4.54 W, is plotted in Figure 2.26(*b*). The far field at all currents was dominated by a diffraction-limited central lobe with a FWHM of 0.11°. The corresponding near-field pattern was observed to be substantially flat with an approximately 10% modulation random intensity ripple. The small sidelobes in the far field adjacent to the central lobe were caused by edge diffraction from the near-rectangular aperture. These sidelobes were also observed to become slightly more prominent with increasing output power due to flattening of the near-field pattern from near-Gaussian to near-rectangular, which resulted from gain saturation.

In summary, broad-area tapered amplifiers have emitted up to 4.5 W cw in a near-diffraction-limited far-field pattern. The design of the tapered amplifier is such that its output can be saturated by a low-power (0.15 W), single-mode laser master oscillator, and is suitable for monolithic integration with a single-mode distributed Bragg reflector (DBR) laser. In the next section, monolithically integrated flared-amplifier master oscillator power amplifiers will be discussed.

2.5 Integrated flared-amplifier MOPAs

In the preceding sections it was shown that high-power diffraction-limited output beams have been achieved in MOPA structures when an external optical signal was injected into the optical amplifier. In this section the topic of monolithically integrated flared-amplifier master oscillator power amplifers (MFA-MOPA) will be discussed[48–50]. As will be presented, the MFA-MOPA has demonstrated, at a heatsink temperature of 5 °C, single lobed diffraction-limited operation to greater than 3 W cw, the highest single spectral and spatial mode output power demonstrated from a monolithic diode emitter. Much of the basis of discussion of this section is built upon the discussion of discrete broad-area power amplifiers in Section 2.4. The characteristics of the discrete power amplifier are similar to that of the integrated MOPA discussed here.

The design of the MFA-MOPA is presented in Figure 2.27. The structure, similar to other M-MOPA designs, consists of a DBR master oscillator coupled to a power amplifier. The DBR laser is fabricated in a two-step metalorganic chemical vapor deposition (MOCVD) process, Intermediate to the first and second growth, a second-order grating is fabricated. The grating is holographically exposed and chemically etched

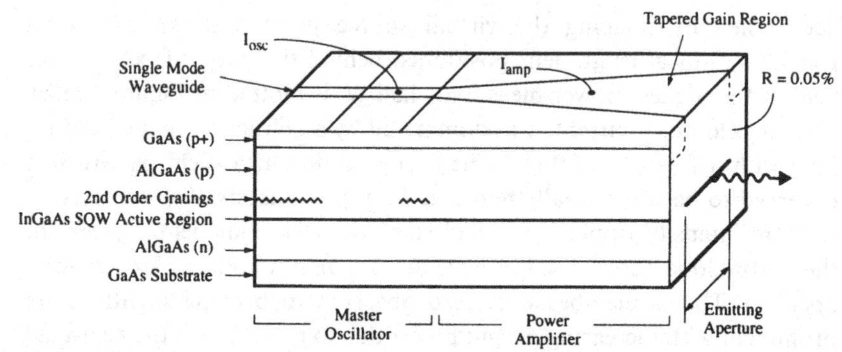

Figure 2.27. Schematic diagram of the MFA-MOPA. (After Ref. 50, © 1993 IEEE.)

into the surface of the AlGaAs p-cladding layer. The output of the master oscillator is coupled to a gain-guided flared amplifier. The light diffracts from the output of the 4 μm-wide single waveguide as it propagates through the power amplifier. As the mode propagates along the length of the tapered gain region the modal intensity remains approximately fixed; however, due to the increasing aperture size the overall power is increased[43–45]. The output power, to first order, scales linearly with amplifier width at a set current density.

The light output as a function of the input current is presented in Figure 2.28 for a device in which the amplifier length is 2 mm and output aperture width is 250 μm. In this configuration the oscillator current is held constant at 150 mA while the light output is plotted as a function of the current to the power amplifier at a heatsink temperature of 5 °C. The maximum output power presented in Figure 2.28 is 2 W cw at an input current of 3.5 A to the power amplifier. The differential efficiency of the power amplifier is approximately 50% and the total conversion efficiency of the MFA-MOPA at an output power of 2 W cw is 29%. When the oscillator is turned off the amplifier operates in a superluminescent mode to a current level of approximately 2.7 A. Above this drive level the amplifier begins to self-oscillate, as is shown in the *L–I* curve. Under injection from the master oscillator, the gain of the amplifier is reduced and therefore the self-oscillation of the power amplifier is quenched.

As previously discussed, an advantage of the MFA-MOPA architecture is that powers as low as 5–10 mW are sufficient to saturate the gain in the flared amplifier. The saturation of the gain of the flared amplifier is characterized in Figure 2.29, which gives the output power from an MFA-MOPA as a function of current to the DBR laser master oscillator.

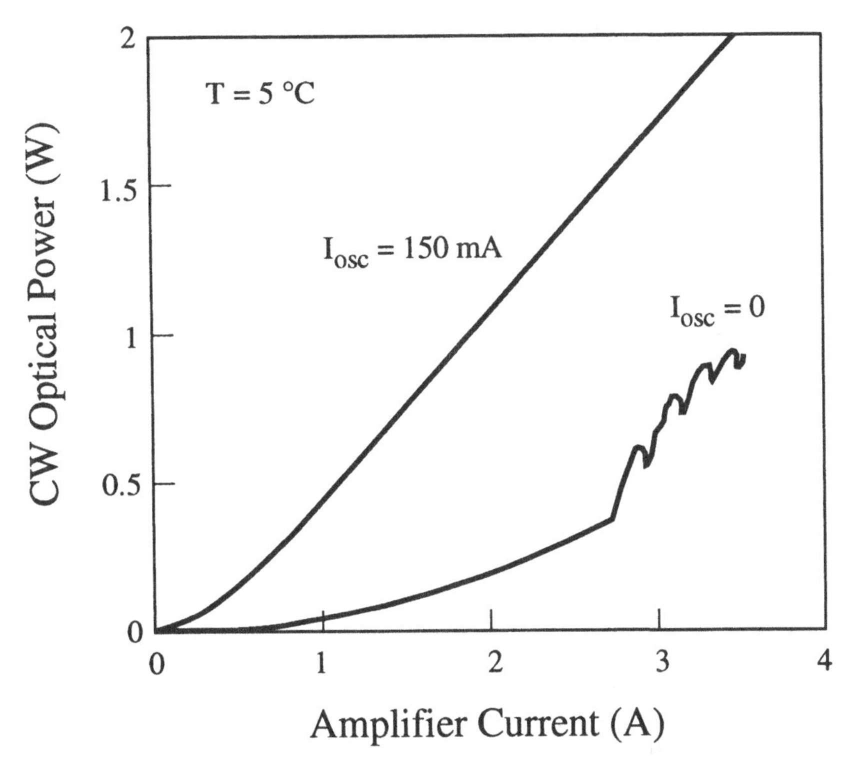

Figure 2.28. Light–current curve of the MFA-MOPA under maximum injection ($I_{osc} = 150$ mA) and zero injection. (After Ref. 50, © 1993 IEEE.)

These data are taken with a current of 2.5 A into the flared amplifier at a temperature of 10 °C. With $I_{osc} = 0$ approximately 250 mW of spontaneous noise is generated by the flared amplifier. The gain begins to saturate with as little as 10–15 mA of drive current to the oscillator. At a drive current of 35 mA the gain of the flared amplifier is completely saturated. The integrated nature of the DBR laser master oscillator makes it difficult to determine exactly how much power is injected into the flared amplifier; however, separate experiments with cleaved, AR-coated DBR lasers indicate that the injected power at a drive current of 35 mA should be of the order of 15–20 mW.

The light propagating through the power amplifier region is affected by the non-uniform saturation of the gain across the width of the amplifier and by the thermal gradient induced by the dissipated power[36,47]. To first order, the effect of the carrier and thermal gradients of the amplifier can be characterized by an effective lens which images the diffracting wave

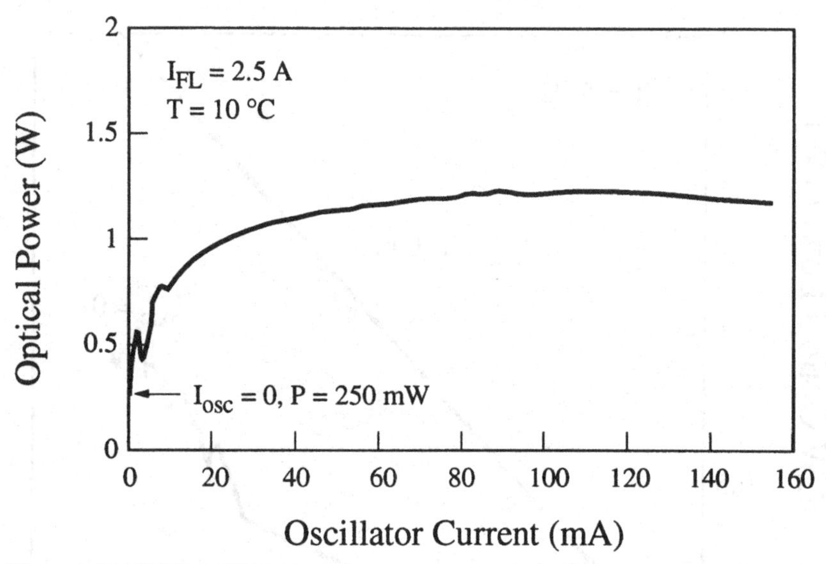

Figure 2.29. MFA-MOPA output power vs master oscillator current. (After Ref. 50, © 1993 IEEE.)

from the 4 μm-wide aperture of the master oscillator. This lensing results in a virtual source within the amplifier with a different waist size and position from that which is injected from the master oscillator. Since the lens power of the amplifier is not known, and therefore the virtual source size and location are not known, the diverging beam at the output of the amplifier cannot be directly related to a known diffraction-limited output. In order to measure the beam quality of the amplifier relative to a known source, the output is imaged through a 14 mm-focal-length lens where the lens is positioned at a focal length away from the emitting aperture. Under these conditions the resultant beam waist that is formed some distance away from the lens can be correlated to the width of the near-field emission of the output aperture of the amplifier. The width of the beam waist can then be characterized relative to the ideal diffraction-limited operation for an emitter of the same width. In the following discussion the beam waist can also be referred to as the far-field pattern from the emitting aperture by transforming the waist size in micrometers to an angle by measuring the angle imaged from the lens.

The radiation pattern of the MFA-MOPA, presented in Figure 2.30, is measured as discussed above. The resultant beam waist represents the Fourier transform of the emitting aperture of a flared amplifier of width 250 μm. To obtain these measurements the detector is positioned and

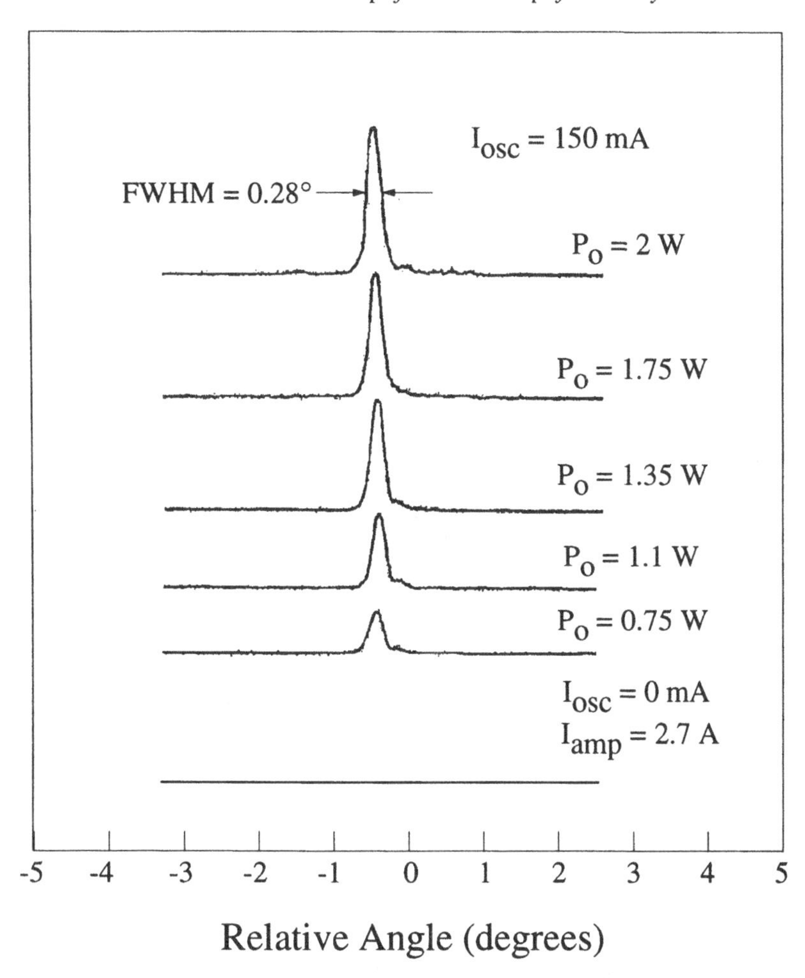

Figure 2.30. MFA-MOPA far field pattern at 5 °C (see text discussions) as a function of power output. (After Ref. 50, © 1993 IEEE.)

adjusted for each power measurement at the waist of the radiation beam. The far-field pattern is predominantly single-lobed with greater than 93% of the power emitted in a diffraction-limited output lobe. The width of the far-field pattern is 0.28° and is independent of drive current to an output power greater than 2 W cw (at 5 °C), correlating to a diffraction-limited output beam.

When the oscillator current is turned off, the far field is quenched to better than 25 dB extinction in the regime where the amplifier is below the self-oscillation threshold. The power that is radiated from the amplifier in the amplified spontaneous emission mode (oscillator off and amplifier below the self-oscillating threshold) is radiated at high angles and is not imaged at the beam waist position, and therefore does not overlap with the far-field pattern. Above the self-oscillation threshold of the amplifier there is a residual lobe in the far field that overlaps with the signal under injection from the oscillator, which results in a 15 dB attenuation of the far-field pattern above the self-oscillation threshold of the amplifier.

Slight variations in carrier-induced and thermal-induced phasefront curvature (i.e. lensing) with increasing output power result from gain saturation and the added dissipated power respectively. The result is a slight change in the virtual source position with output power. Figure 2.31 shows the virtual source position, normalized to the Rayleigh range, as a function of output power. The source position changes by approximately one Rayleigh range associated with the measured beam waist when the output power is increased from 0.4 to 1.3 W cw. In addition, these data show that the position of the virtual source changes only slightly at the higher output powers, which is consistent with the results of rectangular broad-area amplifiers[42,47] which show that carrier-induced lensing is dominant over thermal lensing. At high output powers ($P_0 > 1$ W), gain saturation effectively clamps the carrier population, resulting in minimal lensing. For example, the position of the virtual source changes by less than 1/10 of a Rayleigh range when the output power is increased from 1 to 1.3 W.

The beam quality of an MFA-MOPA has been quantified to an operating power of 1.2 W cw by measuring the beam quality factor, M^2 [51]. To characterize the quality of the beam emitted from an MFA-MOPA the beam is first circularized through the use of an objective lens with a focal length of 6.5 mm and a cylindrical lens with a focal length of 200 mm. The numerical aperture of the objective lens is 0.62. The resulting beam diameter is then approximately 7.5 mm. A variable slit is placed at the far-field plane, after the objective lens, which allows the output beam to be spatially filtered.

The beam quality factor, M^2, is measured using Coherent's MM-2 Modemaster. These data have been verified to be accurate to within 7% by manually determining M^2 by recording the beam diameter as a function of position along the optic axis in the vicinity of a beam waist. It should be noted that M^2 is a measure of beam quality with respect to a Gaussian-shaped beam. For example, a perfect diffraction-limited Gaussian beam has $M^2 = 1.0$. However, single-mode diode lasers do not

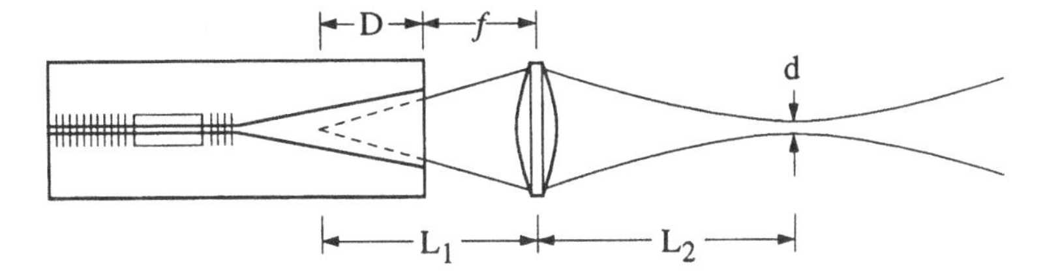

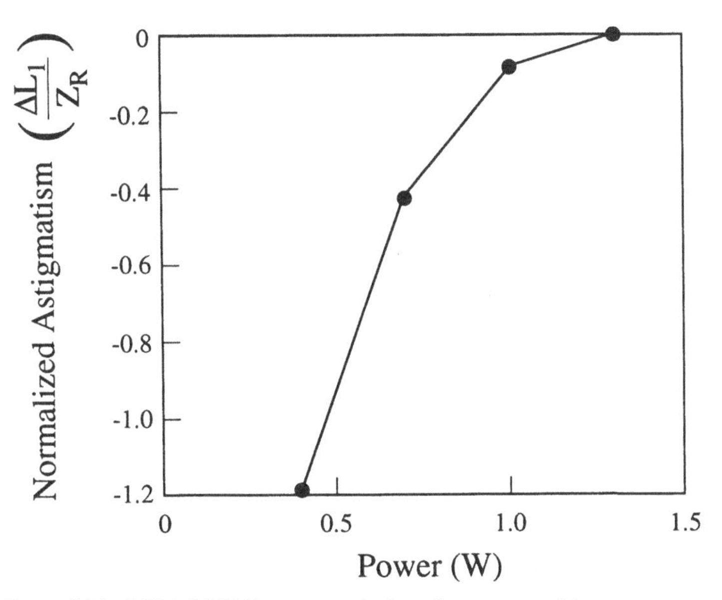

Figure 2.31. MFA-MOPA measured virtual source position vs output power.

produce ideal Gaussian-shaped beams and typically result in M^2 values of $M_{\parallel}^2 = 1.1$ and $M_{\perp}^2 = 1.5$, where $M_{\parallel}^2$ and $M_{\perp}^2$ refer to the directions parallel and perpendicular to the junction. These higher M^2 values reflect the non-Gaussian shape of the beams that can be produced by single-mode diode lasers.

Experimentally measured values of $M_{\parallel}^2$ and $M_{\perp}^2$ for an MFA-MOPA between 0.16 W and 1.1 W cw are shown in Figure 2.32, where the output power is measured after the variable slit. The variable slit is used to transmit the main diffraction lobe in the direction parallel to the junction. For higher output powers, the worst case transmission is 93%. Then $M_{\perp}^2$ remains constant at a value of 1.5 to a power in excess of 1.1 W through the aperture. This value is identical to that measured on single-mode

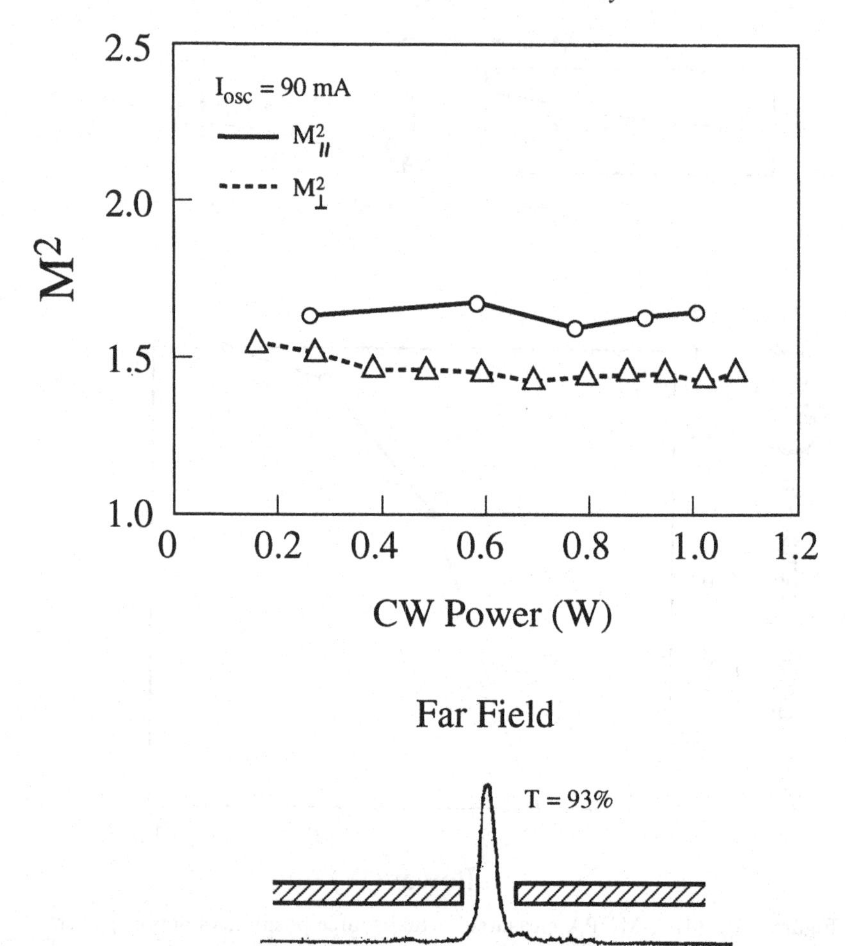

Figure 2.32. Measured M^2 factor parallel and perpendicular to the junction, as a function of MFA-MOPA output power.

single-stripe lasers, indicating high-quality diffraction-limited operation in the fast axis. As a function of output power $M_{\parallel}^2$ is also constant at a value of 1.6 over the operating region. In all of these measurements the current to the oscillator is constant at 90 mA while the current to the flared amplifier is varied to obtain the desired output power. Separate experiments have shown that $M_{\parallel}^2$ and $M_{\perp}^2$ are independent of the oscillator drive current. The power loss incurred by the use of a spatial filter varies from a few percent at low powers to a maximum of 7% at an output power of 1.3 W cw.

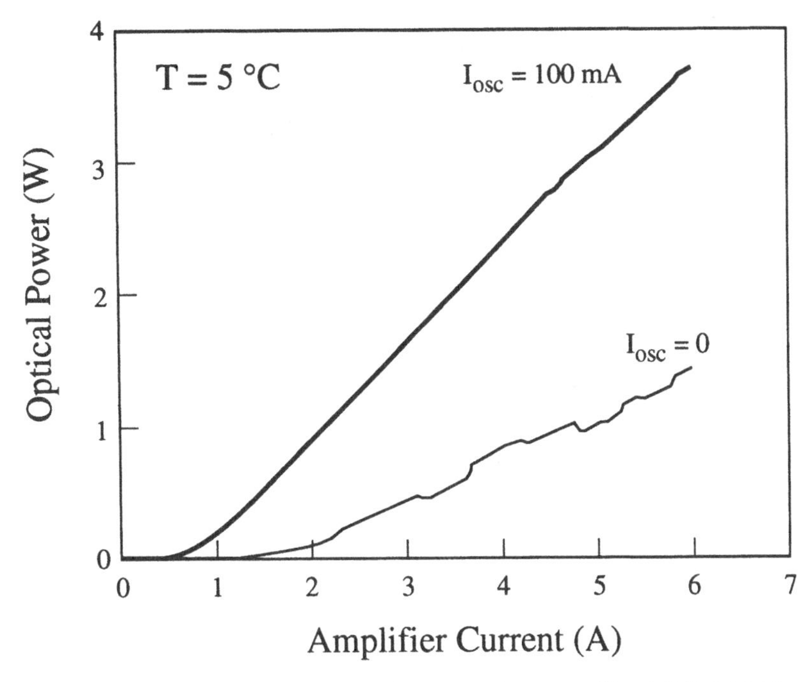

Figure 2.33. Light–current curve of 3.25 mm-long amplifier MFA-MOPA.

Higher output powers have been achieved from the MFA-MOPA by increasing the length of the amplifier. Figures 2.33 and 2.34 present the *L–I* and far-field pattern of an MFA-MOPA in which the amplifier length was increased from 2 mm to 3.25 mm and the output aperture was increased from 250 μm to 400 μm. The maximum output power shown here exceeds 3 W cw at 5 °C, and the corresponding far-field pattern of the source at 3.0 W output indicates a predominantly single-lobed diffraction-limited radiation pattern.

The spectral output of the 3.25 mm-length MFA-MOPA, similar to that of Figure 2.30, is a single longitudinal mode which replicates that of the DBR master oscillator. As discussed in a previous article[52], DBR lasers which are almost identical to the one integrated into the MFA-MOPA operate in a single longitudinal mode with a linewidth less than 4 MHz and a side mode suppression ratio greater than 25 dB. Most of the operating regime of the MFA-MOPA reflects that of the discrete DBR laser; however, in narrow regimes the spectral stability becomes uncertain on the order of 100 MHz due to the residual resonances of the amplifier. The wavelength of the MFA-MOPA shifts slightly with increasing current

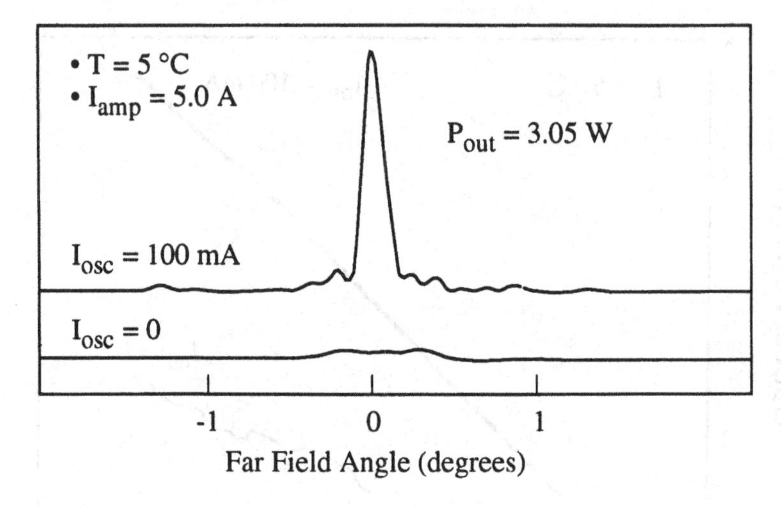

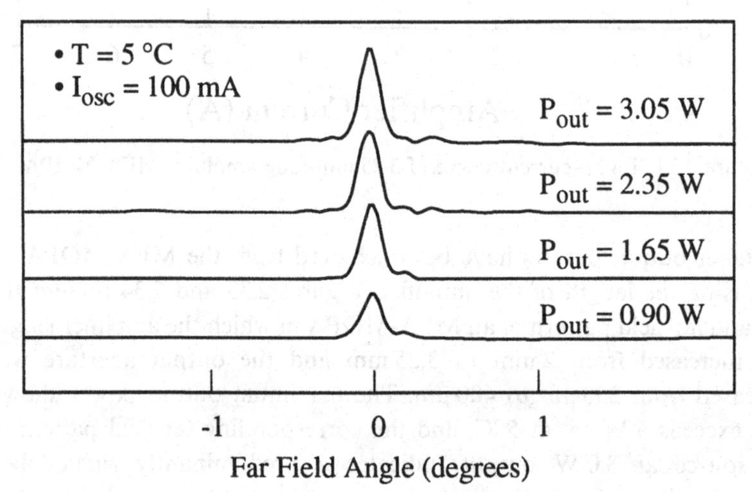

Figure 2.34. Far-field pattern of 3.25 mm-long MFA-MOPA.

to the flared amplifier, approximately 0.1 nm/A, as a result of the heating of the master oscillator due to greater power dissipation from the power amplifier.

The spectral output exhibits one of the advantages of the MFA-MOPA, that of minimum wavelength chirp while operating over a large range in output power. As a result the MFA-MOPA can be modulated in several ways. First, the output power can be modulated by varying the current to the master oscillator. In this scenario the output power in the far field can be digitally modulated with an extinction ratio greater than 25 dB. The modulation of the amplifier output via the master oscillator current can only be performed in a regime below the self-oscillation threshold of the power amplifier, as shown in Figure 2.28. Alternatively, in order to minimize the wavelength chirp of the MFA-MOPA, the master oscillator current can be operated under cw conditions and the amplifier current can be varied to change the output power of the amplifier region. As mentioned previously, in this configuration the wavelength is nearly constant over the operating range from 0 to 2 W cw.

The characteristics of the MFA-MOPA, at a heatsink temperature of 5 °C, represent the highest-power diffraction-limited cw operation from a monolithic semiconductor source. In addition, the output is both a single diffraction-limited lobe and single spectral mode, which is suitable for most applications of coherent laser diodes.

2.6 Conclusions

Over the last several years, semiconductor laser diodes have advanced from low-power, low-efficiency devices to high-power, high-efficiency and coherent lasers. Although many of the laser array coupling approaches studied initially seemed promising for the fabrication of a coherent high-power 2-D array, the uniformity requirements of an array of series-coupled oscillators ultimately limits their scalability. The development of the M-MOPA class of devices alleviates most of the problems associated with the series-coupled oscillators.

The M-MOPA class of devices is immune to mode competition and non-uniform spatial gain saturation. Consequently, the M-MOPA functions in a particular single mode throughout the operation of the device.

The M-MOPA with cascaded amplifiers[14,15] demonstrates temporal coherence to output powers up to 4.5 W pulsed and the scalability of the technology to 2-D[27]. As a result, in this regime, the M-MOPA technology is scalable to higher output powers. The M-MOPA output is a single longitudinal mode and the oscillator linewidth has been measured to be less than 1.65 MHz.

The MAG-MOPA technology extends the characteristics of the M-MOPA class of devices to include diffraction-limited far-field radiation patterns[16,29]. The MAG-MOPA utilizes a uniformly saturated power amplifier to minimize the phase distortions. As a result coherent, single-lobed, diffraction-limited operation has been demonstrated to pulsed powers of 805 mW. More recently it has been demonstrated that the

MAG-MOPA technology is capable of operation to 1.3 W of temporally coherent output.

In addition the M-MOPA class of devices is capable of high-speed electronic beam steering[53]. Devices have been fabricated that demonstrate steering over 1.4°. Coupling these devices with the 0.012° divergence of the MAG-MOPA would correspond to >100 resolvable spots.

The culmination of the M-MOPA technology is summarized in the characteristics of the tapered-amplifier MFA-MOPA[48–50], which has demonstrated high cw power, coherent, diffraction-limited and single-longitudinal-mode output, which can be small-signal-modulated at high speeds with power gain.

Finally, it appears that it is possible to increase the coherent output power available from a single amplifier element significantly by virtue of the demonstration of up to 400-element parallel-injected amplifier arrays[19]. Such amplifier arrays offer the advantages of high mutual coherence and a distributed heat load; however, independent phase control of each element is required to achieve diffraction-limited performance. Nevertheless, the replacement of single-mode waveguide amplifiers by broad-area, tapered-contact amplifiers in array configurations should result in multi-watt cw coherent semiconductor sources.

Acknowledgment

The authors gratefully acknowledge the hard work and dedication of the members of the research group at Spectra Diode Laboratories (D. Doyle, S. Ogarrio, K. Dzurko, B. Tsai, R. Geels, T. Nguyen, R. Lang, N. Biligiri, S. O'Brien, T. Tally, J. Major, J. Parks, R. Parke, T. Johnson, R. Waarts, D. Nam and D. Scifres), financial support of the US Air Force (Phillips Laboratories) and NASA, and the support of and collaboration with the Naval Research Laboratory (L. Goldberg, D. Hall, M. Surette and J. Weller).

References

1. D. Botez, *IEE Proc. J., Optoelectronics*, **139**, 14–23 (1992).
2. J. Major, D. Welch and D. Scifres, *Appl. Phys. Lett.*, **59**, 2210–12 (1991).
3. J. Hohimer, G. Hadley, D. Craft, T. Shiau, S. Sun and C. Schaus, *Appl. Phys. Lett.*, **58**, 452–4 (1991).
4. L. Mawst, D. Botez, M. Jansen, T. Roth, C. Tu and C. Zmudminski, *Electron. Lett.*, **27**, 1586–7 (1991).
5. D. Mehuys, J. Major, Jr, and D. F. Welch, *SPIE Proc. OE-LASE '93*, Vol. 1850, pp. 2–12, Los Angeles, CA (1993).

6. L. Mawst, D. Botez, M. Jansen and J. Rozenbergs, *Electron. Lett.*, **27**, 369–71 (1991).
7. K. Dzurko, D. Scifres, A. Hardy, D. Welch, R. Waarts and S. O'Brien, *Electron. Lett.*, **28**, 1477–8 (1992).
8. L. Goldberg and J. Weller, *Appl. Phys. Lett.*, **50**, 1713–15 (1987).
9. L. Pang, E. Kintzer and J. Fujimoto, *Opt. Lett.*, **15**, 728–30 (1990).
10. R. Lang, *IEEE J. Quantum Electron.*, **QE-18**, 976–83 (1982).
11. E. Kintzer, J. Walpole, S. Chinn, C. Wang and L. Missagia, *IEEE Photon. Tech. Lett.*, **5**, 605–8 (1993).
12. D. Mehuys, L. Goldberg, R. Waarts and D. Welch, *Electron. Lett.*, **29**, 219–21 (1993).
13. L. Goldberg and D. Mehuys, *Appl. Phys. Lett.*, **61**, 633–5 (1992).
14. D. Welch, D. Mehuys, R. Parke, R. Waarts, D. Scifres and W. Streifer, *Electron. Lett.*, **26**, 1327–29 (1990).
15. N. Carlson, J. Abeles, D. Bour, S. Liew, W. Reichert, P. Lin and A. Gozdz, *Photon. Tech. Lett.*, **2**, 708–10 (1990).
16. D. Mehuys, D. Welch, R. Parke, R. Waarts, A. Hardy and D. Scifres, *Electron. Lett.*, **27**, 492–4 (1991).
17. N. Carlson, P. Gardner, M. Renna, J. Andrews, R. Stolzenberger, A. Triano, E. Vangieson, D. Bour, G. Evans, S. Liew, J. Kirk and W. Reichert, *Photon. Tech. Lett.*, **4**, 988–90 (1992).
18. D. Welch, R. Parke, D. Mehuys, A. Hardy, R. Lang, S. O'Brien and D. Scifres, *Electron. Lett.*, **28**, 2011–13 (1992).
19. D. Krebs, R. Herrick, K. No, W. Harting, F. Struemph, D. Driemeyer and J. Levy, *Photon. Tech. Lett.*, **3**, 292–5 (1991).
20. U. Koren, B. Miller, G. Raybon, M. Oran, M. Young, T. Koch, J. DeMiguel, M. Chien, B. Tell, K. Brown-Goebeler and C. Burrus, *Appl. Phys. Lett.*, **57**, 1375–7 (1990).
21. U. Koren, R. Jopson, B. Miller, M. Chien, M. Young, C. Burrus, C. Giles, H. Presby, G. Raybon, J. Evankow, B. Tell and K. Brown-Goebeler, *Appl. Phys. Lett.*, **59**, 2351–3 (1991).
22. J. Whiteaway, G. Thompson, A. Goodwin and M. Fice, *Electron. Lett.*, **27**, 2252–4 (1991).
23. D. Mehuys, A. Hardy, D. Welch, R. Waarts and R. Parke, *Photon. Tech. Lett.*, **3**, 342–4 (1991).
24. D. Welch, R. Parke, A. Hardy, R. Waarts, W. Streifer and D. Scifres, *Electron. Lett.*, **26**, 757–8 (1990).
25. D. Mehuys, R. Parke, R. Waarts, D. Welch, A. Hardy, W. Streifer and D. Scifres, *IEEE J. Quantum Electron.*, **QE-27**, 1574–81 (1991).
26. D. Mehuys, D. Welch, R. Waarts, R. Parke, A. Hardy and W. Streifer, *IEEE J. Quantum Electron.*, **QE-27**, 1900–9 (1991).
27. R. Parke, D. Welch and D. Mehuys, *Electron. Lett.*, **27**, 2097–8 (1991).
28. D. Welch, R. Waarts, S. Mehuys, R. Parke, D. Scifres, R. Craig and W. Streifer, *Appl. Phys. Lett.*, **57**, 2054–6 (1990).
29. N. Carlson, *IEEE J. Quantum Electron.*, **QE-28**, 1884 (1992).
30. W. Chow and R. Craig, *IEEE J. Quantum Electron.*, **QE-26**, 1363–8 (1990).
31. R. Parke, D. Welch, D. Mehuys, W. Plano and D. Scifres, unpublished manuscript.
32. J. Abeles, P. York, N. Carlson, T. Andrews, W. Reichert, J. Kirk, N. Hughes, S. Liew, J. Connolly, G. Evans and J. Butler, *Conf. Digest Thirteen IEEE International Semiconductor Laser Conference*, Takamatsu, Japan, pp. 78–9 (1992).

33. J. Andrews, *Appl. Phys. Lett.*, **48**, 1331–3 (1986).
34. S. Macomber, P. Akkapeddi and A. Montroll, *SPIE Proceedings*, **723**, 36–9 (1986).
35. L. Goldberg, J. Weller, D. Mehuys, D. Welch and D. Scifres, *Electron. Lett.*, **27**, 927–9 (1991).
36. J. Andrews, *J. Appl. Phys.*, **64**, 2134–8 (1988).
37. M. Tamburrini, L. Goldberg and D. Mehuys, *Appl. Phys. Lett.*, **60**, 1292–4 (1992).
38. D. Mehuys and L. Goldberg, *SPIE Technical Digest*, **1634**, 31–8 (1992).
39. A. E. Siegman, *Lasers*, p. 301, University Science Books, Mill Valley, CA (1986).
40. D. Mehuys, L. Goldberg, D. Welch and J. Weller, *Appl. Phys. Lett.*, **62**, 544–6 (1993).
41. L. Goldberg, M. Tamburrini and D. Mehuys, *Electron. Lett.*, **27**, 1593–5 (1991).
42. L. Goldberg, D. Mehuys and D. Hall, *Electron. Lett.*, **28**, 1082–4 (1992).
43. J. Walpole, E. Kintzer, S. Chinn, C. Wang and L. Missagia, *Appl. Phys. Lett.*, **61**, 740–42 (1992).
44. D. Mehuys, D. Welch and L. Goldberg, *Electron. Lett.*, **28**, 1944–6 (1992).
45. J. Jacob, M. Rokni, R. Klinkowstein and S. Singer, *Appl. Phys. Lett.*, **48**, 318–20 (1986).
46. M. Tilton, G. Dente, A. Paxton, J. Cser, R. DeFreez, C. Moeller and D. Depatie, *IEEE J. Quantum Electron.*, **QE-27**, 2098–100 (1991).
47. D. Hall, L. Goldberg and D. Mehuys, *IEEE Photon. Tech. Lett.*, **5**, 922–5 (1993).
48. D. Welch, R. Parke, D. Mehuys, A. Hardy, R. Lang, S. O'Brien and D. Scifres, *Electron. Lett.*, **28**, 2011–13 (1992).
49. J. Abeles, R. Amantea, N. Carlson, J. Andrews, P. York, J. Connolly, R. Rios, W. Reichert, J. Kirk, T. Zamerowski, D. Gilbert, S. Liew, N. Hughes, J. Butler, G. Evans, S. Narayan and D. Channin, *Thirteenth IEEE Semiconductor Laser Conf.*, Paper PD-12, Takamatsu, Japan (1992).
50. R. Parke, D. Welch, A. Hardy, R. Lang, D. Mehuys, S. O'Brien, K. Dzurko and D. Scifres, *IEEE Photon. Tech. Lett.*, **5**, 297–300 (1993).
51. T. Johnson, *Laser Focus World*, May issue, pp. 173–83 (1990).
52. S. O'Brien, R. Parke, D. Welch, D. Mehuys and D. Scifres, *Electron. Lett.*, **28**, 1272–3 (1992).
53. R. Parke, D. Welch, D. Mehuys, R. Waarts, R. Craig and D. Scifres, *Electron. Lett.*, **26**, 1076–7 (1990).

3

Microoptical components applied to incoherent and coherent laser arrays

JAMES R. LEGER

The preceding chapters of this book have described diode laser arrays with a variety of geometries and coherence properties. Edge-emitting geometries were shown to be capable of high power and high efficiency, whereas surface-emitting geometries permitted fabrication of large two-dimensional arrays. In addition, many of these arrays were designed to contain various degrees of mutual coherence between lasing apertures. In the current chapter, we consider several of the systems aspects of these laser arrays. In particular, we will consider external optical components and systems that manipulate the laser output light to satisfy the requirements of specific applications. We will describe optical systems that act as interfaces between the laser array and other electrooptical components, as well as external laser cavities that improve the performance of the laser array itself.

The recent advances in microoptical components permit unprecedented control of light from laser arrays. Many optical functions are now possible that were previously difficult or impossible to perform with conventional optical components. Microlenses and microlens arrays can be used to collimate and expand individual lasers in an array, making it possible to change the fill factor of a given array. Astigmatism, spherical aberration, and other optical aberrations, can be removed with aspheric and anamorphic microlenses. Multiple planes of microlenses can be used to construct microtelescope arrays and other microoptical systems. Microprisms and diffractive elements make it possible to reconfigure light fields by directing portions of the beam to different locations. This chapter will explore the application of these components to diode laser arrays.

The chapter is divided into four sections. The first section reviews some of the fundamental requirements of optical systems interfaced to diode laser arrays. The specific constraints imposed by the numerical aperture and possible astigmatism and asymmetric divergence of an individual diode laser are reviewed. In the second half of the section, the special attributes of laser arrays are considered. Specifically, we describe the effects of mutual coherence and array fill factor.

The second section describes various types of microoptical elements and their fabrication methods. These elements generally fall into two categories: diffractive and refractive. We discuss many of the benefits and liabilities of these two technologies. In this context, we show some simple applications of these optical elements for collimation of light from laser arrays.

In the third section, several applications of external optics to coherent systems are described. We review some optical systems for superimposing mutually coherent laser beams, as well as external optical systems that utilize coherence of the light field to increase the array fill factor. We end this section with a description of external laser cavities that establish or enhance the mutual coherence of a diode laser array.

The final section describes applications of external microoptics to arrays of mutually incoherent lasers. The limits imposed by the radiance theorem are reviewed, and systems for optimal light concentration are considered. We conclude by describing several applications of these optical systems.

3.1 Requirements of incoherent and coherent systems

In this chapter, we assume for simplicity that the laser array consists of individual lasing elements, each producing a single lowest-order spatial mode with perfect spatial coherence. Optical elements that intercept light from a single laser aperture are not affected by the mutual coherence of the array, and their design follows the conventional rules of Gaussian optical beams. We briefly review some of the requirements for these optical elements in the first part of this section.

The occurrence of multiple apertures in an array presents additional design considerations. Two of the most important aspects of the array are the mutual coherence from one laser to another and the array fill factor. The second part of this section describes the effect of coherence on the array radiance and the resulting consequences in systems applications. In the third part, we discuss the role of the fill factor in a coherent array and describe a simple optical system to increase it.

3.1.1 Single-laser optics

Applying separate lenses to each individual laser in an array imposes some strict requirements. First, the lens diameter can be no larger than the spacing of the lasers. This spacing can range from less than ten microns to several hundred microns. Second, we may require adjacent lenses to be placed as close to each other as possible to prevent gaps in the resulting wavefront. Third, the shape of the lens aperture may have to be noncircular to match the asymmetric divergence of the diode laser. Fourth, the numerical aperture of the lens must match the numerical aperture of the diode laser, often requiring a lens with a low f number to collect the laser light efficiently. In addition, diffractive and absorptive losses must be kept low. Finally, each lens should ideally produce an aberration-free beam. The lens must contain both low intrinsic aberrations and have the ability to correct for any aberrations in the original laser beam.

3.1.2 Effects of array coherence

As light propagates from the face of a laser array, the light beams from the individual lasing apertures start to overlap. The resulting field is then greatly affected by the mutual coherence and the phase distribution across the laser array.

It is convenient to describe the output of an array by its radiance, or power per unit area per unit solid angle[1]. Radiance is often the most important performance measure of a laser array, since both power density and divergence are considered simultaneously. For example, the amount of power that can be coupled into a fiber is governed both by the size of the fiber core and its numerical aperture. Thus, it is the total power that can simultaneously be concentrated into the core area and the acceptance cone angle that is important.

3.1.2.1 Radiance theorems

Two important theorems govern radiance in an optical system. The first, called the radiance theorem[2], states that the radiance of the light distribution produced by an imaging system cannot be greater than the original source radiance (assuming that both object and image spaces have the same index of refraction); the radiances are equal only when the losses in the optics are negligible. The second, related, theorem states that the radiance of a collection of mutually incoherent (but otherwise identical) sources cannot be increased by a passive linear optical system to a level greater than the radiance of the single brightest source.

There are certain cases where the radiance of a laser array can exceed the radiance of the individual elements. First, if mutual coherence is established across the laser array, the entire source behaves as a single spatial mode, and the phase and amplitude can, in principle, be made uniform. Adding additional *mutually coherent* lasers to the array does increase the radiance, and the resulting total radiance can be equal to the sum of the individual laser radiances.

A second way to increase radiance is to use lasers with different average properties (such as wavelength or polarization). Passive optical elements such as diffraction gratings and polarizing beam splitters can be used to multiplex the beams.

3.1.2.2 Consequences of the radiance theorem

The fact that the radiance of a mutually incoherent (but otherwise identical) laser array does not increase with source number has direct consequences for many applications. For example, if we are interested in the intensity at the center of the far field, the theorem states that there is no radiometric benefit from using more than one laser in the array. Using additional sources will, of course, increase the far-field intensity proportional to the number of incoherent sources N, as shown in Figure 3.1(a); the area of the source is increased by a similar amount, however, resulting in the same radiance. Alternatively, a single source expanded to this same size will also increase the intensity by N because of the decreased divergence.

In free-space optical communication, the receiver mirror intercepts only the center of the far-field pattern, and the intensity at this point (integrated over the mirror area) determines the amount of power received. Thus, it would appear that there is no apparent advantage to using more than one mutually incoherent source; there are obvious disadvantages associated with the increased electrical power consumed by multiple lasers and the inability to use coherent communication schemes. However, there are several practical reasons for using more than one incoherent source for direct-detection, free-space communication. First, the system is more robust because of the redundancy afforded by the multiple sources. Failure of a single diode in the array only decreases the received power slightly, rather than the catastrophic failure associated with a single-diode source. Second, the increased width of the incoherent beam makes the job of pointing and tracking easier. Finally, the collimation optics in the

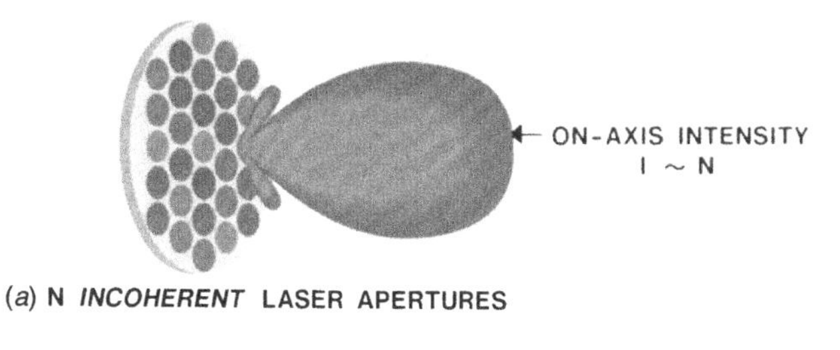

(*b*) N *COHERENT* LASER APERTURES

Figure 3.1. Effect of coherence on the far-field pattern: (*a*) the intensity in the center of the far-field pattern is proportional to the number of lasers N in an incoherent array; (*b*) the intensity is proportional to N^2 in a coherent array.

incoherent case must be diffraction-limited only over the small region illuminated by a single-laser aperture; the collimating optics for the equivalent single-laser system must be diffraction-limited over an aperture of the same size as the entire incoherent array. For large optics such as telescope mirrors, this implies that the mirror have sufficient thickness and supporting structures to maintain a diffraction-limited figure. The cost and weight associated with this more precise optic element can easily offset the advantages of using a single laser.

Applications that require the light to be focused to a small spot are governed by the same considerations as discussed above. Again, if the application requires a high peak intensity at the center of the focused spot, there is no radiometric advantage in using an array of incoherent emitters. On the other hand, if the requirement is to deposit as much power as possible in an area that is larger than the diffraction-limited spot size, the increased power from an array can be used to advantage. We will examine several of these cases in Section 3.4.3.

By establishing coherence across the N lasers in the array, the radiance of the individual sources is summed. Equivalently, the on-axis intensity increases by N^2, as shown in Figure 3.1(*b*). This results from the simultaneous increase in the total power by a factor of N and decrease in solid-angle divergence by an additional factor of N as compared to a single source.

3.1.3 Fill-factor considerations

The array fill factor is defined as that fraction of the array occupied by the lasing apertures. From the discussion in the preceding section, it is apparent that the average radiance from an incoherent array is maximized when the source area and/or divergence are minimized, or equivalently when the fill factor is maximized. We will discuss this case in more detail in Section 3.4.3. For a coherent array, the situation is somewhat more complex due to the constructive and destructive interference between the sources. In this case the Strehl ratio gives a better description of the array performance. The Strehl ratio as defined in Ref. 3 compares the on-axis far-field intensity of a specific array to the on-axis far-field intensity of a uniformly illuminated constant-phase light field of the same power and area. This ratio, often used as a measure of the quality of the laser light, will be less than one if the light field is not uniform in amplitude and phase, or if the coherence across the light field is not perfect.

It can be shown that the Strehl ratio of a coherent array is proportional to the array fill factor[3]. Although it is sometimes possible to place the laser apertures sufficiently close to each other to obtain a relatively high fill factor, high-power arrays usually contain significant nonlasing regions to reduce the heat load. In addition, some surface-emitting geometries have inherently small fill factors. In these cases, an external optical system is required to increase the effective fill factor and thus the Strehl ratio. This can be achieved in principle by providing each laser element with a beam expanding telescope, where the resultant beams fill the array aperture as uniformly as possible. In practice, the laser apertures of many arrays are sufficiently small that diffraction expands each of the beams. After a small propagation distance, the adjacent cones of laser light intersect. Collimating each laser beam at this point with a single lens for every aperture produces a wavefront with the desired uniform phase and quasi-uniform intensity. This application of microlenses will be reviewed in Section 3.3.2.

3.2 Microoptics

Light rays can be bent by using refraction, reflection or diffraction. Microoptics have been fabricated based on each of these effects. In this section, we review some of the techniques for fabricating both refractive and diffractive lenses. Reflective structures can be produced by simply coating the appropriate refractive or diffractive structures with a high-reflectivity coating.

3.2.1 Refractive microoptics

Refractive lenses can be divided into two broad categories. The first type is produced from a homogeneous material and uses the shape or figure of the lens to bend the light. Spherical shapes are the easiest to produce, and are the most common. Aspheric shapes offer the potential to reduce or eliminate spherical and other aberrations. The second type of refractive lens uses a material with a varying index of refraction. These lenses (often called gradient index or GRIN lenses) can be made perfectly flat; the light bending power is provided by the index variation in the material.

The spherical figure of a conventional macroscopic lens is obtained by grinding and polishing. Some attempts have been made to fabricate microlens arrays using these conventional methods, as well as pressing and moulding[4,5]. Other methods have relied on surface tension of liquids to form accurate spherical surfaces. One such method uses a photosensitive glass as a starting substrate[6,7]. The glass is exposed to UV light through a mask consisting of opaque dots on a clear background. Upon thermal development ($\approx$480–600 °C), the glass exposed to the UV light forms noble metal particles that serve as nucleation sites for a microcrystalline phase. This new crystalline structure has a higher density than the homogeneous glass, and contracts upon heating. When the temperature of the thermal development cycle rises above the softening temperature of the unexposed glass, the contraction of the exposed rings squeezes the soft undeveloped glass disk above the surface, where surface tension forms a spherical shape. Since these lenses are defined by photolithography, accurate arrays can be fabricated. The radius of curvature variation between lenses has been measured to be $\pm$0.1 μm, and the surface figure deviates from a perfect sphere by less than 0.1 μm. Figure 3.2 shows an array of microlenses fabricated using this technique.

A second spherical lens fabrication method uses photoresist as the lens material[8]. Figure 3.3 illustrates the process. The substrate is coated with aluminum and patterened to form aperture stops for each of the micro-lenses. Pedestals are formed by patterning conventional positive photo-resist, and are hardened using deep-UV light and hard baking. A photoresist formulated for thick coating is deposited on top of the pedestals and patterned to form 15 μm-high cylinders. A flood exposure to near-UV light lowers the melting point of the cylinders such that heating at 140 °C causes each cylinder to melt and flow to the edge of the pedestal. As before, surface tension causes each lens to form into a spherical shape. The variation in focal length from lenses within the same

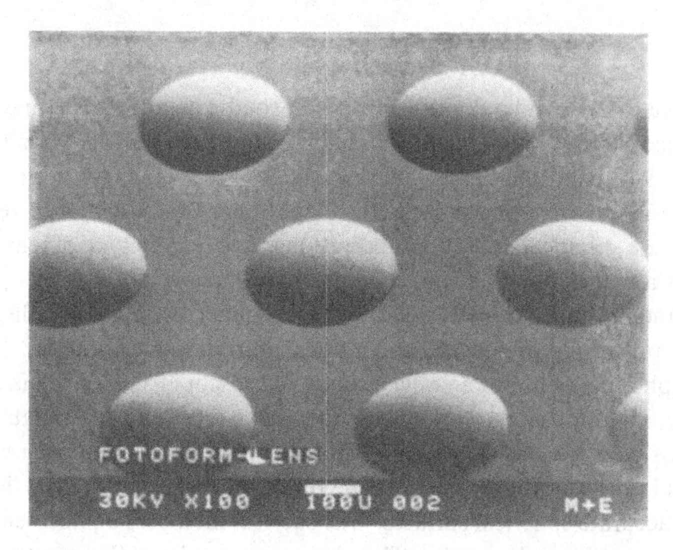

Figure 3.2. Photomicrograph of refractive lenslet array fabricated by photolytic technique. (Ref. 6.)

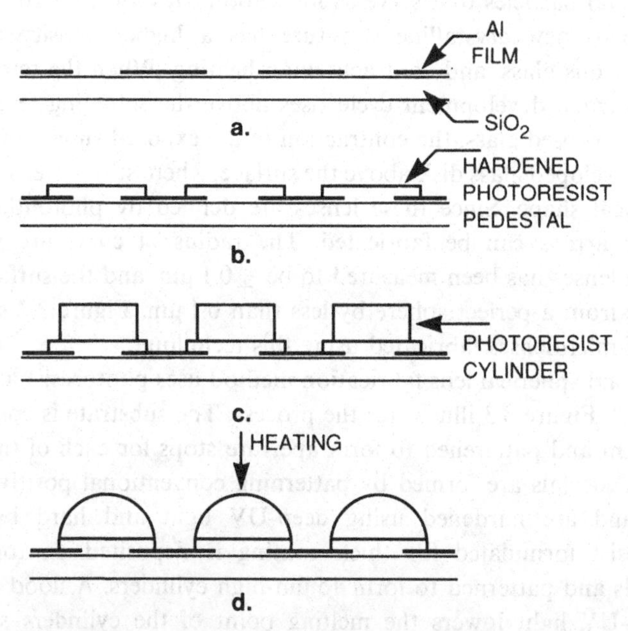

Figure 3.3. Procedure for fabricating photoresist microlenses: (*a*) etch aperture stops in aluminum film; (*b*) fabricate hard photoresist pedestals; (*c*) pattern soft photoresist cylinders on top of pedestals; (*d*) melt soft photoresist cylinders to form a spherical shape by surface tension. (Ref. 8.)

group was measured to be ± 0.5 μm. Spot sizes of 1.2 μm (full-width-at-half-maximum) were obtained from lenses with numerical apertures of NA = 0.32.

The two preceding techniques used surface tension to form highly accurate spherical surfaces. Such lenses suffer from spherical aberration at high numerical apertures. Diode lasers that behave as ideal point sources require lenses with aspheric surfaces for proper collimation. In addition, diode lasers with substantial astigmatism require aspheric lenses with anamorphic focusing properties. Aspheric refractive optical elements suitable for collimating large numerical aperture diode lasers have been fabricated by mass transport in both InP and GaP[9,10]. The process is based on the thermally induced morphological changes that occur below the melting point of the crystal. At these temperatures, some dissociation occurs at the surface of the semiconductor, resulting in free metallic atoms on the surface. When there is a variation of curvature across the surface, the free metallic atoms will diffuse from regions of large curvature to regions of smaller curvature. This mass transport of metallic atoms tends to smooth out any sharp features. However, unlike surface tension, large, slowly varying features are preserved since the transport time varies as the fourth power of the spatial Fourier component[11].

The lens fabrication method consists of generating a multi-level mesa structure shown in Figure 3.4(*a*) by photolithography and wet chemical etching. The diameters and step heights are chosen to produce the desired lens profile based on the assumption that mass is conserved over a local region during mass transport. The substrates are then placed in an H_2/PH_3 atmosphere and heated for several hours (800–880 °C for InP and >1000 °C for GaP). The surface profile of the resulting lens is compared with the ideal aspheric lens in Figure 3.4(*b*). Microlenses of 130 μm diameter with 200 μm focal lengths have been shown to have near-diffraction-limited performance.

Cylindrical optics are often required for anamorphic optical systems. An interesting microcylindrical lens can be fabricated from an optical fiber[12]. The fiber is held along the array axis of a one-dimensional edge-emitting laser array and collimates the light in the transverse direction. A simple cylindrical fiber does not have the ideal figure for collimation and so suffers from cylindrical aberration. This aberration can be eliminated by producing a fiber with a noncircular cross-section[13]. A fiber preform (which is macroscopic in size) is precisely ground by numeric control techniques. When the preform is drawn into a thin fiber, the diameter is reduced by a factor of 50–70 ×, yet the cross-sectional shape

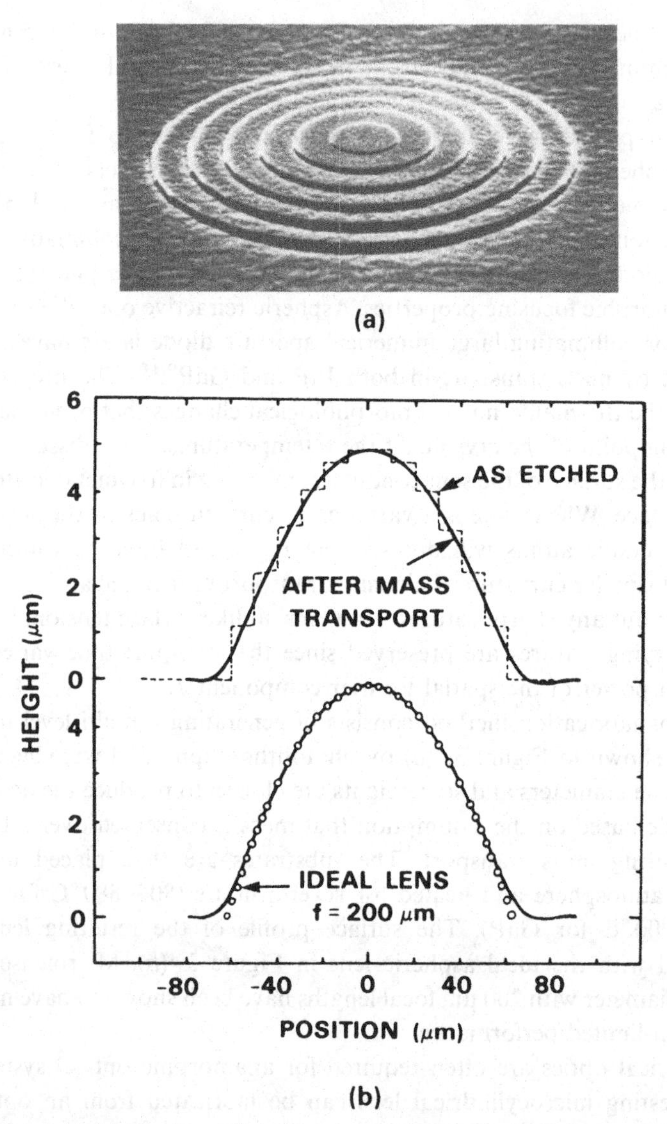

Figure 3.4. Refractive lens fabrication by mass transport: (*a*) photomicrograph of mesa structure produced by masking and chemical etching; (*b*) measured mass-transported surface figure compared to an ideal lens. (Ref. 10.)

of the preform is retained. Figure 3.5 shows an SEM photomicrograph of a typical cylindrical lens with a figure chosen to eliminate cylindrical aberration. The focal length is 220 μm and the diffraction-limited numerical aperture is approximately 0.6.

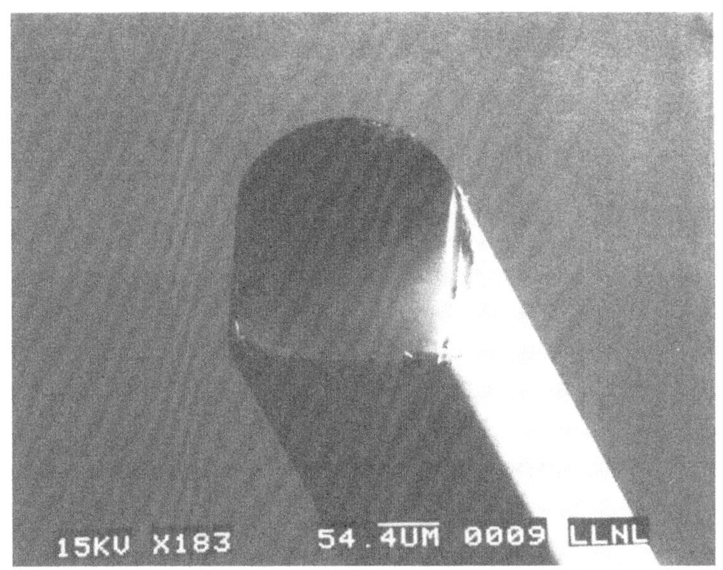

Figure 3.5. Scanning electron photomicrograph of a fiber cylindrical lens with optimized figure for collimation in one axis. (Ref. 13.)

The focal power of the lenses described above comes from refraction from a curved surface. Gradient index, or GRIN, lenses achieve their focal power by creating a continuous variation in the index of the lens material. Focusing and imaging properties can thus be obtained with flat substrates. One of the most common types of GRIN lenses is the rod lens, where the index of refraction varies approximately quadratically in the radial direction. The index gradient is most commonly created by exchanging dopant ions from a molten salt bath with ions present in the original glass rod. Off-axis light rays propagate along the rod in a sinusoidal fashion. By choosing the index gradient and rod length correctly, these lenses can be designed to collimate light from a diode laser. Arrays of these rod lenses have been fabricated by stacking individual rods accurately into a matrix[14].

An alternative GRIN technique employs selective diffusion into a planar substrate[15–17]. A thin film of metal is evaporated on to a glass substrate, and an array of small holes is produced in the metal by photolithography and metal etching. The masked substrate is then immersed in a molten salt at approximately 500 °C to create an ion exchange between monovalent cations in the glass substrate and a different monovalent cation in the molten salt. An electric field can also

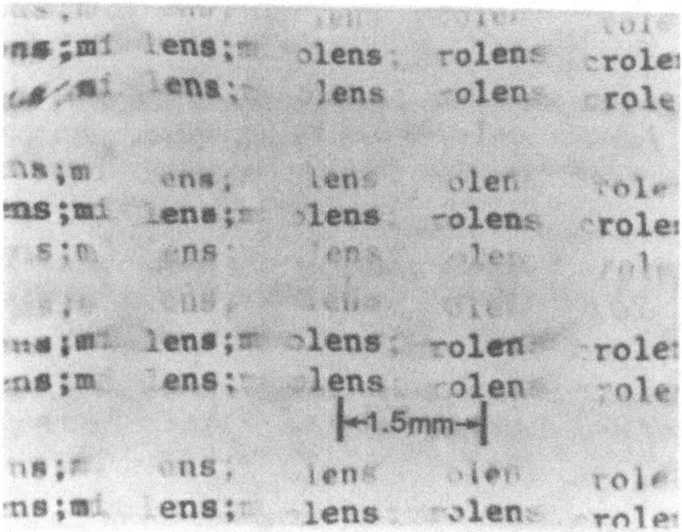

Figure 3.6. Multiple real images produced by an array of planar GRIN microlenses. (Ref. 17.)

be applied to decrease the diffusion time by electromigration[18]. The diffused dopant can be chosen to produce a higher refractive index in the center of each aperture, resulting in an array of positive GRIN microlenses. Unlike the rod lens, the index of refraction is a function of both radial and axial directions. Lenses with diameters ranging from 10 μm to 1000 μm have been fabricated with numerical apertures as large as 0.3. Diffraction-limited performance has been observed in lenses with numerical apertures as large as 0.19. Figure 3.6 shows the real images produced by an array of planar GRIN microlenses.

The maximum numerical aperture achievable by a planar GRIN lens is limited by the index difference between the glass substrate and the exchanged ions. One method of achieving higher numerical apertures is to combine the power of two lenses. Numerical apertures as high as 0.54 have been demonstrated by using this technique[19]. A second method takes advantage of the diameter difference between the exchanged ions. By replacing small-diameter ions with larger ones, the volume of the glass can be increased locally, causing a swelling in the lens area. The resulting bulge creates a spherical shape that adds to the refractive power of the GRIN lens. Numerical apertures of 0.57 have been achieved for lens diameters of less than 100 μm[20].

3.2.2 Diffractive microoptics

One common type of diffractive lens uses a surface-relief pattern to focus and collimate light. The ideal surface-relief pattern shown in Figure 3.7 can be obtained by subtracting integer wavelength amounts from a refractive lens. The resulting profile consists of piece-wise continuous phase values between 0 and 2π. At the design wavelength, this lens performs identically to its refractive counterpart (neglecting the finite thickness of the element).

Diffractive lenses have several important advantages. Since they are planar structures, they can often be fabricated on thin, lightweight substrates. Aspheric lenses and lenses containing aberration correction can be fabricated as readily as simple spherical ones. In addition, large arrays of closely packed lenses can be fabricated to match the spacing of an edge-emitting or surface-emitting diode laser array. Advanced lithographic and etching techniques are used to transfer these patterns into a variety of substrates with the sub-micron accuracy required for high-speed lenses.

Fabricating the piece-wise continuous surface-relief pattern of Figure 3.7(*b*) is difficult due to the tolerances required on the surface height

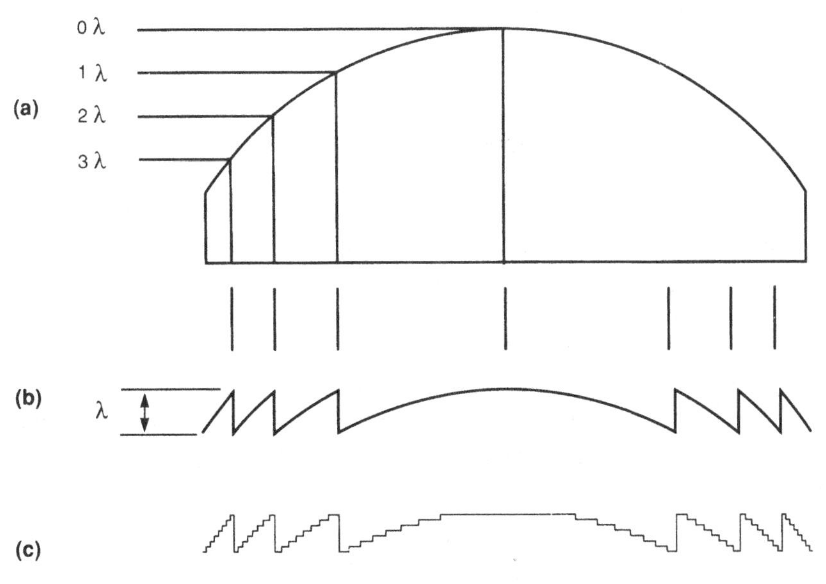

Figure 3.7. Comparison of refractive and diffractive lenses: (*a*) conventional refractive lens; (*b*) piece-wise continuous diffractive lens; (*c*) multi-level diffractive lens.

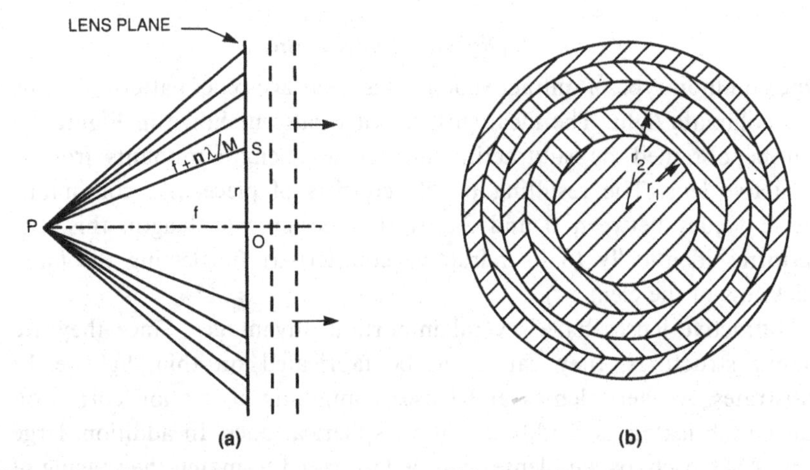

Figure 3.8. Fresnel phase zones of a multi-level diffractive optical element: (*a*) geometry for calculating zone boundaries; (*b*) diffractive lens consisting of concentric rings of constant phase.

and positional errors ($\ll \lambda/4$) of a single continuous region, and the small size of these regions ($\approx 2\lambda$ at the edge of an $f/1$ lens). Continuously blazed lenses have been fabricated by varying the beam current of an e-beam writing system to expose the e-beam resist by different amounts[21–23]. Since the etch rate of the resist depends on the electron dose, a continuous phase pattern can be etched into the substrate material.

An alternative method approximates the continuous sections of the profile by a series of steps (Figure 3.7(*c*))[24,25]. $M = 2^K$ phase levels are fabricated from K etching steps using K binary masks[26–28]. Since each etching step is binary, the etch depth is easier to control.

The phase pattern of a multi-level diffractive collimating lens can be computed with the help of Figure 3.8. A point source P is placed a distance f in front of the diffractive element. At a radius r on the lens plane, the spherical wave emanating from P must propagate a distance PS given by $\sqrt{(f^2 + r^2)}$. We divide the lens into zones, where a zone represents a phase change of $2\pi/M$ in the spherical wave. The zone boundaries r_n are then given by

$$r_n = \sqrt{\left(\frac{2n\lambda f}{M} + \frac{n^2\lambda^2}{M^2}\right)}. \tag{3.1}$$

The phase of the nth zone can be canceled approximately by a constant phase

$$\varphi_n = -\left(\frac{2\pi n}{M}\right)_{\mathrm{mod}\ 2\pi}. \tag{3.2}$$

A simple multi-level diffractive lens consists of circular phase rings with the M different phase levels given by Eq. (3.2) and radii given by Eq. (3.1). Note that if only two phase levels are used, $M = 2$ and the equations reduce to a conventional phase Fresnel zone plate.

A paraxial approximation is valid when $f \gg \lambda n/2M$. In this case, the second term in Eq. (3.1) is small compared to the first, and the zone boundaries are simply given by

$$r_n \approx \sqrt{\left(\frac{2n\lambda f}{M}\right)}. \tag{3.3}$$

The wavefront exiting the multi-level diffractive lens has residual phase errors due to the quantized nature of the element. The effect of these errors can be assessed conveniently by calculating the Strehl ratio of the element given by[3]

$$S = \left|\frac{1}{\pi R^2}\int_0^{2\pi}\int_0^R \exp[\mathrm{j}\phi(r,\theta)]r\,\mathrm{d}r\,\mathrm{d}\theta\right|^2, \tag{3.4}$$

where $\phi(r, \theta)$ is the phase difference between the impinging spherical wave and the multi-level diffractive lens, and R is the radius of the lens.

We first calculate the integral in Eq. (3.4) across the nth zone only:

$$\int_0^{2\pi}\int_{r_n}^{r_{n+1}} \exp[\mathrm{j}\phi(r,\theta)]r\,\mathrm{d}r\,\mathrm{d}\theta = \int_0^{2\pi}\int_{r_n}^{r_{n+1}} \exp\left(\mathrm{j}\pi\frac{r^2}{\lambda f}\right) \times \exp\left(-\mathrm{j}2\pi\frac{n}{M}\right)r\,\mathrm{d}r\,\mathrm{d}\theta, \tag{3.5}$$

where r_n and r_{n+1} are given by Eq. (3.3). The first term in the integral is a paraxial approximation to an expanding spherical wave, and the second is the constant phase of the nth zone in the lens. Evaluating this integral results in

$$\int_0^{2\pi}\int_{r_n}^{r_{n+1}} \exp\left[\mathrm{j}\pi\left(\frac{r^2}{\lambda f} - \frac{2n}{M}\right)\right]r\mathrm{d}r\,\mathrm{d}\theta = 2\lambda f\exp\left(\mathrm{j}\frac{\pi}{M}\right)\sin\left(\frac{\pi}{M}\right). \tag{3.6}$$

Note that Eq. (3.6) is independent of the zone number n. Consequently, the total integral across all N zones in Eq. (3.4) is simply given by

$$\int_0^{2\pi}\int_0^R \exp[\mathrm{j}\phi(r,\theta)]r\,\mathrm{d}r\,\mathrm{d}\theta = 2N\lambda f \exp\left(\mathrm{j}\frac{\pi}{M}\right)\sin\left(\frac{\pi}{M}\right). \tag{3.7}$$

The radius of the lens R can be related to the total number of zones N by Eq. (3.3):

$$R = r_N \approx \sqrt{\left(\frac{2N\lambda f}{M}\right)}. \tag{3.8}$$

Substituting Eqs. (3.7) and (3.8) into Eq. (3.4) results in a Strehl ratio for a multi-level diffractive lens of

$$\begin{aligned} S &= \left|\frac{M}{\pi}\exp\left(\mathrm{j}\frac{\pi}{M}\right)\sin\left(\frac{\pi}{M}\right)\right|^2 \\ &= \frac{\sin^2(\pi/M)}{(\pi/M)^2} = \mathrm{sinc}^2\left(\frac{1}{M}\right). \end{aligned} \tag{3.9}$$

The Strehl ratio of Eq. (3.9) is sometimes called the diffraction efficiency of the multi-level diffractive lens. A lens with eight phase levels ($M = 8$) is predicted to have a diffraction efficiency of 97%. This equation is based on scalar diffraction theory (requiring features to be much greater than a wavelength of light), and assumes a paraxial light cone from an infinitely thin element. Actual efficiencies of diffractive lenses may differ when these assumptions are not valid.

There are several factors that can reduce the Strehl ratio of a simple refractive lens. Single-element spherical lenses cannot collimate large-angular-divergence sources without a significant reduction in the Strehl ratio from spherical aberration. In addition, the laser sources themselves may contain significant aberrations. Diffractive lenses can correct for all these aberrations by simply modifying the equation that specifies the zone locations. The second term of Eq. (3.1) completely corrects all lens aberrations for an ideal on-axis point source. Astigmatism in the laser beam can be corrected by designing an anamorphic lens. The zones are defined by the locus of points (r, θ) that satisfy the equation

$$\sqrt{(f_1^2 + r^2)} + \frac{r^2\sin^2\theta}{2f_2} = \frac{n\lambda}{M} + f_1 \tag{3.10}$$

for various values of the integers n; f_1 is the focal length of the spherically symmetric component, and f_2 is the focal length of the cylindrical component chosen to remove the astigmatism.

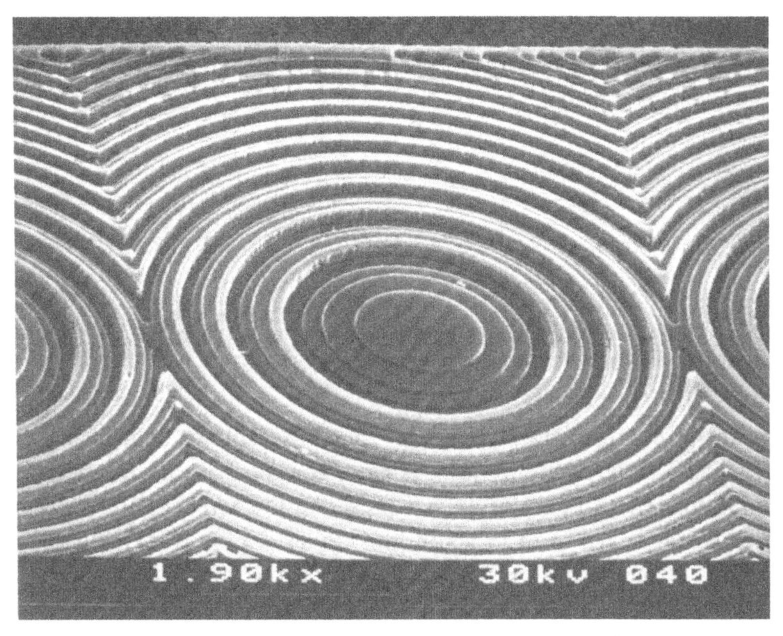

Figure 3.9. SEM photograph of a four-level anamorphic diffractive lens. The lens is designed to remove the astigmatism from a gain-guided laser. (Ref. 26.)

A photograph of an anamorphic multi-level diffractive lens array, obtained from a scanning electron microscope, is shown in Figure 3.9. A two-step etch was performed, producing four discrete etch levels. The theoretical diffraction efficiency of this lens is 81%. Measured diffraction efficiencies ranged from 80% in the central $f/4$ region to about 70% in an $f/2$ region. The lenses were designed to correct for an astigmatic aberration of 1.3 waves in the original laser beams. The elliptically shaped zones given by Eq. (3.10) resulted in an anamorphic lens with a focal length of 100 μm parallel to the array and 69 μm perpendicular to it. The wavefront quality of an $f/2$ region of the microlens is displayed in Figure 3.10(*a*). The flatness of the wavefront corresponds to an RMS phase error of $\lambda/50$. The modulation transfer function in Figure 3.10(*b*) shows nearly diffraction-limited performance, with a Strehl ratio of 0.98.

3.3 Coherent techniques

We saw in Section 3.1.2 that mutual coherence is necessary to increase the radiance of a laser array. To maximize the Strehl ratio, it is also

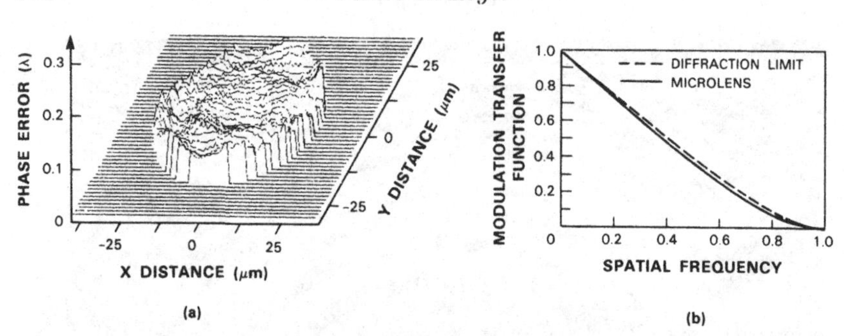

Figure 3.10. Optical performance of anamorphic diffractive lens: (*a*) optical wavefront of collimated beam; (*b*) modulation transfer function. (Ref. 26.)

important to combine the beams such that the resultant light field is as uniform as possible in amplitude and phase. The effect of partial coherence or nonuniformities in amplitude and phase on the Strehl ratio has been described elsewhere[3]. In this section, we will describe several methods of combining *coherent* beams from different lasing apertures to form a quasi-uniformly illuminated aperture with increased radiance. We then describe a method of establishing this coherence using external optics and the Talbot effect. Finally, we show an external optical system for controlling the phase of the array. All of these external cavity optics are inherently two-dimensional, and can be applied to both one-dimensional edge-emitting and two-dimensional surface-emitting laser arrays.

3.3.1 Superposition of coherent laser beams

The radiance theorem described in Section 3.1.2 suggested a natural way to categorize beam combining methods. Laser beam superposition is performed by directing N beams to a common point on a beam-combining element. The element directs the superimposed beams along a single direction. The resultant beam has the size and divergence of a single beam, but a near-field intensity (W/cm^2) N times greater.

The aperture-filling methods described in Section 3.3.2 are performed by rearranging the coherent beams from the array to eliminate the gaps between the individual beams. In this case, the near-field intensity (power per unit area) is the same as for a single expanded beam, but the area is N times greater, and the solid angle divergence is N times less.

These two methods are entirely equivalent from a radiance standpoint. The light distribution from one can be converted into the other by an appropriate afocal telescope. However, there are often practical issues that dictate the use of one over the other. Aperture-filled systems have an advantage in high-power applications since the intensity does not increase with increasing number of lasers. In addition, the low divergence that results can sometimes eliminate the need for a beam expanding telescope. Systems based on superposition are preferred for retrofitting an existing optical system with a higher-power source, since the beam size and divergence are identical to the original single laser.

Perhaps the simplest coherent beam superposition device is a beam splitter. It can be shown that two coherent beams incident on a beam splitter with the correct phase relationship can be combined into a single beam by operating the splitter backwards[3]. By concatenating several beam splitters, N laser beams can be superimposed by $N-1$ beam splitters. Diffractive structures offer an elegant method of performing multi-beam superposition. The first method described here uses a surface-relief grating to convert the nonuniform phase pattern from the superposition of several coherent beams into a near-planar wavefront. The second method is based on the Bragg diffraction properties of thick holograms and crystals.

3.3.1.1 Surface-relief gratings

Laser beam superposition can be accomplished using a specially designed surface-relief grating. The laser beams are made to cross each other at a given point in space, resulting in an interference pattern $|E(x, y)| \exp[j\phi(x, y)]$ that is dependent on the beam angles and the relative phases of the lasers. The grating is placed in the interference plane to convert the field into a single beam. For perfect conversion into a plane wave, the transmittance of the grating would have to be the reciprocal of the interference, or

$$t(x, y) = \frac{1}{|E(x, y)|} \exp[-j\phi(x, y)]. \qquad (3.11)$$

The magnitude of the grating transmittance is restricted to values less than or equal to one. Since any absorption results in an undesirable loss of laser power, the magnitude of the transmittance is chosen to be unity, and the magnitude component of the interference pattern is not corrected. The phase term, however, can be corrected by constructing a grating with the phase of Eq. (3.11). After passing through this conjugate phase grating, the light amplitude coupled into the zero order beam is given by

$$A_0 = \int_{-\infty}^{\infty} \int_{-\infty}^{\infty} |E(x, y)| \, dx \, dy. \qquad (3.12)$$

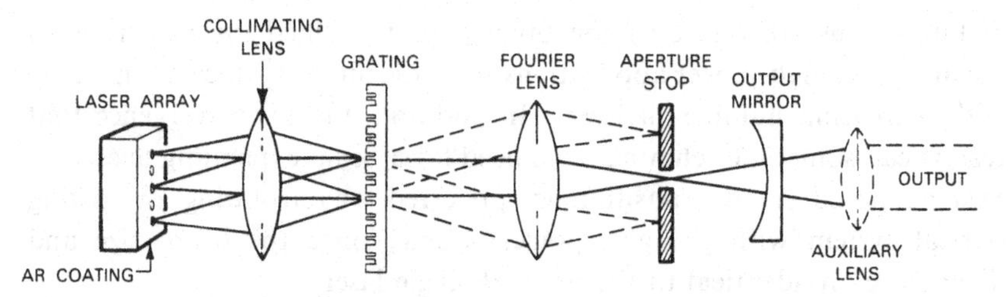

Figure 3.11. Laser beam superposition using binary phase gratings in an external cavity. (Ref. 34.)

The grating coupling efficiency is maximized by choosing the phases of the lasers to produce an interference magnitude $|E(x, y)|$ that maximizes Eq. (3.12). For example, the coupling of six lasers by a binary grating is maximized using laser phases of $(\pi, \pi, 0, 0, \pi, \pi)$; the resulting coupling efficiency from (Eq. 3.12) is 81%[29].

The common cavity configuration is used to establish coherence and proper phasing between lasers (see Figure 3.11). The binary grating inside the cavity couples the laser beams into a single output beam. Any loss of coherence or change in relative phases of the lasers results in lower coupling into the on-axis order, and correspondingly smaller feedback. The lowest lasing threshold occurs when the lasers are both coherent and in the correct phase state for efficient beam combining.

In addition to beam combining, diffraction gratings designed to operate inside a common cavity must distribute the single feedback beam *equally* among the lasers with a minimum of loss to higher orders. A phase grating with the phase transmittance of Eq. (3.11) does not necessarily have this property. By careful modification of the phase transmittance, we can design a grating to split a single beam into N *equal intensity* beams with maximum efficiency[30–32]. It can be shown that this splitting efficiency is identical to the efficiency of combining the N beams into a single beam with the same grating[33]. Thus, the gratings can simply be designed to have good beam-splitting properties.

A one-dimensional binary phase grating designed to couple six lasers was fabricated on a quartz substrate[34]. This grating was placed in the common cavity configuration of Figure 3.11. Six lasers from an anti-reflection-coated AlGaAs diode array were positioned along the ± 1, ± 2 and ± 3 diffraction orders of the grating. The angular plane wave spectrum of the light after passing through the grating is shown in Figure 3.12. The one large on-axis order contains 68.4% of the power from the six lasers.

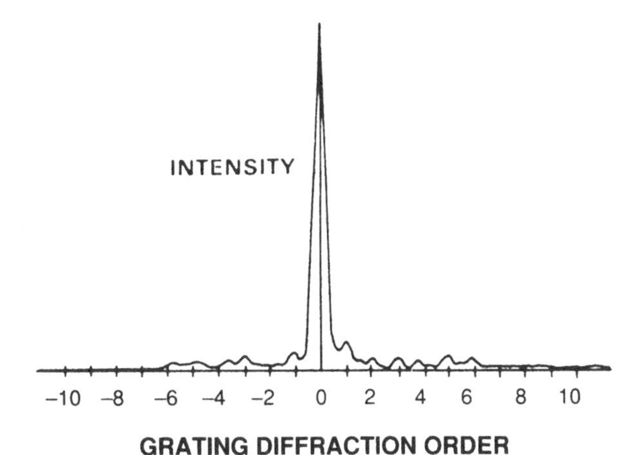

Figure 3.12. Angular plane wave spectrum of light after passing through binary grating: 68.4% of the light is contained in the on-axis order. (Ref. 34.)

3.3.1.2 Volume holograms and photorefractive crystals

If the diffracting medium is sufficiently thick, volume diffractions effects become important. The Q parameter can be used to determine whether these Bragg diffraction effects are significant[35]:

$$Q = \frac{2\pi\lambda_a T}{nd^2}, \tag{3.13}$$

where λ_a is the wavelength of light in air, T is the thickness of the medium, n is the medium index of refraction, and d is the period of the diffracting structure. When $Q > 10$, volume diffraction dominates. This is the case for many holographic recording materials such as thick photographic emulsions and dichromated gelatin, as well as optical diffraction in crystals.

In principle, very high diffraction efficiencies can be obtained by volume diffraction from phase modulated holograms[36]. The surface-relief grating in Figure 3.11 can be replaced by this volume phase hologram, resulting in an efficient superposition of laser beams into a single beam.

Photorefractive crystals such as barium titanate ($BaTiO_3$) and strontium barium niobate (SBN) can be used as volume holograms for beam superposition via two-wave mixing[37]. The principle advantage of using a photorefractive crystal over a prerecorded volume hologram is that the crystal is able to adjust to phase changes in the lasers (subject to the crystal response time) and still maintain high coupling efficiency.

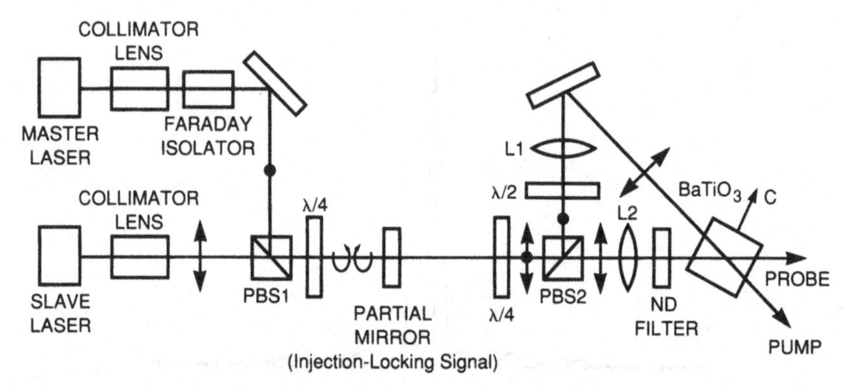

Figure 3.13. Laser beam superposition by two-wave mixing using a photorefractive crystal. (Ref. 38.)

A schematic diagram of an experimental set-up to demonstrate this technique is shown in Figure 3.13[38]. A small amount of light from the master laser is returned to the partial mirror (reflectivity = 8%) and used to injection-lock the slave laser. The polarizing beam splitter PBS1 multiplexes the light from both the master and slave lasers along a common path with different polarizations. The beams are separated again by a second polarizing beam splitter PBS2 and converted to the same polarization. These two coherent beams interfere with one another inside a crystal of $BaTiO_3$. Two-wave mixing transfers the energy from the pump beam to the probe beam. With incident pump and probe beam powers of 13.6 mW and 0.3 mW respectively, the two-wave mixing process results in a depleted pump power of 1.5 mW and an amplified probe power of 6.0 mW.

3.3.2 Aperture filling of a coherent laser array

Aperture-filling optics convert a nonuniform light field into one that is quasi-uniform in amplitude and phase. We start by describing systems that employ microlenses to fill the aperture of a coherent array. (Microlenses applied to incoherent arrays are covered in Section 3.4.2.) The remaining aperture-filling techniques utilize interference effects among different laser apertures and thus can only be applied to coherent arrays.

3.3.2.1 Aperture filling by microlens arrays

Microlenses have been applied to a variety of coherent laser arrays. In one system, a two-dimensional array of lasing apertures was formed by stacking several one-dimensional linear arrays on top of each other[39]. The laser array was operated in a master-oscillator–power-amplifier configuration. The light from each laser aperture expanded with a high divergence transverse to the junction and a lower divergence parallel to the junction. The light was collimated transverse to the junction by placing an array of miniature cylindrical lenses a short distance from the array; the parallel divergence direction was collimated in a similar fashion using longer focal length cylindrical lenses placed behind and orthogonal to the first lens array. Because of the large spacing between lasers and between arrays, the arrays could be assembled from conventional cylindrical optics.

Microlenses have also been employed inside external resonator cavities. There are several reasons why this may be desirable. First, the divergence of the light is reduced, relaxing the requirements on all subsequent optics. Second, increasing the fill factor simplifies the diffraction pattern, and makes it easier to establish coherence by simple spatial filtering with single slits or blocking wires[40,41]. Third, the reduced divergence improves the image-forming properties of a Talbot cavity by reducing edge effects and aberrations. The mode-selecting properties are compromised, however, as discussed in Section 3.3.3. Figure 3.9 shows a diffractive microlens array designed to fill the aperture of a linear array of gain-guided lasers. The lens spacing was 50 μm corresponding to the spacing of the lasers in the gain-guided laser array. An $f/2$ lens speed was selected along the array direction to provide a highly uniform light field without losing excessive amounts of light. The lens speed of $f/1$ was selected transverse to the array to capture as much light as possible. This resulted in the rectangular lens aperture shown.

3.3.2.2 Aperture filling by phase spatial filtering

Phase masks can be used as spatial filters to perform aperture filling[42–44]. The spatial filtering optics shown in Figure 3.14 perform a field transformation on the complex amplitude emitted by a phase-locked laser array. This transformation produces an electric field that is nearly uniform in magnitude and varies in phase. These phase variations are eliminated by a phase-correction plate, so that the final light distribution is nearly uniform in both magnitude and phase. The far-field pattern, therefore, concentrates virtually all its energy in the desired central lobe.

Aperture filling can be performed on a mutually coherent one- or two-dimensional laser array with complex amplitude $\hat{a}(x, y)$ by placing the laser array in the front focal plane of a lens. We assume for simplicity that the phase across the array is uniform. We can express the array

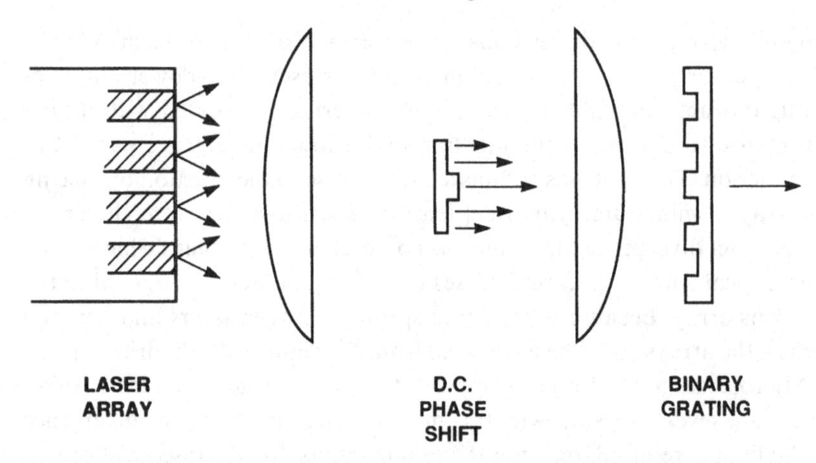

Figure 3.14. Optical configuration for phase spatial filtering. (Ref. 42.)

amplitude as the sum of an average component a_0 and a spatially varying component $a_1(x, y)$:

$$a(x, y) = a_0 + a_1(x, y), \tag{3.14}$$

where

$$a_0 = \frac{1}{XY}\int_0^Y \int_0^X a(x, y)\,\mathrm{d}x\,\mathrm{d}y, \tag{3.15}$$

and X and Y are the length and width of the laser array. An image of the array far field is obtained in the back focal plane of the first lens. The amplitude in the center of the far field is proportional to a_0 given in Eq. (3.15). Consequently, a phase-shifting spatial filter located at this central point can change the phase of a_0 without affecting $a_1(x, y)$. A second lens converts this far-field image back into an image of the laser array. However, the complex field is now given by

$$\hat{a}'(x, y) = a_0 \exp(j\phi) + a_1(x, y), \tag{3.16}$$

where $\exp(j\phi)$ is the phase shift supplied by the spatial filter. The magnitude of $\hat{a}'(x, y)$ is given by

$$|\hat{a}'(x, y)| = \sqrt{[a_0^2 + a_1^2(x, y) + 2a_0 a_1(x, y)\cos(\phi)]}. \tag{3.17}$$

We choose a phase shift ϕ to make the magnitude in Eq. (3.17) as uniform as possible. For example, if the original distribution consists solely of uniformly illuminated apertures of amplitude $A_{\max}$ surrounded by dark space, $a_1(x, y)$ is a binary function:

$$a_1(x, y) = \begin{cases} -a_0 \\ A_{\max} - a_0. \end{cases} \tag{3.18}$$

Equation (3.17) can take on two different values depending on the value of $a_1(x, y)$. Equating these two values and solving for ϕ yields

$$\phi = \arccos\left(\frac{2a_0 - A_{\max}}{2a_0}\right). \tag{3.19}$$

The phase variation in the image is identical in shape to the original binary magnitude variation and has a phase depth of ϕ. Placing a binary phase grating with a phase depth of $-\phi$ in the back focal plane of the second lens cancels the phase variation exactly. The resultant field is uniform in magnitude and phase, and thus has a Strehl ratio of unity. The Gaussian illumination from a real laser array reduces the Strehl ratio only slightly to a value between 0.9 and 0.97.

Phase spatial filtering can be applied to any coherent light distribution. However, Eq. (3.19) can only be completely satisfied when $a_0/A_{\max} \geq 0.25$. This implies that complete aperture filling can only be obtained from coherent laser arrays with two-dimensional fill factors greater than 25%.

Phase spatial filtering was applied to a ten-element AlGaAs Y-guide laser array containing 2.4 μm-wide laser apertures spaced by 6.0 μm[44]. The Y-patterned waveguide configuration established some coherence across the array and ensured that the lasing apertures were mostly in-phase[45]. Figure 3.15(*a*) shows the far-field pattern before aperture filling. Three distinct grating lobes can be seen, with only 51% of the power contained in the central lobe. Figure 3.15(*b*) shows the effect of aperture filling. The off-axis grating lobes are almost entirely eliminated, and their power has been deposited in the main lobe. The intensity in the center of the main lobe has been nearly doubled, with the entire main lobe containing 90% of the total power produced by the array.

3.3.2.3 Aperture filling by the Talbot effect

The preceding aperture-filling method used spatial filters in the back focal plane of a lens to convert an amplitude distribution into a phase distribution. This method is appropriate for low-power applications, but can result in excessive power densities at the spatial filter when used in high-power applications. An alternative method uses simple diffraction to convert amplitude to phase. The phase is corrected via a phase plate as before.

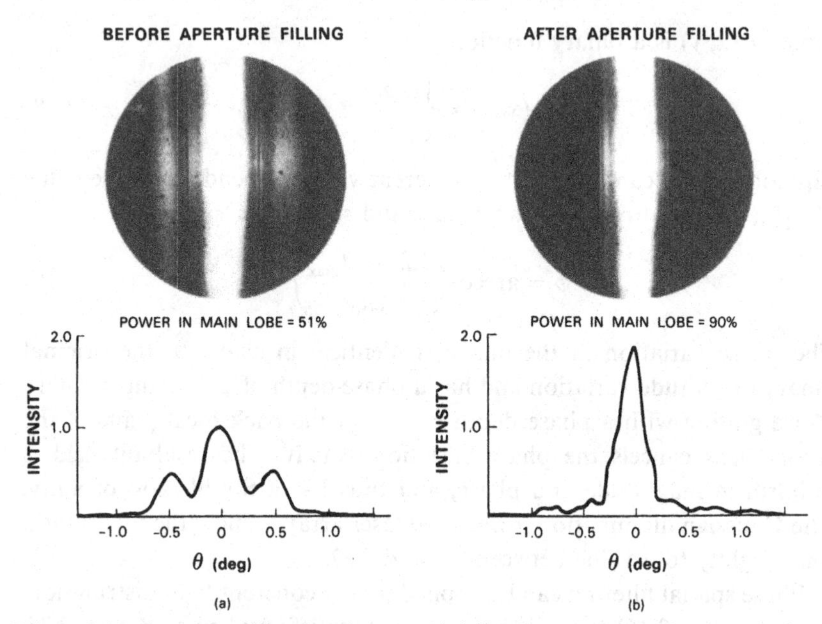

Figure 3.15. Aperture filling by phase spatial filtering: (*a*) far-field pattern before aperture filling; (*b*) far-field pattern after aperture filling. (Ref. 44.)

Many of these diffraction-based aperture-filling systems are based on the fractional Talbot effect[46–48]. The Talbot effect, discussed in the following section, describes the self-imaging properties of a periodic coherent field. Such a field is found to produce an image (correct in amplitude and phase) by free-space diffraction alone at integer multiples of the Talbot distance $Z_T = 2d^2/\lambda$, where d is the period of the array and λ is the wavelength of light. The light distribution at fractional Talbot distances also produces imagelike patterns[49,50]. For example, the plane at $Z_T/4$ produces a doubled image, in which one image is registered with respect to the input object, and the second image is shifted by one-half period. The phase is shifted by $\pi/2$ radians between the two copies of the image. By placing a phase plate in this plane to correct the phase shift, the fill factor of an array can be doubled. This type of optical system has been employed to increase the fill factor of a one-dimensional diode laser array[51].

An extension of the above system to smaller fill factors is possible by using different fractional-Talbot planes[52]. For example, the planes at $Z_T/2n$ $(n \geq 1)$ contain n equally spaced copies of the original aperture in each dimension. These copies are registered with the original aperture for even n and shifted by $1/2n$ period for odd n. A second set of multiple images is formed at the planes corresponding to $Z_T/(2n-1)$ $(n \geq 1)$, where in this case $(2n-1)$ equally spaced and properly registered images result in each dimension. In each case, the phase is constant across a single copy of the aperture but changes from copy to copy within one period. The two-dimensional step-wise phase profile is given by

$$t(I, J) = \exp\left[-j\pi \frac{(I^2 + J^2)}{N}\right], \tag{3.20}$$

where I and J are integers for even N and half-integers for odd N ranging from $-N/2$ to $(N/2-1)$.

Figure 3.16 shows a computer simulation of Fresnel diffraction from a one-dimensional periodic array of apertures with a fill factor of 1/8. Intensity is shown as a solid line, and phase as a dotted line. A single period of the amplitude at the quarter-Talbot plane $(Z_T/4)$ in Figure 3.16(*a*) consists of a doubled image, where the phase between the two copies is $\pi/2$. The amplitudes at $Z_T/8$ and $Z_T/16$ (Figures 3.16(*b* and *c*) respectively) consist of the original image replicated four and eight times respectively. Since the original object is constant over one-eighth of a period, the amplitude at $Z_T/16$ is completely uniform. The phase across a single period of the array consists of a step-wise parabolic profile. If the array consists of Gaussian illuminated apertures rather than uniform ones, the Gaussians are replicated in a similar manner. However, overlap between replicated Gaussians can cause a more complex pattern.

Figure 3.17 shows an experimental demonstration of fractional-Talbot-effect aperture filling. The original distribution consists of a two-dimensional array of uniformly illuminated and coherent apertures with a two-dimensional fill factor of 1/16 (1/4 in each dimension). Fractional Talbot planes at $Z_T/4$, $Z_T/6$ and $Z_T/8$ show two, three and four copies per period. A two-dimensional phase plate with a phase distribution given by Eq. (3.20) was fabricated by etching precision depth steps in fused silica. By placing the phase plate in the $Z_T/8$ Talbot plane, the original distribution in Figure 3.17(*a*) was converted into a plane wave with very high efficiency.

3.3.3 Talbot cavities

A variety of external optical structures have been developed to establish or enhance mutual coherence across layer arrays. Some of these are based on spatial filtering[40,53–62], while others employ diffractive coupling[63–72]. A review of some of these techniques can be found in Refs. 3 and 73. In this section, we will describe one of these methods based on Talbot self-imaging.

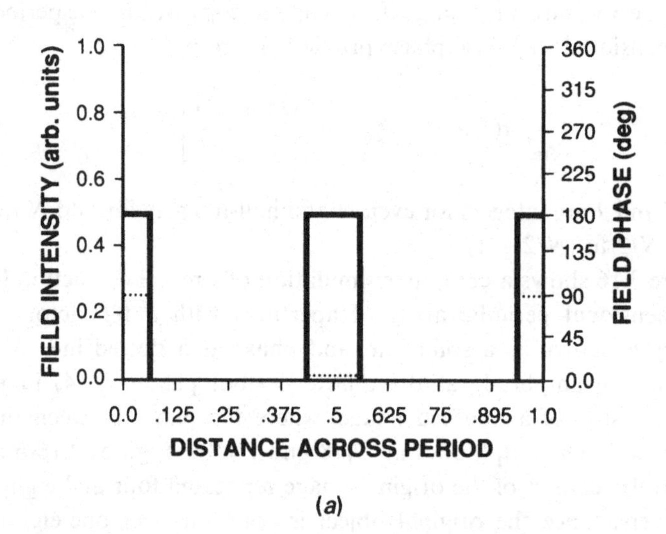

(*a*)

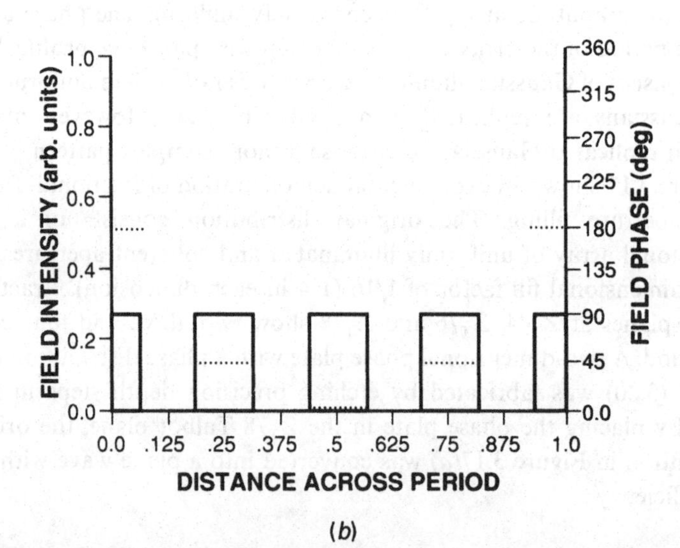

(*b*)

Figure 3.16. Computer simulation of Fresnel diffraction from an infinite period array of apertures with a one-dimensional fill factor of 1/8. The intensity (solid line) and phase (dotted line) across a single period are displayed: (*a*) corresponds to the 1/4 Talbot plane; (*b*) to the 1/8 Talbot plane.

(*continued opposite*)

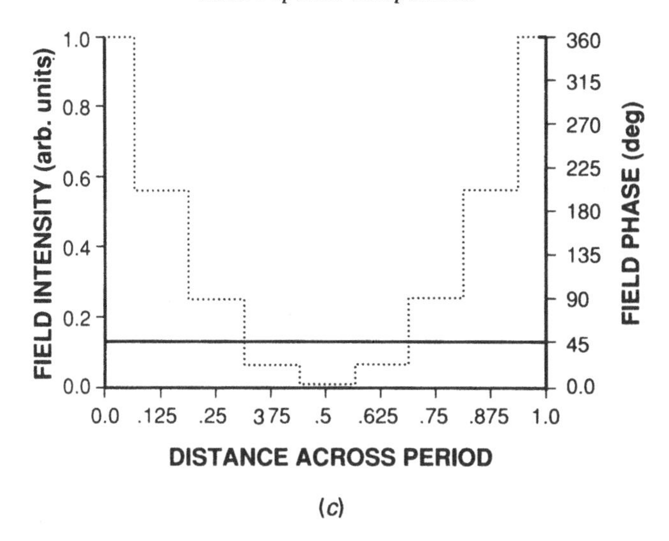

Figure 3.16. (*continued*) (*c*) to the 1/16 Talbot plane. (Ref. 52.)

3.3.3.1 Basic Talbot cavity theory

Talbot cavities employ the self-imaging properties of coherent arrays. The Talbot effect[74,75] can be illustrated quite simply by expressing the periodic light field as a Fourier series:

$$\hat{a}(x, y, z = 0) = \sum_{m=-\infty}^{\infty} \sum_{n=-\infty}^{\infty} b_{mn} \exp\left[j2\pi \frac{mx + ny}{d} \right], \tag{3.21}$$

where b_{mn} are the complex weights of the Fourier components, d is the aperture spacing in both dimensions, and we have assumed the array to be infinite. The Fresnel transfer function for free-space propagation is given by

$$H(m, n, z) = \exp\left(j2\pi \frac{z}{\lambda} \right) \exp\left[-j\pi\lambda z \left(\frac{m^2 + n^2}{d^2} \right) \right], \tag{3.22}$$

where z is the propagation distance, λ is the wavelength of light, and m and n are integers. The field $\hat{a}(x, y, z)$ at a distance z from the array is given by multiplying each Fourier component b_{mn} by the proper phase delay $H(m, n, z)$. Propagation over a distance $z = 2d^2/\lambda$ reduces $H(m, n, z)$ to a constant phase for all values of m and n. Apart from this constant phase, the distribution at this so-called Talbot plane is identical to the original near field of the laser, and hence corresponds to a self-image.

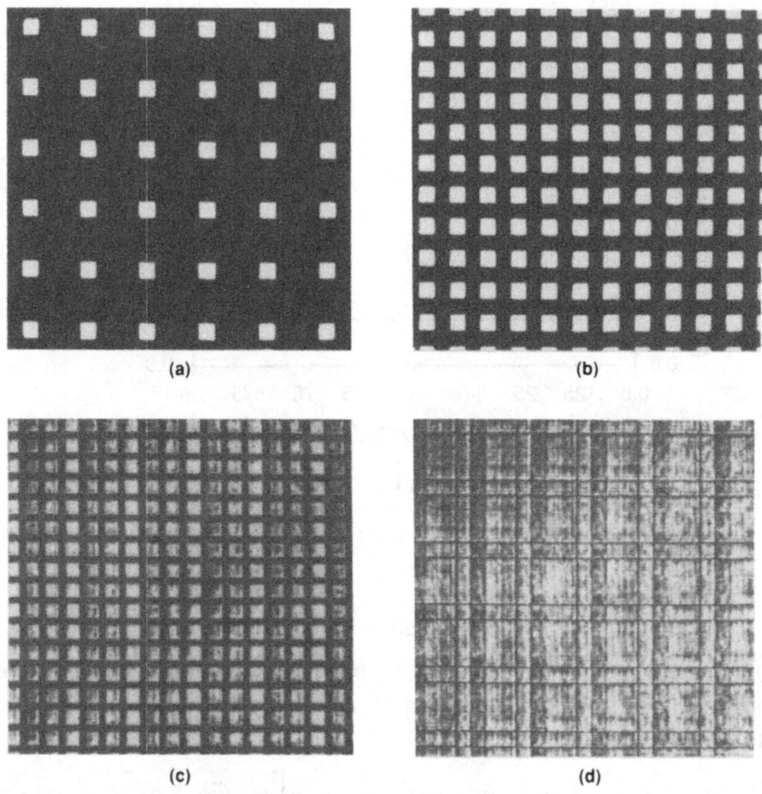

Figure 3.17. Experimental demonstration of light intensity from a periodic object at fractional-Talbot planes: (*a*) shows the intensity in the aperture plane; (*b*), (*c*) and (*d*) are the intensity distributions at the 1/4, 1/6 and 1/8 Talbot planes respectively. (Ref. 52.)

A two-dimensional Talbot cavity is illustrated in Figure 3.18. It consists of a periodic array of laser apertures followed by a flat feedback mirror placed one-half Talbot distance away. The front facets of the diode lasers have been antireflection-coated so that feedback is only provided by the common mirror. Light exits each of the small circular apertures and propagates to the feedback mirror. After reflection from the mirror and propagation back to the array, the light has traveled one complete Talbot distance. If the apertures are all mutually coherent, a self-image will be formed and the light will be efficiently coupled back into the lasing apertures. However, if the array lases with little or no coherence between lasers, Talbot self-imaging does not occur and the return light will be spread out across the entire array. If the apertures have a small fill

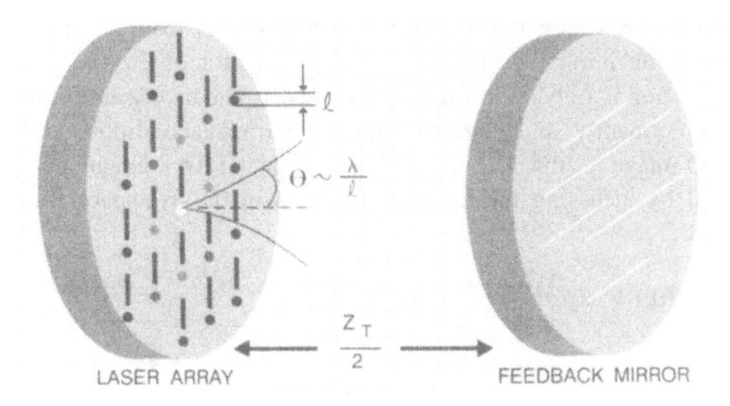

Figure 3.18. A simple two-dimensional Talbot cavity.

factor, most of the light from this incoherent state will not enter the apertures and the mode will have high loss. The Talbot cavity thus promotes lasing in the coherent state by using the laser apertures as a type of spatial filter.

From the above argument, it would appear that it is advantageous to have as small a fill factor as possible so that the incoherent state suffers the largest loss. However, there are several problems with using arbitrarily small fill factors. The most obvious one is that the finite size of a real array produces an imperfect Talbot image at the array edge. The amount of the array affected by this edge effect is inversely related to the fill factor[72,76]. An estimate of this edge effect can be determined by calculating the size of the diffraction pattern from a single aperture at the Talbot distance. For a Gaussian-illuminated aperture with beam waist ω_0 ($1/e^2$ intensity), the beamwidth $\omega(mZ_T)$ at the mth Talbot plane is

$$\omega(mZ_T) = \omega_0\sqrt{[1 + 4(m^2/\pi^2)(d/\omega_0)^4]}, \tag{3.23}$$

where d is the spacing between the apertures. The number of lasing apertures affected by this beam on one side of the array is given by

$$N = \omega(mZ_T)/d. \tag{3.24}$$

If the fill factor is relatively small ($\omega_0/d \ll 1$),

$$N \approx (2m/\pi)(d/\omega_0), \tag{3.25}$$

and the number of apertures affected by the edge effect is inversely proportional to the fill factor. Thus, small arrays with small fill factors will be dominated by edge effects and will have intolerable loss.

A second limitation on the Talbot theory comes from the paraxial approximation implicit in the free-space propagator of Eq. (3.22). This approximation is only valid for paraxial waves. The divergence of diode lasers is frequently much larger than the paraxial condition, and an aberrated image results. The effect of nonparaxial Talbot imaging can be estimated by including the fourth-order term in the free-space transfer function[52]:

$$H(u) = \exp\left\{\frac{j2\pi z}{\lambda}[1-(\lambda u)^2]^{1/2}\right\}$$

$$\approx \exp\left(\frac{j2\pi z}{\lambda}\right)\exp(-j\pi\lambda z u^2)\exp\left(-j\frac{\pi z\lambda^3 u^4}{4}\right), \tag{3.26}$$

where u is the spatial frequency. The first and second terms in Eq. (3.26) are the constant and quadratic phase terms associated with the paraxial approximation. The third term describes an aberration that prevents perfect Talbot imaging. This term is most important for high values of spatial frequency u. We can easily calculate the minimum aperture size and minimum fill factor by requiring the aberration term to be within the Rayleigh limit (that is, less than $\pi/2$ radians). Imposing this constraint and substituting the Talbot distance $2d^2/\lambda$ for z, we have from the third term of Eq. (3.26)

$$u \le \frac{1}{\sqrt{(d\lambda)}}. \tag{3.27}$$

The light diverging from a Gaussian-illuminated aperture with beam waist ω_0 ($1/e^2$ intensity point) is largely contained within a divergence half-angle of $\arctan[\lambda/(\pi\omega_0)]$. Thus, the largest spatial frequency with appreciable power is given by

$$u_{\text{max}} \approx \frac{1}{\pi\omega_0}, \tag{3.28}$$

resulting in a limit on the Gaussian beam width of

$$\omega_0 \ge \frac{\sqrt{(d\lambda)}}{\pi}. \tag{3.29}$$

By defining the fill factor (ff) for Gaussian beams as $(ff) = \omega_0/d$, we can express Eq. (3.29) as

$$(ff) \ge \frac{\lambda}{\pi^2\omega_0}. \tag{3.30}$$

From Eq. (3.30) we see that there is a minimum fill factor allowable for a given aperture size for efficient Talbot imaging. If laser arrays are used that violate this condition, the effective aperture size can be increased by using microlenses so that Eq. (3.29) is satisfied.

3.3.3.2 Modal description of a Talbot cavity

A better understanding of the Talbot cavity can be obtained by considering it as a coupled oscillator. Each aperture forms a single-laser oscillator by reflecting off the output mirror. However, due to the spread of the Gaussian beam from free-space propagation, light from one aperture illuminates adjacent apertures and causes the laser oscillators to become coupled. The overall field from N coupled oscillators can in general be described by the superposition of the N eigenmodes of the coupled system, where each eigenmode has a characteristic loss and frequency associated with it. The extent of the coupling can be estimated by Eq. (3.25), where $2N$ in this case indicates the number of lasers (in one dimension) that are strongly coupled.

To maintain good coherence, the coupled oscillator must be forced to lase in only one eigenmode. A robust system must provide low loss for this one mode while ensuring relatively high loss for all others. A simple Talbot cavity does not satisfy this requirement. It is easy to show from Eqs. (3.21) and (3.22) that an oscillating-phase mode (consisting of apertures with phases alternating between 0 and π radians) has Talbot image planes at $z = md^2/\lambda$, and so will also correspond to a low-loss mode in a standard Talbot cavity[40,76,77]. Several external-cavity spatial filtering techniques have been proposed and demonstrated to provide selective loss to the oscillating-phase mode and promote lasing of the in-phase mode[71,72]. An alternative technique promotes lasing in the oscillating-phase mode instead[72,78]. This can be accomplished by designing an external cavity with a round-trip length of only d^2/λ. This length satisfies the self-imaging condition of the oscillating-phase mode, but corresponds to a high-loss condition for the in-phase mode. Similar techniques have been applied to monolithic coherent arrays by incorporating a diffractive (coupling) section with a length of one-half Talbot distance[79]. The (negative-index) waveguides on either side of the diffractive section are then staggered so that only the in-phase eigenmode will have low loss[79,80].

A complete analysis of the modal thresholds[76,81] supports these basic conclusions. Figure 3.19 shows the modal thresholds of an eight-element laser array in a Talbot cavity as a function of round-trip propagation length. The modes are seen to be degenerate at a propagation distance

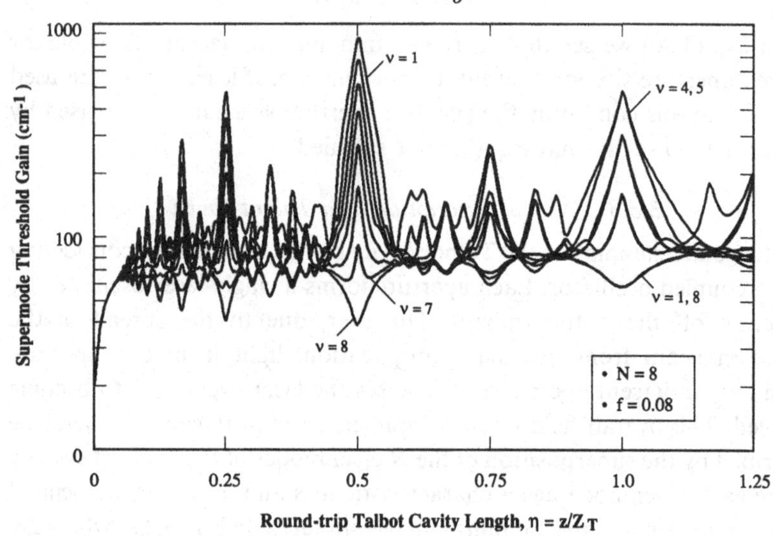

Figure 3.19. Modal threshold gain as a function of round-trip cavity length for an eight-element array. The fill factor is 8%. (Ref. 81.)

of one full Talbot length, with the in-phase ($\nu = 1$) and the oscillating-phase ($\nu = 8$) modes having identical (and lowest) loss. A round-trip distance of one-half Talbot length has a minimum loss for the oscillating-phase mode and a maximum loss for the in-phase mode, as expected from Talbot theory. The modal separation is a function of number of laser elements. For a constant fill factor, increasing the number of elements increases the total number of modes and decreases the separation between adjacent modes.

The effect of fill factor can be seen from Figure 3.20 for a 20-element array[81]. The round-trip propagation distance is $Z_T/2$ and the two modes with the lowest thresholds are $\nu = 20$ and $\nu = 19$. It is clear from the graph that the modal separation becomes poorer with increasing fill factor, again supporting the conclusion of the preceding section obtained from simple filtering arguments.

The results of an experiment to measure the modal thresholds are shown in Figure 3.21[40]. An AR-coated linear diode laser array was placed in an external cavity consisting of an afocal imaging system, a mode-selecting spatial filter, and a movable flat output mirror. The spatial filter was adjusted to permit lasing in only a specific mode, and the lasing threshold was measured as a function of mirror position. The filter was then removed and the new threshold of the array was noted, as well as the far-field pattern. The in-phase and oscillating-phase thresholds cross

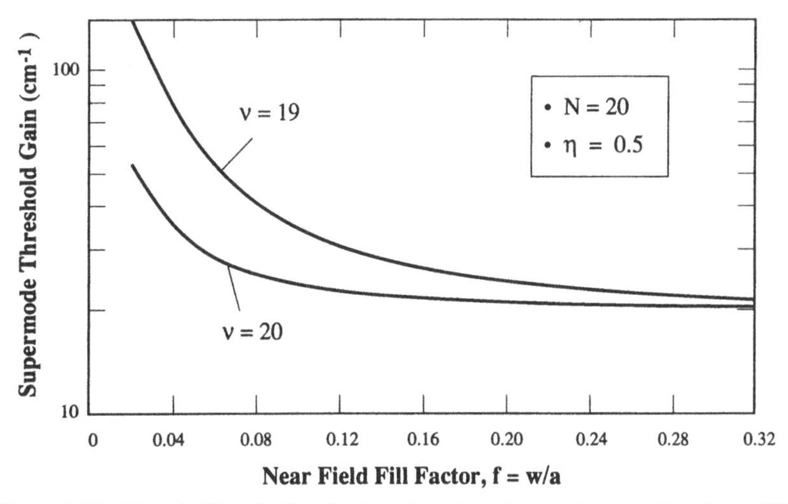

Figure 3.20. Threshold gain for the two lowest-order modes as a function of fill factor. (Ref. 81.)

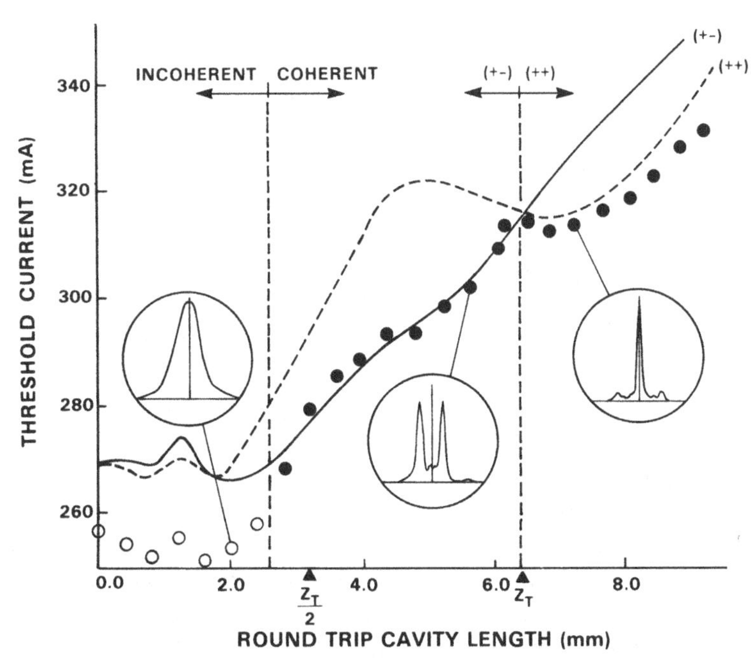

Figure 3.21. Lasing thresholds of in-phase (dashed line) and oscillating-phase (solid line) mode as a function of round-trip cavity length. Circles correspond to thresholds of cavity with no spatial filtering for mode selection. Corresponding far-field patterns are shown. (Ref. 40.)

at one Talbot distance, as expected. At half a Talbot distance, however, the oscillating-phase mode has a lower threshold, again as expected. With the mode-selecting filter removed, the cavity always lased in the mode with the lowest loss (indicated by the circles in the diagram). The general upward trend of all the thresholds is due to loss from the unfocused light perpendicular to the array.

It appears that the small fill factor required for good mode separation is in conflict with the large fill factors and apertures required for low-loss Talbot imaging. In a simple Talbot cavity this is indeed the case, and a trade-off must be made. More advanced designs using (intracavity) rectangular waveguides (described below) and/or mode-selecting mirrors[82] can reduce the fill-factor requirements significantly, permitting large mode discrimination with a low-loss cavity.

3.3.3.3 Talbot cavity experiments

Talbot cavities have been demonstrated using CO_2 laser arrays[65,66] and external cavity semiconductor laser arrays[68,71,83]. They have also been integrated, as spatial filters, within the diode laser array itself[79,84]. As mentioned in the preceding section, a simple Talbot cavity (round-trip distance of one full Talbot length) does not provide separation between the in-phase and oscillating-phase modes. This modal separation can be increased by additional spatial filtering in a fractional Talbot plane. The oscillating-phase mode forms an image at the half-Talbot plane (d^2/λ), where the in-phase mode forms a shifted image[40]. Consequently, the two images are separated and the oscillating-phase mode can be filtered out by an absorbing filter without affecting the in-phase mode[71,72]. A Talbot cavity with a mode-selecting spatial filter is shown in Figure 3.22(*a*) for a linear diode laser array. The cavity was fabricated from a cylindrical substrate, where the curvature was chosen to focus the light transverse to the array. The thickness of the cavity (4.5 mm) was chosen to satisfy the Talbot condition. The laser array consisted of seven elements with a fill factor of approximately 10%. Since Talbot imaging would be dominated by edge effects at this fill factor, microlenses were introduced to expand the beam and reduce the fill factor. Diffractive lenses were etched on to the flat surface of the substrate. A 50% output mirror was deposited on to the curved side, and patterned to match the distribution of the in-phase light at the half-Talbot plane. Results from this cavity are shown in Figure 3.22(*b*).

A second approach to mode separation is to use a round-trip propagation distance that breaks the degeneracy of the modes[72,83]. From

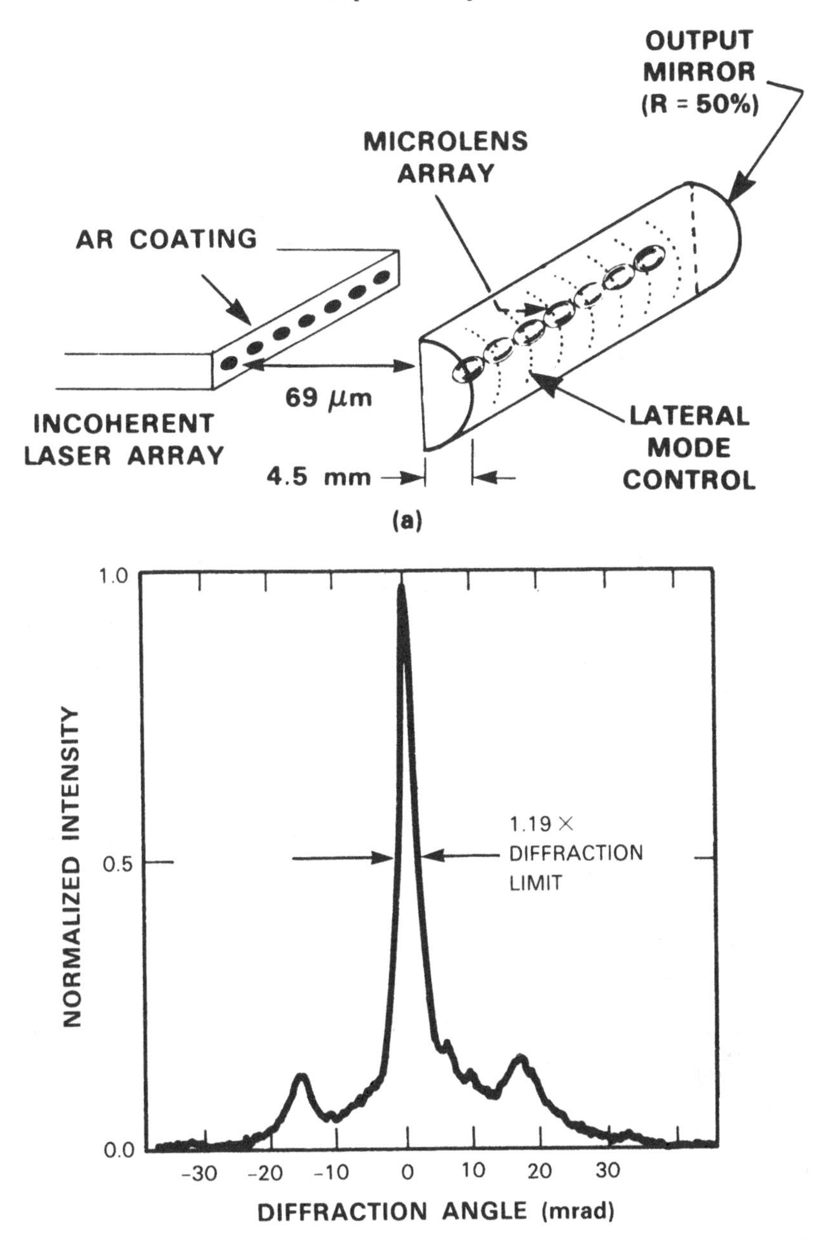

Figure 3.22. Miniature Talbot cavity: (*a*) shows the cavity fabricated from a cylindrical quartz substrate. Diffractive microlenses are etched on the flat side. The curved side contains a patterned 50%-reflectivity mirror. (*b*) shows the far-field diffraction pattern from a seven-element array. (Ref. 72.)

Figure 3.19 we can see that a round-trip distance of one-half a Talbot length appears to have maximum modal separation. This is due to the self-imaging properties of the oscillating-phase mode which are not shared by any other mode. An external Talbot cavity was designed with this cavity length and a 20-element diode laser array[83]. An array fill factor of 8% was chosen as a compromise between good mode discrimination and low edge loss, as discussed in the preceding section. This resulted in a calculated gain discrimination of 15 cm^{-1} between the lowest two modes, and an external cavity loss of 5 cm^{-1}. Since no aperture-filling optics were used, the far-field pattern consisted of several grating lobes. At optical powers of approximately 250 mW, each of these lobes was diffraction-limited. Output powers as high as 900 mW were achieved with this set-up, with a widening of the far-field grating lobes to approximately 1.7 × the diffraction limit.

A two-dimensional Talbot cavity is shown in Figure 3.23[78]. The light source is a two-dimensional array of index-guided InGaAs surface-emitting lasers. The light is generated parallel to the substrate, and then projected through the substrate by a 45° turning mirror. There are several advantages to this design. First, the heat sink can be mounted close to the gain region without interfering with the output light. Second, there is only one end mirror associated with each laser, eliminating the requirement for an AR-coating on the second mirror. Third, the high index of the substrate makes it possible to interface optics to it with reduced numerical aperture requirements. The fill-factor constraint resulting from edge effects has been eliminated by using a rectangular waveguide to propagate the

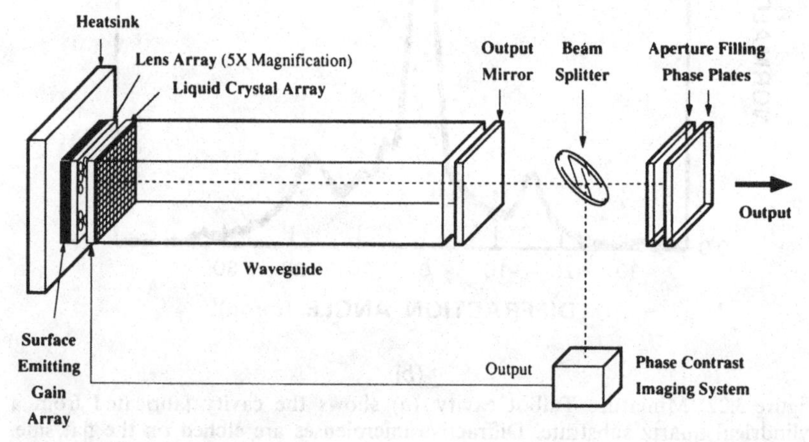

Figure 3.23. Two-dimensional Talbot cavity interfaced to a two-dimensional surface-emitting array. (Ref. 78.)

light to the output mirror. The effect of the waveguide is to place mirrors on all four sides of the array, producing reflection images and making the array appear infinite. The π phase shift that occurs upon reflection requires this device to be operated in the oscillating-phase mode. Consequently, the round-trip propagation distance was chosen to be one-half of a Talbot length to select this mode. The other limitation on fill factor results from the paraxial laser divergence requirement. To account for this, an array of refractive microlenses was placed directly after the laser array to increase the effective aperture size and decrease the divergence. The resulting divergence was chosen to provide maximum mode separation while keeping the aberrations of the Talbot self-image to an acceptable level. The fill factor of the array is increased outside the cavity by using the fractional Talbot effect and a set of phase plates (see Section 3.3.2). Finally, the phase of the array is controlled by using a liquid crystal array and phase contrast imaging system (described below).

3.3.4 External phase sensing and control

The preceding sections have addressed the problem of establishing coherence across a diode laser array and increasing its fill factor by beam superposition or aperture filling. The third requirement for high Strehl ratio performance is phase uniformity. It can be shown that, for uncorrelated phase errors uniformly distributed between $-\alpha$ and α, the expected value of the Strehl ratio is given by[3]

$$\mathscr{E}\{S\} = \left[\frac{\sin(\alpha)}{\alpha}\right]^2. \tag{3.31}$$

Phase variations that are uniformly distributed over a quarter-wave (corresponding to the Rayleigh limit) produce a Strehl ratio of 81%, with larger phase variations causing a rapid deterioration of the Strehl ratio. Some diode laser arrays have sufficient phase uniformity to operate at high Strehl ratios without any phase correction. High-power operation, however, can cause phase distortions that lower the Strehl ratio and broaden the divergence. In addition, scaling of these systems to large arrays makes phase uniformity more difficult to achieve without an active phase control system.

Phase shifting of individual laser apertures in an external cavity has been demonstrated using liquid crystals[85–87]. The phase shifter consisted of an array of nematic liquid crystal elements that were electrically addressable. Application of a voltage to a specific cell caused the liquid crystal molecules to reorientate and change the optical path length presented to linear polarized light. The cells were designed to modulate the phase without altering the polarization of the light. In one demonstration[87], the liquid crystal phase shifting system was able to correct for phase distortions of a high-power (600 mW) Talbot cavity, resulting in far-field lobes that were within 1% of the diffraction limit.

The above phase shifting system has been interfaced to a phase sensing system for automatic control of the laser array light. Phase sensing is performed by imaging the array with a modified Zernike phase contrast imaging system[88]. Phase contrast imaging converts the phase information into intensity by spatial filtering. The technique is basically the reverse of the aperture-filling method described in Section 3.3.2 for converting amplitude variations into phase. For small phase errors $\phi(x)$, we can approximate the complex amplitude from the laser array as

$$A(x, y) = \exp[j\phi(x, y)] \approx 1 + j\phi(x, y), \tag{3.32}$$

where we have assumed a constant intensity here for simplicity. The spatial filtering system shifts the phase of the spatially varying component $\phi(x, y)$ without affecting the phase of the constant. If the phase is shifted by $\pi/2$ radians, the resultant intensity will be given by

$$I(x, y) \approx |1 - \phi(x, y)|^2 \approx 1 - 2\phi(x, y). \tag{3.33}$$

The intensity is seen to be linearly related to the phase error for small errors. In a real system, intensity variations from laser to laser also affect the output. These variations must be measured and compensated for.

Equation (3.33) is only valid for small phase errors. However, by connecting the phase sensor and the liquid crystal phase shifter together in a system, arbitrarily large phase errors can be sensed and corrected in an iterative fashion. The system was found to converge to the phase-corrected solution in only a few iterations.

3.4 Incoherent techniques

Although the radiance of an incoherent array is no greater than the radiance of its single brightest source, there are numerous applications for which the radiance requirements can be relaxed. In these cases, high-power incoherent arrays can be used to great advantage. We begin this section with a description of the concentration limits of incoherent arrays, followed by a discussion of the optical design considerations for optimum concentration. The section concludes with a few applications of incoherent arrays.

3.4.1 Light concentration limits

The étendue of an optical source, defined as the product of the source area and the solid-angle divergence, measures the approximate volume of four-dimensional radiance space containing significant radiance. Mutually incoherent sources of the same polarization must occupy unique volumes in radiance space. Overlap of these volumes would require an increase in radiance that is forbidden by the second radiance theorem. Since the étendue of a single Gaussian beam[89] (assuming small-angle approximation) is λ^2, an array of N mutually incoherent single-spatial-mode sources has a *minimum* étendue of $N\lambda^2$. Smaller étendues can be obtained only by establishing varying degrees of coherence across the array, with a minimum étendue of λ^2 resulting when full coherence is established.

The concept of étendue can also be extended to describe the illumination requirements of the receiving element. The light concentration area is usually defined by the geometry of the receiving element. The maximum solid-angle divergence is also sometimes determined by the receiving element. For example, light coupled to an optical fiber must be deposited in an area defined by the core size and an angular cone given by the fiber numerical aperture. Similarly, end-pumping of an optical crystal requires the pump light to be focused to a small region in the crystal while maintaining a sufficiently small divergence to ensure good collimation throughout the volume of interest. In other simpler systems such as spot illuminators, the maximum divergence angles are determined by the focusing optics and can often be chosen to satisfy the system requirements. It now becomes a simple matter to calculate the maximum number of mutually incoherent source lasers that can be used to illuminate a particular receiving element. Counting both polarizations, the total number of sources N is simply given by

$$N = 2\frac{E_l}{E_p}, \tag{3.34}$$

where E_l is the étendue of the receiver, and E_p is the minimum étendue of a single-mode source.

As an example, we calculate the number of mutually incoherent Gaussian pump beams that can be coupled into a multi-mode fiber with core radius a and numerical aperture, NA. The fiber étendue E_l is given by

$$E_1 = \pi^2 a^2 (NA)^2, \tag{3.35}$$

where we have used small-angle approximation to calculate the solid angle. Since the étendue of a single Gaussian beam is λ^2, the total number of Gaussian pump beams N is given by

$$N = 2\frac{\pi^2 a^2 (NA)^2}{\lambda^2} = \frac{V^2}{2}, \tag{3.36}$$

where V is the V-number of the fiber given by

$$V = \frac{2\pi a (NA)}{\lambda}. \tag{3.37}$$

Note that N is simply equal to the number of modes supported by the multi-mode fiber[90].

As a second example, consider a simple illumination system that must concentrate light into a spot of radius ω_0. If a focusing system with numerical aperture (NA) is used, the étendue of the entire system is $\pi^2\omega_0^2(NA)^2$ and the maximum number of incoherent sources that can contribute to illuminating the desired area is

$$N = \frac{2\pi^2\omega_0^2(NA)^2}{\lambda^2}. \tag{3.38}$$

The minimum focusing lens numerical aperture required to produce a spot of radius ω_0 is given by the diffraction limit as

$$(NA)_{\mathrm{dl}} = \frac{\lambda}{\pi\omega_0}. \tag{3.39}$$

Consequently, the number of incoherent sources is given by

$$N = \frac{2(NA)^2}{(NA)_{\mathrm{dl}}^2}. \tag{3.40}$$

Thus, by choosing a sufficiently large lens aperture, it is often possible to utilize many incoherent sources to illuminate a given area.

3.4.2 Design of light concentrating systems

The preceding section described the theoretical maximum number of incoherent sources that can contribute to a specific application. In this section, we review the optical requirements for achieving optimal light concentration, and describe the effect of several optical systems.

3.4.2.1 Requirements for optimum light concentration

Aberrations from a nonideal source can reduce the beam radiance and increase its effective étendue significantly. The first requirement is to convert each single-spatial-mode laser output into a diffraction-limited distribution with minimum étendue. Astigmatism is a common aberration prevalent in gain-guided lasers. The astigmatic wavefront cannot be collimated with a circularly symmetric lens. An ideal optical system must contain an anamorphic collimating lens to remove this aberration from each laser aperture. Other aberrations should be removed in a similar manner.

The second requirement is to pack the individual emitters in radiance space so that there is no wasted space. Typical cw arrays contain significant inactive regions to reduce the heat load. A simplified linear diode laser source is shown in Figure 3.24(*a*), consisting of emitting apertures separated by large dark regions. Figure 3.24(*b*) shows a two-dimensional slice through radiance space along the x and θ_x axes. Each individual laser is radiating into the same angular cone, but the radiating apertures are separated by substantial inactive regions. The extent of the receiving element étendue is shown as a dotted line, where the integrated power falling in this region is coupled into the receiving element. Since the divergence of the beams is large, only about one-third of each laser beam falls within this dotted region.

The effect of various optical systems on the laser distribution can be displayed in the x–θ_x diagram. The divergence of the beams can be reduced by the magnification system shown in Figure 3.24(*c*), but the spatial separation is increased. Figure 3.24(*d*) shows that a single laser beam is now completely contained within the receiver étendue. The other two beams are imaged outside the region of interest, however, and the power coupled into the receiving element is roughly the same. Note that the total area containing substantial radiance in Figure 3.24(*d*) is unchanged (as required by the radiance theorem) and that the overall area required to contain all the beams is also unchanged.

Figure 3.24(*e*) shows a focusing system where the pump laser has been placed in the front focal plane of the lens, and the light is observed in the back focal plane. The laser beams now occupy a common area, but are incident at different angles. The roles of space and angle have been reversed, and the effect shown in Figure 3.24(*f*) is simply to rotate the

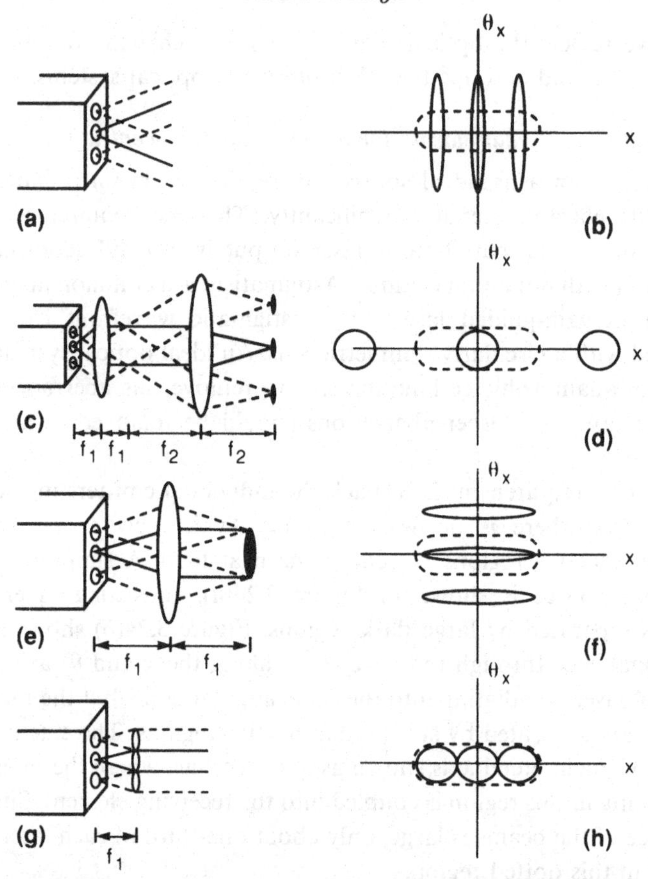

Figure 3.24. Effect of coupling optics on power distribution in radiance space. (*a*) and (*b*) show distribution with no coupling optics. (*c*) and (*d*) show the effect of optical magnification. (*e*) and (*f*) show the effect of focusing optics. (*g*) and (*h*) show the effect of individual microoptic collimating lenses. (Ref. 89, © 1992 IEEE.)

diagram in Figure 3.24(*b*) by 90°. As in the preceding case, only a single laser beam couples into the receiving element, and the useful power is unchanged. Again, the area in the diagram is preserved, and the dark spaces between pump beams have not been removed.

The third optical system, shown in Figure 3.24(*g*), contains collimating microoptics. Each laser beam is allowed to expand to an optimal size, and is then collimated individually. The effect is to perform the magnification illustrated in Figures 3.24(*c–d*) on each beam independently. An increase in beam size to cover the gaps between beams is accompanied by a decrease in divergence. The overall spacing of the laser beams, however,

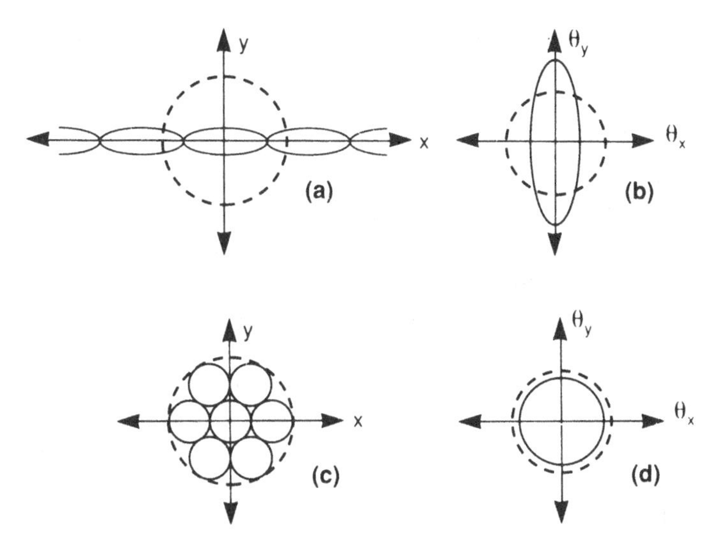

Figure 3.25. Distribution of power in space and angle. (*a*) shows the spatial distribution of a linear laser source, and (*b*) shows its divergence. (*c*) illustrates the advantages of converting the one-dimensional source to a two-dimensional virtual array. (*d*) shows that proper divergence has been achieved by symmetrizing the beams and optimizing the beam waists. (Ref. 89, © 1992 IEEE.)

does not change. Figure 3.24(*h*) shows that this optical system has densely packed three complete laser beams within the receiver étendue. Note that the total area containing light in the diagram is still unchanged, but the area required to contain all the beams is reduced. As with coherent arrays, we see that aperture-filling considerations play an important part in a well-designed, incoherent focusing system.

The third requirement for optimal light concentration is to place laser beams throughout the four-dimensional receiver étendue. This corresponds to fully utilizing the spatial and angular extent of the receiving element. Figure 3.25 illustrates the amount of radiance space used by a linear diode laser array with conventional imaging. A different two-dimensional slice through radiance space containing the x and y axes is shown in Figure 3.25(*a*), and a corresponding slice containing the θ_x and θ_y axes is shown in Figure 3.25(*b*). The extent of the receiver étendue is again shown as a dotted line. We consider an application (such as end-pumping of a solid-state laser) where the receiver étendue is circularly symmetric in both space and divergence. A geometric transformation is required to convert the diode laser distribution in Figure 3.25(*a*) into the ideal symmetric distribution shown in Figure 3.25(*c*). The asymmetric divergence in

Figure 3.25(*b*) can be converted into the required symmetric divergence of Figure 3.25(*d*) by anamorphic imaging.

3.4.2.2 Optical systems for light concentration

The preceding section pointed out the need for aperture filling to ensure optimal light concentration. Since the arrays are incoherent, many of the aperture-filling methods described in Section 3.3.2 are not appropriate. However, the use of individual lenses to expand and collimate individual lasers or groups of lasers in an array does not depend on the coherence, and is an effective method of increasing the fill factor. Both macroscopic[91] and microscopic[26,89] lens systems have been used for aperture filling.

Linear diode laser arrays usually require some geometric transformation of the near-field light distribution for optimal coupling to a receiving element. One simple method of performing a geometric transformation is to use a fiber bundle[92,93]. Each fiber end is butt-coupled to a different portion of the linear array (or separate arrays) such that the fiber receives light from a small group of lasers. The other end of the bundle can be formed into the desired two-dimensional shape. The multi-mode fibers produce a symmetric cone of light with a divergence given by the numerical aperture of the fiber. Consequently, the spatial geometry and divergence can both be symmetrized. A limitation of this technique results from the mismatch between the one-dimensional array and the two-dimensional fiber aperture, resulting in an inefficient utilization of radiance space, and a resulting radiance that is considerably lower than the theoretical limit.

An optical system that utilized radiance space in an optimal manner was designed to concentrate the light from a linear cw diode laser array[89]. The laser array (SDL3480) had a fill factor of 20% to reduce the heat load during cw operation, substantial astigmatism and a nonsymmetric divergence. The optical system consisted of two planes of microoptics shown in Figure 3.26. Each plane contained an array of microlenses and microprisms fabricated as diffractive optics (see Section 3.2.2). Together, the two planes formed an array of imaging systems capable of introducing tilts in both directions. The divergence of each beam was symmetrized and the astigmatism removed by using anamorphic imaging. Aperture filling and geometric transformation were performed by designing the first plane of optics to direct the different laser beams into a two-dimensional geometry. The second plane of optics introduced a complementary tilt that redirected all the beams along the optical axis. A final refractive lens was used to focus the light into a small spot.

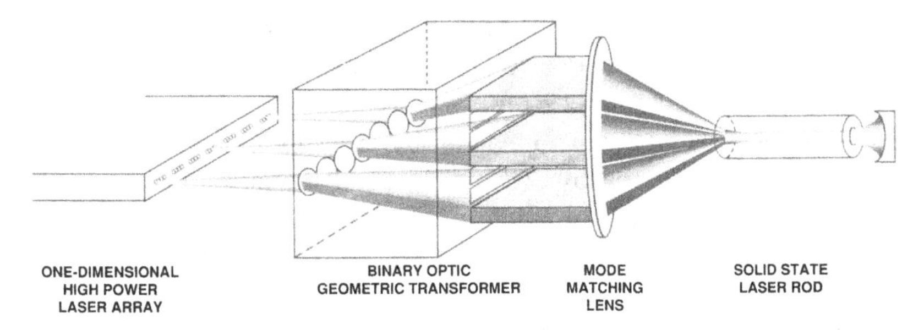

Figure 3.26. Optical system for performing a geometrical transformation on a one-dimensional diode laser array. Diffractive optical elements are contained in two separate planes. (Ref. 89, © 1992 IEEE.)

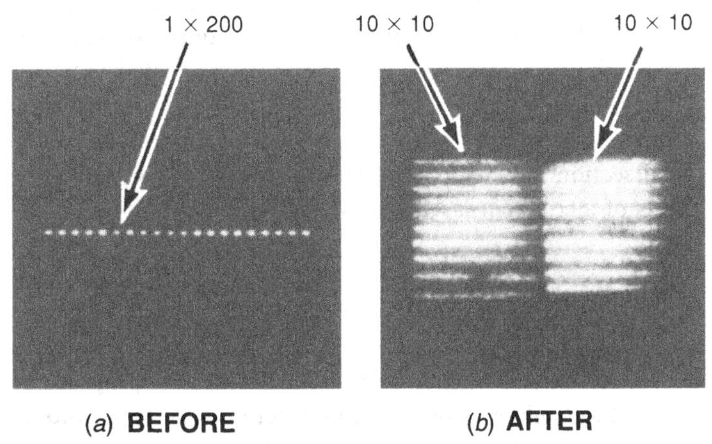

Figure 3.27. Light patterns before and after the geometrical transformation optics: (*a*) a 1 × 200 array of lasers; (*b*) two symmetric sets of 10 × 10 lasers. These two square arrays are superimposed by polarization-multiplexing optics. (Ref. 89, © 1992 IEEE.)

The light intensity before and after the diffractive optics is shown in Figure 3.27. The light from the laser array consisted of a linear array of 200 lasers. Each spot in the diagram consists of ten closely spaced lasers separated by large nonradiating regions. The output from the optical system consists of two arrays of 10 × 10 lasers each. Since polarization is not critical for many applications, these two square arrays were polarization-multiplexed to form a single symmetric array.

The far-field divergence was symmetric, with a divergence angle of approximately 1.5 mrad in each direction. Since the cross-sectional area of the polarization-multiplexed square array was only $0.25\,\text{cm}^2$, the estimated étendue was 1.8×10^{-6} cm^2 Sr. The smallest étendue possible for 200 single-spatial-mode laser beams (polarization-multiplexed) is 6.5×10^{-7} cm^2 Sr, or only 2.7 times smaller. The slightly larger experimental étendue can be accounted for by the sub-optimal packing of Gaussian beams in the array. The efficiency of the uncoated diffractive optics and the polarization-multiplexer limited the optical throughput to about 20%. Nevertheless, the 5 watt laser array was converted into a symmetric distribution with an average radiance of 0.6 MW/cm^2 Sr. This represents an intensity exceeding 270 kW/cm^2 at the focus of an $f/1$ lens.

3.4.3 Incoherent applications

Incoherent diode laser arrays are ideal light sources for many applications. They possess high electrical-to-optical efficiency, and have demonstrated very high reliability. Large laser arrays can be fabricated that produce very high pulsed and cw power levels. In addition, their narrow spectrum offers an advantage in applications such as optical pumping of crystals. In the final section of this chapter, we will review a few of the possible applications of incoherent laser arrays.

3.4.3.1 Illuminators

Illumination of large areas, lines and spots is a natural application of diode laser arrays. Large-area illumination requires only very simple optics to reduce the divergence and direct the light appropriately. The incoherence of the light source can be an asset in some applications since it reduces the effects of interference and speckle.

Linear incoherent diode laser arrays have limited spatial coherence only in the direction of the array. Their coherence transverse to the array is excellent, allowing them to be focused to the diffraction limit in this direction. Consequently, light from these sources can be projected to form thin lines with widths as small as a wavelength of light. Applications include machine vision and optical measurement devices.

Often, it is desirable to illuminate a small area with as much power as possible. As was shown in the preceding section, multiple incoherent sources can be used only when the étendue of the illumination optics is larger than λ^2. In a simple illuminator, the solid-angle divergence is given by the numerical aperture of the focusing lens. This lens can often be chosen to concentrate a large number of incoherent sources in the desired area.

Eq. (3.38) represents an upper limit on the number of usable incoherent sources. In most arrays, the geometry and divergence of the sources is not matched to the desired illumination area, and the fill factor is often much less than unity. In these cases, simple lenses will not be able to concentrate the light in an optimal manner. The light concentration obtainable from a linear array with simple macroscopic optics can be estimated by calculating a one-dimensional étendue along the array direction only. For an array of Gaussian sources with a one-dimensional fill factor $(ff) = \omega_0/d$, we have a linear source étendue E_l given by

$$E_l = \frac{4N\lambda}{\pi(ff)}, \tag{3.41}$$

where we have used the full beam width $(2\omega_0)$ and full divergence angle of the Gaussian beam in the definition. If a simple lens is used to concentrate this linear array into the smallest possible line, the étendue of the receiver is simply given by the product of the lens numerical aperture (full angle) and the desired linear concentration (l), or $E_r = 2(NA)l$. Equating these two étendues and solving for the number of sources N yields

$$N = \frac{2\pi(ff)(NA)l}{4\lambda}, \tag{3.42}$$

where we have not included polarization multiplexing here. Comparing Eq. (3.42) with Eq. (3.38), we see that the effect of the fill factor is to decrease the number of sources that can be concentrated in the desired space. In addition, the number of sources increases linearly with numerical aperture rather than as the square for a two-dimensional source.

3.4.3.2 Fibers and fiber lasers

Eq. (3.36) describes the number of incoherent sources that can be coupled into a fiber. We found that this number was numerically equal to the number of modes supported by the fiber (including the factor of two for both polarizations). Consequently, single-mode fibers can be illuminated with, at most, two mutually incoherent lasers. This number can be much larger for multi-mode fibers. For example, a typical multi-mode fiber with a core diameter of 50 μm and numerical aperture of 0.3 operating at a wavelength of 0.8 μm has a V-number of 58.9, allowing for a coupling of $V^2/2 = 1735$ single-spatial-mode incoherent lasers into the fiber core.

A circular fiber core has a symmetric spatial geometry and divergence. The number calculated above assumes that the diode laser array has similar symmetries and a fill factor of 100%. When conventional optics are used, the geometry of linear diode laser arrays severely limits this number. If we assume for simplicity that the one-dimensional array fill factor is unity, the fiber core radius a must be equal to $M\omega_0$ to accept M laser beams, where ω_0 is the beam waist. Since the divergence of the beam cannot be greater than the numerical aperture of the fiber, we have a maximum number of incoherent laser beams from a linear sub-array given by

$$M = 2\frac{a\pi(NA)}{\lambda} = V, \tag{3.43}$$

where we have again counted both polarizations. Since the theoretical maximum number of laser beams is given by $V^2/2$, simple coupling to a linear laser array reduces the potential étendue and radiance by a factor of $V/2$. In the above example, the total number of lasers coupled by conventional optics is limited to only 59 incoherent sources. A 20% fill factor characteristic of many cw pump arrays would further reduce this number to approximately 12 sources. When we compare this with the theoretical limit of 1735 sources, it is clear that microoptics for geometrical transformation, fill-factor enhancement and divergence equalization can enhance the system potential considerably.

Fiber lasers provide a potential application for laser concentration optics. A fiber laser must be optically pumped by coupling light from a diode laser into the fiber. Eq. (3.36) predicts the number of incoherent single-spatial-mode sources that can be coupled, where the wavelength used is the pump wavelength. Since most fiber lasers are designed with a doped single-mode core and a simple cladding, the pump energy must be deposited into the single-mode core. At the lasing wavelength, only two incoherent beams can be concentrated into the core (one for each polarization). Since the pump wavelength is shorter, it is theoretically possible to achieve a somewhat higher number of pump beams. However, the requirement of concentrating the light into the single-mode core makes it impossible to pump this laser with large numbers of incoherent sources. A coherent array is necessary to achieve very high pump powers.

The number of incoherent pump beams can be increased by using a fiber laser with two claddings[94]. The laser light in the single-mode core is guided by the inner cladding. A second outer cladding is introduced to guide light from the inner cladding, effectively making a multi-mode waveguide. The pump beams can thus be distributed throughout the inner cladding, with the guided pump light being slowly absorbed by the doped core. Double-clad structures can be pumped with, at most, $V_p^2/2$ lasers, where V_p is the V-number of the first cladding guided by the second at the pump wavelength. For one particular double-clad fiber laser[95] with a 45 × 110 μm rectangular inner cladding and $NA = 0.4$, the theoretical number of pump beams is 7775. Considering 65 mW per pump beam, this represents a potential pump power of over 500 watts injected into the fiber laser.

3.4.3.3 End-pumping of solid-state lasers

Solid-state laser end-pumping requires the pump power to be concentrated in a small region of space to match the mode area of the solid-state laser, and simultaneously requires a small divergence angle to allow the power to be absorbed by the crystal over an extended length. The number of pump beams that can be coupled into a laser rod can be determined by calculating the étendue of the solid-state laser mode over a fixed length. A simple approximate expression can be obtained by referring to Figure 3.28. The laser mode has a waist of ω_ℓ, and the crystal is chosen of length L to absorb the majority of the pump beam. The divergence of the laser mode is assumed to be small so that diffraction within the crystal is negligible. We also assume that the pump-beam waist ω_{po} is small compared to ω_ℓ and therefore that the number of pump beams is large. We seek to find the area and solid angle which will completely contain the parallel propagating pump beams within the laser mode throughout the crystal. From the diagram, we see that pump beams can be placed out a distance R from the center, and still be contained within the laser mode as long as we satisfy

$$R \leq \omega_\ell - \omega_{\mathrm{p}}(L/2), \tag{3.44}$$

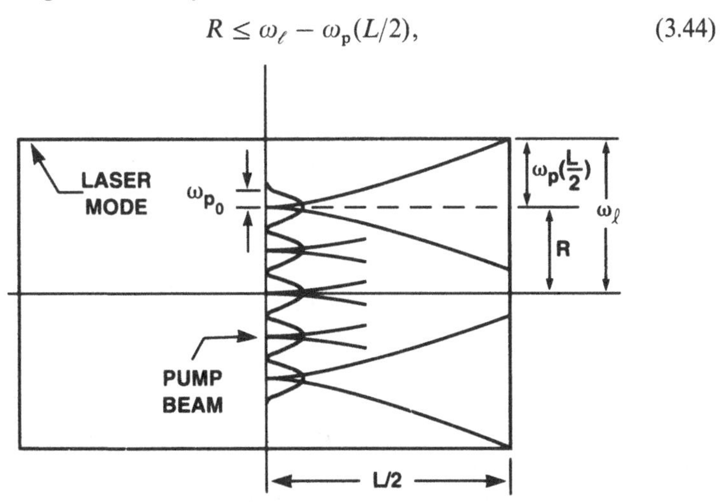

Figure 3.28. Calculating the étendue of an end-pumped rod laser. Parallel-propagating pump beams are shown inside the pump volume. (Ref. 89, © 1992 IEEE.)

where $\omega_p(L/2)$ is the spot size of the pump beam at the end of the crystal. Since ω_{po} is small,

$$\omega_p(L/2) \approx \frac{\lambda_{eff} L}{2\pi\omega_{po}}, \tag{3.45}$$

where λ_{eff} is the effective wavelength of the pump light in the laser rod. The area of the region described by Eq. (3.44) is given by

$$\pi R^2 = \pi\left(\omega_\ell - \frac{\lambda_{eff} L}{2\pi\omega_{po}}\right)^2. \tag{3.46}$$

The solid-angle divergence (using a small-angle approximation) from a pump beam of waist ω_{po} is given by

$$\Omega = \frac{\lambda_{eff}^2}{\pi\omega_{po}^2}, \tag{3.47}$$

resulting in a solid-state laser étendue E_ℓ of

$$E_\ell = \left(\omega_\ell - \frac{\lambda_{eff} L}{2\pi\omega_{po}}\right)^2 \frac{\lambda_{eff}^2}{\omega_{po}^2}. \tag{3.48}$$

It is easy to show that E_ℓ is maximized when

$$\omega_{po} = \frac{\lambda_{eff} L}{\pi\omega_\ell}, \tag{3.49}$$

resulting in a maximum étendue of

$$E_\ell = \frac{\pi^2\omega_\ell^4}{4L^2}. \tag{3.50}$$

Since the étendue of a single Gaussian pump beam $E_p = \lambda_{eff}^2$, the maximum number of mutually incoherent pump beams N that can be packed into the laser pump volume (volume of the laser mode where the pump light is absorbed) is given by

$$N = \frac{\pi^2 n^2 \omega_\ell^4}{2L^2\lambda_p^2}, \qquad N \gg 1 \tag{3.51}$$

where we have substituted the free-space wavelength of the pump light $\lambda_p = \lambda_{eff}/n$, n = refractive index of the laser rod. Note that the number of mutually incoherent pump beams increases as the fourth power of the laser mode waist. This indicates that end-pumping is inherently scalable if the pump volume is utilized efficiently[96]. The number of pump beams increases as the absorption coefficient increases (resulting in a smaller crystal length L). The number can be made even higher, of course, if any mutual coherence that may exist in the laser array is exploited.

Eq. (3.51) is accurate only for large N. The equation for small N as L approaches infinity is given by dividing the étendue of the laser mode itself by the étendue of a pump beam:

$$N = 2\left(\frac{\lambda_\ell}{\lambda_p}\right)^2, \qquad N \approx 1, \tag{3.52}$$

where λ_ℓ is the free-space wavelength of the laser light, and $\lambda_\ell > \lambda_p$.

A high-brightness microoptical system was used to end-pump a solid-state laser rod with a 1 cm linear diode laser array. The 200-laser pump array was converted into a two-dimensional source using the geometrical transformation microoptics described in Section 3.4.2. The solid-state laser was designed with a mode waist $\omega_\ell = 210\ \mu m$. Eq. (3.51) can be used to calculate the expected number of pump lasers that can be coupled into the solid-state laser mode. Using an index $n = 1.8$, an absorption length $L = 6$ mm, a pump wavelength $\lambda_p = 0.808\ \mu m$, and accounting for imperfections in the resulting fill factor of about 70% in both axes, Eq. (3.51) predicts a capacity of 612 pump beams. This value is well above the 200 pump beams used, indicating that virtually all the pump power must be deposited within the fundamental mode volume. No higher-order modes were observed, and the TEM_{00} mode slope efficiency of 56% was demonstrated, indicating efficient coupling from the pump to the lasing mode.

3.5 Conclusion

The application of external microoptical systems to diode laser arrays has been surveyed in this chapter. We have reviewed some of the requirements of these optical systems, and described several methods of fabricating refractive and diffractive microoptical components. The fill factor was shown to have a large impact on the system performance of diode laser arrays. This fill factor can be increased in a variety of ways. Beam expansion by microlenses can enhance the fill factor of both coherent and incoherent arrays. Other methods requiring array coherence are phase spatial filtering, fractional-Talbot phase plates and beam superposition by surface-relief gratings or volume diffraction.

A coherent external cavity technique based on Talbot self-imaging was described. The modes of the cavity are influenced by the fill factor and the number of lasers in the array. A low fill factor increases the gain separation between modes, but also increases the losses due to edge effects and nonparaxial imaging. Microoptics are often required to optimize these effects. Several experiments demonstrated the modal properties and coherent power potential of this technique.

The light concentration properties of specific incoherent systems were analyzed with the help of the étendue concept. Many applications such as multi-mode fiber illumination and end-pumping of solid-state lasers are theoretically capable of accepting light from a large number of incoherent sources. In practice, this number is often reduced considerably due to differences in geometry and divergence between the source and receiver, and because of a low source fill factor. A microoptical system was presented that converts a low fill-factor linear array with asymmetric divergence into a high fill-factor two-dimensional array with symmetric geometry and divergence. This system was used to end-pump a 210 μm solid-state laser mode with a 1 cm bar of diode lasers.

Microoptics are seen to provide a powerful tool for controlling, modifying and adapting light from laser diode arrays. Many current laser array application areas already benefit substantially by employing external microoptics. As optical systems become more and more complex, future applications are likely to become increasingly dependent on this enabling technology.

References

1. F. W. Sears, *Optics*, Addison-Wesley, Reading, Massachusetts (1949).
2. R. W. Boyd, *Radiometry and the Detection of Optical Radiation*, Wiley, New York (1983).
3. J. R. Leger, External methods of phase locking and coherent beam addition, Chapter 8 of *Surface Emitting Diode Lasers and Arrays*, G. Evans and J. Hammer (eds), Academic Press, New York (1993).
4. R. H. Anderson, *Appl. Opt.*, **18**, 477–84 (1979).
5. I. N. Ozerov, V. M. Petrov, V. A. Shishkina and V. M. Shor, *Sov. J. Opt. Technol.*, **48**, 49–50 (1981).
6. N. F. Borrelli, D. L. Morse, R. H. Bellman and W. L. Morgan, *Appl. Opt.*, **24**, 2520–5 (1985).
7. N. F. Borrelli and D. L. Morse, *Appl. Opt.*, **27**, 476–9 (1988).
8. Z. D. Popovic, R. A. Sprague and G. A. N. Connell, *Appl. Opt.*, **27**, 1281–4 (1988).
9. Z. L. Liau, V. Diadiuk, J. N. Walpole and D. E. Mull, *Appl. Phys. Lett.*, **52**, 1859–61 (1988).
10. Z. L. Liau, V. Diadiuk, J. N. Walpole and D. E. Mull, *Appl. Phys. Lett.*, **55**, 97–9 (1989).
11. Z. L. Liau and H. J. Zeiger, *J. Appl. Phys.*, **67**, 2434–40 (1990).
12. T. M. Baer, D. F. Head and M. Sakamoto, *Conf. on Lasers and Electrooptics: Digest of technical papers*, Optical Society of America, p. 416 (1989).

13. J. J. Snyder, P. Reicher and T. Baer, *Appl. Opt.*, **30**, 2743–7 (1991).
14. K. Matsushita and K. Ikeda, *Proc. SPIE*, **31**, 23–35 (1972).
15. M. Oikawa and K. Iga, *Appl. Opt.*, **21**, 1052–6 (1982).
16. M. Oikawa, K. Iga, S. Misawa and Y. Kokubun, *Appl. Opt.*, **22**, 441–2 (1983).
17. Masahiro Oikawa, Kenichi Iga, Takeshi Sanada, Noboru Yamamoto and Kouichi Nishizawa, *Jpn J. Appl. Phys.*, **20**, L296–8 (1981).
18. M. Oikawa, K. Iga and T. Sanada, *Electron. Lett.*, **17**, 452–4 (1981).
19. S. Misawa, M. Oikawa and K. Iga, *Appl. Opt.*, **23**, 1784–6 (1984).
20. M. Oikawa, H. Nemoto, K. Hamanaka and T. Kishimoto, *SPIE*, **1219**, 532–8 (1990).
21. T. Fujita, H. Nishihara and J. Koyama, *Opt. Lett.*, **7**, 578–80 (1982).
22. T. Shiono, K. Setsune, O. Yamazaki and K. Wasa, *Appl. Opt.*, **26**, 587–91 (1987).
23. M. Tanigami, S. Ogata, S. Aoyama, T. Yamashita and K. Imanaka, *IEEE Photonics Tech. Lett.*, **1**, 384–5 (1989).
24. L. d'Auria, J. P. Huignard, A. M. Roy and E. Spitz, *Opt. Comm.*, **5**, 232–5 (1972).
25. V. P. Koronkevich, V. P. Kiriyanov, F. I. Kokoulin, I. G. Palchikova, A. G. Poleshchuk, A. G. Sedukhin, E. G. Churin, A. M. Shcherbachenko and Y. I. Yurlov, *Optik*, **67**, 257–66 (1984).
26. J. R. Leger, M. L. Scott, P. Bundman and M. P. Griswold, *Proc. SPIE*, **884**, 82–9 (1988).
27. G. J. Swanson and W. B. Veldkamp, *Opt. Eng.*, **28**, 605–8 (1989).
28. J. M. Finlan and K. M. Flood, *Proc. SPIE*, **1052**, 186–90 (1989).
29. J. R. Leger, G. J. Swanson and W. B. Veldkamp, *Appl. Opt.*, **26**, 4391–9 (1987).
30. H. Dammann and K. Görtler, *Opt. Commun.*, **3**, 312–15 (1971).
31. U. Killat, G. Rabe and W. Rave, *Fiber Integr. Opt.*, **4**, 159–67 (1982).
32. W. H. Lee, *Appl. Opt.*, **18**, 2152–8 (1979).
33. W. B. Veldkamp, J. R. Leger and G. J. Swanson, *Opt. Lett.*, **11**, 303–5 (1986).
34. J. R. Leger, G. J. Swanson and W. B. Veldkamp, *Appl. Phys. Lett.*, **48**, 888–90 (1986).
35. W. R. Klein, *Proc. IEEE*, **54**, 803–4 (1966).
36. H. Kogelnik, *Bell Syst. Tech. J.*, **48**, 2909–47 (1969).
37. M. Cronin-Golomb, A. Yariv and I. Ury, *Appl. Phys. Lett.*, **48**, 1240–42 (1986).
38. W. R. Christian, P. H. Beckwith and I. McMichael, *Opt. Lett.*, **14**, 81–3 (1989).
39. M. S. Zediker, H. A. Appelman, T. E. Bonham, D. A. Bryan, A. E. Chenoweth, B. G. Clay, J. R. Heidal, J. L. Levy, D. J. Drebs, R. G. Podgornik, B. D. Porter and R. A. Williams, *SPIE Proc.*, **1059**, 162 (1989).
40. J. R. Leger, *Appl. Phys. Lett.*, **55**, 334–6 (1989).
41. V. Diadiuk, Z. L. Liau, J. N. Walpole, J. W. Caunt and R. C. Williamson, *Appl. Phys. Lett.*, **55**, 2161–3 (1989).
42. G. J. Swanson, J. R. Leger and M. Holz, *Opt. Lett.*, **12**, 245–7 (1987).
43. M. Holz, J. R. Leger and G. J. Swanson, *Conf. on Lasers and Electro-optics: Digest of technical papers* (Optical Society of America, New York, 1987), p. 356 (1987).
44. J. R. Leger, G. J. Swanson and M. Holz, *Appl. Phys. Lett.*, **50**, 1044–6 (1987).
45. D. F. Welch, P. Cross, D. Scifres, W. Streifer and R. D. Burnham, *Electron. Lett.*, **22**, 293–4 (1986).
46. O. R. Kachurin, F. V. Lebedev, M. A. Napartovich and M. E. Khlynov, *Sov. J. Quantum Electron.*, **21**, 351–4 (1991).
47. W. Cassarly and J. M. Finlan, Paper PD3, OSA Annual Meeting, Orlando (1989).

48. Liu Liren and Zhao Yiyang, *Chinese Physics*, **9**, 811–14 (1989).
49. J. T. Winthrop and C. R. Worthington, *JOSA*, **55**, 373–81 (1965).
50. B. Packross, R. Eschbach and O. Bryngdahl, *Opt. Commun.*, **56**, 394 (1986).
51. F. X. D'Amato, E. T. Siebert and C. Roychoudhuri, *Proc. SPIE*, **1043**, 100–6 (1989).
52. J. R. Leger and G. J. Swanson, *Opt. Lett.*, **15**, 288–90 (1990).
53. E. M. Philipp-Rutz, *Appl. Phys. Lett.*, **26**, 475–7 (1975).
54. R. H. Rediker, R. P. Schloss and L. J. Van Ruyven, *Appl. Phys. Lett.*, **46**, 133–5 (1985).
55. K. K. Anderson and R. H. Rediker, *Appl. Phys. Lett.*, **50**, 1–3 (1987).
56. L. Goldberg and J. F. Weller, *Electron. Lett.*, **25**, 112–14 (1989).
57. J. Yaeli, W. Streifer, D. R. Scifres, P. S. Cross, R. L. Thornton and R. D. Burnham, *Appl. Phys. Lett.*, **47**, 89–91 (1985).
58. J. Berger, D. Welch, W. Streifer and D. R. Scifres, *Appl. Phys. Lett.*, **52**, 1560–2 (1988).
59. C. J. Chang-Hasnain, D. F. Welch, D. R. Scifres, J. R. Whinnery, A. Dienes and R. D. Burnham, *Appl. Phys. Lett.*, **49**, 614–16 (1986).
60. C. J. Chang-Hasnain, J. Berger, D. R. Scifres, W. Streifer, J. R. Whinnery and A. Dienes, *Appl. Phys. Lett.*, **50**, 1465–7 (1987).
61. C. J. Chang-Hasnain, A. Dienes, J. R. Whinnery, W. Streifer and D. R. Scifres, *Appl. Phys. Lett.*, **54**, 484–6 (1989).
62. L. Goldberg and J. F. Weller, *Appl. Phys. Lett.*, **51**, 871–3 (1987).
63. N. G. Basov, É. M. Belenov and V. S. Letokhov, *Sov. Phys. Solid State*, **7**, 275–6 (1965).
64. N. G. Basov, É. M. Belenov and V. S. Letokhov, *Sov. Phys. Tech. Phys.*, **10**, 845–50 (1965).
65. A. F. Glova, Yu. A. Dreizin, O. R. Kachurin, F. V. Lebedev and V. D. Pis'mennyĭ, *Sov. Tech. Phys. Lett.*, **11**, 102–3 (1985).
66. V. V. Antyukhov, A. F. Glova, O. R. Kachurin, F. V. Lebedev, V. V. Likhanskiĭ, A. P. Napartovich and V. D. Pis'mennyĭ, *JETP Lett.*, **44**, 78–81 (1986).
67. S. A. Darznek, M. M. Zverev and V. A. Ushakhin, *Sov. J. Quant. Electron.*, **4**, 1272–4 (1975).
68. J. R. Leger, M. L. Scott and W. B. Veldkamp, *Appl. Phys. Lett.*, **52**, 1771–3 (1988).
69. J. R. Leger and M. Holz, *LEOS'1988 Annual Meeting Proceedings*, Santa Clara, CA, 2–4 Nov. (IEEE/LEOS, Piscataway, NJ, 1988), pp. 468–71 (1988).
70. C. Roychoudhuri, E. Siebert, F. D'Amato, R. Noll, S. Macomber, E. Kintner and D. Zweig, *LEOS'1988 Annual Meeting Proceedings*, Santa Clara, CA. 2–4 Nov. (IEEE/LEOS, Piscataway, NJ, 1988), pp. 476–9 (1988).
71. F. X. D'Amato, E. T. Sievert and C. Roychoudhuri, *Appl. Phys. Lett.*, **55**, 816–18 (1989).
72. J. R. Leger and M. P. Griswold, *Appl. Phys. Lett.*, **56**, 4–6 (1990).
73. V. V. Likhanskii and A. P. Napartovich, *Sov. Phys. Usp.*, **33**, 228–52 (1990).
74. W. H. F. Talbot, *Philos. Mag.*, **9**, 401–7 (1836).
75. Lord Rayleigh, *Philos. Mag.*, **11**, 196–205 (1881).
76. A. A. Golubentsev, V. V. Likhanskiĭ and A. P. Napartovich, *Sov. Phys. JETP*, **66**, 676–82 (1987).
77. J. Z. Wilcox, W. W. Simmons, D. Botez, M. Jansen, L. J. Mawst, G. Peterson, T. J. Wilcox and J. J. Yang, *Appl. Phys. Lett.*, **54**, 1848–50 (1989).
78. R. G. Waarts, D. W. Nam, D. F. Welch, D. Mehuys, W. Cassarly, J. C. Ehlert, J. M. Finlan and K. M. Flood, *Proc. SPIE*, **1634**, 288–98 (1992).

79. L. J. Mawst, D. Botez, T. J. Roth, W. W. Simmons, G. Peterson, M. Jansen, J. Z. Wilcox and J. J. Yang, *Electron. Lett.*, **25**, 365–6 (1989).
80. P. D. Van Eijk, M. Reglat, G. Vassilieff, G. J. M. Krijnen, A. Driessen and A. J. Mouthaan, *J. Lightwave Technol.*, **9**, 629–34 (1991).
81. D. Mehuys, W. Streifer, R. G. Waarts and D. F. Welch, *Opt. Lett.*, **16**, 823–5 (1991).
82. J. R. Leger and G. Mowry, *Appl. Phys. Lett.*, **63**, 2884–6 (1993).
83. R. Waarts, D. Mehuys, D. Nam, D. Welch, W. Streifer and D. Scifres, *Appl. Phys. Lett.*, **58**, 2586–8 (1991).
84. D. Botez, *IEE Proc. J., Optoelectronics*, **139**, 14–23 (1992).
85. J. M. Finlan, S. M. Hamilton and J. R. Leger, *Proc. SPIE*, **1219**, 377 (1990).
86. W. J. Cassarly, J. C. Ehlert, S. H. Chakmakjian, D. Harnesberger, J. M. Finlan, K. M. Flood, R. W. Waarts, D. Nam and D. Welch, *Proc. SPIE*, **1634**, 299–309 (1992).
87. W. J. Cassarly, J. C. Ehlert, J. M. Finlan, K. M. Flood, R. W. Waarts, D. Mehuys, D. Nam and D. Welch, *Opt. Lett.*, **17**, 607–9 (1992).
88. M. Born and E. Wolf, *Principles of Optics*, Pergamon Press, Oxford, p. 424 (1980).
89. J. R. Leger and W. C. Goltsos, *IEEE J. Quantum Electron.*, **28**, 1088–1100 (1992).
90. D. Gloge, *Appl. Opt.*, **10**, 2252–8 (1971).
91. T. Y. Fan, A. Sanchez and W. E. DeFeo, *Opt. Lett.*, **14**, 1057–9 (1989).
92. J. Berger, D. F. Welch, W. Streifer, D. R. Scifres, N. J. Hoffman, J. J. Smith and D. Radecki, *Opt. Lett.*, **13**, 306–8 (1988).
93. Spectra Diode Labs., *1991 Product Catalog*, model SDL3450-P5.
94. E. Snitzer, H. Po, F. Hakimi, R. Tumminelli and B. C. McCollum, Double clad, offset core Nd fiber laser, *Optical Fiber Sensors '88*, PD5, New Orleans (1988).
95. H. Po *et al.*, *1989 Optical Fiber Communication Conf.*, Houston, Texas, PD-07 (1989).
96. T. Y. Fan and A. Sanchez, *IEEE J. Quantum Electron.*, **26**, 311–16 (1990).

4

Modeling of diode laser arrays

G. RONALD HADLEY

4.1 Introduction

Diode laser arrays were conceived and fabricated in the late 1970s as an attempt to overcome the inherent power limitations of single-aperture diode lasers. Increasing the output power by simply increasing the lateral aperture size from the usual $\approx 5\ \mu m$ to values of 50–100 μm had already been tried with disappointing results. These so-called 'broad-area' devices exhibited such poor modal characteristics that the power focusable into a diffraction-limited spot barely increased at all[1,2]. Arrays were thus seen as a means of increasing output power by phase-locking several diode lasers together so that they operated as a single (hopefully) diffraction-limited source. The last decade has seen a tremendous research effort directed towards the design and fabrication of high-power arrays exhibiting good mode control, with (as usual) mixed results. Today's multi-watt arrays certainly provide more focusable laser power than could be obtained by simply stacking together individual diode lasers. On the other hand, achieving the desired mode control and coherence from arrays has proved considerably more difficult than was originally envisioned. As a result, virtually all the high-power arrays commercially available at present emit their radiation into two broad far-field lobes instead of the desired single diffraction-limited lobe.

This situation has arisen because the first simple types of diode arrays that were easily understandable and relatively easy to fabricate have been shown to exhibit poor mode discrimination[3,4], so that even devices that lased in the desired in-phase mode near threshold became multi-mode at higher currents. The resulting multi-lateral mode operation has in turn resulted in emission into a wide range of angles that is many times the diffraction limit, and hence is difficult to focus. Thus, several more complex designs have been proposed and tested including Y-coupled arrays[5], diffraction-coupled arrays[6] and antiguided arrays[7], to name just a few.

Again, these attempts have met with only partial success. It is clear that commercially practical arrays that emit high-power diffraction-limited radiation are still to be achieved, and will require yet more novel (and probably complex) designs.

It is also clear that the role of the modeler in producing successful designs will continue to be a crucial one. Although the evaluation of certain kinds of devices may at present be limited more by difficulties of fabrication than by lack of understanding, still our capability of modeling some of the newer device types is certainly in need of improvement. For example, a comprehensive model of the newly conceived vertical-cavity surface-emitting arrays would involve the calculation of current and heat flow through multiple material layers, carrier diffusion in multiple quantum wells and a treatment of the waveguiding properties of a complicated three-dimensional structure. Such a model does not exist at present and, if it did, would certainly require significant computational resources to implement. There is thus ample room for the development of new modeling techniques if we are to succeed in designing high-power arrays emitting into a high-quality optical mode.

The purpose of this chapter, then, is to provide an overview of the modeling techniques that have been successfully utilized to describe array behavior, while at the same time indicating promising future directions. While the modeling of arrays is similar in many respects to the modeling of single-stripe lasers, we will limit our scope and concentrate our attention in this chapter specifically on arrays. We will also limit our treatment to include only quasi-steady-state behavior, i.e., we will treat only cw operation or pulsed operation with pulses whose length is many cavity round trip times. Although this choice results in the omission of a large and interesting body of material concerning transient array behavior, space restrictions do not allow its inclusion here.

Since the primary issue with arrays has long been lateral-mode selectivity, the emphasis in modeling activities has correspondingly been the calculation of the gain and shapes of various array modes, given some active medium with known properties. In accordance with this trend, the emphasis in this chapter will be on the calculation of array modes, with less attention being paid to self-consistency or the modeling of the gain medium. Since array-mode calculations have depended historically upon the use of several approximate techniques (such as coupled mode theory or the effective index method), one of the chief objectives of this chapter will be to place these methods in perspective and to give the reader some appreciation of their regimes of validity.

4.2 Edge-emitting arrays

Semiconductor lasers are fabricated by the epitaxial growth of multiple layers of various III–V compounds, interspersed with fabrication steps that may define waveguiding structures, current flow or cavity boundaries[8,9]. Variations of these growth and fabrication steps can then lead to a wide variety of device types. The first diode laser arrays to be successfully fabricated were of the 'edge-emitter' variety, so-called because the lasing cavity is defined by two cleaved facets, both normal to the growth direction, with lasing emission occurring from the edges of the crystal. For this type of laser, the emission area is small compared with the overall device area, in contrast with the more recently developed 'surface-emitting' lasers, in which emission occurs along the growth direction and the emission area is essentially equal to the device area.

A schematic diagram of a generic edge-emitting array is given in Figure 4.1. Current confinement and gain structuring are accomplished by patterning the upper metallization as shown, and waveguiding structures may be built into the device by using various fabrication techniques such as etching and regrowth. The resulting device may be modeled as three basic processes: (1) The device is pumped when electrons and holes flow from the contacts through the doped semiconductor materials to the active region (probably a quantum well or wells). Once there, the carriers

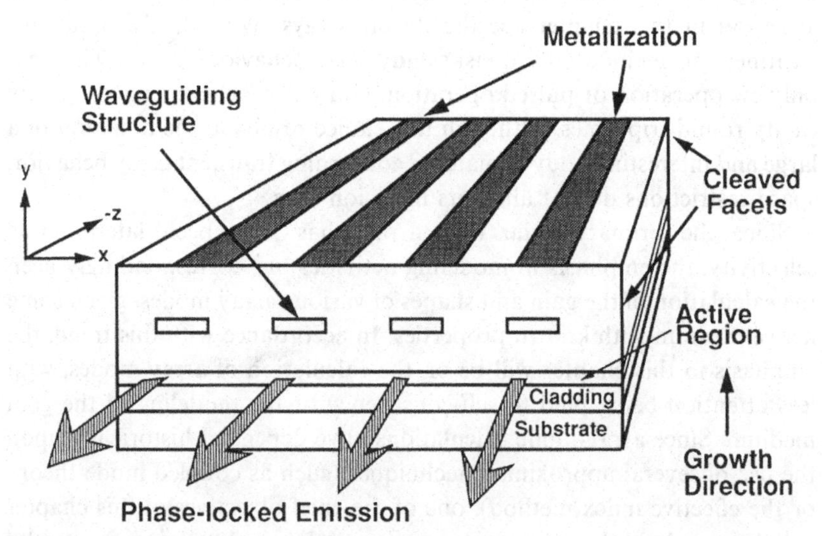

Figure 4.1. Schematic diagram of an edge-emitting diode laser array showing major components (not to scale).

diffuse laterally, combine non-radiatively to produce heat, and radiatively to provide optical gain for the laser. Their presence also lowers the index of refraction due to free-carrier and bandedge-shift effects. (2) Heat generated through both ohmic losses in the cladding layers and non-radiative recombination in the active region flows through these layers towards a heat sink. The resulting temperature rise in the active region changes the index of refraction and thus impacts waveguiding. (3) The electromagnetic fields associated with the cavity modes that are above threshold propagate back and forth between end facets, setting up a standing wave in the cavity and saturating the available gain. In a real device, these three processes couple together, and so a realistic model should be self-consistent, i.e., a calculation including all three should be iterated until convergence is obtained. However, considerable insights can be obtained about device behavior by assuming reasonable gain and index profiles and then solving for the optical modes individually. This is the approach that we will concentrate on, with a brief discussion of active cavity models and thermal effects included later in the section.

4.2.1 Waveguide modeling

Here we intend to solve for the dominant cavity modes of a laser cavity of the general type depicted in Figure 4.1. Starting with Maxwell's equations in the absence of free charge, we assume that the magnetic permeability μ is the vacuum value, and that the electric permittivity ε is constant in space and time. This latter assumption is valid within any single homogeneous material; boundaries between materials will be dealt with later. Assuming the harmonic time dependence $\exp(-i\omega t)$, both the electric and magnetic fields satisfy[10]

$$\nabla^2 \boldsymbol{F} = -\mu\varepsilon\omega^2 \boldsymbol{F}. \tag{4.1}$$

At this point we introduce the first major approximation by assuming that the primary spatial dependence of all field components is upon z (see Figure 4.1 for coordinate conventions) and is of the form

$$\boldsymbol{F} = \boldsymbol{f} \exp(ikz), \tag{4.2}$$

where the field $\boldsymbol{f}$ is taken to depend weakly upon z in a sense that will soon be made clear. Inserting (4.2) into (4.1) for the case of the magnetic intensity $\boldsymbol{H}$ results in

$$2ik\frac{\partial \boldsymbol{H}}{\partial z} + \frac{\partial^2 \boldsymbol{H}}{\partial z^2} + \nabla_\perp^2 \boldsymbol{H} = (k^2 - \mu\varepsilon\omega^2)\boldsymbol{H}. \tag{4.3}$$

We now make the above statement about the weak dependence of $\boldsymbol{H}$ upon z precise by neglecting its second derivative. This is the 'paraxial' approximation, so-called because the resulting equation will only describe fields propagating at small angles with respect to the z axis (the axis of propagation). The resulting equation is then

$$2\mathrm{i}k\frac{\partial \boldsymbol{H}}{\partial z} + \nabla_{\perp}^{2}\boldsymbol{H} = (k^{2} - \mu\varepsilon\omega^{2})\boldsymbol{H}. \tag{4.4}$$

Notice that a solution of Eq. (4.4) for H_x and H_y specifies all fields, since H_z and $\boldsymbol{E}$ may be determined from the Maxwell equations

$$\nabla \cdot \boldsymbol{B} = 0, \tag{4.5}$$

$$-\mathrm{i}\varepsilon\omega\boldsymbol{E} = \nabla \times \boldsymbol{H}. \tag{4.6}$$

Since Eq. (4.4) is only valid within a single material, application of the appropriate boundary conditions at dielectric boundaries is necessary to effect a solution of the overall problem. In order to simplify the discussion of boundary conditions, we assume that all dielectric boundaries making up the waveguide consist of planes that are normal to either the x axis (vertical boundaries) or the y axis (horizontal boundaries). This approximation is well satisfied even for z-dependent structures provided that the structure changes occur slowly. The resulting boundary conditions that apply may be derived directly from Maxwell's equations[11], and are listed in Figure 4.2. Because the structures of diode laser arrays are typically characterized by length scales that differ dramatically in the two transverse dimensions, application of the boundary conditions described in Figure 4.2 leads to solutions of Eq. (4.4) that tend to fall into two general categories: (1) TE modes in which H_y is the dominant component with $H_x \ll H_y$; and (2) TM modes in which H_x is the dominant component and $H_y \ll H_x$. However, the characteristics of these two modes are nearly identical because the relative variations in ε are usually very small, with the resulting derivative discontinuity predicted from the boundary conditions in Figure 4.2 being small as well. (Differences in gain leading to a preference for TE-mode operation result from subtle effects such as differences in facet reflectivity[12], or, for quantum-well materials, polarization-dependent transition selection rules[13].) Consequently, a scalar formulation for arrays, in which only one field component is included, makes good sense. Henceforth, we will employ such a formulation, so that the working equation for the dominant TE-mode component H_y

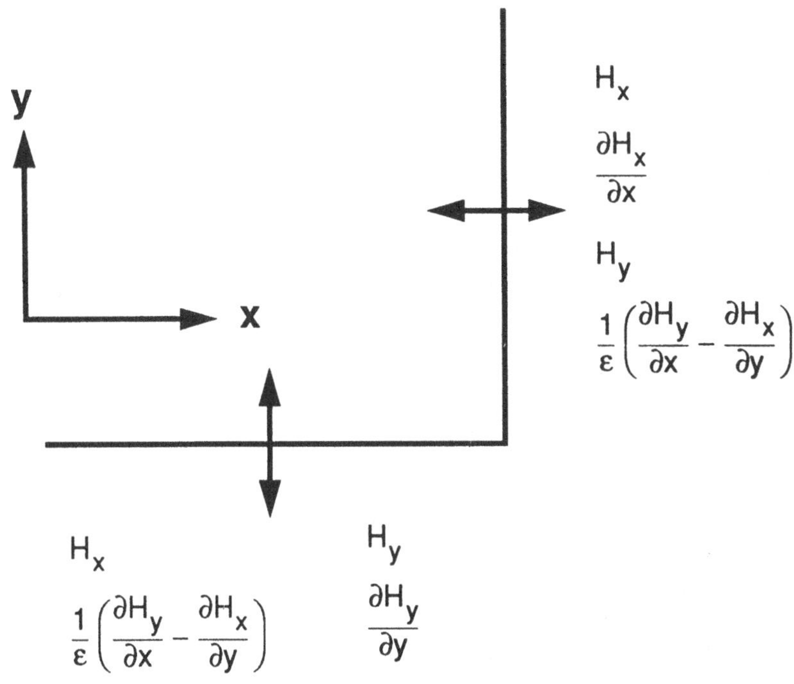

Figure 4.2. Boundary conditions on various components of the magnetic field for horizontal and vertical dielectric boundaries. Shown adjacent to each type of boundary are the quantities that are continuous across that boundary.

becomes

$$2ik\frac{\partial H}{\partial z} + \nabla_\perp^2 H = (k^2 - \mu\varepsilon\omega^2)H, \tag{4.7}$$

where we have dropped the subscript on the field for clarity of notation.

4.2.1.1 Effective index method techniques

EIM derivation. Later on we will discuss methods for solving Eq. (4.4) as it stands. First, however, we present an extremely popular method for reducing the dimensionality of this equation by one. Due to the aforementioned dissimilarity in length scales between the two transverse dimensions, one might imagine that the y dependence is somewhat insensitive to the presence of a lateral structure. In that case, H might be approximately separable, i.e.

$$H(x, y, z) \approx H(x, z)g(y). \tag{4.8}$$

Inserting Eq. (4.8) into Eq. (4.4) and dividing through by g gives

$$2ik\frac{\partial H}{\partial z} + \frac{\partial^2 H}{\partial x^2} = \left[k^2 - \mu\varepsilon\omega^2 - \frac{1}{g}\frac{d^2 g}{dy^2}\right]H. \tag{4.9}$$

Since all y-dependence in the above equation is contained within the square brackets, the quantity within must be at most a function of x. Defining the 'effective' index of refraction through the equation

$$g'' = k_0^2(n_{\mathrm{eff}}^2 - \tilde{\varepsilon})g, \tag{4.10}$$

where $\tilde{\varepsilon}$ is the relative permittivity (the permittivity divided by the vacuum value), allows Eq. (4.9) to be written in the useful form

$$2ik\frac{\partial H}{\partial z} + \frac{\partial^2 H}{\partial x^2} = k_0^2(\bar{n}^2 - n_{\mathrm{eff}}^2)H. \tag{4.11}$$

In the above equation we have defined the overall modal index $\bar{n}$ through the relation $k^2 = k_0^2\bar{n}^2$.

This formulation is known as the 'effective index method' or EIM[14], and has widespread application. In practice, Eq. (4.10) is solved for the effective index at each lateral position x that corresponds to a different vertical structure, resulting in a complete effective index profile. A typical example of the application of this method is illustrated for the simple buried waveguide structure in Figure 4.3(a). Equation (4.11) is then solved by any of several different means, some of which we will examine below. Strictly speaking, the validity of the EIM is dependent upon the degree to which H is separable. In fact, however, this method often gives qualitatively good results even when that assumption is clearly violated, an observation that probably accounts in part for the method's popularity.

Equation (4.11) is treated using one of several approaches depending on the application. If only the array eigenmodes are of interest, the first z-derivative term in Eq. (4.11) is dropped, and the resulting eigenvalue problem solved for the modal index $\bar{n}$. If, on the other hand, information is desired about the exchange of power between eigenmodes, then the first term is retained and an initial solution (usually made up of many eigenmodes) is advanced in z so as to model the propagation characteristics of the array. Propagation techniques may also be used to find the

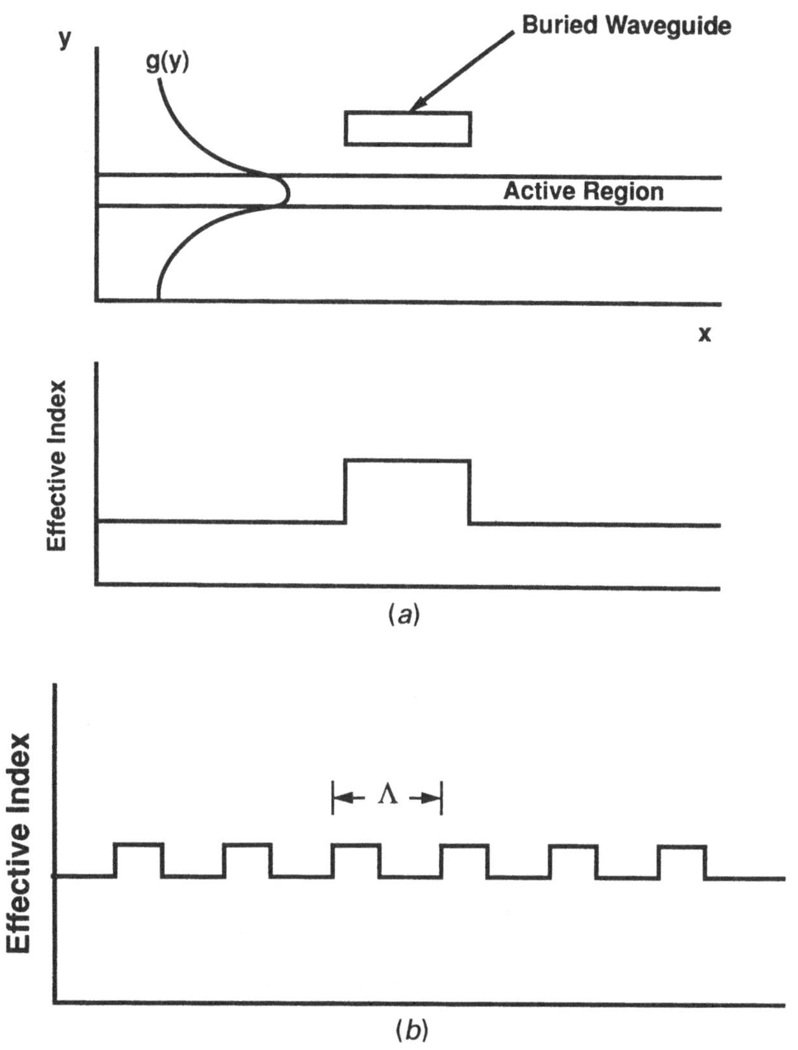

Figure 4.3. Pictorial illustration of the effective index method as it might apply to (*a*) a problem consisting of a single buried waveguide. The upper illustration shows the problem geometry and the field amplitude $g(y)$. The lower illustration shows the resulting effective index profile. For an index-guided array, the effective index profile might appear as in (*b*), with periodicity Λ.

highest gain eigenmode, provided that the modes are well-separated in gain. These methods will be discussed at some length later on in this section. In the meantime we will investigate two popular eigenmode solution techniques by dropping the z-derivative term in Eq. (4.11) and solving the resulting eigenvalue problem:

$$\frac{\partial^2 H}{\partial x^2} = k_0^2(\bar{n}^2 - n_{\text{eff}}^2)H. \tag{4.12}$$

Coupled mode theory. The first technique of eigenmode solution for arrays follows from an expansion of the desired solution in terms of modes of the individual array elements[15]. If an array is comprised of N individual weakly coupled waveguides, as shown in Figure 4.3(*b*), then we may write

$$H(x, z) = \sum_{j=1}^{N} a_j H_j(x), \tag{4.13}$$

where the single waveguide solution $H_j(x)$ satisfies Eq. (4.12) with the effective index restricted to only the contribution from waveguide j:

$$\frac{\partial^2 \mathrm{H}_j}{\partial x^2} = k_0^2(\bar{\varepsilon}_j - \varepsilon_j(x))\mathrm{H}_j. \tag{4.14}$$

In Eq. (4.14) we have introduced the modal dielectric constant $\bar{\varepsilon} = \bar{n}^2$ and the 'effective' dielectric constant $\varepsilon = n_{\text{eff}}^2$ in order to simplify notation. These dielectric constants are dimensionless and should not be confused with the previous usage of the same symbol in Eq. (4.9). Inserting Eq. (4.13) into Eq. (4.12) then results in

$$\sum_{j=1}^{N} (\bar{\varepsilon} - \bar{\varepsilon}_j - (\varepsilon(x) - \varepsilon_j(x)))a_j H_j(x) = 0. \tag{4.15}$$

The above equation is not easy to solve unless the solutions H_j are at least approximately orthogonal. However, if the individual waveguides are well separated (weakly coupled) then we expect their eigenmodes to barely overlap and we may multiply Eq. (4.15) by $H_i^*(x)$ and integrate over x. This results in[16]

$$\sum_{j=1}^{N} (s_{ij} - \kappa_{ij})a_j = 0, \tag{4.16}$$

where

$$s_{ij} = \int (\bar{\varepsilon} - \bar{\varepsilon}_j)H_i^*(x)H_j(x)\,\mathrm{d}x \tag{4.17}$$

and

$$\kappa_{ij} \equiv \int (\varepsilon(x) - \varepsilon_j(x))H_i^*(x)H_j(x)\,\mathrm{d}x. \tag{4.18}$$

As mentioned previously, the above formulation is most useful when the individual waveguide eigenfunctions are approximately orthogonal. In that case κ_{ij} has only off-diagonal elements since the two dielectric constant functions in Eq. (4.18) are identical in the region where the eigenfunctions have appreciable overlap; and s_{ij} has only diagonal elements. Since in this picture only nearest neighbor waveguides couple, the resulting matrix appearing in Eq. (4.16) will be tridiagonal. The eigenvalues are obtained from the requirement that the determinant of this matrix must be zero if non-zero coefficients a_j are to result.

As an example, we consider the simple case of three identical coupled waveguides with real dielectric constants. In this case $\kappa_{12} = \kappa_{21} = \kappa_{23} = \kappa_{32} = \kappa$ and the eigenvalues are determined from

$$\begin{vmatrix} \bar{\varepsilon} - \bar{\varepsilon}_1 & -\kappa & 0 \\ -\kappa & \bar{\varepsilon} - \bar{\varepsilon}_1 & -\kappa \\ 0 & -\kappa & \bar{\varepsilon} - \bar{\varepsilon}_1 \end{vmatrix} = 0, \tag{4.19}$$

with solutions

$$\bar{\varepsilon} = \bar{\varepsilon}_1,\ \bar{\varepsilon}_1 \pm \sqrt{(2)}\kappa. \tag{4.20}$$

These eigenvalues are the modal dielectric constants of the well-known 'array modes' of the three-waveguide system[17,18].

In addition to the present applications involving coupled waveguides, coupled mode theory has proved useful in a number of problems such as acoustooptic coupling and reflection from gratings. The method can also treat directional couplers and other problems involving z-dependent power transfer between modes by employing a different derivation similar to that given above but starting from the more general Eq. (4.11)[15]. The strength of coupled mode theory lies in the qualitative insights it provides by relating solutions of a large formidable problem to known solutions of a simpler problem. For all its advantages, however, it should be remembered that coupled mode theory, in its most common form, is still essentially a first-order perturbation theory. Extending it beyond this limit involves keeping other than nearest neighbor interactions in Eqs. (4.17) and (4.18), an approach whose difficulty in comparison with other more accurate methods probably destroys its utility. Consequently, the use of coupled mode theory for quantitative predictions requires caution unless weak coupling is assured.

Bloch function analysis. Another useful analytic technique makes use of the fact that, for arrays, the index (or gain) perturbation is periodic, as shown in Figure 4.3(*b*). Thus, we may borrow solution methods from the field of solid state physics, where the solution to the problem of an electron moving in a periodic potential is already well understood. In order to clarify the analysis, we will assume for the moment that the periodic perturbations in the refractive index (or gain) shown in Figure 4.3(*b*) can be rounded off and thus approximated by trig functions. (For the case of gain-guided arrays, this approximation is actually quite good, since carrier diffusion smooths out the abrupt changes in gain.) Then Eq. (4.12) may be written as

$$\frac{\partial^2 H}{\partial x^2} + k_0^2\left\{n_0^2 + \frac{\Delta n^2}{2}\left[\exp(2\pi i x/\Lambda) + \exp(-2\pi i x/\Lambda)\right]\right\}H = k_0^2\bar{n}^2 H, \quad (4.21)$$

where n_0 and Δn are considered to be complex. According to Bloch's theorem[19], the solution may be written in the form

$$\begin{aligned} H(x) = {} & A\exp(iqx)[1 + r\exp(2\pi i x/\Lambda)] \\ & + B\exp(-iqx)[1 + s\exp(-2\pi i x/\Lambda)], \end{aligned} \quad (4.22)$$

where the functions within brackets have the periodicity of the dielectric constant variation in Eq. (4.21). If we now insert the above form into Eq. (4.21) and equate coefficients of the exponentials in Eq. (4.22), expressions result that relate r, s, q and $\bar{n}$. Because of the symmetry incorporated into Eq. (4.22), a specification of the parity of the mode together with normalization determines the coefficients A and B. At this point we are still modeling infinite arrays. For an array with a finite number of stripes, the above solution must be matched (both value and derivative) with an appropriate exponential solution for the constant properties' region at each edge of the array. This condition then leads to a discrete set of eigenfunctions and eigenvalues $\bar{n}$.

The character of the resulting eigenmodes may be seen to depend on the value of the Bloch wave vector q. For example, when the latter is near zero, Eq. (4.22) describes the in-phase mode, which appears as a mostly uniphase, slowly varying envelope with perturbations of amplitude r or s at the stripe periodicity. Values of q near $-\pi/\Lambda$ correspond to the out-of-phase mode whose periodicity is half that of the stripes. For this mode, r and s are near -1 and the solution resembles a sinusoid that changes sign at each stripe location. This out-of-phase mode may also be viewed as resulting from a lateral wave vector that satisfies the Bragg condition for reflection from the grating produced by the index (or gain) perturbation. Thus, the sinusoid is the standing wave resulting from interference between the two reflected plane waves. The familiar twin-lobed far-field pattern emitted by gain-guided arrays merely reflects the propagation direction of these two plane waves.

With regard to its use in diode laser array modeling, the Bloch-function technique was first applied to gain-guided arrays[20], where the modulation of the real refractive index due to antiguiding was also included. (In this work, a different plane wave vector Γ was used that is related to the above variable q by $\Gamma = q + (\pi/\Lambda)$.) Results of this calculation for several eigenmodes of interest are shown in Figure 4.4, and compared with direct numerical solutions. As can be seen, this analytic technique is quite capable of accurately predicting the modes of a device whose interelement coupling is strong, in contrast to the coupled mode theory method described previously. This success is understandable once we realize that in this context the Bloch-function method takes on the character of a plane wave analysis. Thus, instead of viewing the device as a series of weakly coupled single-stripe lasers, we are effectively viewing it as a broad-area laser with gain and index perturbations[21,22]. Thus, Bloch-function analysis may in a certain sense be seen as a complementary technique to coupled mode theory. Unlike the latter, it does not depend on the validity of a weak coupling assumption. It is an approximation, however, since non-resonant terms resulting from the insertion of Eq. (4.22) into Eq. (4.21) are neglected. Thus, the accuracy of the method degrades as the number of stripes decreases.

In spite of its resemblance to a plane wave treatment, however, Bloch-function analysis is also applicable to strongly index-guided arrays, as shown by more recent analyses of arrays of antiguides[23,24]. For this application, additional higher spatial frequencies must be used in Eq. (4.22) in order to describe the eigenmodes of these devices that oscillate more rapidly than the stripe periodicity[24].

Analytic matching techniques. A more accurate method of solving Eq. (4.12) is available if the dielectric constant can be approximated as a piecewise constant function. In that case, within a region where $\varepsilon_{\text{eff}} = n_{\text{eff}}^2$ is constant, the solutions to Eq. (4.12) are just complex exponentials:

$$H(x) = A\exp(\lambda x) + B\exp(-\lambda x), \tag{4.23}$$

with

$$\lambda = k_0 \begin{cases} \mathrm{i}\sqrt{(\varepsilon_{\text{eff}} - \bar{\varepsilon})} & \bar{\varepsilon} < \varepsilon_{\text{eff}} \\ \sqrt{(\bar{\varepsilon} - \varepsilon_{\text{eff}})} & \bar{\varepsilon} > \varepsilon_{\text{eff}} \end{cases}. \tag{4.24}$$

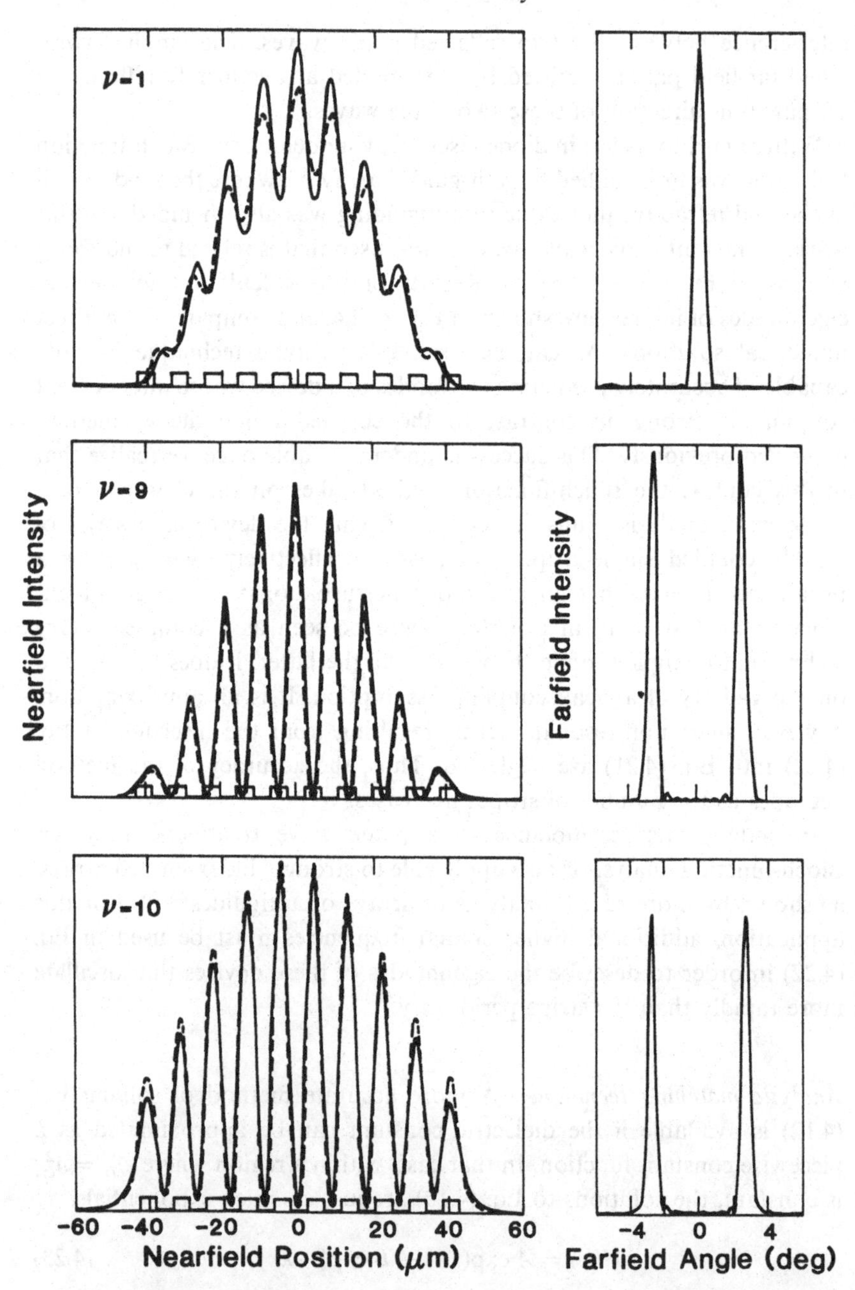

Figure 4.4. Several eigenmodes of interest for a ten-stripe gain-guided array, as determined from Bloch-function analysis (dashed lines) and direct numerical simulation (solid lines). (From D. Mehuys and A. Yariv, *Optics Lett.*, **13**, 571, 1988.)

The requirement of continuity of both $\boldsymbol{H}$ and its derivative at each region interface then provides a matrix linking the corresponding pairs of coefficients A and B. These matrices may in turn be assembled into a large matrix multiplying a column vector made up of all the layer coefficients. Setting the determinant of this matrix equal to zero provides the desired eigenvalues. Although this procedure is straightforward in theory, problems can arise in practice due to some of the exponential functions becoming too large. A numerical procedure that avoids this problem is described in detail in the Appendix, p. 222.

The principal advantage of analytic matching over coupled mode theory is that a weak coupling assumption is not required. Consequently, the only assumption that must be justified for its use (other than those leading up to Eq. (4.12)) is that of piecewise constant dielectric constants. This assumption is not overly restrictive for problems relating to diode laser arrays, since these devices are usually constructed of discrete regions with constant properties. In fact, this technique is quite useful for solving Eq. (4.10) to provide the effective index profile, as well as the final solution of Eq. (4.12) for the array eigenmodes[25].

Beam propagation – Fox–Li methods. For non-self-consistent EIM modeling of the optical modes of a diode laser array, the methods already described are quite adequate. If, however, a more complete model is desired that includes effects such as gain saturation by several modes, together with longitudinal dependence, eigenmode techniques are no longer adequate. For such models, the solution of Eq. (4.11) by beam propagation algorithms is preferable, as it allows a longitudinally dependent mode shape to change with varying gain or index profiles. Multimode operation may also be simulated by propagating several fields simultaneously.

The use of beam propagation, together with reflection between facets, was first introduced by Fox and Li[26] as a technique for determining the eigenmodes of a gas laser cavity; thus it is sometimes referred to by the name 'Fox–Li iteration'. With this technique, an arbitrary initial field (which possesses the symmetry of the desired solution but is generally composed of many eigenmodes) is propagated back and forth between facets until a steady state is reached. So long as a non-uniform gain profile is employed, modal gain discrimination will be expected to favor one mode over the others, and eventually the solution will settle down to this highest-gain mode.

The primary advantage of this technique is that, for self-consistent problems, the shape of the highest-gain mode is allowed to evolve as the gain and index profiles change. The governing equation for beam propagation is a rearranged version of Eq. (4.11):

$$\frac{\partial H}{\partial z} = \frac{\mathrm{i}}{2k}\left[\frac{\partial^2 H}{\partial x^2} + k_0^2(\varepsilon_{\mathrm{eff}}(x, z) - \bar{\varepsilon})H\right]. \tag{4.25}$$

Two primary techniques are commonly used for solving Eq. (4.25). The first is a straightforward finite difference technique. If we denote the finite difference analog of the second spatial derivative operator by

$$D_x H_i \equiv H_{i+1} + H_{i-1} - 2H_i, \tag{4.26}$$

then the finite-differenced form of Eq. (4.25) may be written

$$H_i^{n+1} - H_i^n = \frac{ik_0 \Delta z}{4\bar{n}} \xi_i^{n+1/2}(H_i^n + H_i^{n+1}) + \frac{\mathrm{i}\Delta z}{4k\Delta x^2} D_x(H_i^n + H_i^{n+1}), \tag{4.27}$$

where

$$\xi_i^n \equiv \varepsilon_{\mathrm{eff}}(x_i, z^n) - \bar{\varepsilon}, \tag{4.28}$$

and where subscripts refer to the x-direction, and superscripts to the propagation plane (z-direction). Eq. (4.27) is easily seen to be tridiagonal in the values of H at the new $(n + 1)$ propagation plane, and may consequently be solved quickly using the well-known Thomas algorithm[27]. This particular algorithm is unconditionally stable regardless of the size of the propagation step used. It is also unitary if $\varepsilon_{\mathrm{eff}}$ is real, i.e., energy is exactly conserved for all Fourier components.

The second technique[28] involves the use of the Fast Fourier Transform, more commonly known as the FFT. We start with a formal 'operator' solution of Eq. (4.25):

$$H_i^{n+1} = \exp[(\mathrm{i}\Delta z/2k\Delta x^2)D_x + N]H^n, \tag{4.29}$$

where the exponentiated operator expression is defined in terms of a power series, and

$$N \equiv \frac{ik_0}{2\bar{n}} \int_{z^n}^{z^{n+1}} (\varepsilon_{\mathrm{eff}}(x, z) - \bar{\varepsilon})\,\mathrm{d}z \approx \frac{ik_0 \Delta z}{2\bar{n}} (\varepsilon_{\mathrm{eff}}(x, z^{n+1/2}) - \bar{\varepsilon}). \tag{4.30}$$

Since D_x and N do not commute in general, an exact evaluation of Eq. (4.29) is not possible. Instead, the so-called 'gain sheet' approximation[28] is employed, in which the actions of the two non-commuting operators are split into separate steps:

$$H^{n+1} \approx \exp[(\mathrm{i}\Delta z/4k\Delta x^2)D_x] \exp(N) \exp[(\mathrm{i}\Delta z/4k\Delta x^2)D_x]H^n. \tag{4.31}$$

The first and third exponential operators that describe free-space diffraction are evaluated via the approximate expression

$$\exp[(\mathrm{i}\Delta z/4k\Delta x^2)D_x] \approx \frac{1 + (\mathrm{i}\Delta z/8k\Delta x^2)D_x}{1 - (\mathrm{i}\Delta z/8k\Delta x^2)D_x}, \tag{4.32}$$

resulting in the algorithm

$$\left(1 - \frac{\mathrm{i}\Delta z}{8k\Delta x^2} D_x\right)H_i^{n+1} = \left(1 + \frac{\mathrm{i}\Delta z}{8k\Delta x^2} D_x\right)H_i^n. \tag{4.33}$$

This equation is solved efficiently by expressing H in terms of its discrete Fourier transform:

$$H_k^n = \sum_{j=1}^{M} a_j^n \exp[(2\pi\mathrm{i}/M)(j-1)(k-1)]. \tag{4.34}$$

Inserting Eq. (4.34) into Eq. (4.33) and using the completeness relation

$$\sum_{k=1}^{M} \exp[(2\pi\mathrm{i}/M)(n-m)(k-1)] = M\delta_{mn} \tag{4.35}$$

results in the simplified propagation step in Fourier space:

$$\frac{a_j^{n+1}}{a_j^n} = \frac{1-(\mathrm{i}\Delta z/2k\Delta x^2)\sin^2[\pi(j-1)/M]}{1 + (\mathrm{i}\Delta z/2k\Delta x^2)\sin^2[\pi(j-1)/M]}. \tag{4.36}$$

The algorithm thus proceeds by transforming the initial field, multiplying the transform by the vector given in Eq. (4.36), and finally performing an inverse transform. These transforms can be performed with impressive speed, particularly if the number of mesh points M is an exact power of two, due to a special algorithm first developed by Hockney[29]. After the free-space propagation step, the middle step denoted in Eq. (4.31), namely multiplication by a vector, is performed. This algorithm is also unitary, or energy conserving, for real dielectric constants, because both the free-space FFT propagation and the vector multiplication step are unitary. Although the appearance of Eq. (4.31) is that of a three-step algorithm, the free-space propagation portion for back-to-back steps may be combined into a single propagation step, yielding in effect a two-step algorithm.

It is worth noting that the FFT approach is inherently periodic in nature, and thus the boundary conditions must also be periodic. The expression given in Eq. (4.36) was derived assuming zero Dirichlet boundary conditions at each end. More complicated boundary conditions may also be employed, but *the same boundary condition must be used at each problem boundary*! This stands in sharp contrast to the tridiagonal solution approach described earlier, in which different boundary conditions can be used at the two end points. In general, the FFT method is most suitable for problems with simple boundary conditions.

Although the primary use of the beam propagation method has been to find the highest gain mode, the addition of eigenvector solution techniques such as the Prony method[30] also allows the computation of lower gain modes. The Prony method makes use of dot products between computed fields after consecutive round trips to extract information regarding prominent eigenvectors. This technique has been employed to good advantage in the study of higher-order modes of gain-guided laser arrays[31]. Figure 4.5 shows a comparison between theoretical and

Experimental $\nu = 13$ Mode

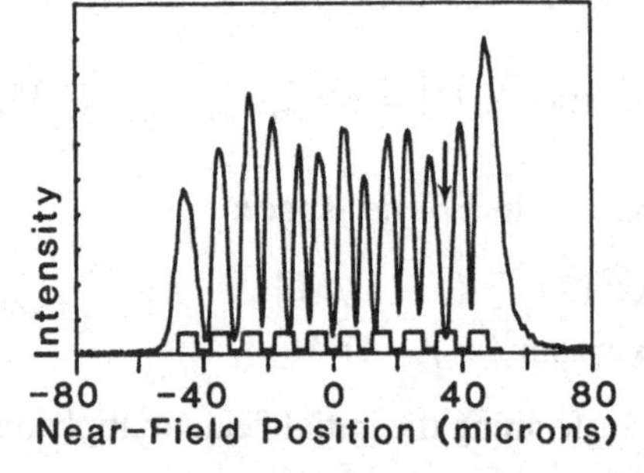

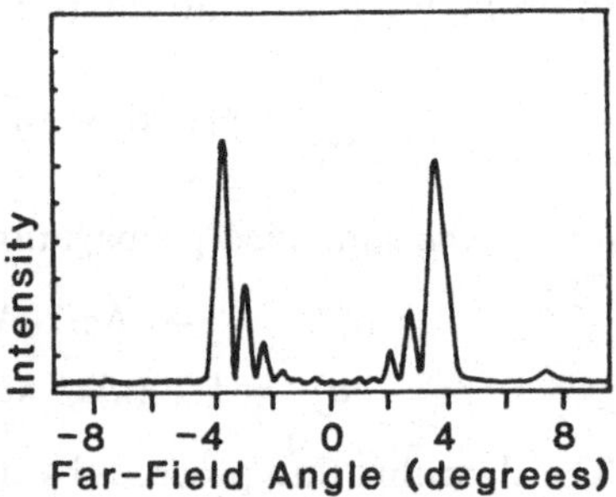

Theoretical $\nu = 13$ Mode with $\Delta T = 10$ K

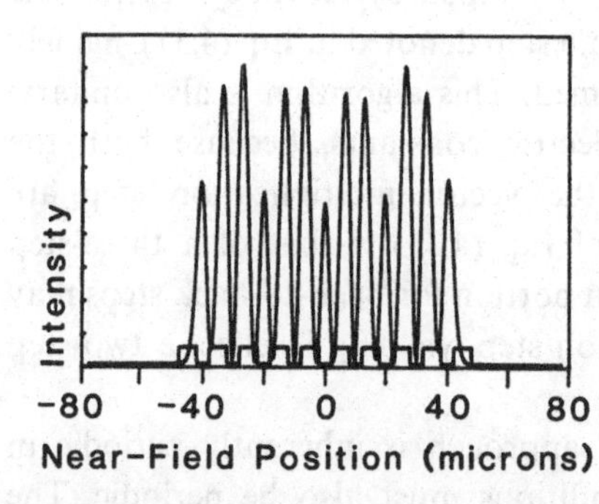

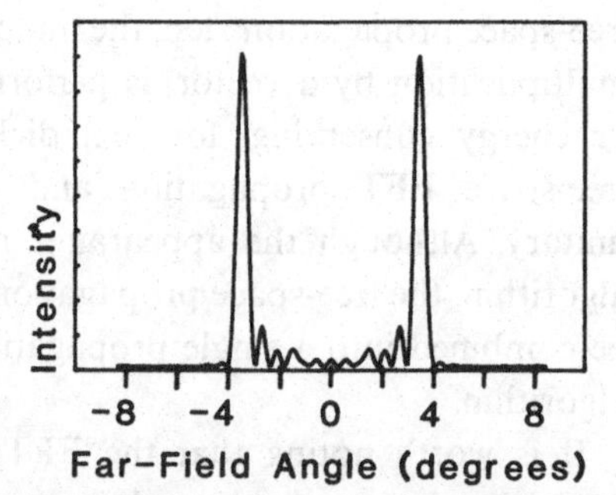

Figure 4.5. Near- and far-field intensity patterns of the 13th order mode of a ten-stripe gain-guided array as determined from experimental measurements (upper curves) and calculation, utilizing the beam propagation method (lower curves). (From Ref. 31, © 1987 IEEE.)

experimental near-field and far-field intensity profiles for the 13th order mode of a ten-stripe gain-guided array. These calculations provided some of the earliest insights into the character of higher-order modes in gain-guided arrays, including particularly the importance of including thermal effects in device modeling.

4.2.1.2 Two-dimensional models

Despite the immense popularity of the effective index method, cases arise where the approximate separability described in Eq. (4.8) is not valid. When this situation occurs, Eq. (4.7) must be solved in its entirety, keeping derivative terms in both x and y. The solution of this equation is in general numerically intensive, and we will discuss several approximate methods designed to minimize the numerical effort involved, as well as one direct method.

Two-dimensional analytic matching. This method is similar in spirit to the method of the same name described previously for solution of the simpler, one-dimensional Eq. (4.12). As before, we set the z-derivative term in Eq. (4.7) equal to zero in order to solve for the eigenmodes of some two-dimensional structure, resulting in the equation

$$\nabla_\perp^2 H = k_0^2(\bar{\varepsilon} - \varepsilon(x, y))H. \tag{4.37}$$

Next we decompose the more general structure into a number of adjacent slabs in which solutions of Eq. (4.37) may be obtained with minimum effort, as illustrated in Figure 4.6. The solutions of the two-dimensional equation may then be constructed from appropriate sums of the one-dimensional solutions by matching the field and its derivative at each slab boundary[32]. Each slab is assumed to be made up of a series of rectangles in each of which all properties are constant. We then assume that H is separable *within each slab*:

$$H_j(x, y, z) = \alpha_j(x)\beta_j(y). \tag{4.38}$$

Inserting this relationship into Eq. (4.37) results in

$$\frac{\alpha_j''}{\alpha} + \frac{\beta_j''}{\beta} = k_0^2(\bar{\varepsilon} - \varepsilon_j(y)) \tag{4.39}$$

for the jth slab. Since α depends only upon x, and β depends only upon y, Eq. (4.39) may be written as the two separate equations

$$\frac{\mathrm{d}^2\beta_j^i}{\mathrm{d}y^2} = k_0^2(\delta_j^i - \varepsilon_j)\beta_j^i \tag{4.40}$$

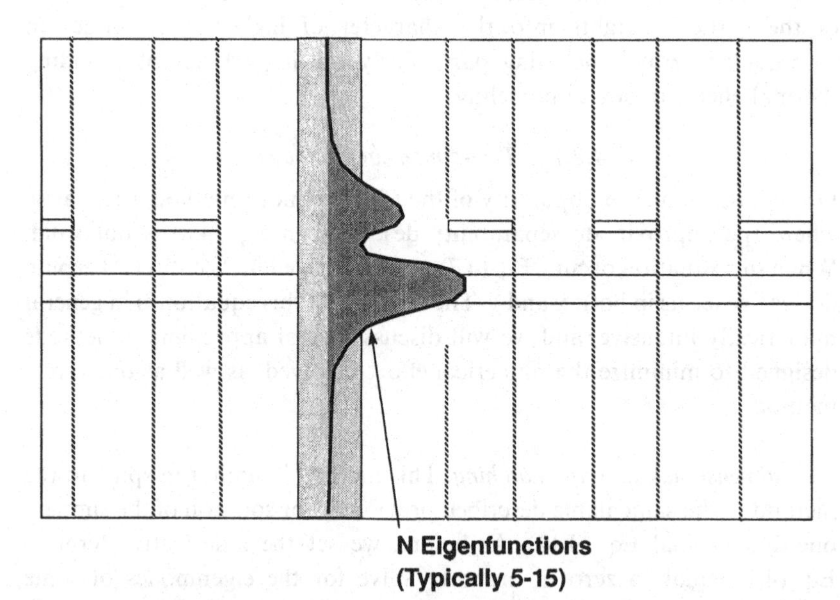

Figure 4.6. Schematic illustrating the 2-D analytic matching technique as it might apply to an index-guided array consisting of buried waveguides.

for the ith eigenvector β_j^i, and

$$\frac{d^2\alpha_j^i}{dx^2} = k_0^2(\bar{\varepsilon} - \delta_j^i)\alpha_j^i \tag{4.41}$$

for the x dependence in slab j due to the ith eigenvector. Note that the above equations are identical to the formulas for the effective index method provided that β_j^i is approximately the same for all slabs. In that case, only a single eigenmode is used (usually the fundamental waveguide mode), and δ_j^i is the square of the effective index for the chosen mode.

Having solved Eq. (4.40) for the first N eigenmodes, we then write the field in slab j as

$$H_j(x, y) = \sum_{i=1}^{N} [a_j^i \exp(h_j^i x) + b_j^i \exp(-h_j^i x)]\beta_j^i, \tag{4.42}$$

where

$$h_j^i = k_0^2\sqrt{(\bar{\varepsilon} - \delta_j^i)}. \tag{4.43}$$

For TE-like modes, the matching conditions at the interface between slabs j and $j + 1$ (see Figure 4.2) are

$$H_j(x_j, y) = H_{j+1}(x_j, y). \tag{4.44}$$

$$\frac{\partial H_j}{\partial x}(x_j, y) = \frac{\varepsilon_j(y)}{\varepsilon_{j+1}(y)} \frac{\partial H_{j+1}}{\partial x}(x_j, y). \tag{4.45}$$

At this point Eq. (4.42) is substituted into the above matching equations, both resulting equations are multiplied by β_{j+1}^{k*} and integrated over y. This procedure results in the matrix equation

$$\begin{pmatrix} a_{j+1}^i \\ b_{j+1}^i \end{pmatrix} = \boldsymbol{R}_j \begin{pmatrix} a_j^i \exp(h_j^i x_j) \\ b_j^i \end{pmatrix}, \tag{4.46}$$

where

$$(R_j)_{ki} = \begin{pmatrix} (B_j)_{ki} & (B_j)_{ki} \\ \dfrac{h_{j+1}^i}{k_0}(A_j)_{ki} & -\dfrac{h_{j+1}^i}{k_0}(A_j)_{ki} \end{pmatrix}^{-1} \begin{pmatrix} (C_j)_{ki} & (C_j)_{ki}\exp(-h_j^i x_j) \\ \dfrac{h_j^i}{k_0}(C_j)_{ki} & \dfrac{-h_j^i}{k_0}(C_j)_{ki}\exp(-h_j^i x_j) \end{pmatrix}. \tag{4.47}$$

In arriving at Eq. (4.47) we have redefined $x = 0$ to be the left-hand boundary of each slab, just as was done for the one-dimensional analytic matching technique (see Appendix). The auxilliary matrices in Eq. (4.47) are defined by

$$(A_j)_{ki} \equiv \int \frac{\varepsilon_j(y)}{\varepsilon_{j+1}(y)} \beta_{j+1}^{k*} \beta_{j+1}^i \, \mathrm{d}y. \tag{4.48}$$

$$(B_j)_{ki} \equiv \int \beta_{j+1}^{k*} \beta_{j+1}^i \, \mathrm{d}y. \tag{4.49}$$

$$(C_j)_{ki} \equiv \int \beta_{j+1}^{k*} \beta_j^i \, \mathrm{d}y. \tag{4.50}$$

In a similar manner to the one-dimensional case, a large matrix equation may be constructed from the several interface equations (Eq. (4.46)) and the modal dielectric constants (eigenvalues) for each mode found by setting the determinant of the resulting matrix equal to zero. Also needed to construct the overall matrix are boundary conditions at the left and right boundaries. Once again, factoring out the exponentials $\exp(hx)$ that have positive real parts keeps the overall matrix well-conditioned for problems in which thick outer slabs contain solutions of exponential character.

The 2-D analytic matching procedure just described is similar in many respects to another technique commonly known as the 'method of lines'[33].

In this latter technique, the problem is discretized in one lateral dimension (say the y direction), resulting in a differential equation in the remaining variable that permits an analytic solution provided that the dielectric constant depends only upon that variable. Thus this method also assumes slab-like geometries in which ε is piecewise constant in a direction normal to the slab boundaries. There are, however, two major differences between these seemingly similar methods:

(1) Due to the discretization along the slab length, the number of eigenvectors employed in the expansion for the method of lines is always equal to the number of mesh points used. Therefore, the matching conditions between slabs are exact at each mesh point. By contrast, the technique described above typically utilizes fewer eigenvectors, and the matching is accomplished only in an averaged sense (thus the integrals in Eqs. (4.48)–(4.50)). Thus, the matching conditions are more accurate using the method of lines.
(2) Because the method of lines utilizes a finite difference solution in one dimension, its accuracy is limited by the mesh size. This discretization error can be considerable, with certain problems in which the field peaks sharply in a material layer. The 2-D matching technique, however, is analytic in both lateral dimensions, and so no discretization error is incurred.

This analytic matching technique has been employed to good advantage in the study of antiguided (leaky-mode) arrays[34], for which the use of the effective index method is not strictly justifiable. This kind of array is characterized by waveguide elements (such as the buried waveguides depicted in Figure 4.1) having high indexes of refraction, thus leading to large 'effective' index steps in comparison with conventional index-guided arrays. Although these devices often include complicated geometries, Figure 4.7(*a*) shows a 'rectangularized' version in which boundaries between different materials have been changed so as to be either horizontal or vertical. Gain and loss were introduced by adding an imaginary part to the dielectric constant in the active region and in the lossy GaAs layers, thus allowing a prediction of modal gain discrimination. Radiation boundary conditions were imposed at the right boundary (necessary for 'leaky' modes) by setting the coefficients in Eq. (4.42) equal to zero if the corresponding complex exponent was found to have a negative imaginary part. The resulting simplified structure was then analyzed[34] using 2-D matching. Figure 4.7(*b*) shows intensity contours for the lowest-order leaky mode of the structure for $\varepsilon = 12.2$, using 15 eigenmodes in each

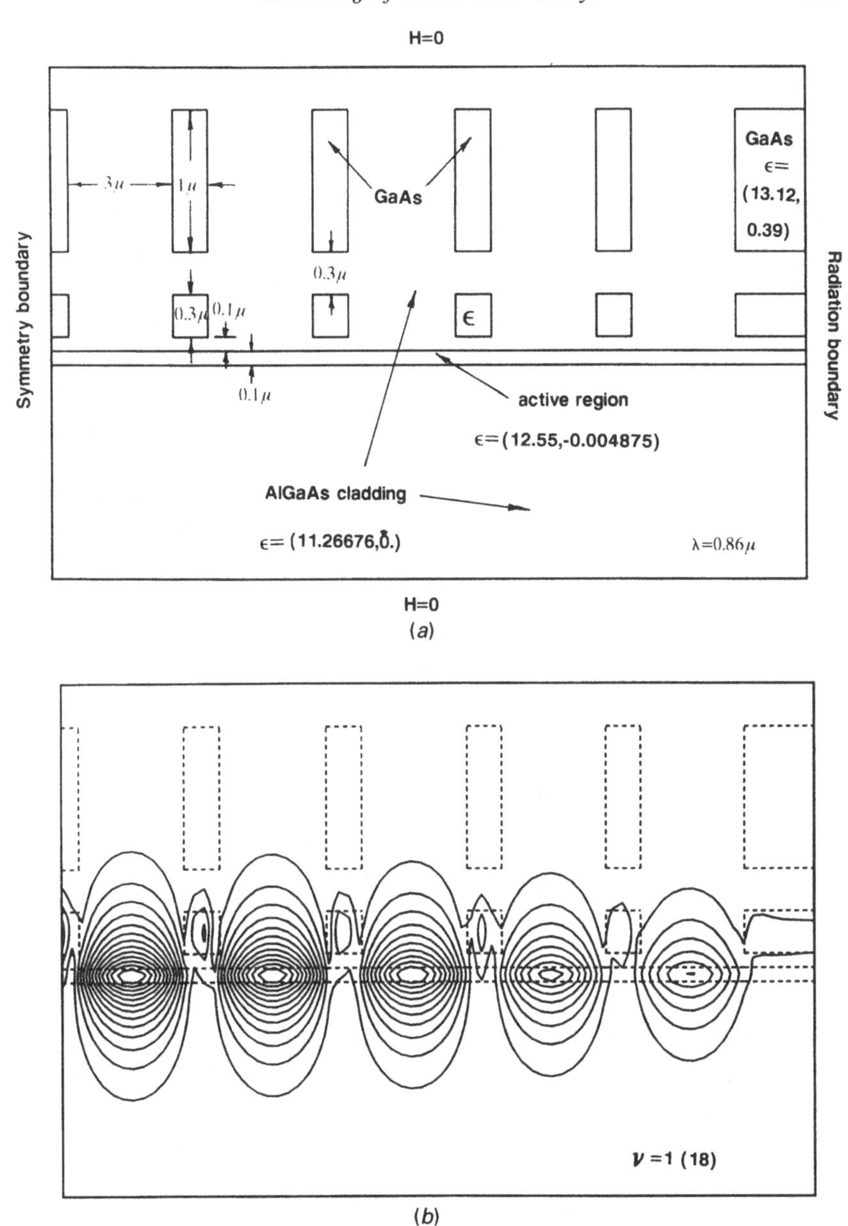

Figure 4.7. Sample problem for the use of 2-D analytic matching. (*a*) Geometry and dielectric constant values for the right half of a ten-channel leaky-mode array. The dielectric constant boundaries have been steepened to make the problem rectangular. The dielectric constant of the buried waveguides is considered variable. (*b*) Resulting near-field intensity contours of the lowest-order in-phase mode for $\varepsilon = 12.2$. (From Ref. 34, © 1991 IEEE.)

vertical slab. This calculation was performed in about 45 seconds on an IBM RS/6000 workstation.

Triangular mesh finite difference method. The preceding technique is limited to structures whose material boundaries are horizontal or vertical lines. A more general (but more computationally intensive) method[34] results from directly discretizing Eq. (4.37) on a non-regular mesh that can conform to fit boundaries of arbitrary shape. Originally developed to solve Poisson's equation for the fields present in accelerator magnets[35], this differencing scheme was recently adapted for the solution of the Helmholtz equation.

In order to have a well-defined solution algorithm, it is important that the mesh be regular topologically, even though it may be non-uniform spatially. Consequently, we construct a specialized triangular mesh[36] in which each interior mesh point is surrounded by exactly six neighboring points, as illustrated in Figure 4.8. If we assume the field to vary linearly over each triangle, then we may integrate Eq. (4.37) over the region bounded by the dotted line in Figure 4.8, resulting (for points not on a

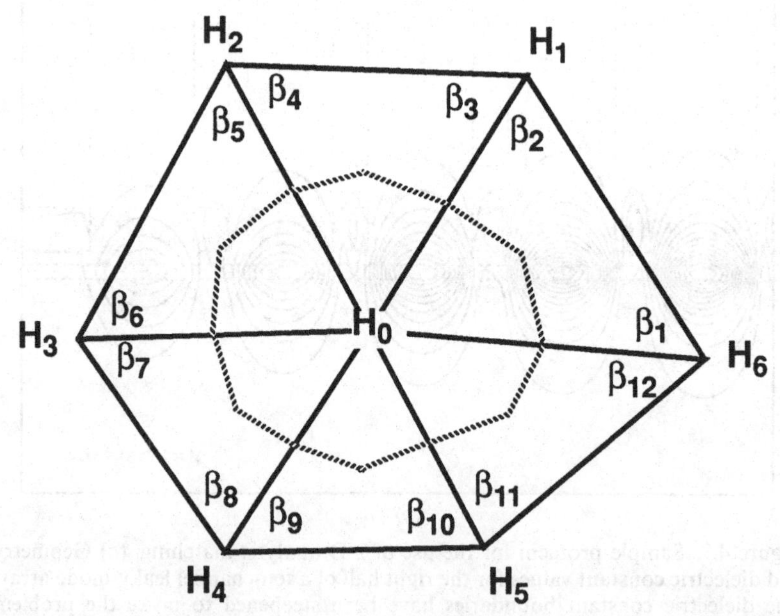

Figure 4.8. Mesh topology for the triangular mesh finite difference method. The dashed lines joining triangle centroids with segment midpoints enclose the region of integration.

dielectric boundary) in the difference equation

$$H_0 \sum_{i=1}^{6} \omega_i - \sum_{i=1}^{6} \omega_i H_i = \frac{k_0^2}{3}(\varepsilon - \bar{\varepsilon}) A H_0, \tag{4.51}$$

where ε is the dielectric constant at the center mesh point, A is the area of integration, and the coupling factors ω are given by

$$\left.\begin{aligned} \omega_1 &\equiv \tfrac{1}{2}(\cot \beta_1 + \cot \beta_4) \\ \omega_2 &\equiv \tfrac{1}{2}(\cot \beta_3 + \cot \beta_6) \\ &\vdots \\ \omega_6 &\equiv \tfrac{1}{2}(\cot \beta_2 + \cot \beta_{11}) \end{aligned}\right\} \tag{4.52}$$

and depend only upon the triangle shapes.

For the case where the central mesh point is on a dielectric boundary, a similar difference equation may be derived starting from the more general Maxwell equation

$$\nabla_\perp^2 \boldsymbol{H} - \frac{1}{\varepsilon}\frac{\partial \varepsilon}{\partial x}\frac{\partial \boldsymbol{H}}{\partial \boldsymbol{x}} = k_0^2(\bar{\varepsilon} - \varepsilon)\boldsymbol{H}, \tag{4.53}$$

which is valid so long as $\boldsymbol{H}$ has only a y component, and ε is independent of z. Note that the derivative of the dielectric constant is non-zero (a delta-function) only on the boundary, and we average $\partial \boldsymbol{H}/\partial x$ over the two adjacent triangles.

With the inclusion of appropriate boundary conditions, the above difference equations may be expressed as a banded matrix equation with a bandwidth of approximately twice the number of mesh points in the x or y direction (whichever is smaller). As before, the modal dielectric constant is determined by finding the zeros of the determinant of this matrix, with the latter being calculated by commercially available band matrix routines. The solution of a specific problem also necessitates determination of the mesh to be used. This is accomplished by first assigning coordinates for the material boundaries on a grid of regular equilateral triangles. A mesh generator then 'relaxes' the mesh in some optimum way to ensure gradual changes in mesh size throughout the problem region. More detail concerning the mesh generator and the triangle algebra may be found in the original paper by Winslow[36].

As an example of the ability of this formalism to solve for the optical modes of very complex structures, we revisit the leaky-mode array problem discussed above, but this time without the rectangularization needed in order to accommodate the 2-D matching technique[34]. The

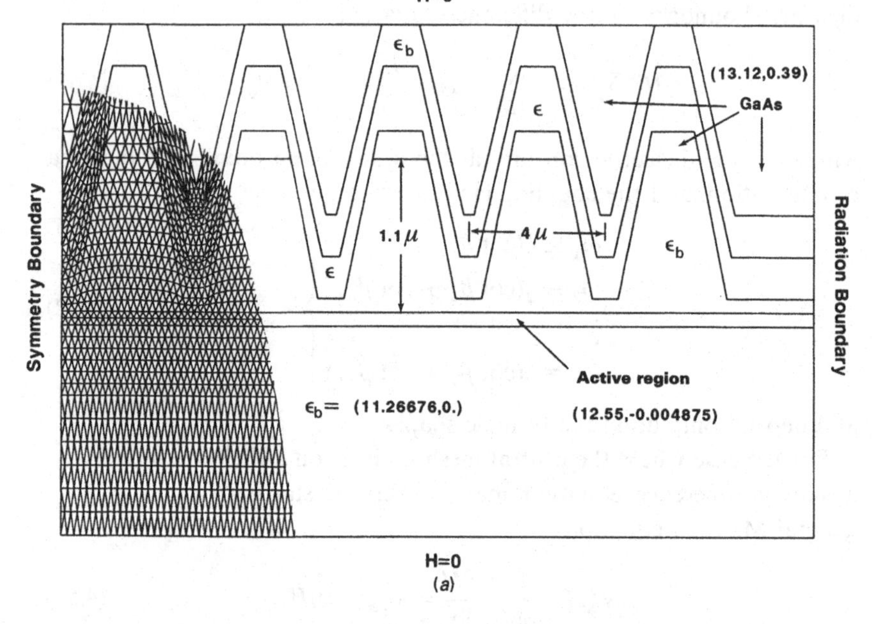

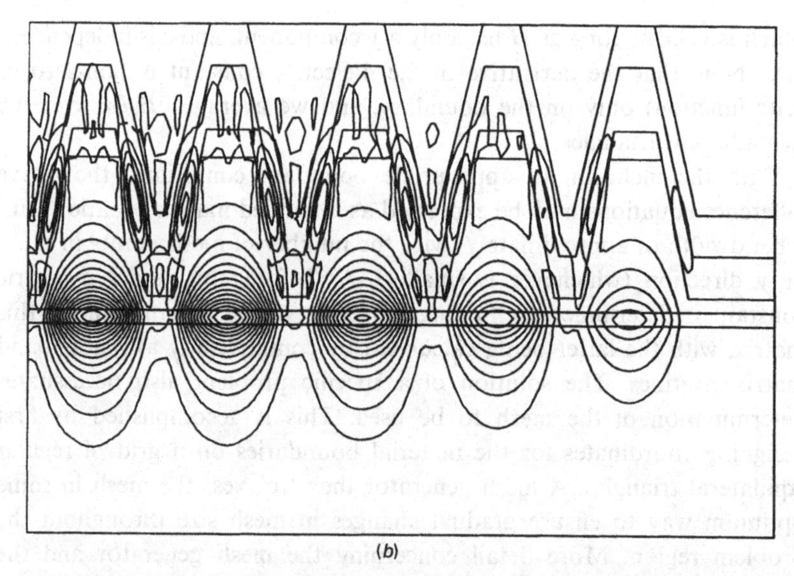

Figure 4.9. Sample problem illustrating the use of the triangular mesh finite difference method. (*a*) Geometry and dielectric constants for the right half of a ten-channel leaky-mode array as determined from SEM photos. At the left is shown a portion of the mesh used for this problem. (*b*) Resulting near-field intensity contours of the lowest-order in-phase mode for $\varepsilon = 12.11$. (From Ref. 34, © 1991 IEEE.)

resulting device structure, together with a portion of the output from the triangular mesh generator, is shown in Figure 4.9(*a*). A solution for the same fundamental mode as shown in Figure 4.7(*b*) (but for a different value of waveguide dielectric constant ε, i.e. 12.11) is shown in Figure 4.9(*b*). The obvious complexity of this mode and the resulting sharp resonances in the modal gain curves[34] could not be reproduced using the simpler rectangular model.

4.2.2 Active cavity models

The role of an active cavity model is to simulate the flow of carriers and heat throughout the device, ultimately resulting in a prediction of the complex dielectric constant to be used as input to the waveguide calculation. Of course, the optical field is itself a source or sink for both carriers and heat, so that, ideally, a self-consistent formulation is required[28,37]. This self-consistency is usually achieved by underrelaxed iteration, in which a new optical field causes changes in the carrier and temperature distribution, which then give rise to a new field, etc. Although these effects are actually three-dimensional in nature, they may be treated as two-dimensional and merely solved repetitively along various planes normal to the axial (z) direction for most devices whose structures change slowly in z. This approximation ignores, for example, both current spreading and carrier diffusion in the z direction.

4.2.2.1 Current spreading

Virtually all diode laser arrays employ either patterned metallization, ion bombardment or current blocking layers to restrict current flow to certain desired areas on the growth surface (the epitaxial side) of the wafer, as shown in Figure 4.10. However, the current density reaching the active region is still strongly affected by current spreading, and so Laplace's equation

$$\frac{\partial^2 \phi}{\partial x^2} + \frac{\partial^2 \phi}{\partial y^2} = 0 \tag{4.54}$$

must be solved for the potential in the cladding layer. Because the substrate metallization is so far ($\approx 100\ \mu\text{m}$) from the active region, patterning the substrate would be ineffective due to current spreading. Also, the low resistivity of the n-doped substrate and the high electron mobility imply that the current density reaching the active region is primarily determined by the hole transport in the upper cladding.

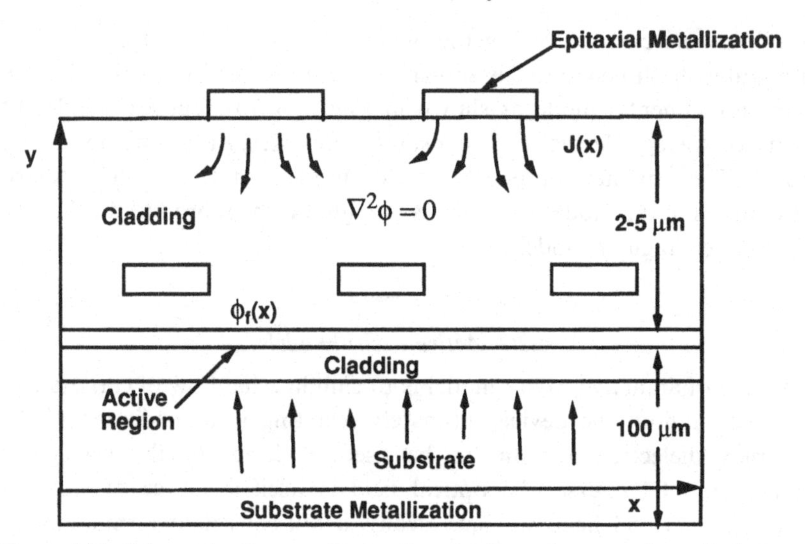

Figure 4.10. Schematic diagram illustrating the flow of current in a diode laser array (not to scale). Laplace's equation is solved for the potential in the upper cladding with current density fixed at the upper boundary and known potential at the lower boundary.

Consequently, we solve Eq. (4.54) subject to a Neumann boundary condition at the upper surface (current density specified) and a Dirichlet boundary condition at the active region given by[38]

$$\phi_f(x) = \frac{kT}{q}\left(2\ln\left(\frac{N}{\sqrt{(N_c N_v)}}\right) + A_1 N\left(\frac{1}{N_c} + \frac{1}{N_v}\right)\right), \tag{4.55}$$

where N is the carrier density (electrons or holes since their densities have been set equal in order to derive Eq. (4.55)), N_c and N_v are the conduction and valence band density of states, respectively, A_1 is a known constant, k is Boltzmann's constant, T the temperature and q the electronic charge. Eq. (4.54) may be solved efficiently using fast Fourier transforms[37]. The presence of the lower boundary condition described by Eq. (4.55) reflects the filling of available states by carriers in the conduction and valence bands, thereby raising the band gap. This in turn has the effect of discouraging current flow into the active region at those positions where the boundary potential is high, and thus leads in effect to further current spreading[39]. Note that the band gap voltage has been omitted from Eq. (4.55) since the addition of a constant potential would not affect the solution of Eq. (4.54) or the resultant current density at the active region.

4.2.2.2 Carrier diffusion

Once they reach the active region, carriers may diffuse laterally or recombine. The carrier density appearing in Eq. (4.55) is thus the solution of the carrier diffusion equation[37]:

$$D\frac{\partial^2 N}{\partial x^2} = -\frac{J}{qd} + \frac{N}{\tau_{\mathrm{nr}}} + BN^2 + CN^3 + \frac{g\Gamma|H|^2}{h\omega d}. \tag{4.56}$$

The current spreading calculation above results in a current density J that acts as a source for the carrier diffusion equation. The carriers diffuse with coefficient D, recombine non-radiatively in a time τ_{nr}, recombine and emit spontaneously with coefficient B, recombine via an Auger process with coefficient C, or recombine via stimulated emission as described by the last term. In this term, g is the power gain, Γ the confinement factor (overlap of the optical mode intensity with the active region), and d the thickness of the active region. This last term leads to gain saturation since the presence of a large optical field acts to suppress the carrier density and thus lower the gain. Although Eq. (4.56) is non-linear, it may be solved by linearization of the non-linear terms, and iteration. Values for the various coefficients may be found in Ref. 37.

4.2.2.3 Active region gain and linewidth enhancement factor effects

The presence of electrons and holes in the active region provides gain through stimulated emission, and also effects changes in the index of refraction. Near threshold, the power gain coefficient may be described by a simple linear relationship of the form $g = a(N - N_t)$ where N_t is the transparency density, i.e., that density at which there is neither gain nor loss. Likewise, changes in carrier density near threshold lead to linear changes in the index of refraction. These are expressed in terms of the so-called 'linewidth enhancement factor'[40] defined by

$$R \equiv -2k_0\frac{\partial n/\partial N}{\partial g/\partial N}. \tag{4.57}$$

With R approximately constant and g given by the linear relationship described above, we obtain

$$\frac{\partial n}{\partial N} = -\frac{aR}{2k_0}. \tag{4.58}$$

With values for R ranging from 3 to 5 for bulk GaAs, the resulting lateral variations in the refractive index across an array turn out to be substantial, and this effect was responsible for filamentation in early broad-area lasers.

The simple picture presented above changes as more carriers are injected into the active region, and the device is brought above threshold. These new carriers occupy states of higher momentum, and thus cannot participate in gain at the frequency corresponding to the band edge. Also, the Coulomb interaction between carriers influences the gain expression[41]. Thus the gain rolls off, and g is more properly expressed as a polynomial in N[42]. This non-linear behavior also influences the linewidth enhancement factor, so that it is no longer constant.

In addition, the use of quantum-well active regions leads to more complicated expressions for the gain and the linewidth enhancement factor. This results from the density of states function, which is different for quantum wells as compared with bulk material. In general, the rollover of the gain with carrier density is more severe for quantum-well devices, but the linewidth enhancement factor is somewhat lower[43]. The latter difference implies that quantum-well devices operate more stably, with less tendency towards filamentation[42]. We refer the interested reader to the rather extensive literature on this subject[40–44].

4.2.2.4 *Thermal effects*

Changes in device temperature affect both the threshold-current density and the index of refraction. The former is found empirically to scale as $J \approx \exp(T/T_0)$ over a limited temperature range, with $T_0 \approx 150$ K for quantum-well GaAs material. This increase in threshold is due primarily to increased losses from spontaneous emission and non-radiative recombination, so that more current is required to maintain the same inversion[45]. This effect is even more pronounced for long-wavelength lasers, where values of T_0 values below 100 K are common[46].

Changes in the index of refraction are approximately linear with temperature, and have been found to have a surprisingly large effect on mode selectivity in some arrays[37,47–49]. With a coefficient $\mathrm{d}n/\mathrm{d}T$ of 4×10^{-4} K^{-1}[50], a 10° temperature variation across the device results in an index change of 0.004. Changes of this magnitude are sufficient to cause thermal lensing, and can sometimes have a determining effect on the cw or quasi-cw operation mode of arrays.

In order to model this effect, a predictive capability for the temperature distribution in an array is necessary. Such a predictive capability is possible without undue effort provided some simplifying assumptions are made. Although reabsorption of radiation and ohmic heating in the

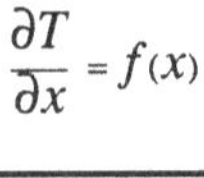

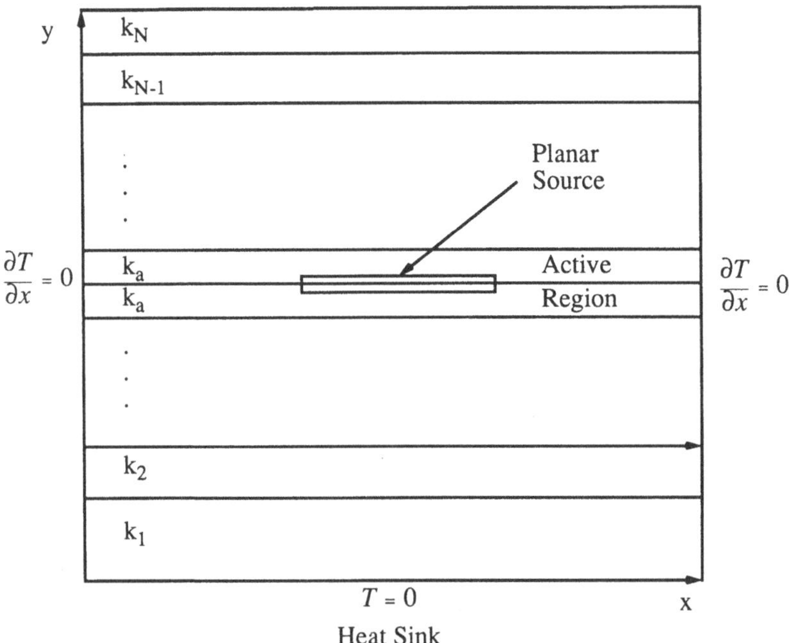

Figure 4.11. Schematic diagram depicting heat flow and the thermal model (not to scale). The primary heat source (due to non-radiative recombination) is in a plane bisecting the active region, as shown. External heating may also be applied at the upper boundary.

cladding layers are known to be sources of heat, we restrict ourselves to considering only non-radiative recombination in the active region[37]. In that case, with the continued neglect of any z-dependence, we may solve for two-dimensional heat flow using a slab model, as shown in Figure 4.11. Here the source of heat is considered to be confined to a plane bisecting the active region as shown. General boundary conditions allow for the presence of additional heat flux terms at the upper boundary. The resulting steady-state temperature profile is determined by solving Laplace's equation for the temperature using fast Fourier transforms in a manner that is very similar to the current flow problem discussed above.

The current flow, carrier diffusion and thermal models presented above have been used together in a self-consistent calculation to model an experiment designed to examine the effects of temperature on the free-running mode of a broad-area laser[37]. In the experiment, a 60 μm-wide

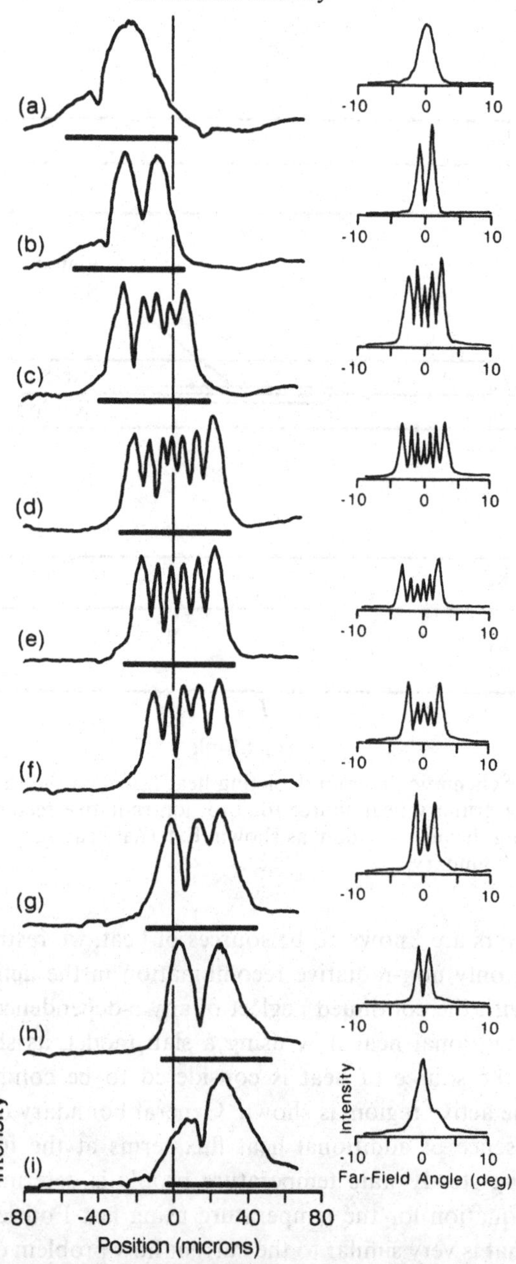

Figure 4.12. Experimentally measured near- and far-field intensity profiles for a 60 μm-wide broad-area laser free-running with external heating applied in the form of a narrow stripe at the upper metallization. For each of the nine patterns the position of the heating stripe relative to the upper metallization is depicted by the vertical line. (From Ref. 37, © 1988 IEEE.)

broad-area laser was mounted p-side-up on a heat sink, and the upper metallization illuminated with light from an argon ion laser focused into a single stripe running parallel to the lasing axis. The resulting measured near- and far-field intensity patterns are shown in Figure 4.12 for different lateral positions of the heating beam. As is clear from the diagram, the mode order could be varied from 1 to 7 by adjusting the position of the external heat source, thus demonstrating the controlling influence of thermal effects on the operation of broad area lasers.

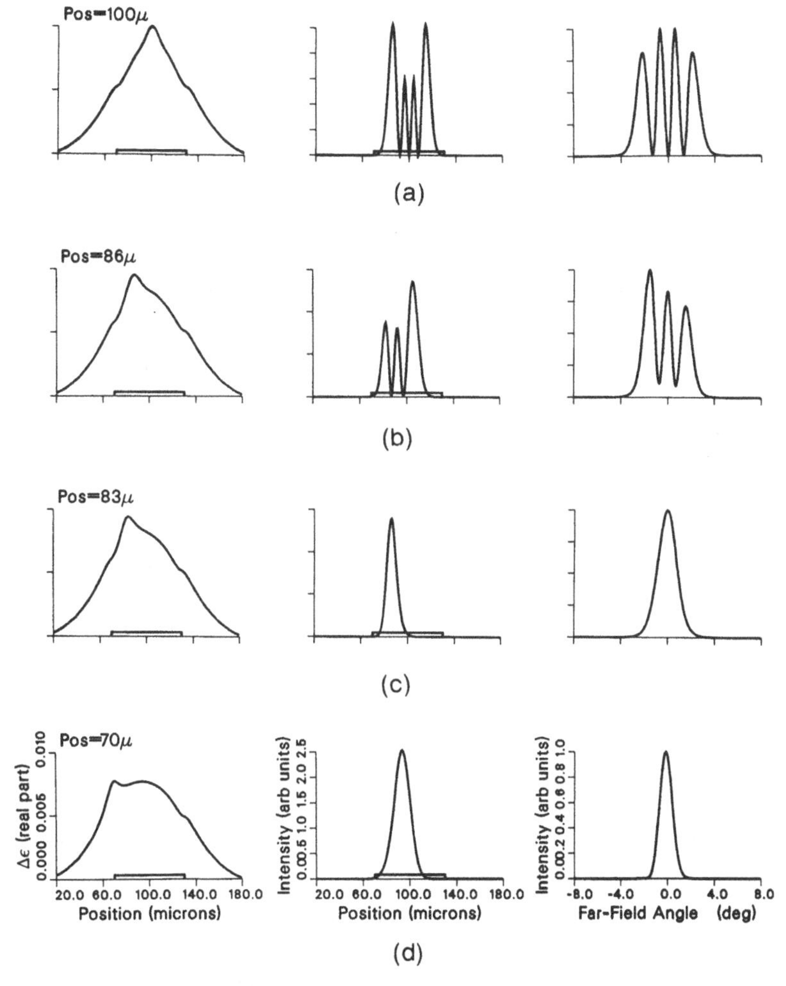

Figure 4.13. Numerical simulation of the data shown in Figure 4.12. The position of the heating stripe can be seen as the peak of the dielectric constant curves at the left. (From Ref. 37, © 1988 IEEE.)

The simulation utilized the beam propagation approach with coupled optical, electronic and thermal models as described above. (A simple linear gain approximation was used. Other details of the model may be found in Ref. 37.) The results of the calculation are presented in Figure 4.13. The relative position of the heating stripe and the effect of temperature on the dielectric constant can be seen in the far left column in the diagram. The accompanying near- and far-field profiles demonstrate good qualitative agreement with the data, predicting an increase in mode order as the heating stripe is moved towards the center of the device.

4.3 Vertical-cavity surface-emitting arrays

4.3.1 Introductory discussion

Vertical-cavity surface-emitting lasers (VCSELs) have been actively pursued because of their potential as a low-threshold laser source whose emission is normal to the growth planes[51–56]. Although the output power from these devices is typically only a few milliwatts, other desirable characteristics such as their low thresholds (sub-mA) and astigmatic emission has encouraged continued research. Two-dimensional arrays of such lasers have been pursued for a number of applications. If the individual lasers are widely separated (not phase-locked), they may be addressed individually for use in optical computing. Arrays of lasers spaced within a few microns of each other may be phase-locked by evanescent wave overlap, reflective coupling or leaky-mode coupling. Two-dimensional phase-locked VCSEL arrays have thus emerged naturally as a means of scaling up the available power without sacrificing coherence or focusability.

Because of the difficulty involved in fabricating these devices, only a few varieties have been successfully tested so far. These have all been of the reflection modulation type, illustrated schematically in Figure 4.14, employing both optical and electronic pumping. Fabrication of such a device usually begins by growing a broad-area VCSEL structure consisting of an active region containing gain (usually one or more quantum wells) sandwiched between two mirror stacks. These stacks are made up of a few dozen layers of alternating dielectric constant with indices and thicknesses carefully chosen to maximize reflectivity at the operation wavelength. The resulting wafers are then processed in order to pattern the upper reflector in some manner. This patterning has been performed in at least two different ways. In one approach, the upper contact

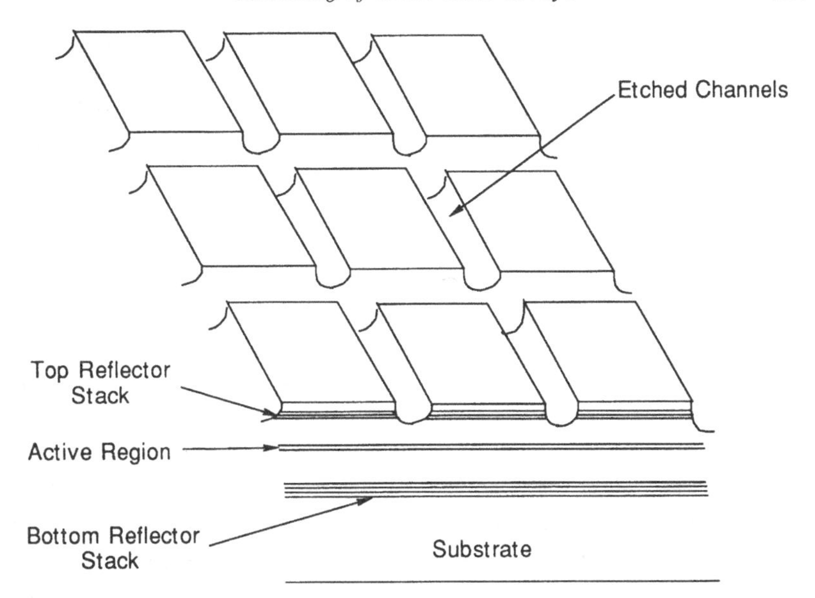

Figure 4.14. Schematic drawing of a 2-D VCSEL array fabricated by etching partway through the upper mirror stack. Similar devices have been fabricated by patterning the upper metallization.

metallization layer was ion milled to provide a spatially modulated reflectivity[57,58]. A slightly different approach involved growing only the bottom mirror stack and patterning the contact metallization, followed by electron beam evaporation of a dielectric mirror[59]. Yet a different approach[60] (shown in Figure 4.14) has been to etch part way through the upper mirror stack, thus lowering the reflectivity underneath the etched regions. The overall goal of a reflection modulated device is to break the gain degeneracy of the optical modes so as to achieve single array mode operation.

Despite the present limitations in fabrication technology, one can also envision other designs, which might, for example, involve a much deeper etch than that portrayed in Figure 4.14, followed by a regrowth of high-index material so as to result in a truly index-antiguided array. The potential advantages of such a device will become clear as we examine the model predictions.

4.3.2 Modeling

Due to the extremely short cavity length of these devices, we expect few changes in the field between successive reflections. Consequently, taking the propagation axis to be normal to the epitaxial growth planes, we propose a distributed gain model in which both the active region gain and reflective losses (output coupling) are averaged over the cavity length. Then the cavity may be considered effectively infinite and uniform along the propagation axis, and eigenmode solutions of Eq. (4.37) can be sought. Although Figure 4.14 suggests rectangular arrays, this approach is applicable to other geometrical arrangements as well, but solutions of Eq. (4.37) for these cases require considerably more effort. Non-rectangular arrays have been investigated theoretically using coupled mode theory to provide insights into the weak coupling limit[61]. Our treatment here allows strong coupling, but we will limit ourselves to the case of rectangular arrays.

Having proposed a model that reduces to solutions of Eq. (4.37), our problem would seem to be solved satisfactorily, since we have already demonstrated two methods of solution for this equation in Section 4.2.1.2 above. We will indeed pursue this approach a little later, but first it is prudent to ask whether or not a useful relationship exists between the solutions we desire and the well-known modes of a geometrically similar edge-emitting array. The answer is in the affirmative, and so we will investigate this simpler model first. It is not only numerically less taxing but, more importantly, provides some very useful insights into two-dimensional phase locking.

4.3.2.1 Separability approximation

We consider a somewhat specialized array consisting of rectangular pixels positioned on a rectangular lattice, as shown in Figure 4.15. In keeping with the distributed gain model described above, we allow the dielectric constant to take on just two values as shown. (For the device illustrated in Figure 4.14, for example, ε and ε_r would differ only in their imaginary parts.) Consider the approximation[62] for the array dielectric function

$$\varepsilon(x, y) \approx \varepsilon_r + \Delta\varepsilon[u(x) + v(y) - 1], \tag{4.59}$$

where the functions u and v are defined to be unity within a pixel and zero otherwise, and $\Delta\varepsilon \equiv \varepsilon - \varepsilon_r$. A cursory examination of Eq. (4.59) reveals that it is exact everywhere except in the region of intersection of the channels, one of which is shown as a hatched area in Figure 4.15. In those areas it is low by an amount $\Delta\varepsilon$. This error is acceptable because the channel intersections typically see relatively low optical field. Thus, proceeding with the analysis, the use of Eq. (4.59) then allows the variable

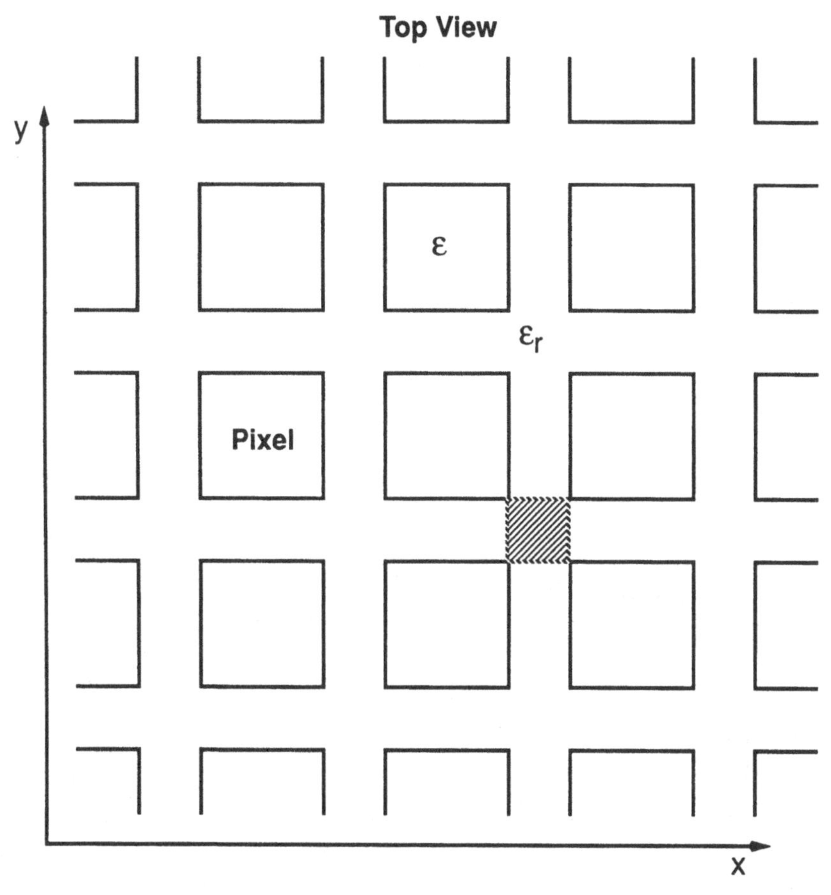

Figure 4.15. Geometry and dielectric constants for modeling VCSEL arrays. The 1-D separability approximation model fails in the regions indicated by the hatched area.

separation

$$H_{ij}(x, y) = \psi_i(x)\varphi_j(y), \tag{4.60}$$

where ψ_i and φ_j are eigenfunctions of the 1-D equations

$$\frac{\partial^2 \psi_i}{\partial x^2} = k_0^2[\bar{\varepsilon}_i - \varepsilon_r - \Delta\varepsilon u(x)]\psi_i, \tag{4.61}$$

$$\frac{\partial^2 \varphi_j}{\partial y^2} = k_0^2[\bar{\varepsilon}_j - \varepsilon_r - \Delta\varepsilon v(y)]\varphi_j, \tag{4.62}$$

with eigenvalues $\bar{\varepsilon}_i$ and $\bar{\varepsilon}_j$. Eqs. (4.61) and (4.62) describe the 1-D problems obtained by passing planes through the pixels in the x and y directions respectively. The modal dielectric constant for the array is then given by

$$\bar{\varepsilon}_{ij} = \bar{\varepsilon}_i + \bar{\varepsilon}_j - \varepsilon_r - \Delta\varepsilon. \tag{4.63}$$

We may now extract some very useful information from the above formalism. First of all, the far-field intensity pattern for the array is trivially

$$F_{ij}^{2D}(\theta_x, \theta_y) = F_i^{1D}(\theta_x) F_j^{1D}(\theta_y), \tag{4.64}$$

where θ_x and θ_y are the angles of emission measured from the y–z and x–z planes respectively. Also, for the usual case of identical modes in both directions and square pixels, we have for the mode discrimination

$$\Delta\bar{\varepsilon}^{2D} = 2\Delta\bar{\varepsilon}^{1D}. \tag{4.65}$$

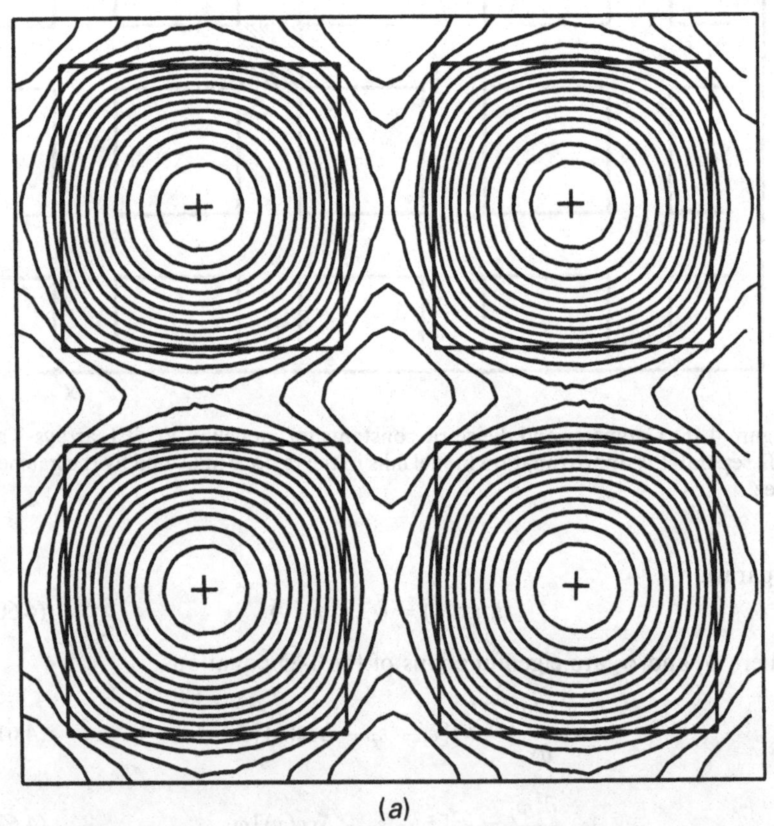

(*a*)

Figure 4.16. Calculated near-field intensity contours for several dominant modes of an infinite 2-D VCSEL array computed using 2-D matching and symmetry boundary conditions: (*a*) in-phase evanescent mode. (*continued*)

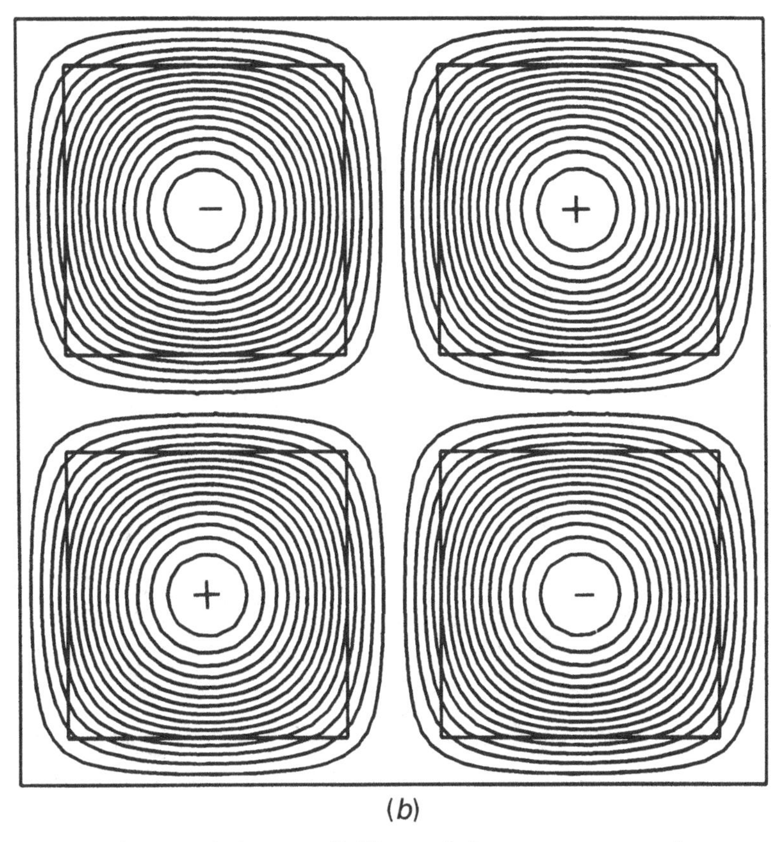

Figure 4.16. (*continued*) (*b*) out-of-phase evanescent mode.

This important result implies that if a 1-D structure (such as a gain-guided array, for example) is known to favor an out-of-phase mode, then the analogous 2-D array will also do so, but with approximately twice the mode discrimination. Furthermore, the far-field pattern of such an array would, on the basis of Eq. (4.64), consist of four off-axis modes shifted by 45° with respect to the array axes. Both these predictions have been observed in the laboratory[58,60] for devices of the general type shown in Figure 4.14. These devices are effectively gain-guided since the averaged gain is lower in the channels between pixels due to the higher outcoupling there, and the index of refraction is identical in both regions.

Conversely, it should be possible to design 2-D arrays that lase with a desirable single-lobed far-field pattern by tailoring the geometry to favor fundamental mode lasing in the associated 1-D array. For example,

Figure 4.16. (*continued*) (*c*) in-phase leaky mode. (*continued*)

regrowth of deeply etched channels with material of higher index of refraction than under the pixels should result in leaky-mode operation of the 1-D array. Such linear arrays have been shown to produce primarily single-lobed output to high powers with good modal gain discrimination[34]. Consequently, we would expect the 2-D array to lase in a fundamental leaky mode with even better modal discrimination. These modes will be calculated using a more rigorous 2-D solution technique in the next section.

4.3.2.2 Two-dimensional modeling

As pointed out previously, the techniques already developed to explore 2-D solutions for waveguiding in edge-emitting arrays may be immediately applied to the modeling of 2-D VCSEL arrays. We assume a rectangular

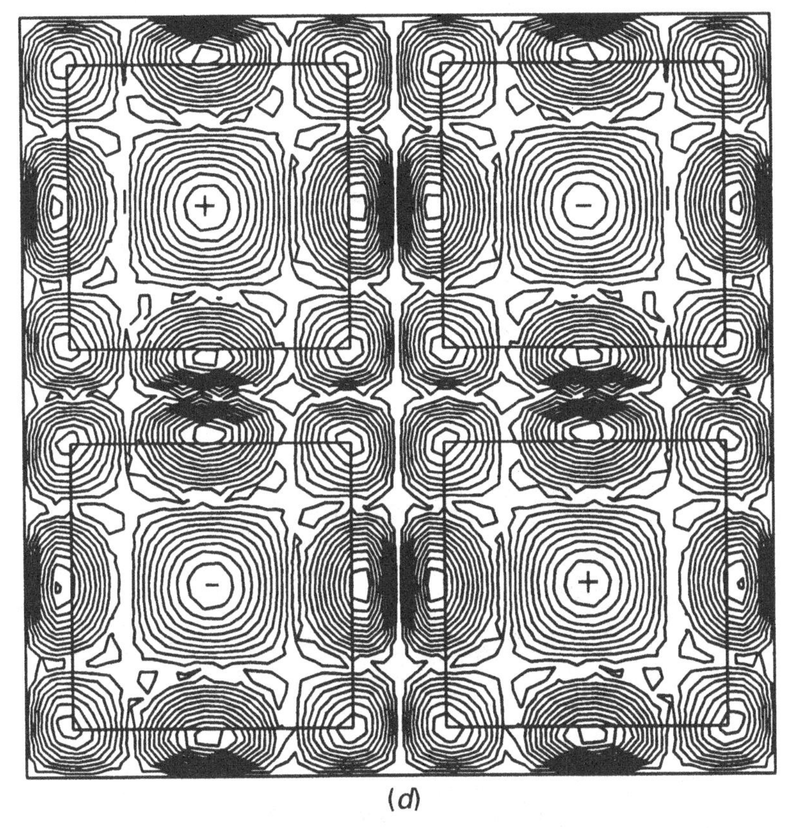

Figure 4.16. *(continued)* (*d*) out-of-phase leaky mode. (From Ref. 62.)

geometry, as shown in Figure 4.15, and apply the 2-D analytic matching technique discussed earlier using a variety of boundary conditions to select different modes. We note in passing, however, that more complicated arrangements could be modeled using the triangular mesh algorithm presented in Section 4.2.1.2.

Figure 4.16 shows calculated near-field intensity profiles for four modes of interest of an index-guided VCSEL structure[62]. In Figure 4.16(*a*, *b*), the index of the channel regions has been set slightly lower than under the pixels as it would be for a linear array. These diagrams depict the in-phase and out-of-phase modes, respectively. These modes are nearly indistinguishable, except for the noticeable null between pixels present in Figure 4.16(*b*) and absent in Figure 4.16(*a*). Notice that for these modes, the intensity in the channel intersection region is indeed low, justifying the

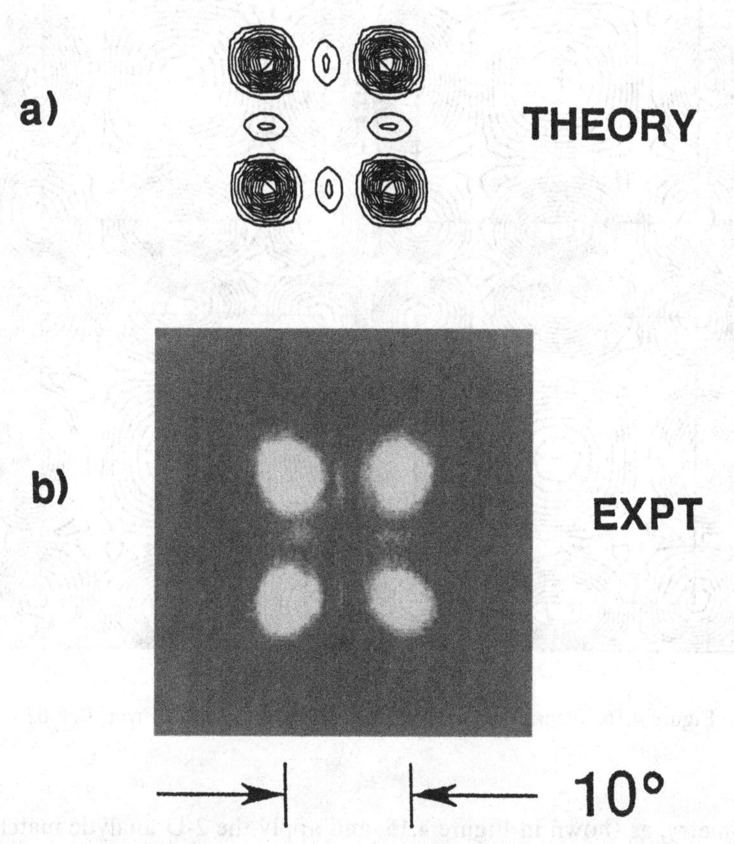

Figure 4.17. Theoretical and experimental far-field patterns for a 3 × 3 reflectivity modulated VCSEL array operating in the out-of-phase mode. (From Ref. 60.)

separability approximation used earlier. For these modes, the 1-D formalism was found to agree quite well with the more exact 2-D solutions. For Figure 4.16(c, d), the index in the channel regions was increased slightly over the value under the pixels in order to investigate the leaky-mode operation. The resulting modes are the leaky in-phase and out-of-phase modes respectively[34]. They differ in appearance by the secondary maxima between pixels, which in turn lead to side lobes in the far-field pattern. The presence of appreciable energy in the channel intersection region decreases the accuracy of the separability approximation for this case, although it has still been verified to be a reasonably good approximation.

We have used the same procedure to model reflection-modulated (gain-guided) 2-D arrays fabricated by etching part way through the upper mirror stack[60]. For a 3×3 array, we present both the calculated and measured far-field patterns in Figure 4.17. The model not only predicts the correct lobe separation (which would be expected since it depends only on the pixel spacing), but also details of the pattern such as the lobe width and the presence of the four minor lobes.

4.4 Conclusion

The modeling of diode laser arrays has progressed considerably in the short time since the first coupled mode theory calculations reported the characteristics of array modes. Waveguiding models have increased in sophistication using analytic matching and beam propagation techniques towards the 2-D models in use today[34]. At the same time, active region models have also become more complete, including the self-consistent coupling of current spreading and active layer diffusion, and, more recently, the addition of thermal modeling[37]. Some treatments have also included a separate transport equation for electrons and holes.

None the less, the fabrication of devices with built-in lateral lensing, Y-guides and Talbot cavities continues to provide the theorist with significant new challenges. A multi-watt cw diode laser array that emits essentially all its radiation into a single far-field lobe has yet to be demonstrated, and so the need for good theoretical models has not diminished. The parameter space for designing such a device is still too large and the fabrication steps too costly for the experimenter to proceed based solely on empirical guidance.

In addition, there is considerable room for improvement in the active region models, particularly in regard to quantum-well devices. The simple linear gain models typically used are highly suspect for quantum-well lasers. Even though a considerable body of work is presently available detailing the complex mechanisms operative in both bulk and quantum-well material[41–44], this information has yet to be incorporated to any appreciable extent into device models. Indeed, it seems that at present most of the effort expended on device modeling has been in the area of waveguiding.

Recently, laser array modeling has taken on a new twist with the advent of 2-D VCSEL arrays. Although similar in some respects to the operation of 1-D arrays, these devices none the less offer an extra dimension of complexity that is just beginning to be explored. Present predictions indicate the possibility of using leaky-mode coupling to couple large numbers of pixels, resulting in an array that operates with good mode discrimination and an astigmatic single-lobed output[62]. However, these predictions will have to await their test until some still formidable fabrication difficulties are solved. Also, it should be noted that significant differences, such as cavity Q and device length, existing between these arrays and the more standard edge-emitting arrays, have not yet been thoroughly examined. These differences may well lead to new insights and designs for future devices.

Appendix: numerical implementation of analytic matching

Consider a composite made up of N layers of material, each characterized by a constant dielectric constant. In this domain we wish to find the solution of the Helmholtz equation:

$$\frac{\partial^2 H}{\partial x^2} = k_0^2(\bar{\varepsilon} - \varepsilon_{\text{eff}})H, \tag{A.1}$$

where ε_{eff} has the constant value ε_i in layer i. We define

$$h_i \equiv k_0\sqrt{(\bar{\varepsilon} - \varepsilon_i)} \tag{A.2}$$

and select the root in such a way that the real part of h_i is always positive. The solution to Eq. (4.12) is then given by

$$H = H_i \exp(h_i x) + H_i' \exp(-h_i x), \tag{A.3}$$

where x denotes the relative distance from the left boundary of each individual layer. Equating the field and its derivative at each interface then yields the matrix equation

$$\begin{bmatrix} H_{i+1} \\ H'_{i+1} \end{bmatrix} = \frac{1}{2h_{i+1}} \begin{bmatrix} h_+ & h_- \exp(-h_i d_i) \\ h_- & h_+ \exp(-h_i d_i) \end{bmatrix} \begin{bmatrix} H_i \exp(h_i d_i) \\ H_i' \end{bmatrix}, \tag{A.4}$$

where

$$h_\pm = h_{i+1} \pm h_i. \tag{A.5}$$

Notice that the exponential whose argument has a positive real part has been factored into the vector, leaving the matrix with only those exponentials having arguments with negative real parts. This procedure guards against scaling problems in the matrix that typically result if a particularly thick layer has solutions with primarily exponential character. Denoting the matrix in Eq. (A.4) (including the prefactor) by $A^{(i)}$, we may now assemble a composite matrix equation for the entire problem:

$$\begin{bmatrix} \exp(-h_1d_1) & B & 0 & 0 & 0 & 0 & 0 & \cdots & 0 \\ -A_{11}^{(1)} & -A_{12}^{(1)} & \exp(-h_2d_2) & 0 & 0 & 0 & 0 & & 0 \\ -A_{21}^{(1)} & -A_{22}^{(1)} & 0 & 1 & 0 & 0 & 0 & & 0 \\ 0 & 0 & -A_{11}^{(2)} & -A_{12}^{(2)} & \exp(-h_3d_3) & 0 & 0 & & 0 \\ 0 & 0 & -A_{21}^{(2)} & -A_{22}^{(2)} & 0 & 1 & 0 & & 0 \\ \vdots & & & & & & & & \\ 0 & 0 & 0 & \cdots & 0 & -A_{11}^{(N-1)} & -A_{12}^{(N-1)} & \exp(-h_Nd_N) & 0 \\ 0 & 0 & 0 & & 0 & -A_{21}^{(N-1)} & -A_{22}^{(N-1)} & 0 & 1 \\ 0 & 0 & 0 & & 0 & 0 & 0 & C & D \end{bmatrix} \begin{bmatrix} H_1\exp(h_1d_1) \\ H_1' \\ H_2\exp(h_2d_2) \\ H_2' \\ H_3\exp(h_3d_3) \\ \vdots \\ H_N\exp(h_Nd_N) \\ H_N' \end{bmatrix} = 0. \tag{A.6}$$

The constants B, C and D are uniquely determined from the desired boundary conditions. Eigenvalues are determined using a root finding algorithm that varies $\bar{\varepsilon}$ in an optimum manner in order to minimize the determinant of the matrix in Eq. (A.6). Both the calculation of the determinant and the solution of the eigenvectors corresponding to known eigenvalues are computed using commercially available band matrix routines.

The above procedure has been found in practice to be superior to the simpler method of multiplying the individual transfer matrices together to form a smaller determinant. Although both methods are equivalent in many cases, problems involving thick intermediate layers in which exponential behavior is present should be adequately treatable using the full matrix, but may cause the simpler matrix to become ill-conditioned.

References

1. G. H. B. Thompson, *Optoelectronics*, **4**, 257 (1972).
2. P. A. Kirkby, A. R. Goodwin, G. H. B. Thompson and P. R. Selway, *IEEE J. Quantum Electron.*, **QE-13**, 705 (1977).
3. E. Kapon, J. Katz, S. Margalit and A. Yariv, *Appl. Phys. Lett.*, **45**, 600 (1984).
4. E. Kapon, Z. Rav-noy, S. Margalit and A. Yariv, *J. Lightwave Tech.*, **LT-4**, 919 (1986).
5. D. F. Welch, P. Cross and D. Scifres, *Electron. Lett.*, **22**, 293 (1986).
6. J. Katz, S. Margalit and A. Yariv, *Appl. Phys. Lett.*, **42**, 554 (1983).
7. D. Botez, L. Mawst, P. Hayashida, G. Peterson and T. J. Roth, *Appl. Phys. Lett.*, **53**, 464 (1988).
8. H. C. Casey and M. B. Panish, *Heterostructure Lasers*, Academic Press, Orlando, Florida (1978).
9. H. Kressel and J. K. Butler, *Semiconductor Lasers and Heterojunction LEDs*, Academic Press, New York (1977).
10. ibid, p. 126.
11. J. D. Jackson, *Classical Electrodynamics*, John Wiley and Sons, New York (1962), p. 217.
12. Ref. 8, p. 79.
13. N. K. Dutta, D. C. Craft and S. G. Napholtz, *Appl. Phys. Lett.*, **46**, 123 (1985).
14. J. Buus, *IEEE J. Quantum Electron.*, **QE-18**, 1083 (1982).
15. A. Yariv, *IEEE J. Quantum Electron.*, **QE-9**, 919 (1973).
16. A. Hardy and W. Streifer, *Opt. Lett.*, **10**, 335 (1985).
17. J. K. Butler, D. E. Ackley and D. Botez, *Appl. Phys. Lett.*, **44**, 293 (1984).

18. E. Kapon, J. Katz and A. Yariv, *Opt. Lett.*, **9**, 125 (1984).
19. N. W. Ashcroft and N. D. Mermin, *Solid State Physics*, Saunders College, Philadelphia, Pa. (1976), p. 133.
20. D. Mehuys and A. Yariv, *Opt. Lett.*, **13**, 571 (1988).
21. G. R. Hadley, A. Owyoung and J. P. Hohimer, *Opt. Lett.*, **11**, 144 (1986).
22. J.-M. Verdiell and R. Frey, *IEEE J. Quantum Electron.*, **QE-26**, 270 (1990).
23. P. G. Eliseev, R. F. Nabiev and Yu. M. Popov, *J. Sov. Las. Res.*, **10**, 449 (1989).
24. D. Botez and T. Holcomb, *Appl. Phys. Lett.*, **60**, 539 (1992).
25. W. K. Marshall and J. Katz, *IEEE J. Quantum Electron.*, **QE-22**, 827 (1986).
26. A. G. Fox and T. Li, *Bell. Sys. Tech. J.*, **40**, 453 (1961).
27. R. D. Richtmeyer and K. W. Morton, *Difference Methods for Initial Value Problems*, 2nd edn, Interscience Publishers, J. Wiley and Sons, New York (1967), p. 199.
28. G. P. Agrawal, *J. Appl. Phys.*, **56**, 3100 (1984).
29. R. W. Hockney, *J. Assoc. for Computing Mach.*, **12**, 95 (1965).
30. A. E. Siegman and H. Y. Miller, *Appl. Opt.*, **9**, 2729 (1970).
31. G. R. Hadley, J. P. Hohimer and A. Owyoung, *IEEE J. Quantum Electron.*, **QE-23**, 765 (1987).
32. M.-C. Amann, *IEEE J. Quantum Electron.*, **QE-22**, 1992 (1986).
33. S. B. Worm and R. Pregla, *IEEE Trans. Microwave Theory and Techn.*, **MTT-32**, 191 (1984).
34. G. R. Hadley, D. Botez and L. J. Mawst, *IEEE J. Quantum Electron.*, **QE-27**, 921 (1991).
35. K. Halbach and R. F. Holsinger, *Particle Accelerators*, **7**, 213 (1976).
36. A. M. Winslow, *J. Comp. Phys.*, **2**, 149 (1967).
37. G. R. Hadley, J. P. Hohimer and A. Owyoung, *IEEE J. Quantum Electron.*, **QE-24**, 2138 (1988).
38. W. B. Joyce and R. W. Dixon, *J. Appl. Phys.*, **49**, 3719 (1978).
39. Ref. 37, esp. Fig. 3.
40. M. Osinski and J. Buus, *IEEE J. Quantum Electron.*, **QE-23**, 9 (1987).
41. C. Ell, H. Haug and S. W. Koch, *Opt. Lett.*, **14**, 356 (1989).
42. W. W. Chow and D. Depatie, *IEEE J. Quantum Electron.*, **QE-24**, 1297 (1988).
43. W. W. Chow and D. Depatie, *Opt. Lett.*, **13**, 303 (1988).
44. W. W. Chow, S. W. Koch, M. Sargent III and C. Ell, *Appl. Phys. Lett.*, **58**, 328 (1991).
45. Ref. 9, section 3.3.
46. A. Mozer, K. M. Romanek, O. Hildebrand, W. Schmid and M. H. Pilkuhn, *IEEE J. Quantum Electron.*, **QE-19**, 913 (1983).
47. G. R. Hadley, J. P. Hohimer and A. Owyoung, *J. Appl. Phys.*, **61**, 1697 (1987).
48. J. P. Hohimer, G. R. Hadley and A. Owyoung, *Appl. Phys. Lett.*, **52**, 260 (1988).
49. J. P. Hohimer, G. R. Hadley, D. C. Craft, T. H. Shiau, S. Sun and C. F. Schaus, *Appl. Phys. Lett.*, **58**, 452 (1991).
50. Ref. 8, p. 31.
51. K. Iga, F. Koyama and S. Kinoshita, *IEEE J. Quantum Electron.*, **QE-24**, 1845 (1988).
52. M. Ogura, W. Hsin, M. Wu, S. Wang, J. R. Whinnery, S. C. Wang and J. J Yang, *Appl. Phys. Lett.*, **51**, 1655 (1987).
53. L. M. Zinkiewicz, T. J. Roth, L. J. Mawst, D. Tran and D. Botez, *Appl. Phys. Lett.*, **54**, 1959 (1989).
54. P. L. Gourley, S. K. Lyo and L. R. Dawson, *Appl. Phys. Lett.*, **54**, 1397 (1989).
55. J. L. Jewell, A. Scherer, S. L. McCall, Y. H. Lee, S. Walker, J. P. Harbison and L. T. Florez, *Electron. Lett.*, **25**, 1123 (1989).

56. C. F. Schaus, M. Y. A. Raja, J. G. McInerney, H. E. Schaus, S. Sun, M. Mahbobzadeh and S. R. J. Brueck, *Electron. Lett.*, **25**, 637 (1989).
57. H.-J. Yoo, A. Scherer, J. P. Harbison, L. T. Florez, E. G. Paek, B. P. Van der Gaag, J. R. Hayes, A. Von Lehmen, E. Kapon and Y.-S. Kwon, *Appl. Phys. Lett.*, **56**, 1198 (1990).
58. M. Orenstein, E. Kapon, N. G. Stoffel, J. P. Harbison, L. T. Florez and J. Wullert, *Appl. Phys. Lett.*, **58**, 804 (1991).
59. D. G. Deppe, J. P. van der Ziel, N. Chand, G. J. Zydzik and S. N. G. Chu, *Appl. Phys. Lett.*, **56**, 2089 (1990).
60. P. L. Gourley, M. E. Warren, G. R. Hadley, G. A. Vawter, T. M. Brennan and B. E. Hammons, *Appl. Phys. Lett.*, **58**, 890 (1991).
61. H.-J. Yoo, J. R. Hayes, E. G. Paek, A. Scherer and Y.-S. Kwon, *IEEE J. Quantum Electron.*, **QE-26**, 1039 (1990).
62. G. R. Hadley, *Opt. Lett.*, **15**, 1215 (1990).

5

Dynamics of coherent semiconductor laser arrays

HERBERT G. WINFUL AND
RICHARD K. DEFREEZ

5.1 Introduction

In this chapter we review our recent work on the dynamics of coherent semiconductor laser arrays. While most of the published literature on laser arrays has focused on the spatial properties of these sources, it is becoming increasingly apparent that their temporal behavior can be marvelously rich and complex. This complexity should not be surprising. A laser is, after all, a nonlinear oscillator. When one creates an array of coupled nonlinear oscillators the resulting dynamical behavior of the system will range from synchronization and phase locking to instabilities and chaos. An understanding of the complex dynamics is essential for the design of stable, compact, high-power sources for applications such as intersatellite optical communications. Because the characteristic time scale of semiconductor laser array dynamics is less than a nanosecond, it took the pioneering streak camera measurements of DeFreez *et al.*, to provide the first indication of temporal instabilities in these lasers[1]. Since then, complex dynamical behavior has been observed in most types of semiconductor laser arrays. Theoretical work has proceeded apace, from the early time-dependent coupled-mode theories[2–12] to more recent partial differential equation models which treat the array as a single entity[13–15]. The models predict both coupling induced instabilities and saturable-absorber induced instabilities, depending on array geometry and material parameters. They also suggest new possibilities for extremely high-speed modulation of laser arrays. In what follows we present the current theoretical understanding of laser array dynamics, as well as a survey of relevant experimental results.

5.2 Coupled lasers

Conventional single-stripe semiconductor lasers are limited in their power output to about 100 mW as a result of the catastrophic damage that

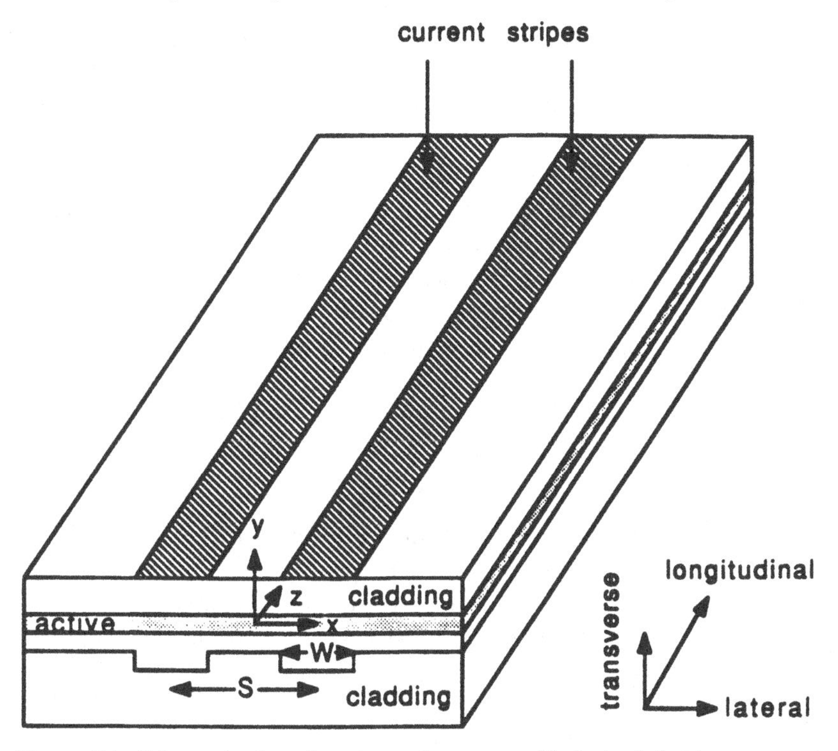

Figure 5.1. Schematic of semiconductor laser array. Each shaded stripe represents one laser in the array. Propagation is in the z-direction.

occurs at high levels of optical flux density. One way to circumvent the damage problem, while increasing the power output, is to create an array of coherently coupled lasers. When the radiation from a coherent array of N lasers is focused to a spot, the power density is proportional to N^2. In this manner, coupled laser arrays have yielded output power values in excess of 1 W under cw operation. For closely spaced semiconductor positive-index waveguide lasers (Figure 5.1), lateral distributed coupling between the lasers is by means of their overlapping evanescent fields. The fields of adjacent lasers can couple with either a 0° phase shift or a 180° phase shift between them. The 0° phase shift configuration ('in-phase mode') is the desirable one since it leads to a single central lobe in the far field. When a 180° phase shift exists between adjacent elements (the 'out-of-phase mode') the far-field pattern exhibits multiple lobes. Experimentally, it has been noted that phased arrays with evanescent coupling tend to operate in the out-of-phase mode. We have shown by means of a stability analysis of an evanescent-wave coupled laser model that this preference for the out-of-phase mode is to be expected for weakly coupled lasers[2]. In addition, we find that, beyond a certain coupling strength, the phase-locked state loses stability through a supercritical Hopf bifurcation. The laser output then exhibits sustained intensity pulsations at gigahertz frequencies. These pulsations can be periodic, quasi-periodic or chaotic, depending on the coupling strength. The instability predictions mentioned here are based on a coupled-mode theory, described in the next section, which works best for index-guided evanescent coupled arrays. For gain-guided, antiguided and Y-guide arrays, a partial differential equation model is more appropriate. Such a model is presented in Section 5.4.

5.3 Coupled-mode theory for laser array dynamics

5.3.1 The coupled-mode equations

In the time-dependent coupled-mode theory, the electric field of the guided mode in the jth laser is taken as $E_j(t)\exp(-\mathrm{i}\omega_0 t)$, where the complex amplitude $E_j(t)$ varies slowly compared to the optical frequency ω_0. Assuming nearest-neighbor coupling, the evolution of the mode amplitude (E_j) and the population (N_j) in the jth laser is described by the equations[2–4]

$$\frac{\mathrm{d}E_j}{\mathrm{d}t} = \frac{1}{2}\left[G(N_j) - \frac{1}{\tau_\mathrm{p}}\right](1 - \mathrm{i}\alpha)E_j + \mathrm{i}K(E_{j+1} + E_{j-1}) \times \exp(-\mathrm{i}\psi) + \mathrm{i}(\omega_0 - \omega_j)E_j, \tag{5.1}$$

$$\frac{\mathrm{d}N_j}{\mathrm{d}t} = P - \frac{N_j}{\tau_\mathrm{s}} - G(N_j)|E_j|^2, \tag{5.2}$$

where G is the gain, τ_p (≈ 1 ps) is the photon lifetime, τ_s (≈ 2 ns) is the lifetime of the active population, P is the pump rate, K is the magnitude of the coupling strength between adjacent lasers and ψ is the coupling phase. The uncoupled laser frequencies are given by ω_j. The parameter α is known as the linewidth enhancement factor in semiconductor lasers and is a measure of the carrier-density-dependent refractive index. For operation not too far from the lasing threshold of the uncoupled lasers, the gain may be expressed as $G(N_j) = G(N_\mathrm{th}) + g(N_j - N_\mathrm{th})$, where N_th is the carrier density at threshold, $G(N_\mathrm{th}) = 1/\tau_\mathrm{p}$, and $g = \partial G/\partial N$ is the differential gain.

It is convenient to transform Eqs. (5.1) and (5.2) into dimensionless form for the normalized magnitude (X_j) and phase (ϕ_j) of the electric field, and the normalized excess carrier density Z_j in the jth laser. These equations read

$$\dot{X}_j = Z_j X_j - \eta[X_{j+1} \sin(\phi_j - \phi_{j+1} + \psi) - X_{j-1} \sin(\phi_{j-1} - \phi_j + \psi)], \tag{5.3a}$$

$$\dot{\phi}_j = \alpha Z_j - \eta[(X_{j+1}/X_j) \cos(\phi_j - \phi_{j+1} + \psi) + (X_{j-1}/X_j) \cos(\phi_{j-1} - \phi_{j+1} + \psi)] + \Delta_j, \tag{5.3b}$$

$$T\dot{Z}_j = p - Z_j - (1 + 2Z_j)X_j^2, \qquad j = 1, 2, 3, \ldots, N, \tag{5.3c}$$

with $X_0 = X_{N+1} = 0$, and η is the coupling coefficient. Here the overdots signify derivatives with respect to a reduced time t/τ_p and we define the following variables and parameters:

$$X_j = (\tfrac{1}{2}g\tau_s)^{\frac{1}{2}}|E_j|, \qquad Z_j = \tfrac{1}{2}gN_{th}\tau_p(N_j/N_{th} - 1),$$

$$p = \tfrac{1}{2}gN_{th}\tau_p(P/P_{th} - 1), \qquad P_{th} = N_{th}/\tau_s,\ \eta = K\tau_p,\ T = \tau_s/\tau_p,$$

$$\Delta_j = (\omega_0 - \omega_j)\tau_p.$$

In the absence of coupling ($\eta = 0$), each laser behaves as a weakly damped oscillator which evolves toward a steady state characterized by $Z_j = 0$, $X_j = \sqrt{p}$, and an arbitrary phase ϕ_j. The approach to steady state is oscillatory with a frequency (the relaxation oscillation frequency)

$$\Omega_r = \left[\frac{2p}{T} - \left(\frac{1 + 2p}{2T}\right)^2\right]^{\frac{1}{2}} \approx \left(\frac{2p}{T}\right)^{\frac{1}{2}} \tag{5.4}$$

and damping rate

$$\gamma = \frac{1 + 2p}{2T}, \tag{5.5}$$

in units of the inverse photon lifetime.

For nonzero coupling ($\eta \neq 0$), the assembly of lasers can organize itself into a phase-locked state with a constant output intensity and a fixed phase difference between adjacent lasers. Phase locking can occur if the detuning Δ_j in the native frequencies of the uncoupled lasers lies within a certain locking range. It can be shown that if the carrier recovery time is sufficiently short and the coupling η is sufficiently weak, the evolution of the coupled laser system may be described by the phase equation (Eq. (5.3b)), which for two lasers reads

$$\dot{\theta} = 2(\alpha\eta' + \eta'') \sin\theta - \Delta, \tag{5.6}$$

where $\theta = \phi_1 - \phi_2$ and $\Delta = (\omega_1 - \omega_2)\tau_p$. The quantities $\eta' = \eta \cos \psi$ and $\eta'' = \eta \sin \psi$ are the real and imaginary parts of the coupling coefficient. They determine the wavelength separation and the gain splitting of the array modes. Solution of the phase equation in steady state yields a locking range

$$|\Delta| \leq 2(\alpha\eta' + \eta''). \tag{5.7}$$

For example, two coupled lasers with $\alpha = 2$, a purely real coupling parameter of $\eta' = 0.1$ and an operating wavelength of 850 nm are predicted to have a locking range of about 40 GHz.

5.3.2 Stability of phase locking

The phase-locked state is not necessarily stable when the carrier dynamics are taken into account. For two coupled lasers with zero detuning, the phase-locked solutions have $\theta = 0$ (in-phase) and $\theta = \pi$ (out-of-phase). For lasers with purely real coupling ($\eta'' = 0$), a linear stability analysis[2] of the coupled-mode equations shows that the out-of-phase solution is stable if

$$\eta < \frac{1 + 2p}{2\alpha T}, \tag{5.8a}$$

while the in-phase solution is stable if

$$\eta > \frac{\alpha p}{1 + 2p}. \tag{5.8b}$$

Figure 5.2 shows the stability boundaries described by Eqs. (5.8) in the plane of the variables η and p. The parameters used were $\alpha = 5$ and $T = 2 \times 10^3$. It can be seen that for $\eta < 10^{-4}$ the stable solution is the out-of-phase one. There is a small region on the η–p plane (imperceptible in Figure 5.2) where both in-phase and out-of-phase solutions are stable and may coexist. Hysteresis effects are possible in this region.

The stability boundary $\eta = (1 + 2p)/2\alpha T$ corresponds to a supercritical Hopf bifurcation. For a given value of p, as the coupling strength is increased, the time-independent out-of-phase mode loses stability in favor of a self-pulsing solution in which all the variables – field amplitudes, phases and carrier densities oscillate periodically in time. Phase oscillations in the near field result in beam steering in the far field[10–12]. At the Hopf

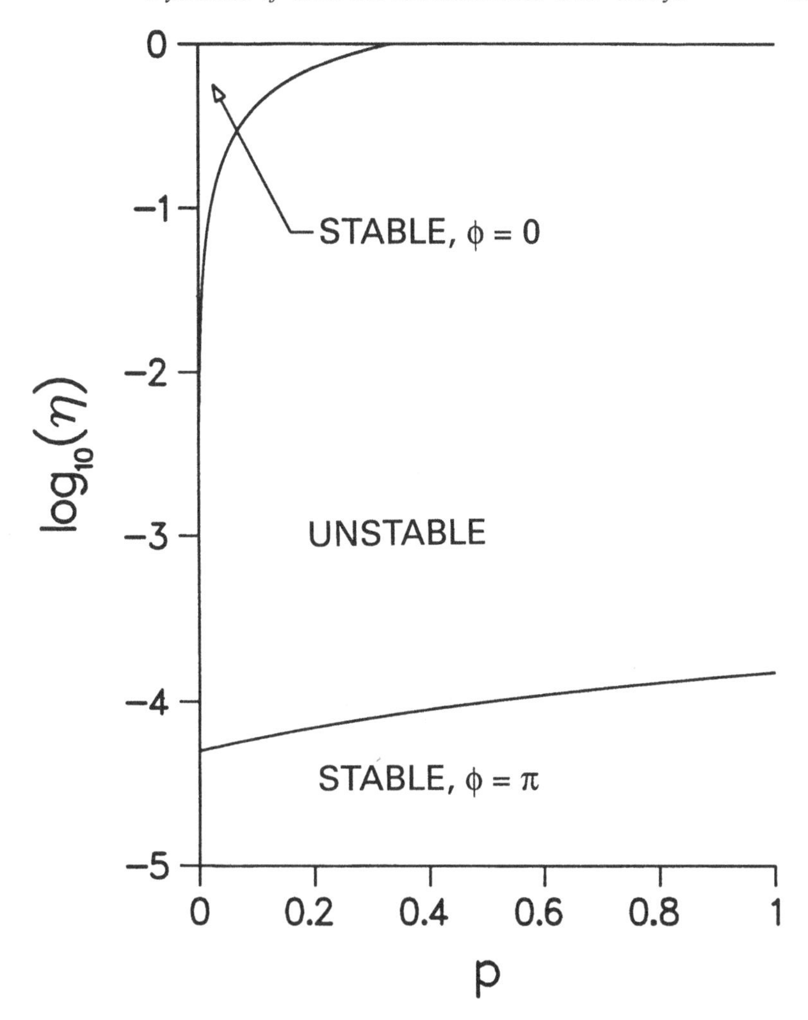

Figure 5.2. Instability domain for two coupled lasers. Here p is the excess pump current and η is the normalized coupling strength.

bifurcation the linear stability analysis yields a pulsation frequency

$$\Omega = \left[\frac{2p}{T} + 4\eta^2\right]^{\frac{1}{2}}. \tag{5.9}$$

For low values of coupling strength, the pulsation frequency is simply the relaxation oscillation frequency $\Omega_r \approx (2p/T)^{1/2}$, while at large coupling the pulsation frequency becomes the rate 2η of energy transfer between the lasers. Figure 5.3 shows some of the variety of temporal behaviors

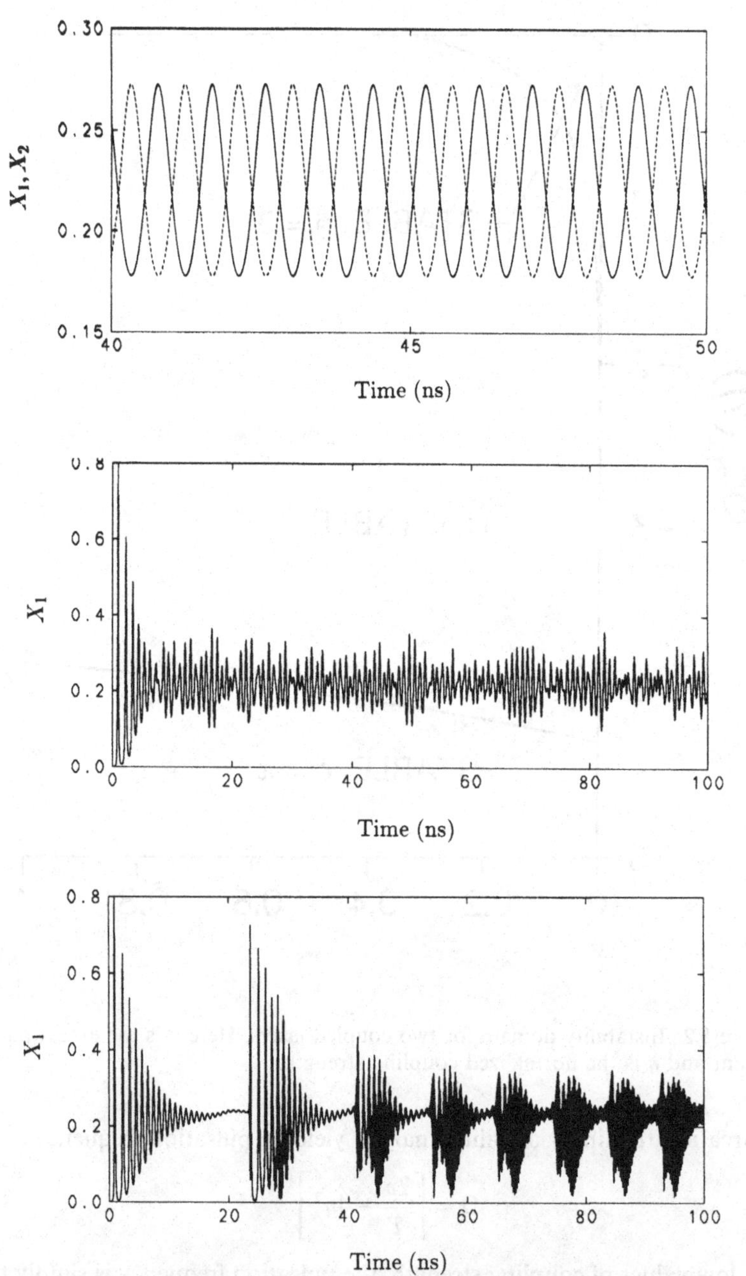

Figure 5.3. Time series of the output field amplitude (X_1) from one of a pair of coupled lasers for different values of coupling strength η. The dotted curve in the top illustration represents X_2. Top: $\eta = 10^{-4}$; Middle: $\eta = 10^{-3}$; Bottom: $\eta = 10^{-1}$.

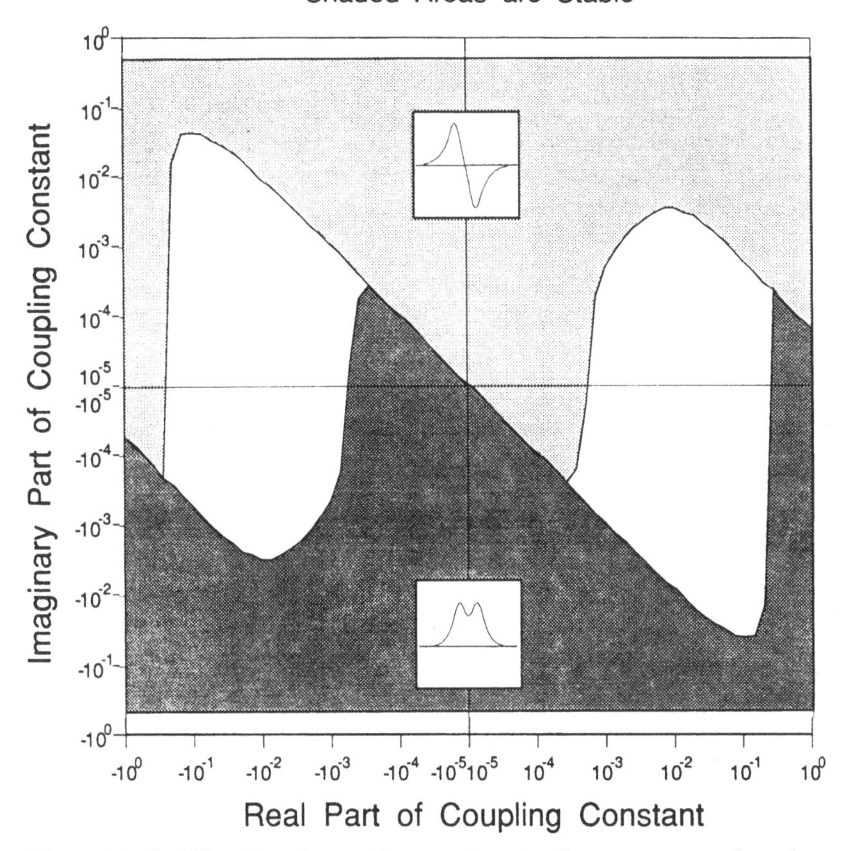

Figure 5.4. Stability domain on the complex coupling parameter plane for a twin-emitter semiconductor laser array. The linewidth enhancement factor is 1, the pumping level is $1.66 \times I_{\text{th}}$.

that can be observed as η is increased. Note that these coupling induced oscillations are in antiphase: the two lasers exchange energy periodically or aperiodically.

For most laser array designs the coupling coefficient is a complex quantity. The stability of the phase-locked state will depend on both the real and imaginary parts of the coupling constants[7,8]. Figure 5.4 shows the stability boundaries of the in-phase and out-of-phase solutions in the complex η plane. The imaginary part of η introduces extra damping into the system and enhances the stability of the locked state.

The stability analysis of evanescently coupled lasers has been extended to N coupled emitters[9]. For an array of N lasers with purely real coupling, the out-of-phase solution loses stability when

$$\eta \approx \frac{1 + 2p}{4\alpha T \cos[\pi/(N + 1)]}. \tag{5.10}$$

This result shows that the array becomes somewhat less stable as the number of emitters is increased. For N greater than 10, the critical coupling strength reaches a limiting value of

$$\eta = \frac{1 + 2p}{4\alpha T}, \tag{5.11}$$

which is independent of the number of emitters. Of course, in very large arrays there will be inhomogeneities, such as thermal gradients, which will cause detunings between emitters and impose a limit on their ability to phase lock.

5.4 A propagation model for laser array dynamics

The coupled-mode theory presented in Section 5.3 is strictly valid only for structures such as index-guided arrays where the modes of the individual lasers are well confined and thus serve as proper basis fields for the coupled-mode description. In index-guided laser arrays, the transverse mode of each laser is confined by a real refractive index step in both the x- and y-directions of Figure 5.1. Gain-guided laser arrays, however, do not have the benefit of strong mode confinement in the lateral direction. There is no built-in refractive index step in that direction. They are best described as a single entity: a broad-area laser with a lateral (x-dependent) variation of pump current, gain, loss and refractive index. Lateral diffusion of carriers may also play an important role in these lasers. Antiguided arrays, where the coupling mechanism is a global interaction among all the emitters by means of lateral propagating waves, can also be described as a single entity. Y-junction arrays are another class of lasers that are not properly described by coupled-mode theory. Here we present a partial differential equation model that is quite general and can be used to describe a wide variety of arrays[13–15]. (There is also a less general approach that is based on interacting lateral modes[16].)

In the geometry of Figure 5.1 the electric field of the light wave is taken as

$$\mathbf{E} \approx \mathrm{Re}\{\Psi(x, z; t)\phi(y)\exp[\mathrm{i}(kz - \omega t)]\}, \tag{5.12}$$

where $\phi(y)$ is the transverse field distribution of the fundamental TE mode, and $\Psi(x, z; t)$ is the lateral field distribution whose amplitude may depend on time. The lateral field satisfies the paraxial wave equation

$$2\mathrm{i}k\frac{\partial \Psi}{\partial z} + \frac{\partial^2 \Psi}{\partial x^2} + k_0^2 \Gamma \Delta\varepsilon \Psi = 0, \tag{5.13}$$

where Γ is the active layer confinement factor, k is the propagation constant, $k_0 = \omega/c$, x is the lateral direction and z is the direction of propagation. The dielectric perturbation $\Delta\varepsilon$ represents the distribution of gain, loss and refractive index in the lateral direction. It depends on the carrier density N through

$$\Delta\varepsilon = -an_1\alpha N/k_0 - \mathrm{i}(n/k_0)a(N - N_0) + \mathrm{i}(n_2/k_0)(1 - \Gamma)\alpha_c/\Gamma, \tag{5.14}$$

where n_1 and n_2 are the refractive indices of the active layer and cladding layers, a is the gain coefficient, α_c is the loss in the cladding layer, α is the linewidth enhancement factor and N_0 is the carrier density required to achieve transparency. The carrier density $N(x, t)$ satisfies the diffusion equation

$$\frac{\partial N}{\partial t} = P(x) - \frac{N(x)}{\tau_s} - \Gamma v_g a(N - N_0)|\Psi(x)|^2 + D_{xx}\frac{\partial^2 N}{\partial x^2}. \tag{5.15}$$

Here P is the pump rate, τ_s is the carrier lifetime, D_{xx} is the diffusion coefficient and v_g is the group velocity.

5.4.1 Gain-guided arrays

We have used this model to simulate the behavior of two-stripe gain-guided laser arrays[14]. The simulations yield stable (time-independent), periodic or chaotic output intensities, depending on the pump current and the separation S between the stripes. Figure 5.5 shows chaotic intensity pulsations for a spacing of $S = 10.0$ μm. The upper trace is the output from stripe 1 and the lower trace is the output intensity from stripe 2. Note that the chaotic pulsations in the two stripes are synchronous with each other. This differs from the behavior of index-guided arrays where the pulsations are in antiphase for most values of coupling parameter.

There is a range of spacing S that gives rise to a train of regular pulses in the simulated output. Figure 5.6 shows the output of lasers 1 and 2

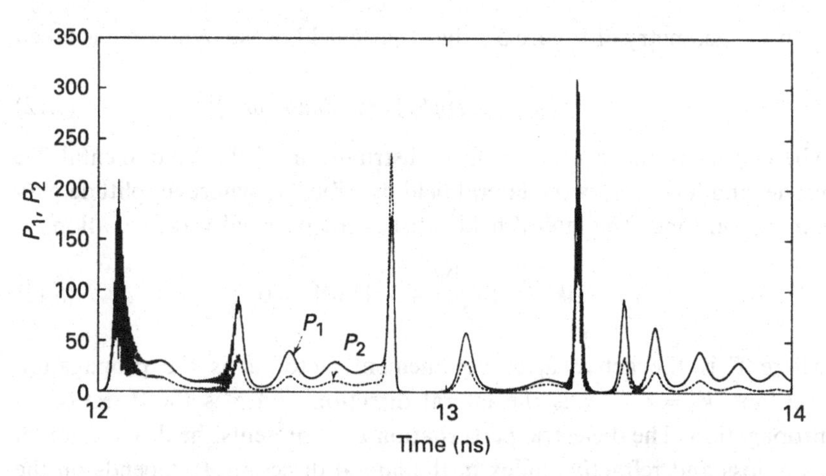

Figure 5.5. Simulated output intensity of a twin-stripe gain-guided array showing chaotic pulsations. Parameters used are listed in Ref. 20.

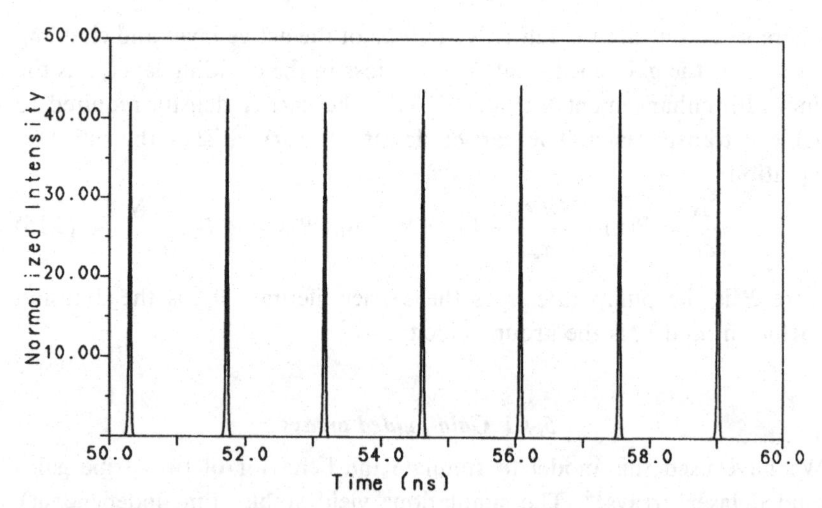

Figure 5.6. Temporal evolution of the light intensity in the near field of a twin-stripe gain-guided laser under regular self-pulsing conditions. (After Ref. 14.)

(superimposed) as a function of time for $S = 16$ μm. These pulsations have the character of repetitively Q-switched spikes of pulse width 40 ps and separation 2 ns (of the order of the carrier lifetime). In this self-Q-switched mode, the output intensity remains in a fixed lateral distribution, as seen in Figure 5.7, and a phase difference of π is maintained between the two lasers. The presence of several high-intensity peaks in the absorbing region

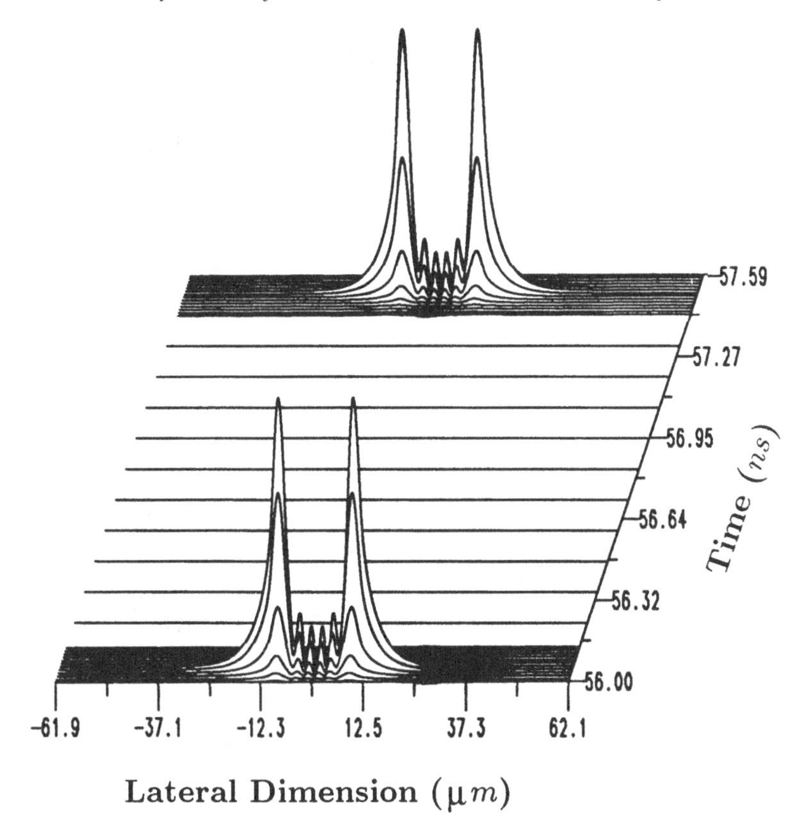

Figure 5.7. Spatiotemporal evolution of the near-field intensity of a twin-stripe gain-guided laser array. (After Ref. 14.)

between the pumped stripes is significant. These fields can saturate the absorption in the unpumped region and lead to repetitive Q-switching. This mechanism underlies the self-pulsations seen in these simulations and in many other array designs that incorporate absorbing regions between gain sections.

5.4.2 Antiguided laser arrays

Resonant antiguided laser arrays have achieved some impressive results in terms of diffraction-limited cw output power[17]. Recent devices have yielded in excess of 0.5 W cw in a diffraction-limited beam. Under pulsed conditions diffraction-limited-beam operation up to 2.1 W has been obtained. The operating principle is illustrated in Figure 5.8. The basic

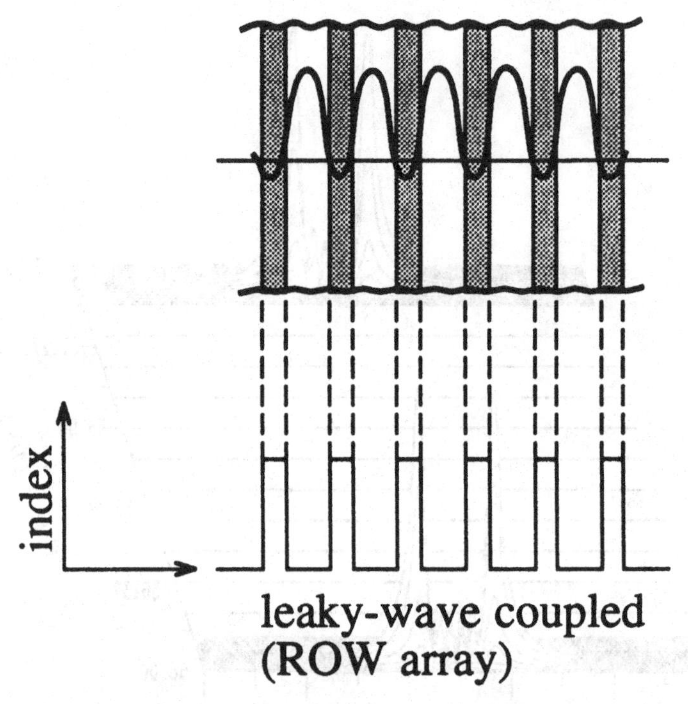

Figure 5.8. Schematic of an antiguided laser array. Typical index-step values are in the 10^{-1} to 10^{-2} range. (After Ref. 17.)

element is a single real-index antiguide consisting of a low-index core surrounded by a higher-index cladding. Such a structure can, of course, only support leaky waves. Gain in an antiguide core compensates for the edge radiation loss of the leaky modes. In an array of antiguides, coupling between the elements is by means of these leaky waves, which are propagating modes as opposed to evanescent modes. Antiguide arrays that satisfy a certain lateral resonance condition provide strong discrimination against out-of-phase modes in favor of the in-phase mode[17].

Because of the long-range nature of the coupling, a propagation model is necessary for describing the dynamics of antiguided arrays. Simulations based on Eqs. (5.14) and (5.15) indicate that antiguide arrays with interelement loss can also exhibit a number of instabilities[18]. Depending on device parameters such as the pump current, the total output power can pulsate in time, with each element in the array oscillating in unison. Figure 5.9 shows a pulsing solution. For these simulations the built-in index step is 4×10^{-2}, the stripe width is 4 μm, while the interelement spacing is 2 μm. The linewidth enhancement factor is 2. Here, too, the

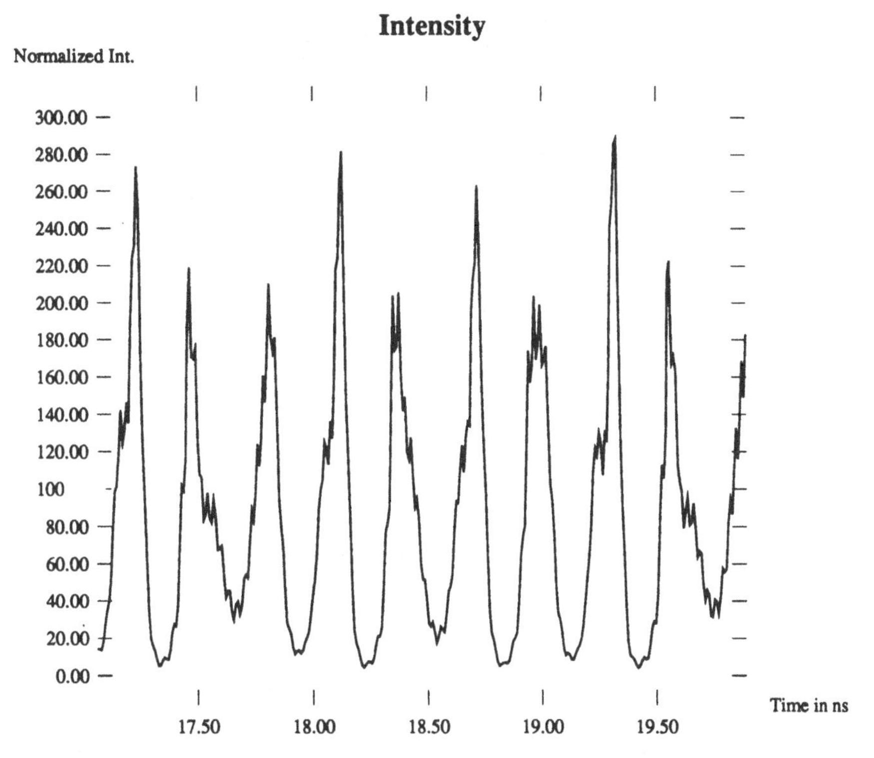

Figure 5.9. Simulated total output power of a ten-stripe antiguided array with significant interelement loss showing sustained pulsations.

self-pulsations are due to the saturable-absorber action of the interelement regions. The unsaturated interelement loss coefficient is 50 cm^{-1}. Interelement losses of this magnitude are sometimes purposely introduced in antiguided laser arrays in order to enhance their mode discrimination ability[17]. If there is little or no interelement loss the arrays *are* stable. Antiguided arrays without interelement loss have been fabricated at a wavelength of 0.98 μm[17], and do not show self-pulsations[27,29].

5.4.3 Y-junction arrays

The Y-junction array is a design that favors the in-phase mode of operation. The principle of the Y-junction laser is shown in Figure 5.10. When the incident modal amplitudes of the two-stripe segment are in-phase, they add constructively to excite the fundamental mode of the output waveguide. When they are out-of-phase they add to form an

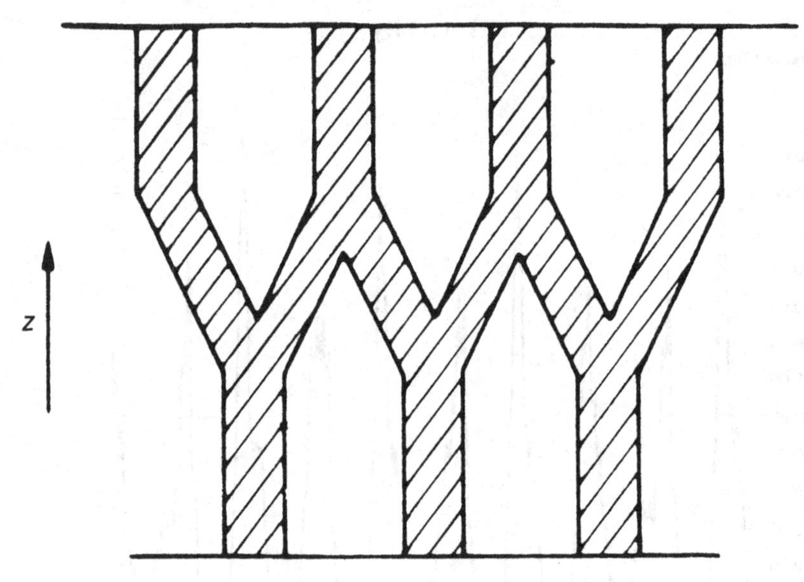

Figure 5.10. Schematic of a Y-junction array.

antiphase mode which is radiated away in the substrate. Because the guides in the two-stripe region of the Y-junction array are widely separated, there is no evanescent coupling. A proper theoretical description of the dynamics of the Y-junction array requires the use of the propagation model.

Computer simulations based on Eqs. (5.14) and (5.15) show that even a single Y-junction laser is capable of rich dynamic behavior[19]. Figure 5.11 shows the output of a Y-junction laser driven by a constant pump current. The laser exhibits sustained self-pulsations at a multi-gigahertz rate. The mechanism for the self-pulsations is as follows: constructive interference at the Y-junction creates an intense in-phase mode which strongly depletes the carriers at the center of the junction. The hole burnt in the carrier distribution means that the out-of-phase mode will have a better overlap with the gain distribution and will begin to grow at the expense of the in-phase mode. However, the out-of-phase mode is not supported by the single-guide section and begins to be radiated away. The output intensity drops as the intense in-phase mode is converted to the lossy out-of-phase mode. This allows the depleted carriers to be pumped up again and the process repeats itself indefinitely. This laser structure is essentially a device with a periodic self-modulation of loss.

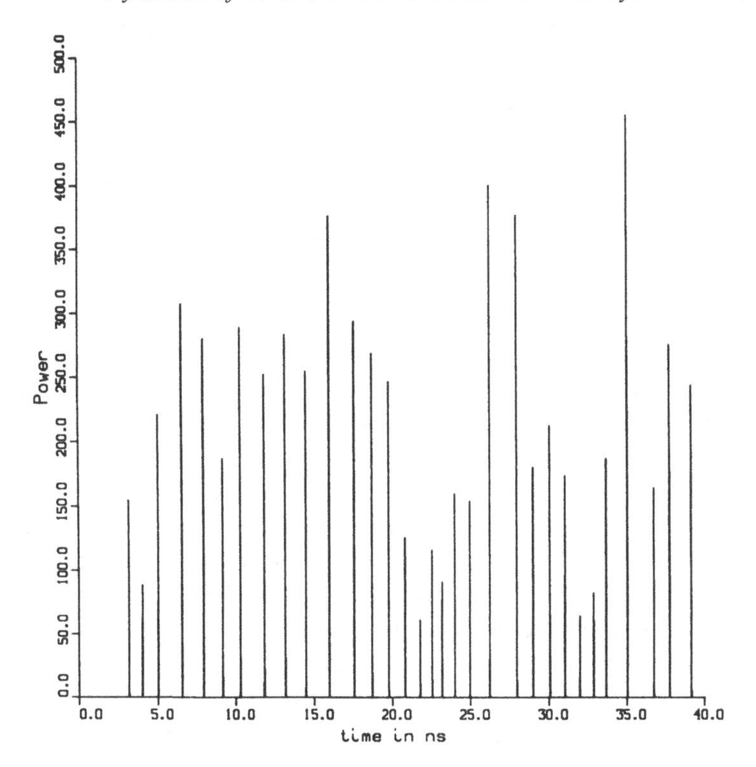

Figure 5.11. Simulated output power of a single Y-junction laser showing sustained pulsations under cw pump current.

Of course, if a single Y-junction laser exhibits self-sustained pulsations, an array of coupled Y-junction lasers should also pulsate. This pulsation, however, has less to do with the coupling between lasers than with the dynamics of the individual lasers.

5.5 Experimental observations

Spatiotemporal instabilities have now been observed in a wide variety of semiconductor laser arrays. These include gain-guided[1,20,21], weakly index-guided[25], antiguided[26,27], Y-coupled[22–24], and two-dimensional grating-surface-emitting arrays. To date, most of the experimental studies on laser array dynamics have employed high-speed streak cameras. A streak camera uses the nearly instantaneous photocathode emission of electrons which are subsequently accelerated and scanned (streaked) across a phosphor to create an image of the optical event of interest.

The streak camera can be thought of as equivalent to a large linear array of photodiodes, each having extremely fast temporal response. Thus, using a streak camera, one can simultaneously probe the spatial and temporal characteristics of radiation from a semiconductor laser array with bandwidths up to several hundred gigahertz.

5.5.1 Measurement techniques

The spatiotemporal evolution of the optical output from various laser arrays has been investigated under both pulsed and continuous (cw) conditions while operating the lasers at various drive currents above lasing threshold. The experimental arrangement in each case was similar to that first demonstrated[1] by DeFreez *et al.*, in the mid-1980s, and consisted, in part, of imaging the near or far field of the laser array on the entrance slit of a streak camera. For the pulsed measurements, the time evolution of the light intensity was monitored with a Hamamatsu C979 streak camera, or a C1587 main-frame equipped with a M1952 high-speed streak unit, in the experimental set-up shown in Figure 5.12. Achievable temporal resolutions were 8 and 1.6 picoseconds respectively. The current pulse and the streak camera were triggered synchronously by a video frame pulse extracted from the free-running video output of the vidicon by a video

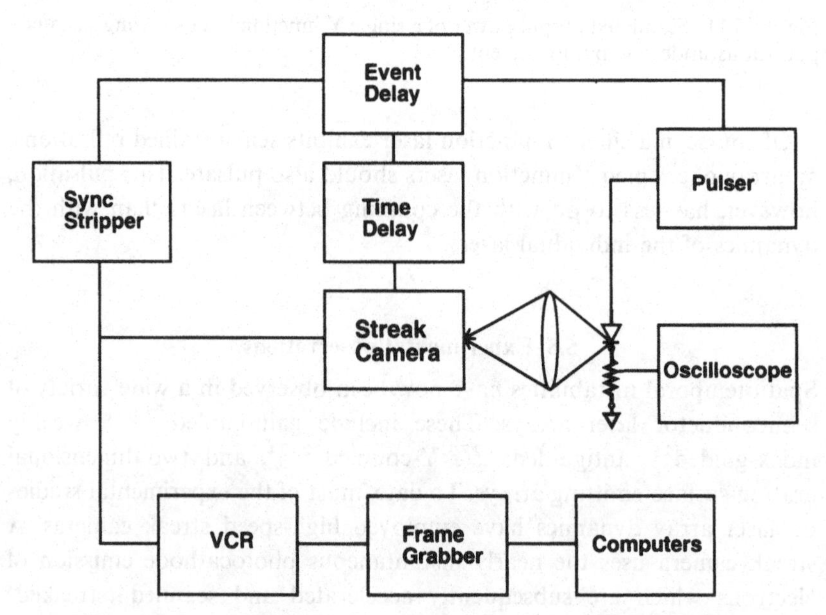

Figure 5.12. Experimental set-up for dynamic measurements on laser arrays.

sync-stripper. The duty cycle was controlled by the digital delay unit by setting the number of video frames between pulses. Another digital delay generator was used to adjust the time delay, making it possible to observe up to 6 nanoseconds of the time evolution of the laser output anywhere within the pulse duration. For cw operation, the set-up was essentially the same, except that the current pulser was replaced by a cw constant current source, the time delay device was removed and only the C1587/M1952 system was used. Streak events were digitized with a video frame grabber and transferred to various computers for data analysis and graphical presentation. In all cases extreme care was taken to avoid external optical feedback into the arrays which could lead to other instabilities. The following sections provide a survey of the rich dynamical behavior observed in the output of various phased array semiconductor lasers.

5.5.2 Gain-guided arrays

Two-stripe gain-guided arrays are the simplest nontrivial arrays and thus are of special interest from both a theoretical and experimental point of view. In order to reach insight into the dynamics of larger arrays we have carried out extensive studies on twin-stripe arrays.

Figure 5.13 displays the near-field optical output intensity from a twin-emitter semiconductor laser array during cw operation at 1.3 times threshold. The data shown here was recorded after the laser had been operating for several minutes. This array exhibits sustained self-pulsations with undamped, regular spikes which are well correlated between the two stripes. The pulsation frequency of 1.5 GHz increased with increasing drive currents, fitting the well-known relationship between the semiconductor relaxation oscillation frequency and the current overdrive. The fact that the two stripes pulsate in synchronism suggests that the saturable-absorber mechanism described in Section 5.4.1 is responsible for the dynamics. (Coupling induced instabilities, on the other hand, generally lead to periodic or aperiodic energy exchange between the emitters.) Indeed, the measured time-averaged near-field profile from the twin-stripe array reveals a substantial high-intensity peak in the absorbing region between the stripes (Figure 5.14). Saturation of the absorption by this peak will result in a repetitive Q-switching effect. Gain-guided laser arrays will generally be susceptible to this instability. We have observed similar self-pulsations in ten-element gain-guided arrays under both cw and pulsed conditions[1,20,21].

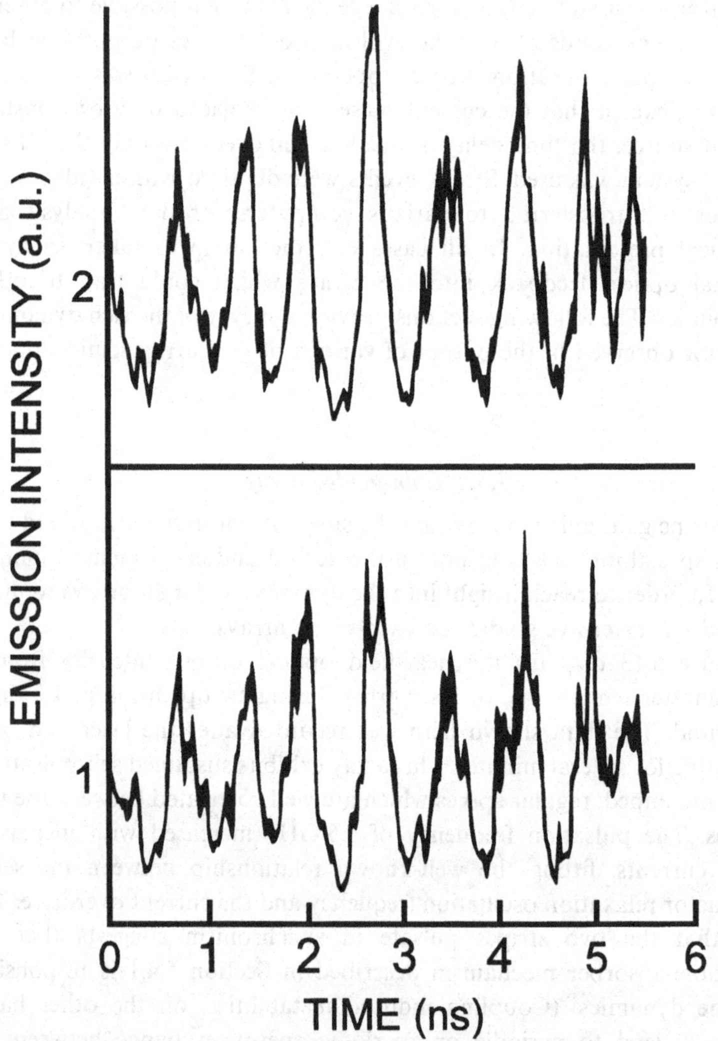

Figure 5.13. Measured temporal evolution of the output intensity of a twin-stripe gain-guided array.

5.5.3 Y-coupled index-guided arrays

Y-coupled index-guided semiconductor laser arrays have also displayed sustained self-pulsations and other interesting dynamics. Figure 5.15 shows the time evolution of the single-lobed far-field radiation pattern of

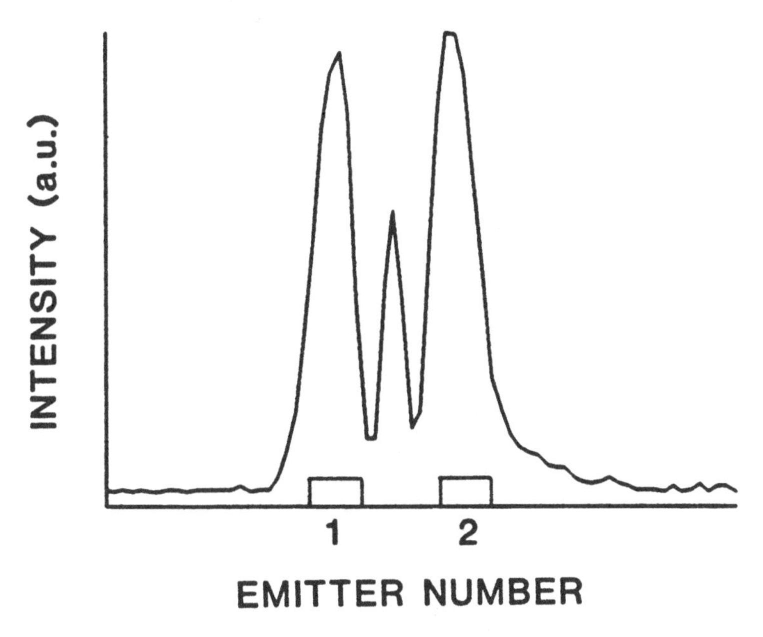

Figure 5.14. Time averaged near field intensity profile of the twin-stripe gain-guided array. (After Ref. 21.)

a ten-stripe Y-coupled array during pulsed operation[22]. Analysis of the diagram and other data taken at faster sweep rates revealed that the formation of the single-lobed pattern occurs within 20 ps of the initiation of lasing action. This implies that the phase difference between the fields in adjacent emitters evolves from a random value to a value near zero within that time interval. Figure 5.15 also shows evidence of beam steering in the far field. This suggests, as discussed in Section 5.3.2 that the phase is entrained but oscillates about a mean value of zero. Under cw operation[23] the output intensity exhibits strong sustained self-pulsations, as shown in the streak camera data of Figure 5.16(*a*). In some elements of the array, the modulation depth approaches 100% (Figure 5.16(*b*)). Because of local inhomogeneities, such as those due to temperature gradients, multi-stripe arrays sometimes break up into subarrays with different oscillation characteristics[24]. The overall dynamic behavior is extremely complicated. As indicated in Section 5.4.3, a single Y-guide laser by itself is capable of sustained pulsations. Saturable absorption can also lead to pulsations in an array. It is thus not clear how much of the observed dynamics of the Y-junction array results from coupling between different emitters.

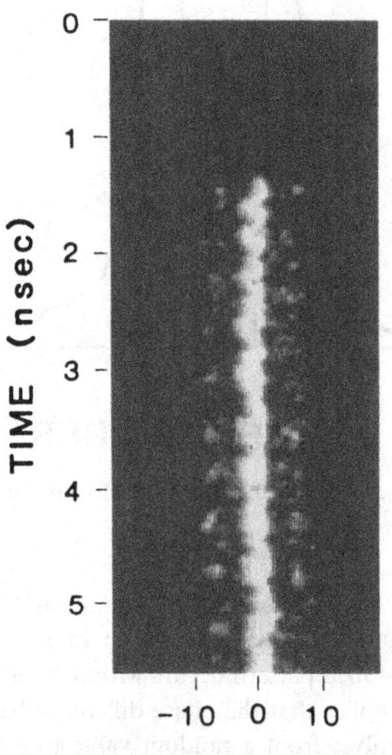

Figure 5.15. Far-field evolution of a ten-stripe Y-junction array showing beam scanning. (After Ref. 22.)

5.5.4 Antiguided arrays

Recent experimental studies have determined that some Resonant Optical Waveguide (ROW) antiguide arrays can exhibit pulsations in their output powers[26,27]. Although generally well phase-locked, with diffraction-limited output to many times above threshold, some arrays have been found to operate stably only at low-output power levels. Figure 5.17 shows the time evolution of the central far-field lobe for a 20-element ROW laser array operating cw at $2.3I_{th}$. Deeply modulated sustained self-pulsations near the relaxation oscillation frequency are clearly visible. All emitters were observed to pulsate in unison in the near field, apparently a consequence of the strong, parallel, resonant optical coupling of the antiguides. Increasingly more erratic, possibly chaotic, dynamics are observed as the injection current is increased, resulting in an extremely broad intensity noise spectrum. Figure 5.18 demonstrates erratic fluctuations in the far-field intensity of a ROW device operating near three times threshold and having strong interelement loss. The self-pulsations are due to interelement saturable absorption, as discussed in Section

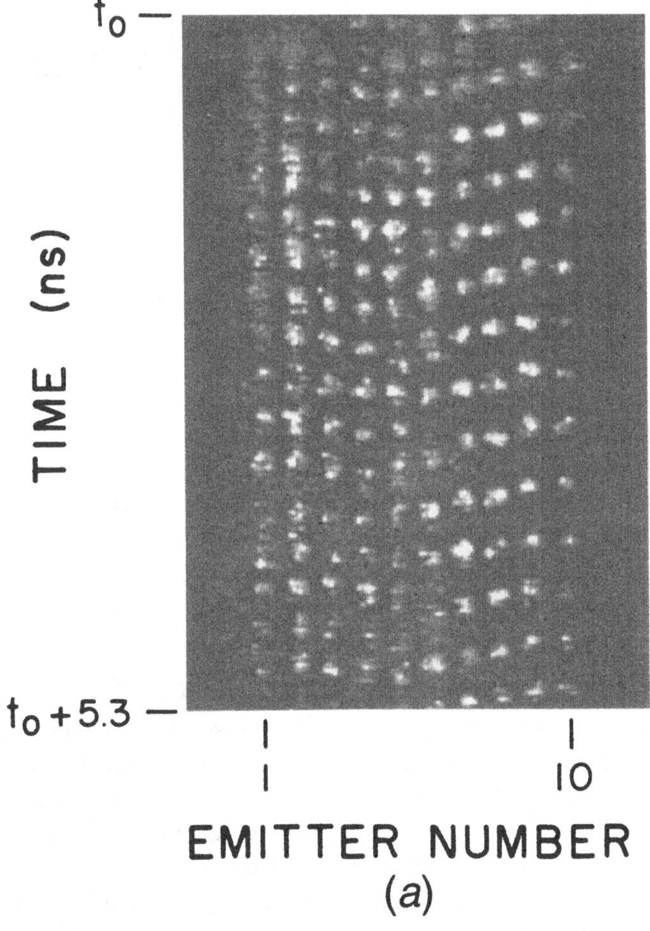

Figure 5.16. (*a*) Streak camera measurement of a near-field evolution of a ten-stripe Y-junction array under cw operation. (*continued*)

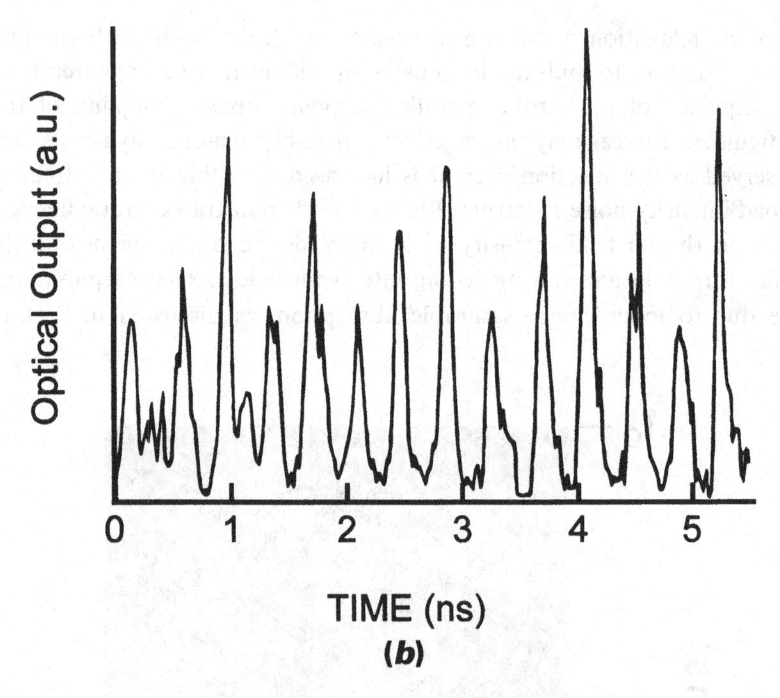

Figure 5.16. (*continued*) (*b*) Output of a single strip (No. 7) in the array. (After Ref. 23.)

5.4.2. In fact, it is well known that periodically distributed absorbers in loss-coupled DFB lasers can lead to highly nonlinear behavior[28]. For antiguided arrays, stabilization may be achieved by eliminating or minimizing the interelement absorption, as successfully demonstrated for loss-coupled DFB lasers[28]. Indeed, very recent experimental results indicate that antiguided arrays with little or no interelement absorption are stable[27,29] to ≥ 0.45 W cw in near-diffraction-limited beams[27].

5.5.5 Two-dimensional grating-surface-emitting (GSE) arrays

Grating-Surface-Emitting (GSE) arrays[30] consist of a number of gain regions connected longitudinally by passive grating regions. The gratings serve three purposes: (i) they provide feedback to each gain region; (ii) they couple different gain sections together through mutual injection; and (iii) they couple useful light vertically out of the plane of the array. In a two-dimensionally coherent GSE array, each gain region in turn consists of an array of laterally coupled index-guided lasers. The lateral coherence is achieved either through evanescent coupling or by the use of Y-branches.

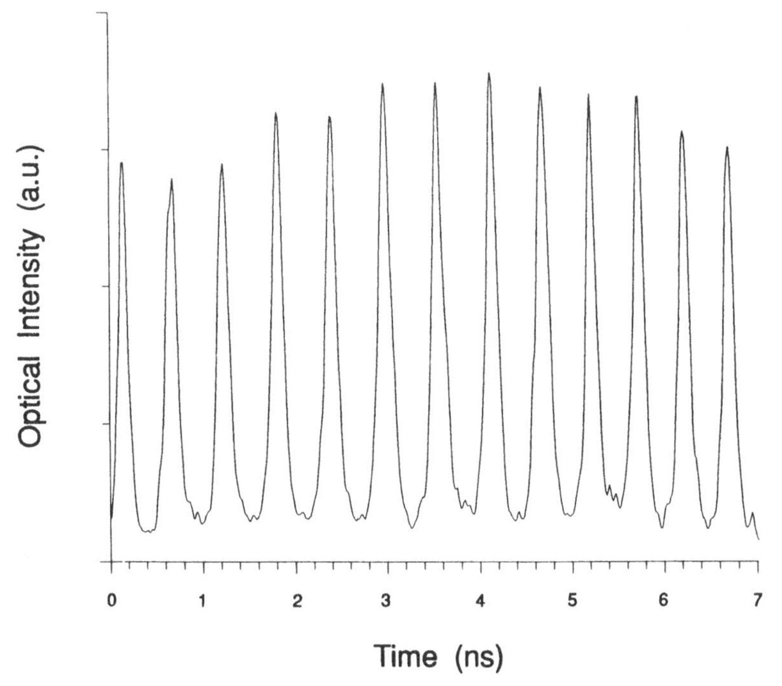

Figure 5.17. Measured time evolution of the central far-field lobe for a 20-element antiguided array with significant interelement loss, operating cw at $2.3I_{th}$. (After Ref. 26.)

As with other array designs, the goal is to achieve stable single-mode operation with very narrow beam divergence. The complexity of the structure, however, suggests the possibility of interesting dynamic behavior on subnanosecond time scales.

Streak camera studies on 2D-GSE arrays have revealed oscillatory instabilities as well as stable, time-independent, single-longitudinal-mode behavior, depending on device parameters and operating conditions[31,32]. The first devices tested were 10×2 arrays – two gain sections, each consisting of ten laterally coupled index-guided lasers. These early devices could only be operated in the pulsed mode because of their relatively high threshold currents. When driven with 200 ns pulses these lasers exhibited strong sustained self-pulsations at the relaxation oscillation frequency of 2 GHz. The origin of the strong self-pulsations is very likely to be the saturable-absorber action of the passive-grating section of the array. Both the passive section and the gain section share the same quantum-well waveguide layer. The absorption in the passive waveguide section is easily

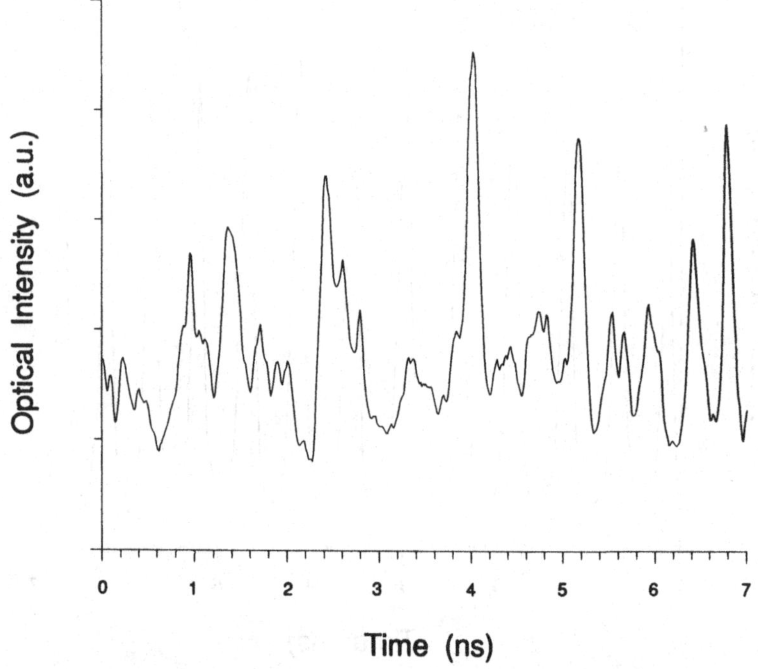

Figure 5.18. Time evolution of the central far-field lobe for a 20-element antiguided array with significant interelement loss, operating cw at $3I_{th}$. (After Ref. 26.)

saturated[33]. Under the appropriate conditions this can lead to the repetitive Q-switching behavior that is typical of lasers with longitudinally inhomogeneous pumping[34].

Recent advances in 2D-GSE laser array technology have resulted in devices that can operate under cw conditions and in a single, narrow ($\approx$35 MHz) longitudinal mode. Under these conditions we have observed stable, time-independent outputs from a 10 × 5 GSE array[32].

We conclude this survey of experimental results with a look at 2D-GSE ring laser arrays[35]. These devices are the only ones we have studied that have shown clear evidence of coupling induced instabilities, as opposed to the saturable-absorber variety. The ring configuration consists of two parallel columns of 2-D GSE arrays coupled at both ends by a pair of total internal reflection 45° turning mirrors (Figure 5.19). This configuration allows multiple columns to be optically coupled in large-aperture coherent arrays. By tuning the currents supplied to the various gain sections, each column can be made to operate in a single longitudinal mode. Within the ring structure we can consider the interaction between

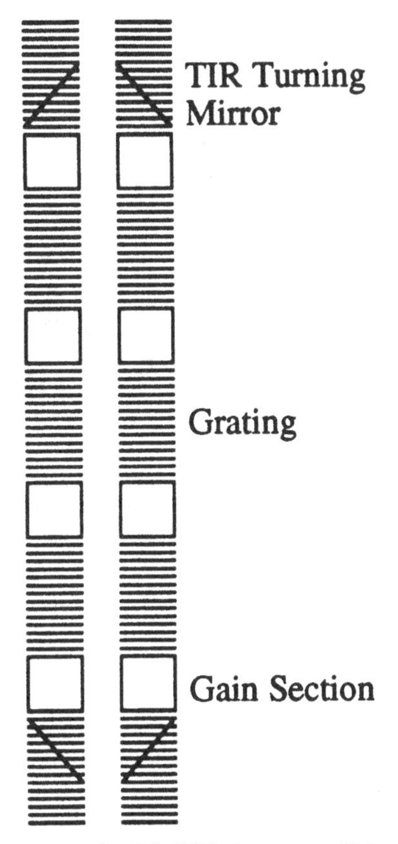

Figure 5.19. Geometry of a 2D-GSE ring array. (After Ref. 35.)

two counter propagating traveling waves coupled by means of the distributed feedback gratings and by the gain media. Such coupling often occurs in ring lasers and can lead to instabilities in which the two waves periodically exchange energy[36]. This energy exchange should manifest itself in antiphase dynamics when the outputs of the two columns are monitored simultaneously.

Figure 5.20 shows the sustained oscillations at approximately 10 GHz from two gratings of a $10 \times 4 \times 2$ GSE ring array. The array is driven by 120 ns current pulses at 1.5 times threshold and operates in a single longitudinal mode. It is clear from the diagram that the oscillations are in antiphase, a signature of coupling induced instability. Measurements on other arrays have revealed antiphase dynamics at frequencies as high as 27 GHz. It has been suggested that the modulation response of

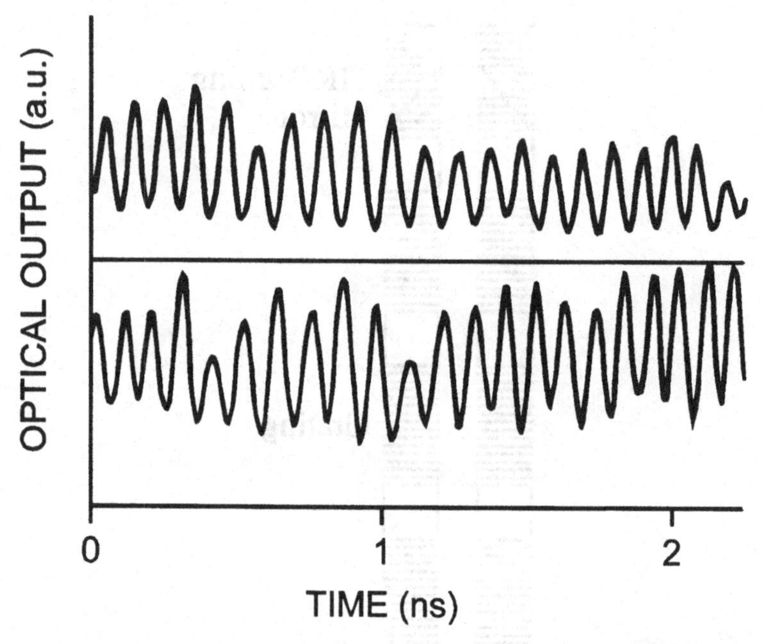

Figure 5.20. Output of two gratings in a 10 × 4 × 2 GSE ring array. Note the antiphase dynamics.

semiconductor lasers can be extended far beyond the relaxation oscillation frequency by taking advantage of this coupling induced resonance[7,8].

5.6 Conclusions

Semiconductor laser arrays are of considerable technological importance as compact sources of high-output power in a well-collimated beam. At a more fundamental level, we suggest that semiconductor laser arrays are ideal structures for the study of self-organization and complex spatio-temporal behavior in extended nonlinear systems. In this review we have presented theoretical and experimental results that show that these arrays can exhibit a number of dynamic instabilities. An understanding of their rich dynamics will aid in the development of stable, compact, semi-conductor light sources for a variety of technological applications.

Acknowledgement

We acknowledge the support of the National Science Foundation under Grant ECS-8906214. Additional support was provided by the US Air Force Office of Scientific Research.

References

1. R. K. DeFreez, R. A. Elliott, T. L. Paoli, R. D. Burnham and W. Streifer, in *Proc. 13th Int. Quantum Electron. Conf.*, 18–21 June 1984, Anaheim, CA, Paper PD-B3; R. A. Elliott, R. K. DeFreez, T. L. Paoli, R. D. Burnham and W. Streifer, *IEEE J. Quantum Electron.*, **QE-21**, 598 (1985).
2. H. G. Winful and S. S. Wang, *Appl. Phys. Lett.*, **53**, 1894 (1988).
3. S. S Wang and H. G. Winful, *Appl. Phys. Lett.*, **52**, 1774 (1988).
4. J. G. Tsacoyeanes, *J. Appl. Phys.*, **64**, 32 (1988).
5. P. K. Jakobsen, R. A. Indik, J. V. Moloney, A. C. Newell, H. G. Winful and L. Rahman, *J. Opt. Soc. Am.*, **B8**, 1674 (1991).
6. H. G. Winful and L. Rahman, *Phys. Rev. Lett.*, **65**, 1575 (1990).
7. G. A. Wilson, R. K. DeFreez and H. G. Winful, *IEEE J. Quantum Electron.*, **QE-27**, 1696 (1991).
8. G. A. Wilson, R. K. DeFreez and H. G. Winful, *Optics. Commun.*, **82**, 293 (1991).
9. H. G. Winful, *Phys. Rev. A.*, **46**, 6093 (1992).
10. S. S. Wang and H. G. Winful, in *OSA Proceedings on Nonlinear Dynamics in Optical Systems*, N. B. Abraham, E. M. Garmire and P. Mandel (eds.), vol. 7, p. 119 (Optical Society of America, Washington, DC, 1991).
11. K. Otsuka, *Phys. Rev. Lett.*, **65**, 329 (1990).
12. Z. Jiang and M. McCall, *Proc. IEE*, Part J, **139**, 88 (1992).
13. S. S Wang and H. G. Winful, in *OSA Proceedings on Nonlinear Dynamics in Optical Systems*, N. B. Abraham, E. M. Garmire and P. Mandel (eds.), vol. 7, p. 114 (Optical Society of America, Washington, DC, 1991).
14. S. S. Wang and H. G. Winful, *J. Appl. Phys.*, **73**, 462 (1993).
15. L. Rahman and H. G. Winful, *Opt. Lett.*, **18**, 128 (1993).
16. P. Mendendez-Valdes, E. Garmire and M. Ohtaka, *IEEE J. Quantum Electron.*, **QE-26**, 2075 (1990).
17. D. Botez, *Proc. IEE*, Part J, **139**, 14 (1992).
18. S. Ramanujan and H. G. Winful, *Conference on Lasers and Electro-Optics*, paper CTu 17, Baltimore, MD, May 1993.
19. P. Subramanian and H. G. Winful, *Integrated Photonics Research Conference*, Palm Springs, CA, March 1993.
20. R. K. DeFreez, N. Yu. D. J. Bossert, M. Felisky, G. A. Wilson, R. A. Elliott, H. G. Winful, G. A. Evans, N. W. Carlson and R. Amantea, *OSA Proc. on Nonlinear Dynamics in Optical Systems*, Afton, 4–8 June 1990, Paper ThA1.
21. N. Yu, PhD Thesis, Oregon Graduate Institute of Science and Technology, Beaverton, Oregon (1990).
22. R. K. DeFreez, R. A. Elliott, K. Hartnett and D. F. Welch, *Electron. Lett.*, **23**, 589 (1987).
23. N. Yu, R. K. DeFreez, D. J. Bossert, R. A. Elliott, H. G. Winful and D. F. Welch, *Electron. Lett.*, **24**, 1203 (1988).
24. R. K. DeFreez, D. J. Bossert, N. Yu, K. Hartnett, R. A. Elliott and H. G. Winful, *Appl. Phys. Lett.*, **53**, 2380 (1988).
25. N. Yu, R. K. DeFreez, D. J. Bossert, G. A. Wilson, R. A. Elliott, S. S. Wang and H. G. Winful. *Appl. Opt.*, **30**, 2503 (1991).
26. D. J. Bossert, Phillips Laboratory, Kirtland AFB, NM, unpublished.
27. S. Ramanujan, H. G. Winful, M. Felisky, R. K. DeFreez, D. Botez, M. Jansen and P. Wisseman, *Appl. Phys. Lett.*, **64**, 827 (1994).

28. Y. Luo, H.-L. Cao, M. Dobashi, H. Hosomatsu, Y. Nakano and K. Tada, *IEEE Photonics Tech. Lett.*, **4**, 692 (1992).
29. M. Jansen, P. Wisseman and D. Botez, private communication.
30. G. A. Evans, N. W. Carlson, J. M. Hammer, M. Lurie, J. K. Butler, S. L. Palfrey, R. Amantea, L. A. Carr, F. Z. Hawrylo, E. A. James, C. J. Kaiser, J. B. Kirk and W. F. Reichert, *IEEE J. Quantum Electron.*, **QE-25**, 1525 (1989).
31. R. K. DeFreez, D. J. Bossert, N. Yu, J. M. Hunt, H. Ximen, R. A. Elliott, G. A. Evans, M. Lurie, N. W. Carlson, J. M. Hammer, S. L. Palfrey, R. Amantea, H. G. Winful and S. S. Wang, *Photon. Technol. Lett.*, **1**, 209 (1989).
32. M. K. Felisky, G. A. Wilson, R. K. DeFreez, G. A. Evans, N. W. Carlson, S. K. Liew, R. Amantea, J. H. Abeles, C. A. Wang, H. K. Choi, J. N. Walpole and H. G. Winful, unpublished.
33. K. Kojima, S. Noda, K. Mitsunaga, K. Kyuma and K. Hamanaka, *Appl. Phys. Lett.*, **50**, 1705 (1987).
34. N. G. Basov, *IEEE J. Quantum Electron.*, **QE-4**, 855 (1968).
35. D. J. Bossert, R. K. DeFreez, H. Ximen, R. A. Elliott, J. M. Hunt, G. A. Wilson, J. Orloff, G. A. Evans, N. W. Carlson, M. Lurie, J. M. Hammer, D. P. Bour, S. L. Palfrey and R. Amantea, *Appl. Phys. Lett.*, **56**, 2068 (1990).
36. N. V. Kravtsov, E. G. Lariontsev and A. N. Shelaev, *Laser Physics*, **3**, 21 (1993).

6

High-average-power semiconductor laser arrays and laser array packaging with an emphasis on pumping solid state lasers

RICHARD SOLARZ, RAY BEACH, BILL BENETT, BARRY FREITAS, MARK EMANUEL, GEORG ALBRECHT, BRIAN COMASKEY, STEVE SUTTON AND WILLIAM KRUPKE

6.1 Introduction

In recent years there has been an evolution in interest in semiconductor laser technology from exploring the performance limits of single devices of inherently low power to exploiting the potential of large arrays of devices of much higher power. While the low-power semiconductor devices are a proven component in many commercial systems, it is clear that a number of applications also exist for high-power arrays and that semiconductor-laser researchers are now poised to tackle them.

The manufacture of large volumes of semiconductor laser arrays with high yield has enabled the fabrication of larger and larger arrays. Rather than mounting single devices on copper submounts, and thus limiting the modest amount of steady state waste heat which can be dissipated, the laser designer is now faced with the challenging task of fabricating increasing numbers of devices in increasingly smaller volumes and operating these units at increased duty cycle or average power. We are faced with the same evolution of technology which IC chip manufacturers have faced for years; however, the semiconductor lasers are far more temperature sensitive than silicon-based electronic circuits and the task is ever more challenging.

Depending on the application, the utility of any laser is governed by its ability to deliver photons to a remote location, which in turn depends on the beam quality. Extensive research has been performed over the last decade in the area of phase locking arrays of semiconductor lasers and amplifiers directly as the route to high-power coherent beams. While significant progress has been made (as reported in chapters 1 and 2), the consensus is that, at least in the near term, the shortest path to combining the output of many individual diode-laser apertures into a single spatially and temporally coherent multiwatt beam is to use a solid state laser as the storage medium for the semiconductor diode lasers at least at powers above several watts. (In this book, Dan Botez's chapter on phase-locked arrays (Chapter 1) and D. Welch and D. Mehuys's chapter on MOPA arrays (Chapter 2) demonstrate that at the watt level diode arrays out-perform diode-pumped solid state lasers.) Thus an additional constraint is placed upon the designer of high average power diode laser arrays, which is that the arrays must be designed to operate in a manner which allows the solid state insulating crystal laser to function properly. As a result the design of the arrays is not simply a matter of packaging as many devices of high power into as small a volume as possible, but it is also defined by average radiance, beam intensity and uniformity, and reliability requirements which are set by each application.

In this chapter we will first set the requirements in Section 6.2 for the fabrication of high average power diode arrays as set by the thermal, electrical and solid state laser performance boundary conditions. In Section 6.3 we will review various approaches to the packaging of high-power arrays. In Section 6.4 we will cite performance from the devices which have been fabricated to date and we will briefly review the reliability or lifetime results to date. In Section 6.5 we will discuss high-power performance limitations and we will also review important work which is currently being performed in the area of improving the lifetime and reliability of very high-power arrays.

6.2 High-power semiconductor laser array requirements

6.2.1 Semiconductor bar package thermal resistivity requirements

Consider a more or less standard semiconductor laser cavity. Garden variety performance from a quantum-well laser in AlGaAs would be to obtain approximately 5 mW per μm of emission aperture of radiation from a laser stripe which is approximately 500 μm in length. If this laser is 50% efficient and has a 50% packing density, this corresponds to a thermal footprint from the device of 2000 W/cm^2 if the laser is operated cw. The characteristic temperature, T_0, which describes the temperature sensitivity of AlGaAs operational parameters such as threshold current, is typically approximately 125–140 K. Thus in order to maintain efficient operation of the semiconductor diodes as well as to maintain good lifetime and reliability the temperature rise at the diode junction should typically be as small as possible and in general should be no greater than approximately 25° above room temperature. In order to run high-power cw devices it is thus necessary to develop a packaging technology with a thermal resistivity of less than $0.025°(\text{cm}^2/\text{W})$[1,2].

6.2.2 *Two-dimensional array peak and average power requirements*

Next consider the requirements from the standpoint of the required fluence and uniformity in intensity as well as in wavelength as needed by the solid state laser. What we will show in the next paragraphs is that an efficient high average power diode pumped solid state laser made from a thermally robust material such as YAG can typically accept approximately 100–300 W/cm^2 average power per square centimeter, depending on the slab thickness, of pumped surface from a diode array which is driving the insulating crystal[3].

The solid state laser must operate at high gain; generally the gain must be approximately 15 times the losses experienced in the laser material itself in order to operate efficiently. Since losses in most well-developed insulating crystals are of the order of 0.25% per centimeter, the laser gain should be of the order of a few percent per centimeter or higher.

This sets the peak irradiance requirements for a diode array which is used to energize a solid state laser. The excited state number density, N^*, is

$$N^* = P_{\text{abs}} \times (1 - f_{\text{loss}}),$$

where P_{abs} is the number of photons absorbed per cm^3 and f_{loss} is the fraction not decayed, which is

$$f_{\text{loss}} = [1 - \exp(-t_{\text{pump}}/t_{\text{fluor}})]\, t_{\text{fluor}}/t_{\text{pump}},$$

where t_{pump} and t_{fluor} are the pump array 'on' time and the storage lifetime of the medium respectively. The solid state laser must be designed so that it absorbs a reasonably large fraction of the array photons where this fraction is

$$f_{\text{abs}} = 1 - \exp(-N\sigma_{\text{a}}t),$$

where the number density of absorbers is N, t is the absorbing medium thickness, and σ_{a} is the absorption cross-section. As an example the peak absorption cross-section for Nd:YAG at 808 nm is 6.0 pm^2 but the average over 3 nm FWHM (a typical diode array wavelength spread) is 3.0 pm^2. Thus at a doping density of 0.75 atom per cent, 70% of the photons from a 3 nm FWHM array are absorbed in 4 mm, and 80% are absorbed in 5 mm, of absorption path.

A given thickness of insulating material, whether used in a rod or slab configuration, can be driven thermally so hard that the material is fractured due to thermally induced stresses. Normally, a laser is designed with a safety margin of operation, b, such that it operates at 20% of this fracture value. The design point for the diode arrays is then

$$P_{\text{outavg}} = 12bR_{\text{T}}/2t\chi\eta f_{\text{abs}}DC,$$

where P_{outavg} is the average power useable from the diode arrays, b is 0.2, R_{T} is the thermal shock parameter (11 W/cm for YAG), t is the material thickness, χ is the waste heat generated per photon absorbed (for YAG measured as 0.38), η is the transport efficiency from the array to the slab, which we arbitrarily take to be 0.9, and DC is the operating duty cycle of the array. The result is that a 5 mm-thick slab will use as much as 100 W/cm^2 average power when pumped cw, and as much as 1000 W/cm^2 when pumped at 10% duty cycle.

Returning to check that the laser is, in any case, still operating efficiently, that is, at a gain to loss ratio sufficiently high that efficient Q-switched extraction is obtained, we calculate the gain, α, as

$$\alpha = 12bR_{\text{T}}\sigma_{\text{e}}t_{\text{pump}}f_{\text{loss}}QD/t^2h\nu\chi,$$

where σ_{e} is the stimulated emission cross-section, QD the quantum defect of 0.76, and $h\nu$ the photon energy of 1.16 eV. We find for 5 mm-thick YAG pumped cw and Q-switched at 5 kHz that the gain is 4.3% cm^{-1}, which exceeds the nominal losses of 0.25% cm^{-1} in YAG by our stated ratio of 15 to 1, but just barely. A laser of this type pumped cw, but Q-switched at higher prf, will become gain-starved, and lasers which are pumped at lower duty cycles with higher peak fluence will always Q-switch efficiently.

To summarize, we have shown that the average power and peak fluence from the diode array are specified by the materials' properties of the solid state laser. For YAG, the most thermally robust material, we have seen that an average power of approximately 100 W/cm^2 from the pump array is the most that can be used for high average power pumping in the cw limit. This figure scales inversely to first order with the operating duty cycle of the arrays.

6.2.3 Two-dimensional array uniformity requirements

Next consider pump wavelength and intensity uniformity requirements. The importance of maintaining some degree of wavelength uniformity and

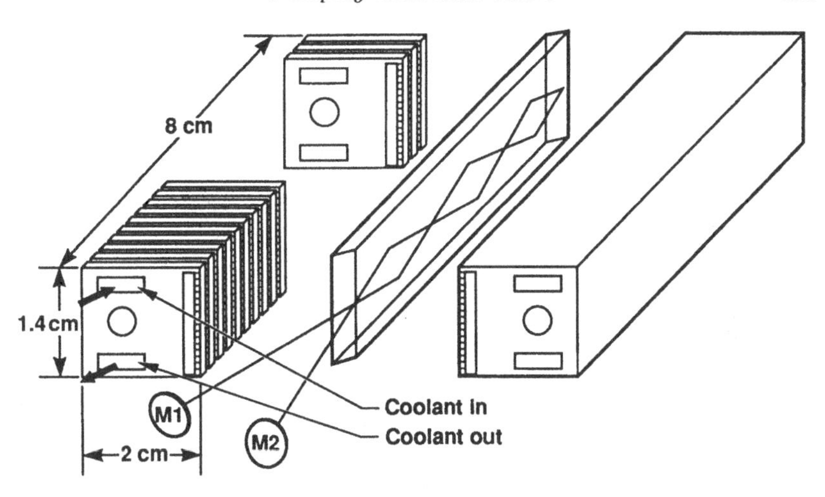

Figure 6.1. A schematic of a diode pumped slab showing a three-diamond optical path. Rays leave mirror 1, enter the slab and propagate down the slab by total internal reflection. The back face of the slab is mirrored, returning the rays on a 'mirrored' path. Such a ray propagation path achieves unity fill.

intensity uniformity from the two-dimensional diode arrays is to deposit stored energy sufficiently uniformly that the wavefront out of the solid-state laser is not degraded. Figure 6.1 displays the so-called three-dimensional optical path, the optical footprint or propagation path which was used to calculate the optical properties of the 1 μm Nd:YAG laser beam. Figures 6.2 and 6.3 shows the differences in wavefront from pumping a zigzag slab amplifier with two different sets of requirements placed on the arrays. In both cases a Nd:YAG slab was pumped which was 4 mm thick, 1.4 cm high, and 8 cm in length. The 1 μm-laser cavity is formed by zigzagging the beam along the length of the laser such that it samples the surfaces which face the diode pump arrays three times in each pass through the amplifier. The result is that the variation on wavefront uniformity is not changed substantially as the specification on diode array uniformity is changed from $\pm 15\%$ in intensity and ± 3 nm FWHM in wavelength (Figure 6.2) to $\pm 20\%$ and ± 5 nm (Figure 6.3). In fact, the variation in wavefront goes from about a half-wave to a wave through the centerline across the long dimension (1.4 cm) of the output aperture and from zero to about a half-wave across the thickness dimension (4 mm) in both cases. Since this level of distortion is not strongly dependent on the number of diamonds traced out within the

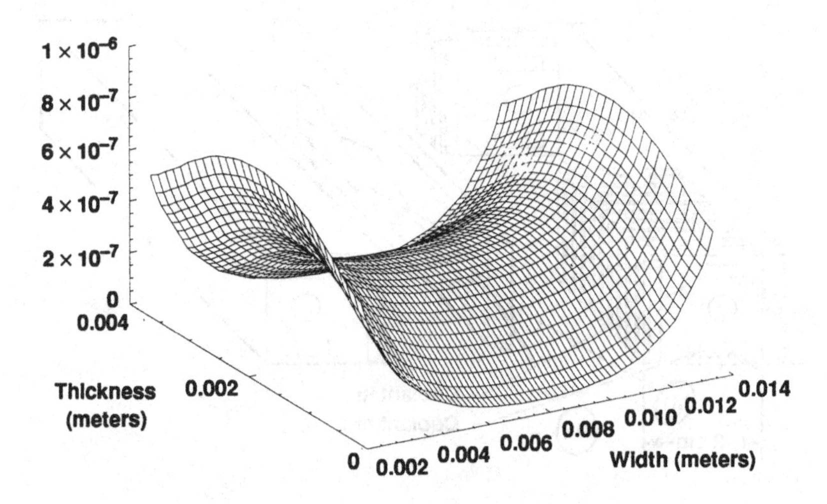

Figure 6.2. A calculated wavefront for a $\pm 15\%$ diode intensity variation and a ± 3 nm FWHM variation in diode wavelength. The uplift of the wavefront at the edges results from a lower temperature due to pump roll-off. Adding heat at the edges can flatten the wavefront in the height direction. Tailoring the source can be used to reduce the wavefront distortion in the thickness direction.

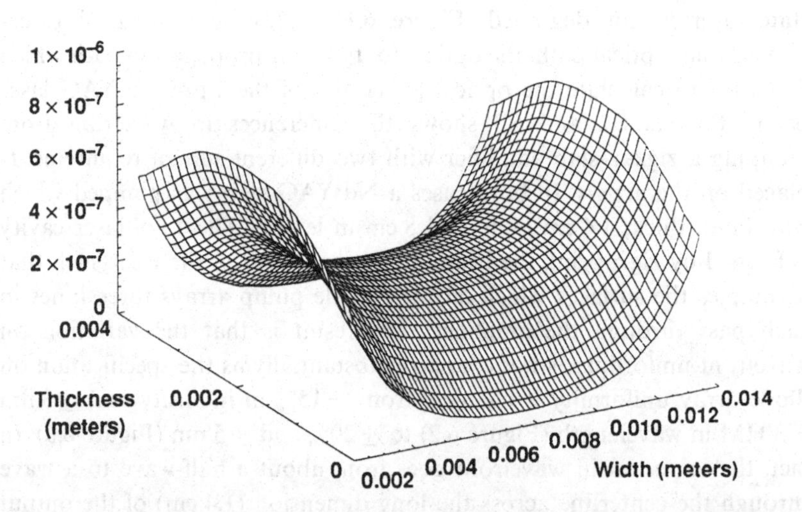

Figure 6.3. A calculated wavefront for the same geometry as used in the Figure 6.2 calculation, both with the intensity variation and wavelength variation increased to $\pm 20\%$ and ± 5 nm respectively. Note that there is negligible change in the wavefront.

slab, we can conclude that end effects dominate the remaining wavefront non-uniformity. The solution to better wavefront uniformity is, then, not tighter specifications on the arrays, but, rather, either to limit the useful aperture of the slabs or tailor the intensity of the arrays to minimize the end effects.

Thus, the final summary of the specifications on the arrays is: (a) the packaging technology should be capable of a thermal resistivity of less than 0.025° (cm^2/W); (b) the average power capabilities of the two-dimensional arrays need not exceed typically 100 W/cm^2 average optical power; and (c) the wavelength uniformity need not exceed typically 5 nm FWHM across the array. Assuming that one-third of this wavelength variation is budgeted against temperature non-uniformity, this means that a random temperature variation of approximately 5° across the array is acceptable. Intensity variations of 20% or so are adequate.

Borrowing a result from Section 5 in this chapter we find that AlGaAs cw arrays with good lifetime performance (of the order of 1000 hours or more) can typically be fabricated to deliver 20 W or more per linear centimeter. (Although, in Section 6.5 we will see that rapid advances now allow the fabrication of reliable devices operating cw at 25 mW/μm of emission aperture – a five- to six-fold improvement over the figure assumed here.) Assuming that the 20 W figure is used, this means that a cw-pumped slab should be excited by arrays of thickness no greater than 2 mm each in order to obtain 100 W/cm^2 cw pumping. The 'stacking pitch' must then be at least five per centimeter. Solid state lasers which are operated at lower duty cycles, say 5%, can be pumped by arrays operating at higher peak power but the same average power. A YAG slab operated at 5% duty cycle can be excited by two-dimensional pulsed arrays which have a stacking pitch of 40 bars per linear centimeter, assuming that the pulsed bars typically deliver 50 W peak power per linear centimeter of bar.

Finally, we comment on the types of arrays needed to pump rod lasers. It has been found that virtually the same arrays operated with the same stacking pitches, uniformity, etc., are needed. More detailed calculations show that extremely good radial excitation uniformity can be achieved, even in the rod geometry. Figure 6.4 shows a cross-sectional cut of a 6 mm diameter yttrium orthosilicate (Nd:YOS) rod which is pumped from four quadrants by 1.8 cm-long bars which are oriented with the fast axis of their beam divergence propagating parallel to the axis of the laser rod. The diodes are coupled on to the rod using an $f/3$ cylindrical coupling optic.

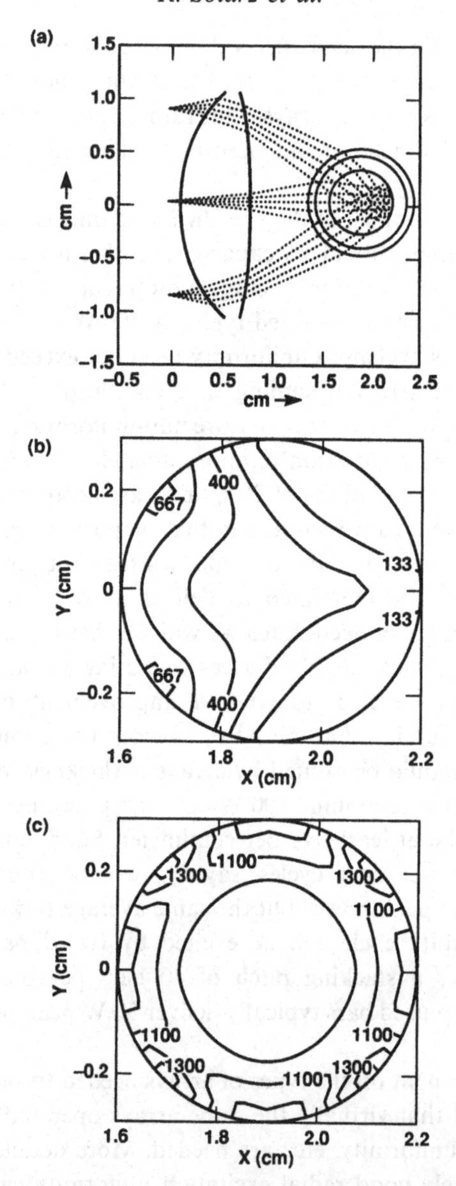

Figure 6.4. Good uniformity of illumination for an insulating crystal laser rod can be achieved by arranging two-dimensional arrays and pumping with their fast axis diverging in and out of the plane of the paper, along the principal axis of the insulating crystal laser rod, as shown in panel (*a*). An $f/3$ coupling optic is used to match the slow axis of the two-dimensional array on to the 0.5 cm diameter rod. Energy deposition from one array pumping from one quadrant is shown in panel (*b*) and the summation of deposited energy pumping with a two-dimensional array from all four quadrants is shown in panel (*c*).

6.3 Approaches to packaging high-average-power two-dimensional semiconductor laser arrays

Three different approaches have emerged for packaging high average power laser arrays. The first to appear commercially was the use of relatively large pieces of copper to serve as heat spreaders which deliver the waste heat to thermoelectric coolers. The other two approaches, both of which have now appeared commercially, are the use of impingement coolers, also known as compact high-intensity coolers (CHIC coolers), and the use of laminar-flow-cooled devices, i.e. silicon microchannel coolers. A variation on the implementation of the latter technology has achieved the lowest thermal resistivities to date and will be commented on in Section 6.4.

We will first briefly review the key elements of heat flow in semiconductor laser bars and their submounts. We will then review the performance of the various packages from the standpoint of hydraulic power, temperature rise, packing densities or pitch. Then we will comment in some detail on the microchannel cooler approach and the various forms in which it has appeared.

Many papers have been published on the subject of heatsinking high-power laser diode arrays. Many of these treatments contain detailed three-dimensional finite element analyses of heat flow in complex structures. However, to first-order the problem is a very simple one and can be broken down analytically into a few separate pieces. Consider the laser diode bar shown in Figure 6.5. The bar is taken to be 1 cm in width, with a laser cavity 500 μm long, and the total bar thickness is taken to be 150 μm with the bar mounted p-side down on to the submount. In this configuration the heat generated at the laser diode junction must typically flow through 2 μm of cap layer GaAs (as well as cladding regions of AlGaAs). Although the bulk thermal conductivity of GaAs is modest, 0.8 W/cm-°C, this term in the thermal resistance, as well as the flow of heat through a 2 μm-thick indium solder layer, are minimal and do not dominate the problem due to the small physical dimensions.

The heat has at this point reached the submount. This submount may either be a thermally conductive carrier of heat to a cooler mounted on the back plane of the two-dimensional diode array, or the submount itself may contain the cooler structure. If the latter is the case the three significant terms still to be encountered in the thermal resistance are the bulk resistance, θ_{bulk}, generated by the flow of heat through a finite thickness of material between the solder and the coolant itself. This term

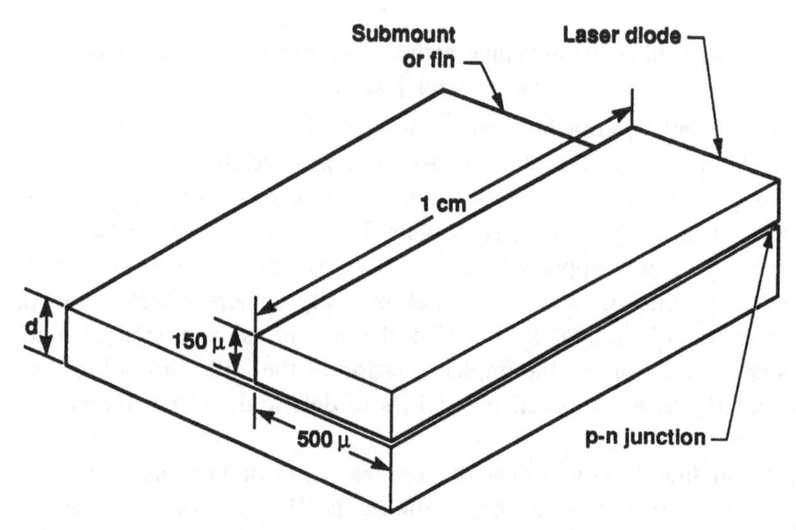

Figure 6.5. Laser diode bar mounted junction side down on a submount or fin. Using microchannels, the cooler can be embedded directly inside the submount.

is to first order, simply

$$\theta_{\text{bulk}} = \delta/\kappa_{\text{b}}A,$$

where δ is the thickness of the material, κ_{b} is the bulk conductivity of the submount material (1.4 W/° cm for silicon and 4 W/° cm for copper), and A is the area through which heat is conducted. The next term is for the boundary layer thickness of the coolant, θ_{bound}, which is calculated as

$$\theta_{\text{bound}} = \delta/\kappa_{\text{c}}\alpha A,$$

where δ is the thickness of the coolant boundary layer, κ_{c} is the thermal conductivity of the coolant (for water 6×10^{-3} W/° cm) and A is the area of the device; α is an effective fin area multiplication factor ($\approx H/W$ if a microchannel fin cooling structure is used). (Here H is the height of the etched channel and W its width.) The final major term in the thermal resistance is the caloric resistance of the coolant itself due to its finite heat capacity, θ_{cal}, where

$$\theta_{\text{cal}} = 1/\rho CfA$$

where ρC is the heat capacity of the coolant (1 cal/gram degree for water); A is again the cooled area; and f is the flow rate of the water (which at 20 psi with 1 cm-long channels which are 75 μm wide by 350 μm deep is 10 cc/s/cm^2).

If a microchannel structure is embedded in the submount, assuming that the channels are 500 μm long, 16 μm wide and 160 μm deep, with 100 μm of silicon bulk between the diode and the microchannels, and assuming a diode bar 1 cm wide and 500 μm long, we calculate simply, with no account taken of the thermal two-dimensional spreading effects, and assuming a flow f of 1 cm^3/s, that

$$\theta_{\text{bulk}} = 0.14\,^{\circ}\text{C/W}$$

$$\theta_{\text{bound}} = 0.27\,^{\circ}\text{C/W}$$

$$\theta_{\text{cal}} = 0.24\,^{\circ}\text{C/W}$$

so that the boundary layer thickness and the caloric heating dominate the problem. In fact, both these terms are overestimates since the boundary layer term does not take credit for thermal spreading in the bulk layer and the caloric term is minimized by going to higher flow rates made possible by shorter channels. Figure 6.6 shows data for microchannels which are 25 μm wide, 150 μm deep, and 1.4 mm in length. Least squares fitting this data shows a linear dependence with a slope of

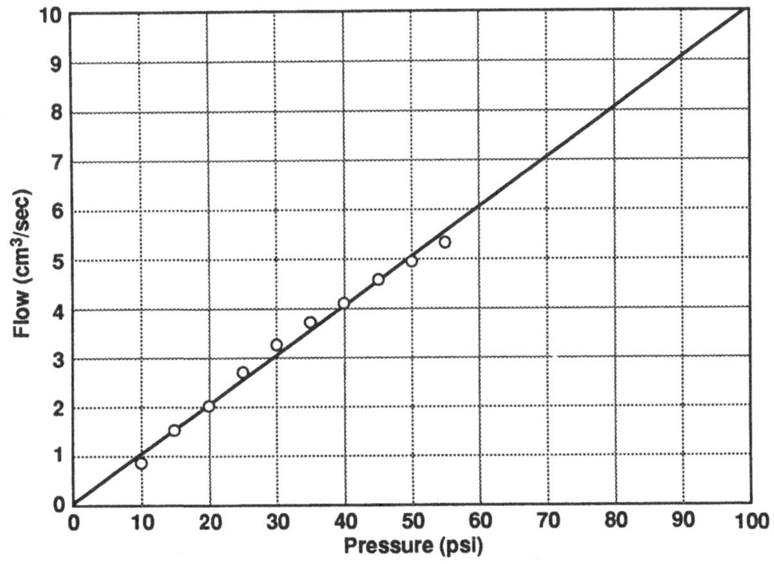

Figure 6.6. Flow rate versus pressure for silicon microchannels which are 25 μm wide, 150 μm deep, and 1.4 mm long. The channels are distributed across a 2 cm-wide silicon submount for a total of 200 channels. See Figure 6.9 for a drawing of the cooler used to obtain this data. (After Ref. 1, © 1992 IEEE.)

0.10 cc/s-psi. Shortening and deepening the channels rapidly improves this slope. As we will see in the next section, high-power microchannel cooled semiconductor packages have been fabricated which show a total measured thermal resistance of 0.23°/W. None the less the above simple arguments clearly delineate the fundamentals.

Consider, instead, that the submount does not contain the coolant but busses the heat from the diode to a cooling structure which is mounted on the back plane of the two-dimensional diode array. This conductive 'fin' will have a bulk thermal resistivity given by the expression for θ_{bulk} above. The thermal conductivity of the fin material may be enhanced over the above example by moving to copper; however, the length of material used to bus the heat to the back plane (often a millimeter or more) more than offsets this advantage. Most serious is the conductive area disadvantage. As one moves to structures with high pitch, say 25 devices per linear centimeter of rack and stack array, the conductive area for a 1 cm bar is 1 cm by approximately 200 μm. Even without taking account of increases in the resistance due to thermal constriction, the fin resistivity is simply calculated to be 2.5°/W. In reality, due to constriction and the lack of ideal and instant spreading of the heat into the full area of the fin, the resistance is significantly larger. Thus, at high stacking pitches, the heat conductivity fin raises the value of the thermal resistance well beyond that of any package which is equally thin and successfully incorporates the cooler into the fin. However, in applications which require low stacking pitch (a few per centimeter) or lower duty cycle (of order 5% with stacking pitches of approximately 30 cm^{-1}), this packaging approach is quite adequate.

We next consider the three types of coolers which can be placed on the back plane of a device if one is allowed to depart from the very aggressive full duty cycle and moderately high-pitch approach achievable with embedded microchannel coolers. Three approaches have been investigated more than others; (a) impingement (CHIC) coolers; (b) microchannel coolers; and (c) thermoelectric coolers.

Thermoelectric coolers have the largest disadvantage in that they typically move only a very small amount of heat per unit area, of the order of a watt per square centimeter of area. Thus the approach used by practitioners of this method has been to spread the heat from a small 1 cm-wide bar through a substantial mass of copper on to the thermoelectric cooler on the back plane. While this approach 'works', it is fair to say it is of no value to high average power two-dimensional stackable diode arrays. This brute force approach is suitable for single high-power cw bars.

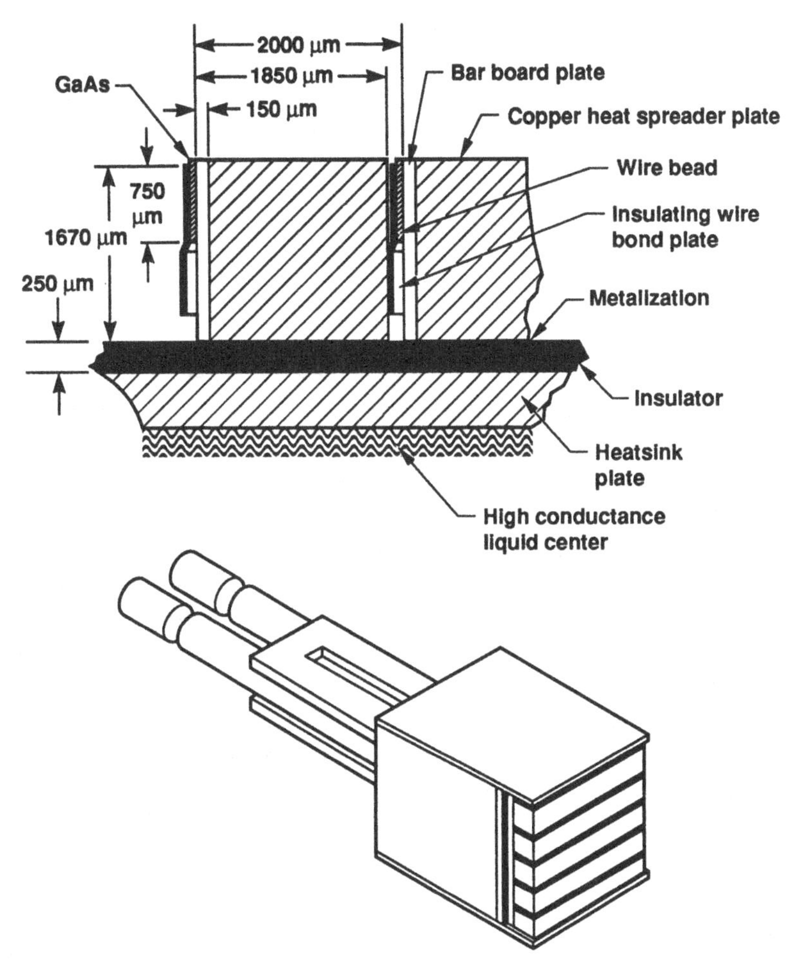

Figure 6.7. 1 cm × 1 cm liquid cooled diode array package with 2.0 mm bar spacing. (Reproduced with permission from Spectra Diode Laboratory.)

The impingement coolers and microchannel coolers represent two opposites. The success of low thermal resistance impingement coolers has been enabled by stacking them in series, as shown in Figure 6.7. Termed multiple surface impingement, this technique allows one to both enhance heat transfer coefficients with relatively low pressure drops and pump power, while also obtaining areal enhancements for the cooling interface. The impingement jets use turbulent flow to minimize coolant boundary layer thickness as opposed to microchannels which use fully developed laminar flow through narrow channels to achieve the same effect. The difference is that the CHIC coolers require a cascade or series of cooling interfaces to achieve low thermal resistivity, while the microchannel coolers can perform this function at a single interface so long as the channels can be kept sufficiently narrow and short to achieve this for a reasonable investment in hydraulic power. We are thus confronted with a parallel and a series cooling technology, a happy dilemma for any engineer or physicist looking for clear design choices.

6.4 Performance comparison of devices fabricated to date

The most complete report on the fabrication and performance of CHIC coolers is Ref. 5. The CHIC consists of an assembly of orifice plates, spacers, end plate (on to which the two-dimensional diode array is mounted), a fluid return channel, outer shell, dowel pins and tubes. The orifice plates and spacers are fabricated using chemical photo-etch methods which allow for accurate placement of the orifices. Very accurate machining is required since the offset of the orifices from one plate to the next is comparable to one orifice width. Electroforming is used to attach the thin-walled outer shell to the stack of plates and spacers. The plates and spacers in the stack are attached to one another using a diffusion bonding technique in one step with dowel pins being used for alignment. It is essential that the joints are uniformly in contact with each other as this series device also depends upon good heat conduction up through successive orifice plates for good overall performance. Pore or orifice sizes of approximately 500 μm diameter are typical of those used in these devices. The attachment of the two-dimensional arrays to the coolers on the front plane is shown in Figure 6.8. Bars of various thicknesses are used to conduct heat to the back plane.

Table 6.1 published by Spectra Diode Laboratory[6], gives performances for these devices when operated at 10% duty cycle, when the laser semiconductor array is varied in pitch from 25 bars/cm (thermal resistance of 5.4°/W) to 6 bars/cm (thermal resistance of 0.86°/W). Note that these values are the laser diode array resistances only and do not include the resistance of the cooler. The cooler resistance must be added to the above figures and can be calculated from the fourth and fifth columns of Table 6.1, assuming the thermal dissipation rates given in the footnotes to Table 6.1. It can be seen that at 15 psi the thermal impedance of the cooler alone is 0.080°/W and that at 30 psi it is only 0.050°/W. Thus, again, the performance of the coolers themselves is excellent and

Table 6.1. *Slab laser junction to coolant temperature rise, 10% duty cycle*

LSA period (mm)	*LSA*-R_{TH} (°C/W)	ΔT_{J-K}[(1)] (°C)	[15 PSI][(2)] ΔT_{J-C} (°C)	[30 PSI][(2)] ΔT_{J-C} (°C)	Peak power density Standard (W/cm²)	Peak power density Condenser (W/cm²)
0.4 mm	5.4	60	82.2	73.9	1500	3750
0.8 mm	1.65	18.3	29.4	25.2	750	1875
case of 50 W/B[(3)]		15.3	24.5	21.0	625	1560
1.2 mm	1.08	12.0	19.4	16.6	500	1250
1.6 mm	0.86	9.5	15.4	13.2	400	1000

[(1)] Junction to cooler temperature rise calculated for 60 W/bar, 35% *E* (@ 70 °C junction), 10% duty cycle, 11.1 W/bar dissipation.
[(2)] Junction to coolant temperature rise calculated for (two) different cooler flow rates.
[(3)] Second case of 50 W/bar is cited to show how 0.8 mm option might be used to 10% D.C. while retaining $\Delta T_{J-C} < 20$ °C. Options to go to >60 W/bar may exist for other packages.

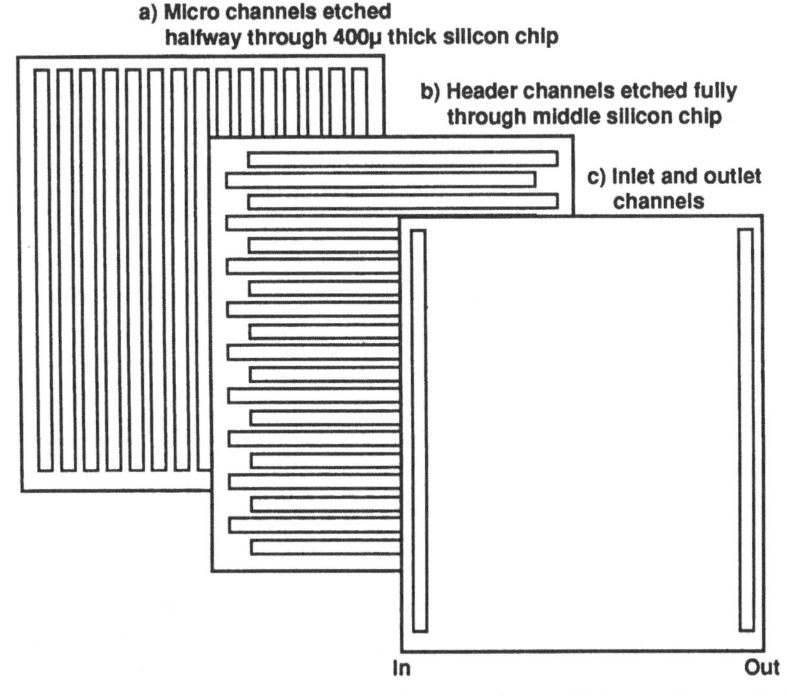

Figure 6.8. Multi-layer microchannel architecture for pack-plan cooling of diode arrays. The center section of staggered header channels is needed to limit the use to hold the length of each segment of the microchannel flow field to approximately 1.5 mm or less so that caloric self-heating does not limit the two-dimensional thermal resistivity.

they are handicapped only by the large thermal resistivity of the conductive cooling submounts attached to the laser diodes.

It is useful to compare the thermal impedance of the CHIC devices mounted on the back plane with microchannel coolers mounted on the back plane. In Ref. 2, 75 μm-wide channels which were 1 cm long were used in this manner and exhibited a measured cooler thermal impedance of 0.04° cm^2/W. Using the narrower channels, cooler thermal impedances of 0.014° cm^2/W were demonstrated in Ref. 1. Again, the advantage of the microchannel is that it can be fabricated in an ultra-thin package and allows the elimination of heat conductive fins, and can thereby be customized to individual diode bars. Currently they are limited to a stacking pitch of 10 or 11 bars/cm and will thus deliver peak powers of less than 750 W/cm^2 without using condensing optics. At peak powers above this value, assuming that duty factors of less than 10% are required, back-plane CHIC or microchannel-cooled two-dimensional diode arrays are currently the architecture of choice, at least until thinner experimental versions of microchannel coolers become fully developed. It should also be noted at this point that back-plane microchannel cooled diode arrays require microchannels which are relatively thin and, most importantly, short in order to compete with CHIC coolers. This can be achieved by moving to an architecture such as that shown in Figure 6.8 in which many layers of channels are used to successively divide the flow regions into finer and finer flow fields. Note that this device is now more parallel in nature and begins to mimic the CHIC devices. Note also that it is more complex than the microchannel coolers customized for individual bars. However, as compared with back-plane cooled CHIC devices, simply sawing 100 μm-wide, 1 cm-long microchannels into a silicon back-plane cooler does not yield a device which is competitive at the 3–10% duty cycle range of operation with >1 kW/cm^2 peak power, at least not using rack and stack technology.

As a brief but important departure, we emphasize that monolithic laser technology, as opposed to rack and stack, does not rely upon heat conductive fins to bus heat to the back plane. Rather, monolithic laser arrays which are mounted junction side down match exceedingly well to CHIC devices or to multilayered parallel architecture microchannel devices. Monolithic arrays, assuming that the performance (efficiency) and real estate consumed by each laser diode is comparable to that of rack and stack lasers, should be capable of eventually generating several kilowatts/cm^2 of areal intensity. For rack and stack devices to compete favorably with cw monolithic devices would require that individual microchannel cooler submounts be thinned to as little as a few hundred μm each or thinner.

While the first report of integrating high-performance coolers with surface emitting truly monolithic lasers was by Mott and Macomber[12], more recently work by Jane Yang and co-workers represents the highest degree of sophistication of packaging of two-dimensional monolithic lasers[37]. The TRW group has fabricated a 0.54 cm^2 unit cell monolithic array mounted on a microchannel heatsink. This device operated at 106 W/cm^2 peak-pulsed optical power, and 46 W/cm^2 cw[38] at a slope efficiency of 0.625 W/A (differential quantum efficiency of 40%) and a wall plug efficiency of 22%. The monolithic laser arrays (see Figure 6.13) have through-the-substrate emission, and are fabricated by reactive-ion-beam-etching of turning mirrors into the active region of the GaAs lasers[37]. SDL has demonstrated 50 W cw power from a 1 cm^2 monolithic InGaAs/AlGaAs ($\lambda = 0.95$ μm) array[39].

The silicon microchannel module pioneered by the LLNL group is shown in Figure 6.9(*a*) and (*b*). It consists of a silicon–glass–silicon

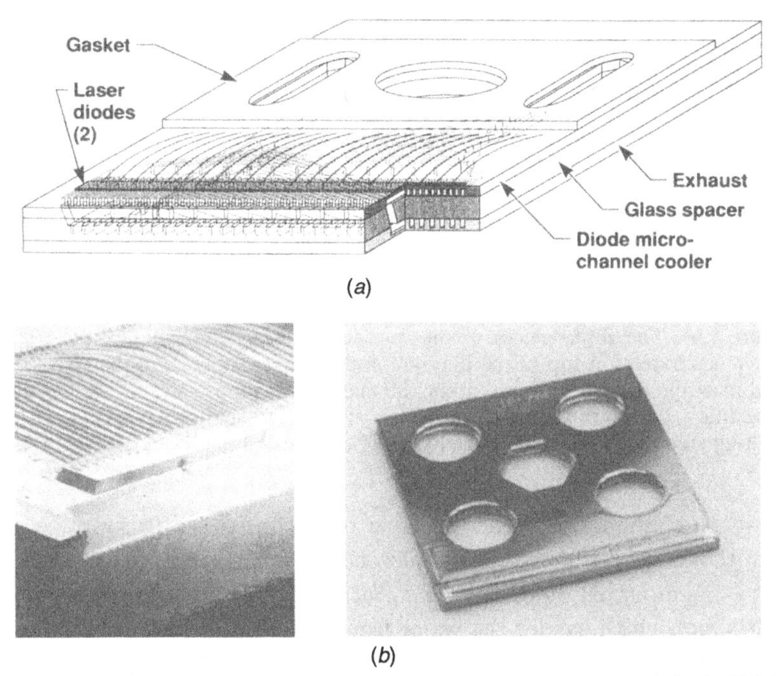

Figure 6.9. (*a*) Modular, stackable microchannel cooler design detail for individually cooling laser diode bars. The total thickness of each cooler (including the thickness of the rubber gasket used to seal the water flow) can be as little as 1 mm. (*b*) Photograph of device. (After Ref. 1, © 1992 IEEE.)

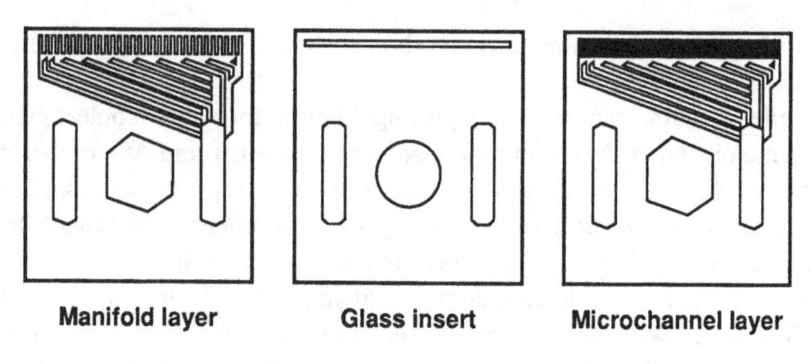

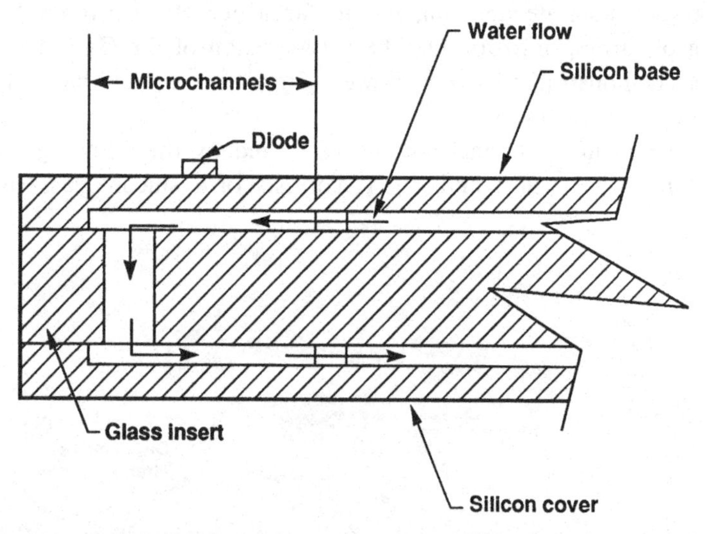

Figure 6.10. The three pieces of the microchannel cooler from Figure 6.9 are shown separately on top of the diagram and after assembly from the side at the bottom of the diagram. To assemble, the manifold is flipped over on to the glass piece and both are translated on to the top of the microchannel layer. Two anodic bonding steps are required. (After Ref. 1, © 1992 IEEE.)

sandwich configuration. The top layer of silicon is used to supply water from the inlet port to the etched microchannels located in the silicon just below the mounted diode bar. The central glass insert is micromachined with a slot which carries the water from the front of the microchannels to the return plenum which is etched into the bottom layer of silicon. Figure 6.10 shows the three individual pieces, as well as the flow path of the coolant through the assembled structure. The thickness of the sandwich is currently 1.0 mm (this includes 0.30 mm for the metal impreg-

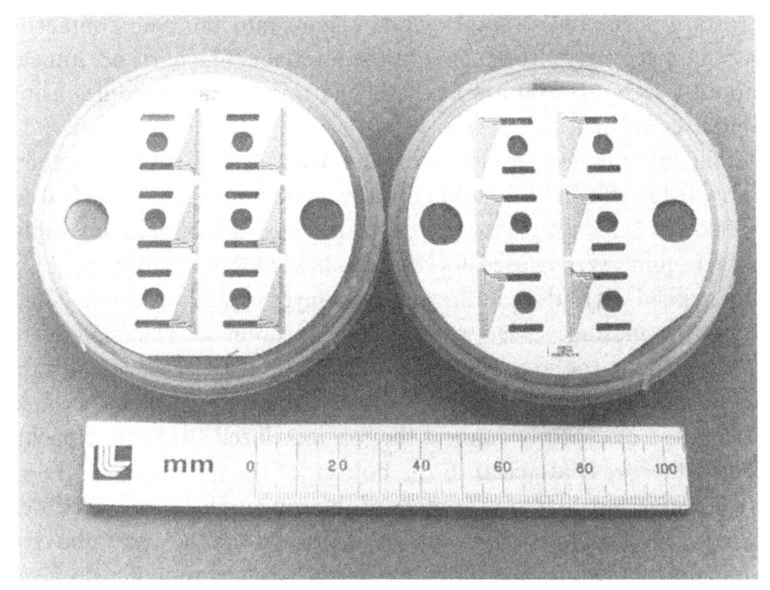

Figure 6.11. Photograph of two 3-inch silicon wafers on which six header manifolds (left wafer) and six microchannel coolers have been etched. (After Ref. 1, © 1992 IEEE.)

nated silicon rubber gasket used to contain the flow of water in stacked two-dimensional arrays while also conducting current in series fashion from bar to bar in much the same way as conventional two-dimensional arrays operate).

The microchannels and the return plenums in the bottom silicon piece are produced at the wafer scale using well-developed anisotropic etch procedures. Three-inch wafers can be used to obtain six modules per wafer, as shown in Figure 6.11. The wafers used are 381 μm thick, (110)-oriented, and have a resistivity of greater than 300 ohm/cm (currently we are evaluating the use of lower resistivity silicon). Si_3N_4 is deposited in an 800 Å-thick layer on the wafers and photolithographically patterned to define those areas of the wafers that are to be etched. The etch is carried out at 35 °C in 44% KOH. The etch rate is approximately 4 μm/h.

The three layers of the module are then joined using an anodic bonding procedure[1]. The three layers are placed in contact with each other in an oven heated to 560 °C, the annealing temperature of the borosilicate glass. The glass is used as the negative electrode and the silicon layers as the positive electrode. An increasing voltage difference is then applied in 100 V increments up to 500 V over approximately 45 minutes. The resulting electrostatic force pulls the glass and silicon into intimate contact and forms an exceedingly strong bond at the interface. The parts are annealed for about two hours and are finally cut into individual modules using a computer controlled dicing saw.

Metallization is next applied to the coolers using a sputtering source. Starting at the silicon and moving to the more outward layers 1000 Å of Ti, 1000 Å of Pt, 9 μm of Au, 1000 Å of Pt and 3 μm of gold are applied. The thick gold layer is used to ensure a lower electrical resistance path for the applied currents, and the thin gold layer is used to form a eutectic bond to the indium solder which is placed upon it. The modules are metallized on the top, front and bottom with a break in the metallization which occurs just below the diode bar and around the package perimeter. This serves to electrically connect the top metallized strip of the package on which the bar is attached to the bottom of the package and gives an isolated metal pad that covers most of the top surface of the module. A 3 μm layer of indium is then vacuum deposited on to a pad above the microchannels. Two 0.9 cm diode bars are then butt-mounted to form a single 1.8 cm-long bar directly above the microchannels. Indium reflow is accomplished in a pure hydrogen atmosphere at approximately 500 °C. The final step is to wirebond from the n-side of the diode bars to the module n-side Au contact pad. Approximately 200 30 μm-diameter gold wires are used.

We have already covered the principles of thermal conductance in microchannel architectures. The key concept is minimizing the boundary layer thickness of the stagnant layer of the coolant by using fully developed laminar flow in very narrow channels. CHIC devices also minimize boundary layer thickness, but by the use of turbulent flow on the impingement plates. At this point it is worth while comparing the hydraulic power expended in the two types of packages. Beach[1] has developed a simple model for the microchannel cooler performance as a function of pressure of the coolant. For 25 μm-wide channels which are 150 μm deep and 1.4 mm long embedded in the silicon submounts this dependence becomes, for the thermal resistance,

$$R_{th} = 0.20°/\mathrm{W} + (1.5/P\text{-psi/W})$$

where P is the header inlet pressure. Normalizing to a laser diode footprint of 1.8 cm by 330 μm and rewriting this equation in terms of the hydraulic power expended to force water through the microchannels gives the thermal impedance (as opposed to the resistance) as

$$Z_{th} = (0.012)\,°\mathrm{C\text{-}cm^2/W} + (2.34 \times 10^{-3}/\sqrt{P_{hyd}})\,°\mathrm{C\text{-}cm^2}/\sqrt{\mathrm{W}}$$

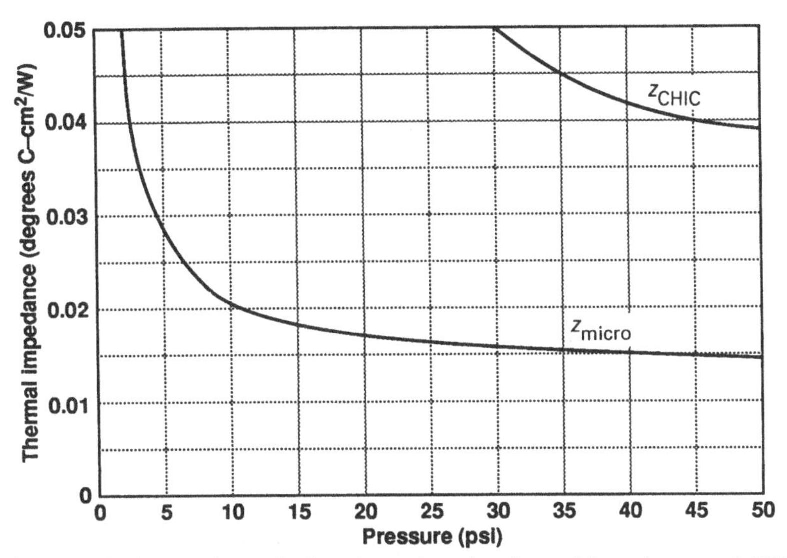

Figure 6.12. Comparison of microchannel cooler thermal impedance and CHIC cooler thermal impedance versus inlet pressure. In this example, a stacking pitch of 10 bars/cm is used and the standard 25 μm-wide by 150 μm-deep and 1.4 mm-long microchannels are used.

Table 6.2. *Impingement cooler/microchannel cooler comparison* (compare 10 bar stack, each bar 1 cm long)

	Package cooling architecture	
	Impingement cooler	Microchannel cooler
Stacking pitch	10/cm	5/cm glass package 10/cm milar package
R_T/bar	1.37°C/W (copper chip carrier)	0.40°C/W*
Flow geometry	Turbulent	Laminar
Required flow rate	37.5 cm^3/sec	12.5 cm^3/sec
ΔP	55 psi	25 psi
Hydraulic power	14.2 W	2.2 W

* Microchannel packages can run at 3.4× the optical output power of impingement cooler packages for the same ΔT.

where P_{hyd} is the hydraulic power supplied to the package. Figure 6.12 plots Z_{th} vs P_{hyd} from this equation and for comparison includes the hydraulic power dependence versus Z_{th} for CHIC devices. For ten-bar stacks (stacking pitch of 10 cm^{-1}) the thermal resistivity of the back-plane cooled CHIC device is 3.5 times that of the in-plane microchannel cooler and requires nearly seven times the hydraulic power to achieve this performance. Table 6.2 gives these comparisons, which illustrate one of the principal reasons why microchannel cooling competes favorably with other technologies.

6.5 Laser diode array performance

Given the various methods of packaging high-power arrays, the proof of each approach is its ability to deliver high average power array performance with useful lifetime. In the following pages we will compare the current status of high-power arrays fabricated from several research groups in the US. Specifically we will review the work of the groups at Lincoln Laboratory, Advanced Optoelectronics, Spectra Diode Laboratory, and Lawrence Livermore National Laboratory. While large arrays of impressive size have also been fabricated at McDonnell–Douglas, this organization has fabricated principally low duty cycle arrays, which are not thermally stressed by high average power.

The work at Lincoln has focused on a hybrid two-dimensional array approach with microchannel cooling on the back plane[4]. The devices appear to be monolithic and should therefore possess the key advantage of not having to bus the heat to the back-plane microchannel cooler. In Ref. 7, Goodhue *et al.*, reported InGaAs/AlGaAs and AlInGaAs/AlGaAs monolithic arrays with over 50% differential quantum efficiencies. In this reference the lasers are horizontal folded-cavity 45°-internal-mirror devices with the top surface containing the output facets for the lasers. No cooling was reported in this paper. However, if a transparent cooler could be placed on to the top layer or output facet, very low thermal impedance could be achieved. An alternative approach would be to employ lift-off technology[8,9] in order to remove the semiconductor laser array from the poorly thermally conducting GaAs substrate and to use Van der Waals bonding technology to bond the monolithic arrays to silicon microchannel coolers. Van der Waals bonding of GaAs arrays has been reported for silicon, glass, sapphire, diamond, $LiNbO_3$, InP and diamond in Ref. 8. TEM micrographs indicate that in most cases roughly a 100 Å-thick region of amorphous material comprise the interface. The impact of these structures, which are not lattice matched, on the lifetime performance of high-power arrays has not been demonstrated, but the approach is an important new development.

More recent work[10a,b,11] by the Lincoln group reports the fabrication of two-dimensional hybrid surface emitting arrays of low thermal conductance. In this work, thermal resistivities as low as 0.70 °C-cm^2/W were reported. In actual practice however, these devices are not high-duty-factor structures. Since the heat flows through the GaAs substrate in this junction-side-up configuration, the dynamics of the heat flow is that initially during a quasi-cw pulse the heat first flows to the copper electrodes mounted near the junctions. As the diodes are turned off, this heat is returned through the junction, through the substrate, and finally into the heatsink. Operation of these devices at high duty cycles is limited by the interpulse accumulation of heat into the substrate. The high duty cycle thermal resistance of these devices is in fact no better than the thermal conductivity of bulk GaAs of 75 μm thickness. Again, lift-off technology would be useful in improving the high duty cycle performance of these devices, but since the hybrid semiconductor bars are fully exposed to the wax used in the lift-off process, it is not at all clear that lift-off technology can be employed with these structures.

In Ref. 12, Mott *et al.*, report on a p-side-down-mounted distributed feedback laser array mounted on a microchannel cooler. With a 4 × 10 array of lasers fabricated on a 0.2 cm^2 chip and linear stacking density of 25/cm, a cw power of 3 W is reported with 27 of the 40 lasers operational. A thermal impedance of 0.038°C-cm^2/W was reported. The thermal resistance is thus 1°/W. Even though very good thermal performance was reported (assisted somewhat by thermal spreading into the much larger heatsink by a copper bonding layer), overall performance was poor due to the relatively high electrical resistance, and thus low laser efficiencies, of the monolithic devices. The entire array had a reported ohmic resistance of 75 mΩ.

Recently, Advanced Optoelectronics introduced a commercial product of back-plane cooled two-dimensional rack and stack diode arrays. In this product silicon channels which are fairly broad and long are used to cool arrays which are mounted at a pitch of up to 25 bars/cm using BeO submounts to bus the heat to the back plane. Not surprisingly, the performance of these arrays is limited to 2% duty cycle at average power densities of 20 W/cm^2. A thermal impedance of 0.3 °C-cm^2/W is quoted. This converts to a per bar thermal resistance of 9 °C/W. Laser Diode Array, Inc. is using microchannels to back-plane-cool two-dimensional arrays of rack and stack diodes implanted in a monolithic BeO substrate. Similar thermal performance is expected as in the results reported above. TRW Inc. has recently demonstrated a monolithic surface-emitting laser structure with back-plane microchannel cooling (Figure 6.13)[37].

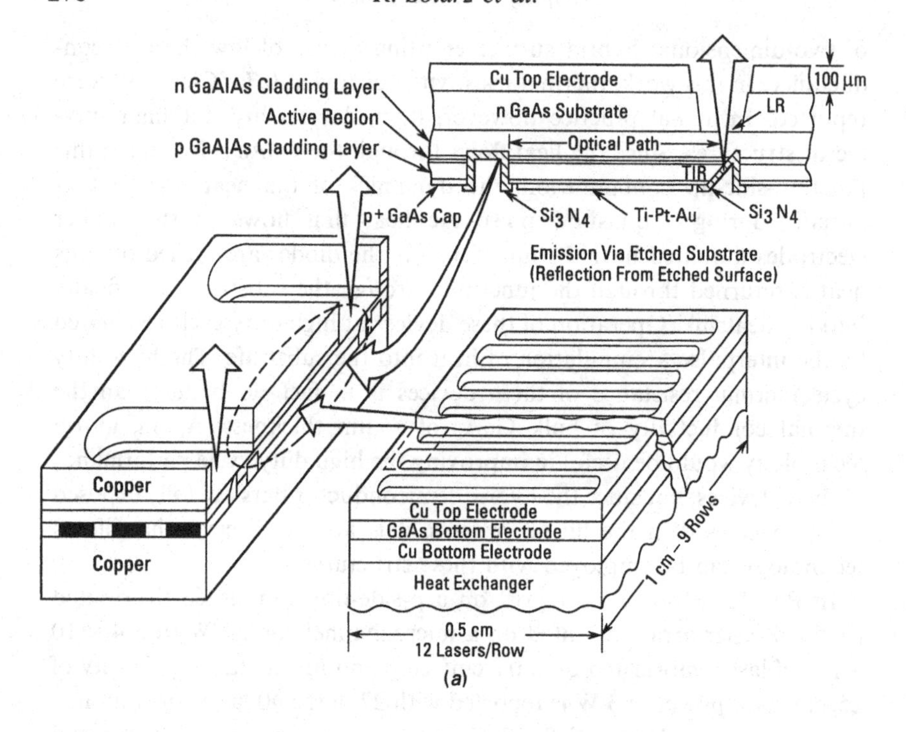

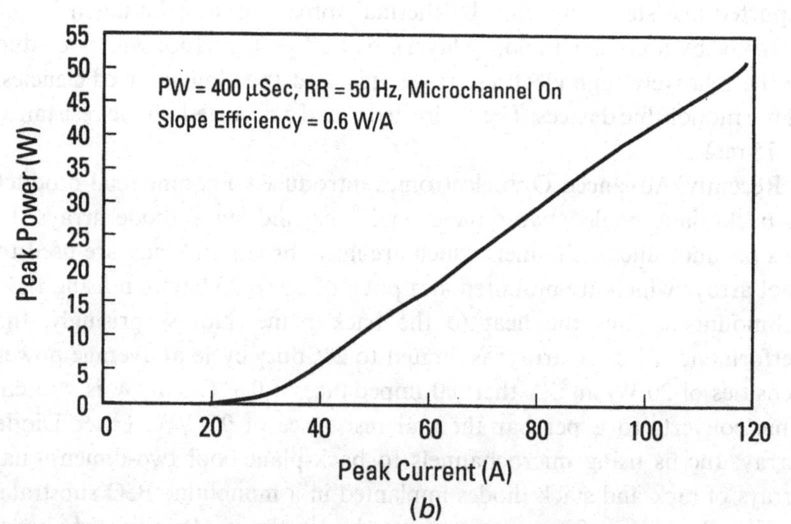

Figure 6.13. Two-dimensional diode arrays fabricated at TRW: (*a*) schematic drawing of monolithic surface-emitting array; (*b*) power versus current characteristics. (After Ref. 37.)

The Livermore group is the only group using stackable heatsinks on two-dimensional diode arrays. Figure 6.14 displays data acquired at low duty factor (10 Hz prf and 100 μs) from one of our modules containing 1.8 linear cm of diode array. The peak optical output of 190 W (106 W/cm) at 225 amps displayed in Figure 6.14(*a*) is supply-limited and typical of the performance of our modules. Figure 6.14(*b*) displays voltage vs drive current data indicating a slope resistance of 7.5 mΩ, which is again a typical value for a module including the two conductive silicon rubber gaskets. The effect of the series resistance is to introduce an I^2R loss term that causes the wall plug efficiency to roll over at high currents.

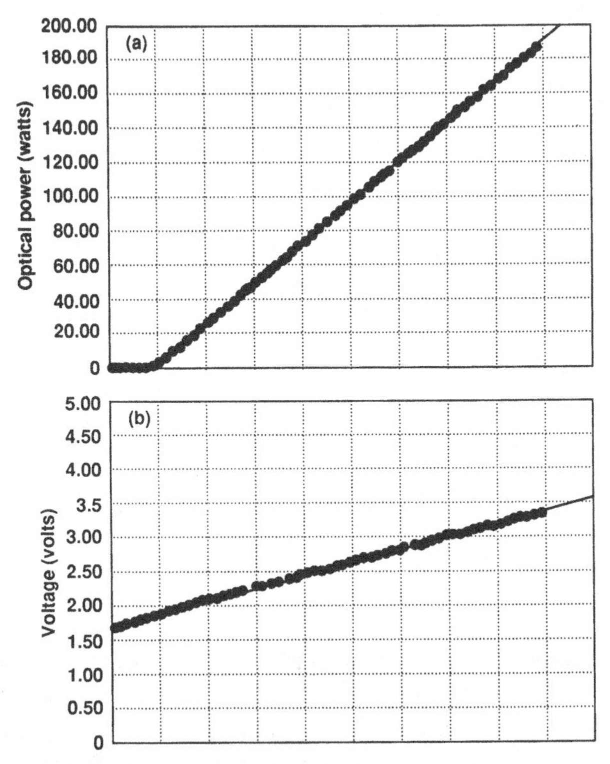

Figure 6.14. Low duty factor characterization data from a module using 10 Hz prf and 100 μs current pulses. (*a*) Peak optical power out vs peak drive current. The maximum output of 190 W (106 W/cm) is supply-limited. The least-square-fit straight line indicates a slope efficiency of 0.95 W/A and a lasing threshold current of 23.7 A. (*b*) Voltage across module (including two conductive silicon rubber gaskets for sealing) vs drive current. The least-square-fit straight line indicates a series resistance of 7.5 mΩ. (*continued*)

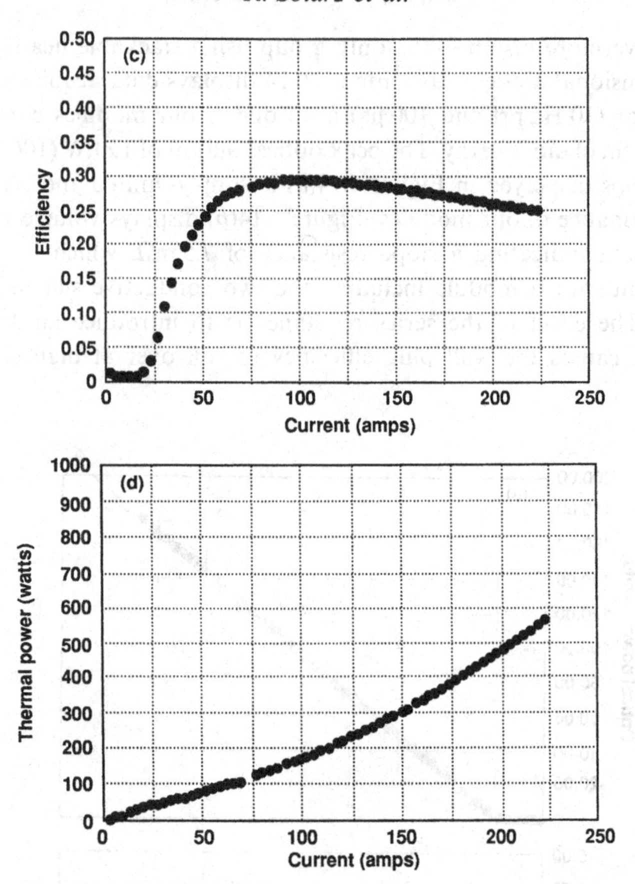

Figure 6.14. (*continued*) (*c*) Measured wall plug efficiency vs drive current. The observed roll off at high drive currents results from I^2R loss. Efficiency optimizes for drive currents between 90 and 115 A. (*d*) Peak thermal power dissipated as a function of drive current. (After Ref. 1, © 1992 IEEE.)

Figure 6.14(*c*) plots the measured wall plug efficiency vs drive current and indicates peak efficiency operation obtained for drive currents between 90 and 115 A. From the standpoint of optimizing the high-average-power operation of diodes it is then desirable to operate at high duty factor in this current range. Figure 6.14(*d*) plots the peak dissipated thermal power for this low duty factor characterization. From these data it is evident that to optimize module efficiency at high duty factor the heatsink must have the capability of dissipating approximately 170 W while holding the active junction at a reasonable operating temperature. Using our measured thermal resistance value of 0.23 °C/W and an inlet water temperature near 0 °C, this module enables cw operation of the laser diodes at a junction temperature of 40 °C and at drive currents that optimize their efficiency.

Figure 6.15(*a*) plots cw light output from a single module as a function of supplied current. The peak optical output is 70 W (38.9 W/cm) at 100 A of input current and is supply-limited at this point. The measured cw slope efficiency is 0.91 W/A and the threshold current is 21.6 A. Figure 6.15(*b*) plots the thermal power dissipated in the same package as a function of the input current. The maximum thermal power dissipation at 100 A of input current is 160 W, which corresponds to a thermal

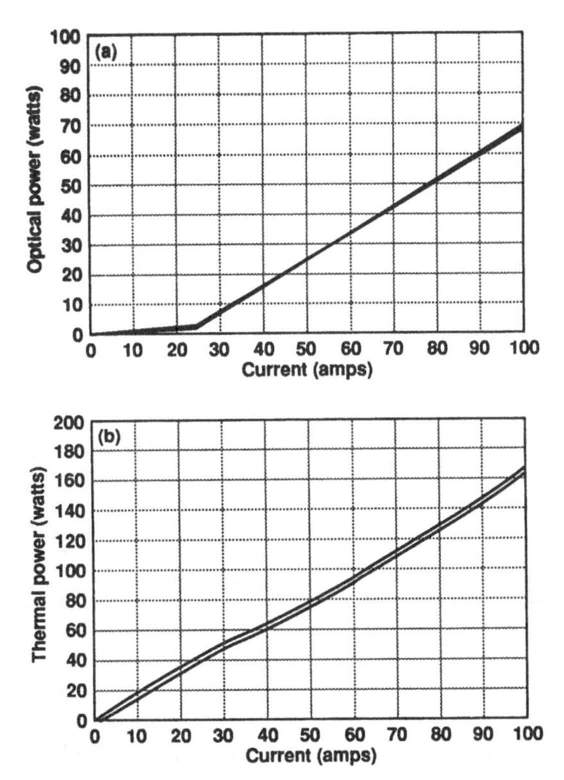

Figure 6.15. (*a*) Optical output power vs drive current from a single module operating cw. The maximum cw optical output is 70 W and is supply-limited. The measured slope efficiency is 0.91 W/A. (*b*) Dissipated thermal power vs drive current. At 100 A the thermal footprint is 2.7 kW/cm^2 and the temperature of the diode junction is approximately 37 °C above the coolant inlet temperature of 15 °C. (*continued*)

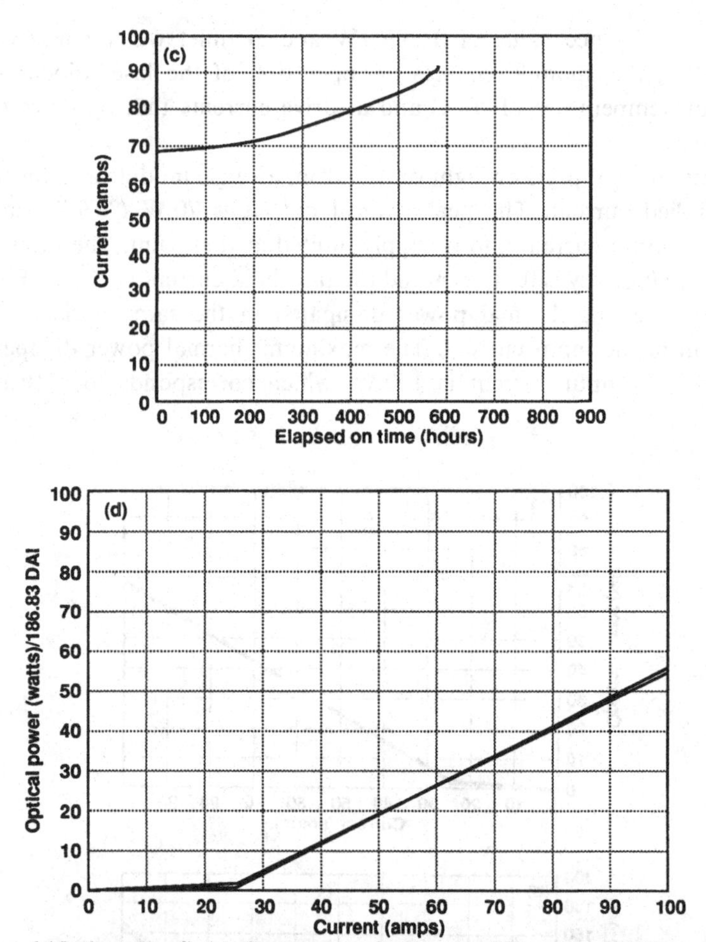

Figure 6.15. (*continued*) (*c*) Constant optical power lifetest results. The current was continually adjusted to maintain a 40 W cw optical output power until the required current increased by 30% after 580 hours. (*d*) An optical power vs drive current characterization performed approximately 400 hours into the lifetest. The data indicate that device degradation is entirely accounted for by the reduced slope efficiency. No degradation in the thermal performance of the heatsink is observed. (After Ref. 1, © 1992 IEEE.)

footprint of 2.7 kW/cm^2. Even at these heat loads, evidence of thermal roll over is only just beginning to be visible. Figure 6.15(*c*) plots reliability data that were acquired in a constant-optical-power lifetest. During this test the bar was operated cw at an output power of 40 W by continuously adjusting the drive current to compensate for device degradation. At the beginning of the lifetest the thermal power dissipated was approximately

110 W and the diode junction temperature was approximately 25 °C above the inlet water temperature of 15 °C. The lifetest was terminated after 580 hours when the required drive current had increased by 30%. Recent cw lifetests of a 20 W cw 9 mm bar to 1800 hours of operation were demonstrated at LLNL with less than 5% degradation during this period. Figure 6.15(*d*) displays an *L–I* characterization curve made 400 hours into the lifetest. The reduced performance of the diode is entirely accounted for by a degradation in the slope efficiency from an initial value of 0.91 W/A to a final value of 0.70 W/A. Again, even at the peak heat load of 2.7 kW/cm^2, evidence of thermal roll over is only barely visible, indicating no degradation in the thermal performance of the heatsink.

The spectral linewidth of the emitted radiation is of the utmost importance to the performance of diode pump arrays used to excite crystalline lasers. This is due to the nm-wide absorption lines that are typically used to couple the diode radiation to the solid state laser. The effective (time-integrated) spectral linewidth of the laser is broadened by the effects of transient heating during a current pulse, an effect commonly referred to as chirp. Depending on the particular pulse width at which the diode is operated, different effects can dominate the observed linewidth. Figure 6.16 plots the measured centroid emission wavelength from one of our modules during a 1 ms pump pulse. The observed chirp results from the transient heating of the diode bar and heatsink during the pulse. These spectral snapshots were acquired in 5 μs windows at increasing delays into the pulse with the package operating at a peak optical power of 100 W. For pump pulses having roughly 100 μs duration, such as would be typical for exciting Nd^{3+}:YAG storage lasers, the effective linewidth of the pump pulse is hardly affected by transient heating and is typically of the order of 3 nm FWHM or less. For pulse widths of the order of several hundred microseconds to a millisecond the emission wavelength of the diode bar is swept toward longer wavelengths as the bar and heatsink heat up. In this regime the transient heating can effectively double the effective emission linewidth to nearly 6 nm FWHM. For very-long-pulse or even cw operation, that part of the emitted light that is strongly chirped due to transient heating during the first millisecond of operation contributes only a small fraction of the total energy to the pulse, and the effective linewidth once again approaches 3 nm FWHM. Figure 6.17 displays spectral line shapes from a module operating at a pulse optical power of 100 W at four different pulse widths: 100 μs, 250 μs, 500 μs and 1 ms. In this case the line shapes are representative of the entire time-integrated pulse and have FWHM of 3.1, 3.8, 4.8 and 5.8 nm respectively.

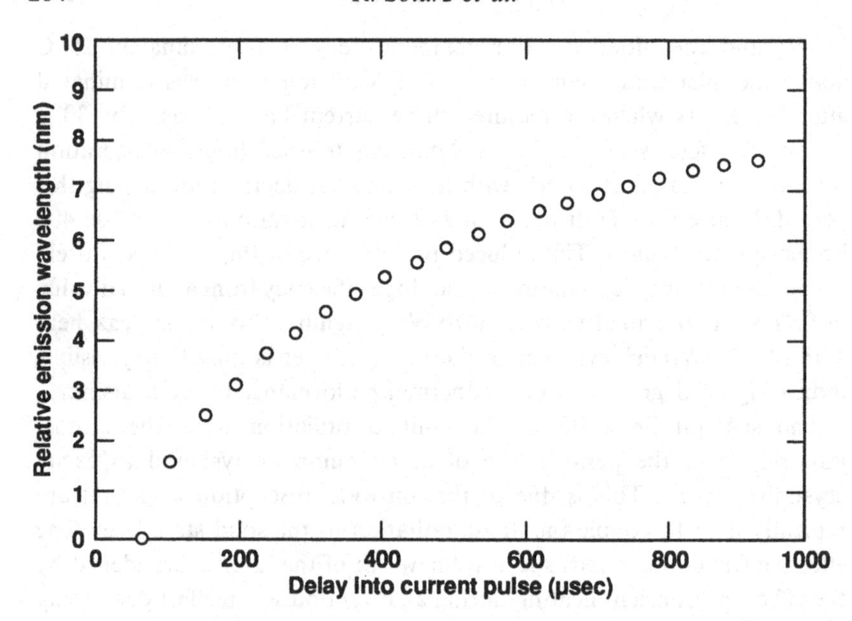

Figure 6.16. Measured pulse chirp through a 1 ms, 100 W diode pulse. The data represent emission wavelength centroids for spectral snapshots acquired in 5 μs-long windows at the indicated delay into the pulse. (After Ref. 1, © 1992 IEEE.)

Another factor impacting emission linewidth is the uniformity of the thermal impedance along the bar. This can be particularly important under conditions of cw operation in which the bar is operating under aggressive thermal loading. Figure 6.18 shows three high-resolution spectral scans taken from different locations on one of the 9 mm bars that make up half of the array on a module. These scans were made with the module operating at a cw optical output power of 20 W (11.1 W/cm) by imaging a small section of the near-field pattern of the bar on to a spectrometer slit. In this case the spectra were acquired from regions at either end and in the center of the bar. The modulation evident in the scans is due to the different longitudinal laser modes. The overlap in wavelength of these three spectra is typical of the observed emission uniformity for our heatsink and is evidence of the excellent thermal control it affords. The linewidths of 2.5 to 3 nm FWHM are also typical of the measured emission linewidths under conditions of cw operation.

Figure 6.19 shows the emission spectrum measured from a 41 module stack. The displayed spectra were acquired using 100 μs pulses at various current levels. To ensure that the emission of the entire array was sampled

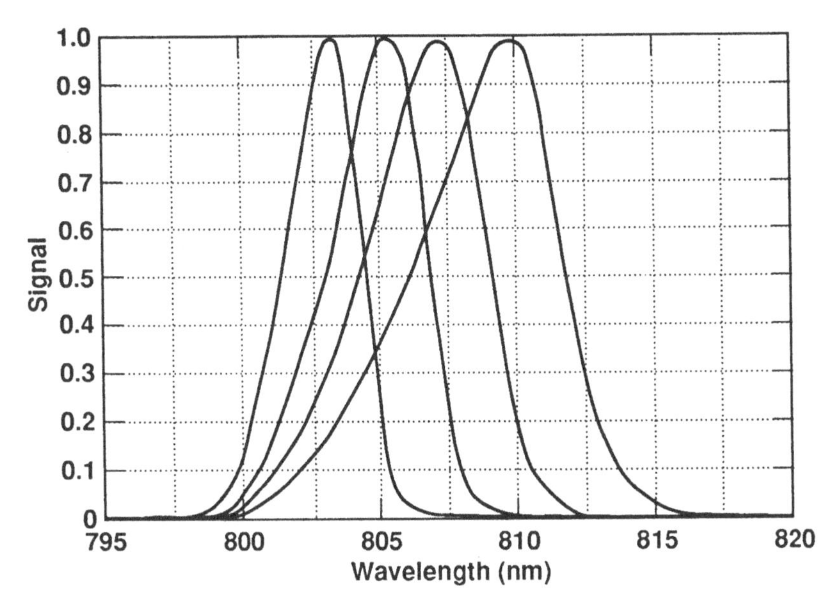

Figure 6.17. Measured linewidths from a module producing 100 W of optical output at four different pulsewidths: 100 μs, 250 μs, 500 μs and 1 ms. The observed linewidths are 3.1, 3.8, 4.8 and 5.8 nm respectively. In this case the emission features are representative of the integrated pulse. Longer pulse widths result in greater emission red shift due to transient heating of the bar and heatsink. (After Ref. 1, © 1992 IEEE.)

equally, the output of the stack was directed to an integrating sphere and the spectral measurement was made on a port at 90° to the entrance. At currents of 40, 80 and 140 A the observed linewidths were 2.5, 3.1 and 3.8 nm FWHM respectively. In this pulse width regime the increasing linewidths are due to transient heating and are indicative of the excellent performance obtainable even in very large pump arrays.

The results on CHIC back-plane-cooled two-dimensional rack and stack arrays as calculated by Spectra Diode Laboratories are given in Table 6.1. At 10% duty cycles, and at a stacking pitch of 25/cm assuming 35% efficient bars, an average power of 150 W/cm^2 optical is anticipated. Arrays now being sold are quoted as being guaranteed up to 4% duty cycle with average powers of 60 W/cm^2 optical. The lifetest data reported by Spectra Diode Laboratories is very good. With single 60 watt bars operating at 20% duty cycle they report only 1.6% drop in output power after one billion 200 μs pulses. This is to be compared with the Livermore group's report above of 5% drop per billion shots for 1 cm-long bars operated at 80 W/cm linear power density, at 25% duty cycle.

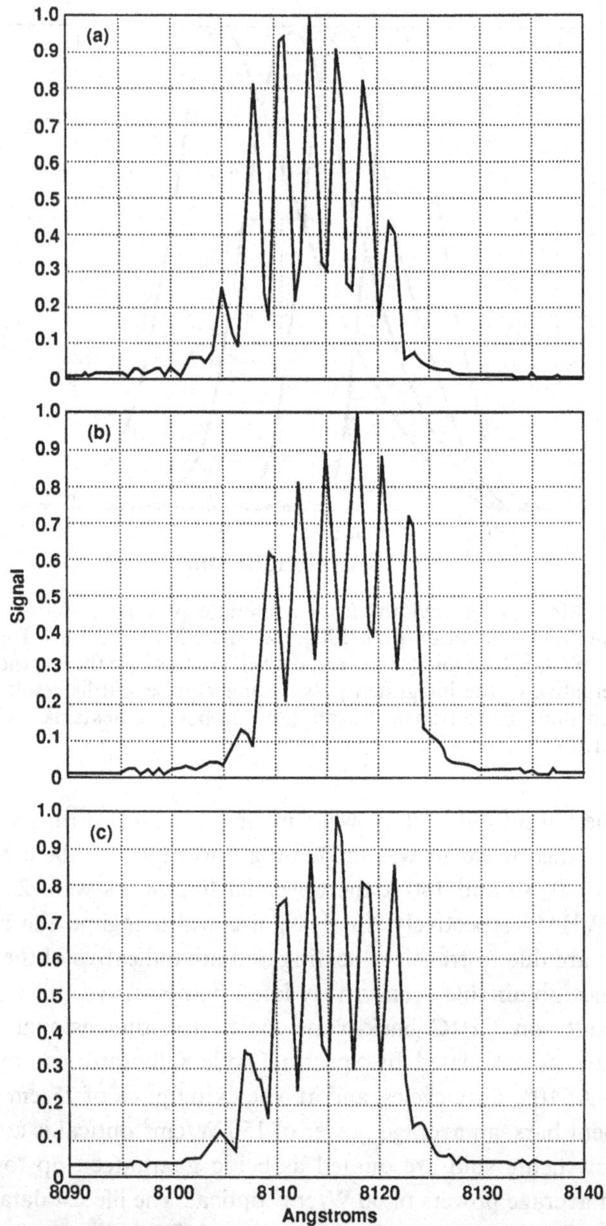

Figure 6.18. High-resolution spectral scans made on one of the 9 mm bars that make up half the linear array on a module. The module was operated cw at a total output of 20 W. The three scans were made from radiation collected at either end and the middle of the bar. The FWHM is 2.5–3.0 nm and the good overlap of the spectra imply a uniform thermal impedance along the length of the bar. (After Ref. 1, © 1992 IEEE.)

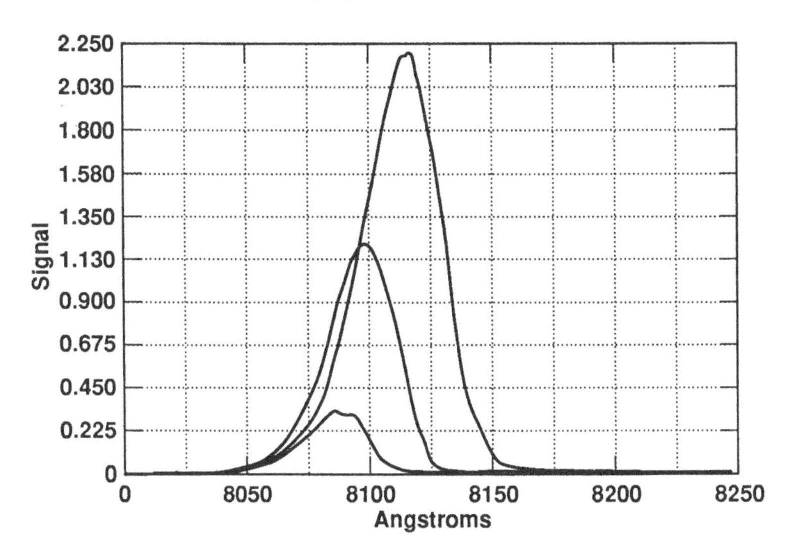

Figure 6.19. Emission spectrum measured from the 41 module stack of Figure 6.4(*b*). The displayed spectra were acquired using 100 μs-long current pulses at 40, 80 and 140 amps. The observed linewidths were 2.5, 3.1 and 3.8 nm FWHM respectively. (After Ref. 1, © 1992 IEEE.)

6.5 Fundamental limits to high-average-power operation and potential improvements

From the foregoing pages we see that much progress has been made in the fabrication and testing of very high power two-dimensional diode laser arrays. Low degradation rates on MOCVD-grown GRINSCH-QW lasers were first reported in 1983[13], followed by studies on the correlation of reliability with V/III ratios[14], with stresses[15], and defect propagation[16]. Only a few short years ago, diode laser arrays which emitted 10–20 W of average power were champion devices[17]. Now two-dimensional diode laser arrays emitting near to 1 kW average power are being reported. Another order of magnitude improvement in average power is anticipated over the next 12–18 months as a result of several programs in the US, all of which are aimed at building kilowatt average power diode pumped solid state lasers.

What is particularly important is that as these high-power arrays become more commonplace, lifetest data will proliferate and higher and higher power devices will be built with increasingly good lifetime performance at high power. Currently, many groups[18–21] are exploring the fundamental limits to high-power operation and much progress is being made. Owing to the lack of availability of good packaging technology, lifetime performance at the very highest powers could not be explored previously and is only now being tackled.

Maximum peak power is believed to be limited by the onset of catastrophic optical damage (COD) at the output facet. Until recently the highest output power density reported was 24 mW/μm of output aperture[22]. This was achieved by using impurity-induced-disordering to produce non-absorbing mirror-facet regions. Once photon absorption takes place, thought to be initiated by the presence of dangling oxide bonds at the cleaved facets, thermal runaway can lead to catastrophic facet damage. Other methods of increasing the catastrophic optical damage limit include increasing the aperture in the dimension normal to the junction[23] and utilizing non-absorbing 'window' regions. In more conventional structures, values near 14 mW/μm have been achieved, with values near 10 mW/μm being typical for reliable operation. Naito *et al.*[24] have reported on the benefits of non-absorbing mirrors for high-power, low-noise operation in recording applications, and Yoo *et al.* recently reported on a chemical passivation technique[25] to enhance the peak power capabilities of semiconductor lasers. Most recently, the IBM Zurich group has reported[21] on a careful study of the thermodynamics of facet damage in cleaved AlGaAs lasers. They report operation to 1000 hours at 40 mW/μm if one uses single quantum wells with quarter-wave Al_2O_3 or Si_3N_4 coatings applied to the cleaved ends. They report more rapid degradation in double-quantum-well material due to the increased area for exposure. Also, recently, Latta *et al.*[26] reported 5000 hour lifetimes from 200 mW single-element diodes at room temperature. These were achieved by protecting the facets using a proprietary procedure. Figure 6.20 is reproduced from Moser *et al.*[21] and shows the MTTF for AlGaAs taken from this reference. Spectra Diode Laboratories recently reported 60%-efficient AlGaAs operating at 125 mW/μm[27], as well as highly reliable quasi-cw operation of 100 W/cm, low-divergence diode arrays[40].

Average power, as opposed to peak power degradation, is now well understood to be correlated to the propagation of dark line defects[15,16] from the cladding regions into the active region of the laser. A number of workers have studied this effect and correlated it to average power, temperature, etc.[7,18,26]. Naturally, good heatsinking, which keeps the active region as close to room temperature as possible, is of great importance.

More important perhaps is the recent realization that the incorporation of indium into short wavelength (0.8–1.0 μm) quantum-well structures mitigates the propagation of dark line defects into and/or along the active

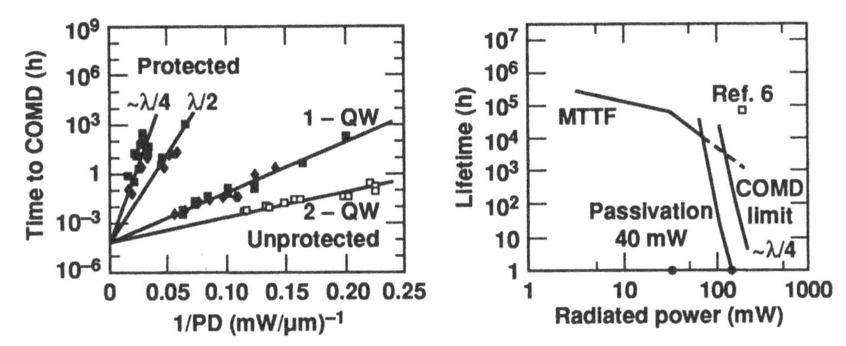

Figure 6.20. Lifetime limitations for AlGaAs lasers. (Taken from Ref. 21 with the permission of the authors). The mean time to failure data for gradual degradation of Al_2O_3-passivated facets, the mean time to failure data, and the square are taken from references cited in Ref. 21.) The region to the left is the low-power regime dominated by gradual failure and the regime on the right is dominated by catastrophic damage for oxygen containing coated facets. The data point labeled Ref. 6 is for treated (oxide-free) facets.

regions of the laser[7,19,20,26,28–33]. It was noted long ago that the longer wavelength InGaAsP lasers were much more reliable than AlGaAs lasers[34] and this was explained by the fact that dislocation glide andclimb rocesses occurred in this material only with difficulty. Recently, Garbuzov*et al.*[31] have reported the growth of high-power InGaAsP/GaAs quantum-well lasers with 100 μm stripe widths and 1 mm cavity lengths. The devices were operated at 0.5–1.0 W each or at 5–10 mW/μm, and during 100–1000 hours of testing a population of 50 devices no sudden failures were observed. The rate of power decrease for many of the devices was less than 3% per hundred hours. Separate experiments revealed that no dark spots in the InGaAsP active regions were observed to grow into larger defects as they had in a similar population of AlGaAs lasers. Garbuzov and colleagues[30] also report operating 100 μm-wide InGaAsP/GaAs lasers at powers up to 5.3 W when mounted junction side down on passive copper heatsinks. Long-term operation of these devices was not reported; however, it should be noted that at 53 mW/ μm these devices are operating well into COD limits and their sudden failure is most certainly due to facet oxidation during operation followed by the rapid onset of COD. With this method of heatsinking it is certain that the junction (and the facets due to the method of operation) was functioning at a very elevated temperature.

Waters and colleagues[29] noted the inhibition of dark line defects in InGaAs/GaAs. AlGaAs and InGaAs populations (12 and 6, respectively) were both intentionally scribed near the active regions to initiate degradation. The lasers were each 60 μm by 600 μm and operated at 30 °C. When operating the InGaAs diodes at either 30 mW or 70 mW for 650 hours no sudden failures were observed. Operating the AlGaAs diodes at 100 mW for 160 hours produced eight failures within 100 hours. In related, more recent, experiments Waters *et al.*[28] intentionally scribed InAlGaAs lasers (again 60 μm by 600 μm laser cavities) to initiate degradation, and after operating cw at 100 mW for 207 hours found total suppression of ⟨100⟩ dislocation climb defects within the InAlGaAs devices; ⟨110⟩ defect growth associated with the glide mechanism was not suppressed, however. Extended testing to 1800 hours still did not produce sudden failures, although the degradation rates were 14%/kh. The authors conclude that since these degradation rates are comparable to those observed in similar AlGaAs devices, the InAlGaAs devices at this point should be considered at least as good as AlGaAs. Recent reports on high-powered, long-lived, visible AlGaInP lasers[35] and high-powered 1.48 μm InGaAsP Fabry-Perot lasers[36] have also appeared.

Very recently, Tang and colleagues[18] have reported on a careful study in which evidence was obtained that the temperature rise at the facets associated with COD correlated more strongly with injected current density then photon flux, at least in the initial stages of gradual degradation of the facet. The authors caution that this mechanism for degradation (non-radiative decay of carriers at surface states) is not necessarily at odds with the picture of eventual absorption of photons at the degraded facets once defect densities have built up. Still, designs which minimize carrier densities near the facet may stall the eventual onset of COD.

Now that astonishing advances are being made in the ability to fabricate very high-power lasers (with the figure of merit being milliwatts of light per micron of emission aperture), it is worth while returning to Sections 6.2 and 6.3 to re-examine our earlier assumptions regarding the requirements for packaging technology. Will the microchannel-cooling technology allow us to operate cw the highest-power 2-D arrays which we are now equipped to assemble? At 40 mW/μm with a 50% bar fill factor we should now be able to manufacture 200 W per linear centimeter arrays. With a thermal resistance of 0.23 °C/W, a 50% efficient 200 W bar will have a temperature rise of 46 °C, where it might operate quite reliably. Note that at a stacking pitch of 10 cm^{-1} a 2-D cw array of this architecture would deliver 2000 W average power from a square centimeter aperture – certainly overkill for side-pumping any solid state laser thicker than 1 or 2 mm.

In summary, tremendous progress has been made by a number of groups worldwide on developing reliable and high average power two-dimensional semiconductor laser arrays. With the new packaging methods developed to allow the devices to be operated at higher and higher average power, and with increasing understanding of the roles and mechanisms of facet damage and dark-line-defect propagation upon semiconductor laser longevity, it is clear that continued progress is assured. As these devices are initially operated at higher powers it is certain that reliable performance will initially be degraded from lower-power performance. Once operated in these new regimes and with their failure modes diagnosed and overcome, however, it is optimistic, but not wildly speculative, to envision devices operating above 1 kW cw per cm^2 of aperture for periods near to 10 000 hours. This kind of performance can be expected from this technology, and applications which require it should begin to drive these important new devices to this level of operation.

Acknowledgment

This work was performed under the auspices of the US Department of Energy by Lawrence Livermore National Laboratory under contract W-7405-Eng-48. The authors would also like to express their appreciation and thanks to E. V. George, J. F. Holzrichter, W. R. Sooy and J. I. Davis for their encouragement and interest in this work. We also thank D. Ciarlo, V. Sperry, L. DiMercurio, J. Hamilton, T. Rodrigues, E. Utterback and S. Mills of LLNL for all their help in the design and fabrication of the devices. Finally, we wish to express the generous support of the US Marine Corps., US Navy, SDIO, and the US Air Force for supporting major portions of this work.

References

1. R. Beach, B. Benett, B. Freitas, D. Mundinger, B. Comaskey, R. Solarz and M. Emanuel, *IEEE J. Quantum Electron.*, **28**, 966–76 (1992).
2. D. Mundinger, R. Beach, W. Benett, R. Solarz, W. Krupke, R. Staver and D. Tuckerman, *Appl. Phys. Lett.*, **53**, 1030–32 (1988).
3. R. Solarz, R. Beach, W. Benett, B. Freitas, B. Comaskey and M. Emanuel, *1991 Diode Laser Technology Proceedings*, 13–15 April, Washington, DC (1991).
4. R. J. Phillips, Microchannel heat sinks, *Lincoln Lab. Journ.*, **1**, 31–48 (1988).
5. T. J. Blanc, R. E. Niggemann and M. B. Parekh, *Soc. Autmot. Eng. Paper No. 831127* (1983).

6. J. Endriz, SDL Technical Note in response to Sol. N66001-91-X-6011, *Laser Diode Array Designs for the NOSC High Efficiency Advanced Solid State Laser*, 1 April (1991).
7. W. D. Goodhue, J. P. Donnelly, C. A. Wang, G. A. Lincoln, K. Rauschenbach, R. J. Bailey and G. D. Johnson, *Appl. Phys. Lett.*, **59**, 632–4 (1991).
8. E. Yablonovich, D. M. Hwang, T. J. Gmitter, L. T. Florez and J. P. Harbison, *Appl. Phys. Lett.*, **56**, 2419–21 (1990).
9. E. Yablonovich, T. Gmitter, J. P. Harbison and R. Bhat, *Appl. Phys. Lett.*, **51**, 2222–4 (1987).
10a. J. P. Donnelly, R. J. Bailey, C. A. Wang, G. A. Simpson and K. Rauschenbach, *Appl. Phys. Lett.*, **53**, 938–940 (1988).
10b. R. Williamson, Planar diode arrays, *Diode Laser Technology Program*, 11 April (1989).
11. L. J. Missaggia, J. N. Walpole, Z. L. Liau and R. J. Phillips, *IEEE J. Quantum Electron.*, **QE-25**, 1988–92 (1989).
12. J. S. Mott and S. H. Macomber, *IEEE Photon. Technol. Lett.*, **1**, 202–4 (1989).
13. R. D. Dupuis, R. L. Hartman and F. R. Nash, *IEEE Electron. Device Lett.*, **EDL-4**, 286–8 (1983).
14. R. G. Waters and D. S. Hill, *J. Electron. Materials*, **17**, 239–41 (1988).
15. M. E. Polyakov, *Sov. J. Quantum Electron.*, **19**, 26–9 (1989).
16. R. G. Waters and R. K. Bertaska, *Appl. Phys. Lett.*, **52**, 1347–8 (1988).
17. D. L. Begley, *Technical Digest of International Conference on Lasers '87*, Paper TI.5, p. 13 (1988).
18. W. C. Tang, H. J. Rosen, P. Vettiger and D. J. Webb, *Appl. Phys. Lett.*, **59**, 1005–7 (1991).
19. M. Okayaso, M. Fukuda, T. Takeshita and S. Vehara, *IEEE Photon. Technol. Lett.*, **2**, 689–91 (1990).
20. R. J. Fu, C. S. Hong, E. Y. Chan, D. J. Booher and L. Figueroa, *IEEE Photon. Tech. Lett.*, **3**, 208–10 (1991).
21. A. Moser, *Appl. Phys. Lett.*, **59**, 522–4 (1991).
22. D. F. Welch, W. Streifer, R. L. Thornton and T. Paoli, *Electron. Lett.*, **23**, 525–7 (1987).
23. R. G. Waters, M. A. Emanuel and R. J. Dalby, *J. Appl. Phys.*, **66**, 961–3 (1989).
24. H. Naito, H. Nakanishi, H. Nagai, M. Yuri, N. Yoshikawa, M. Kume, K. Hamada, H. Shimizu, M. Kazumura and I. Teramoto, *J. Appl. Phys.*, **68**, 4420–25 (1990).
25. J. Yoo, H. Lee and P. Zory, *Photon. Tech. Lett.*, **3**, 202–3 and 594–6 (1991).
26. E.-E. Latta, A. Moser, A. Oosenbrug, C. Harder, M. Gasser and T. Forster, *IBM Zurich Solid State Physics Research Report* (1991).
27. D. Welch, R. Craig, W. Streifer and D. Scifres, *Electron. Lett.*, **26**, 1481–3 (1990).
28. R. G. Waters, R. J. Dalby, J. A. Baumann, J. L. De Sanctis and A. H. Shepard, *IEEE Photon. Tech. Lett.*, **3**, 409–11 (1991).
29. R. G. Waters, D. P. Bour, S. L. Yellen and N. F. Ruggieri, *IEEE Photon. Tech. Lett.*, **2**, 531–3 (1990).
30. D. Z. Garbuzov, N. Y. Antonishkis, A. B. Gulakov, S. N. Zhigulin, A. V. Kochergin and E. U. Rafailov, *Proceedings of OE Lase Conference '91*, Los Angeles, Ca, January (1991).
31. D. Z. Garbuzov, N. Y. Antonishkis, A. D. Bondarev, S. N. Zhigulin, A. V.

Kochergin, N. I. Katsavets and E. U. Rafailov, *Proceedings of OE Lase Conference '91*, Los Angeles, January (1991).
32. K. Itaya, M. Ishikawa, H. Okuda, Y. Watanabe, K. Nitta, H. Shiozawa and Y. Uematsu, *Appl. Phys. Lett.*, **53**, 1363–5 (1988).
33. W. D. Goodhue, J. P. Donnelly, C. A. Wang, G. A. Lincoln, K. Rauschenbach, R. J. Bailey and G. D. Johnson, *Appl. Phys. Lett.*, **59**, 632–4 (1991).
34. P. M. Petroff, in *Semiconductors and Semimetals*, vol. 224, R. K. Willardson and A. C. Beer (eds.), New York, Academic Press, p. 379 (1985).
35. H. B. Serreze and Y. C. Chen, *IEEE Photon. Tech. Lett.*, **3**, 397–9 (1991).
36. S. Oshiba and Y. Tamure, *J. Lightwave Technol.*, **8**, 1350–6 (1990).
37. M. Jansen, S. S. Ou, J. J. Yang, M. Sergant, C. Hess, C. Tu, F. Alvarez and H. Bobitch, *LEOS '92 Conference*, Boston, MA, 16–17 November (1992).
38. M. Jansen, S. S. Ou, J. J. Yang, M. Sergant, C. A. Hess, C. T. Tu, F. Alvarez and H. Bobitch, *SPIE O/E-LASE '93*, Laser Diode Technology and Applications V, Paper 1850–37, Los Angeles, Ca, 18–20 January (1993).
39. D. W. Nam, R. G. Waarts, D. F. Welch and D. R. Scifres, *Tech. Digest CLEO '92*, pp. 346–7, Anaheim, Ca., May (1992).
40. J. M. Haden, M. De Vito, W. Plano and J. G. Endriz, *LEOS '92 Annual Mtg.*, Paper DLTA 5.3, Boston, MA, 16–17 November (1992).

7

High-power diode laser arrays and their reliability

D. R. SCIFRES AND H. H. KUNG

7.1 Introduction

Diode lasers were first demonstrated in 1962[1–4]. These structures were broad area homojunction devices which could only be operated at 77 K with short (< 100 ns) pulse widths. From those early lasers, much progress has been made in increasing the output power, reliability, operating temperature and efficiency of diode lasers and diode laser arrays. For example, at present, monolithic laser diode arrays have been operated at temperatures in excess of 100 °C[5,6] at output power levels in excess of 100 W cw[7] and at power conversion efficiencies of greater than 50% with multiwatt output powers[8]. This chapter will describe the history of the improvements made in high-power diode laser research and the failure mechanisms responsible for the limitations in output power of high-power diode lasers. We also present the results of reliability testing at high output power for a variety of semiconductor laser structures and review environmental testing for laser array products. The design considerations for long life operation at high power are also presented.

7.1.1 History of high-power diode laser arrays

Following the early introduction of homojunction[1–4] diode lasers, single heterostructure[9–11] and double heterostructure (DH) diode lasers[12] were demonstrated in the late 1960s. The double heterostructure laser was the first laser to exhibit cw room temperature operation[13,14] This DH laser offered two major improvements over the original homojunction device. These improvements were (1) good carrier confinement of the electrical charge to a thin 'active' region, and (2) optical waveguiding.

As shown in Figure 7.1, the laser active region (in this case shown to be composed of GaAs with an energy gap of ≈1.4 eV) is surrounded

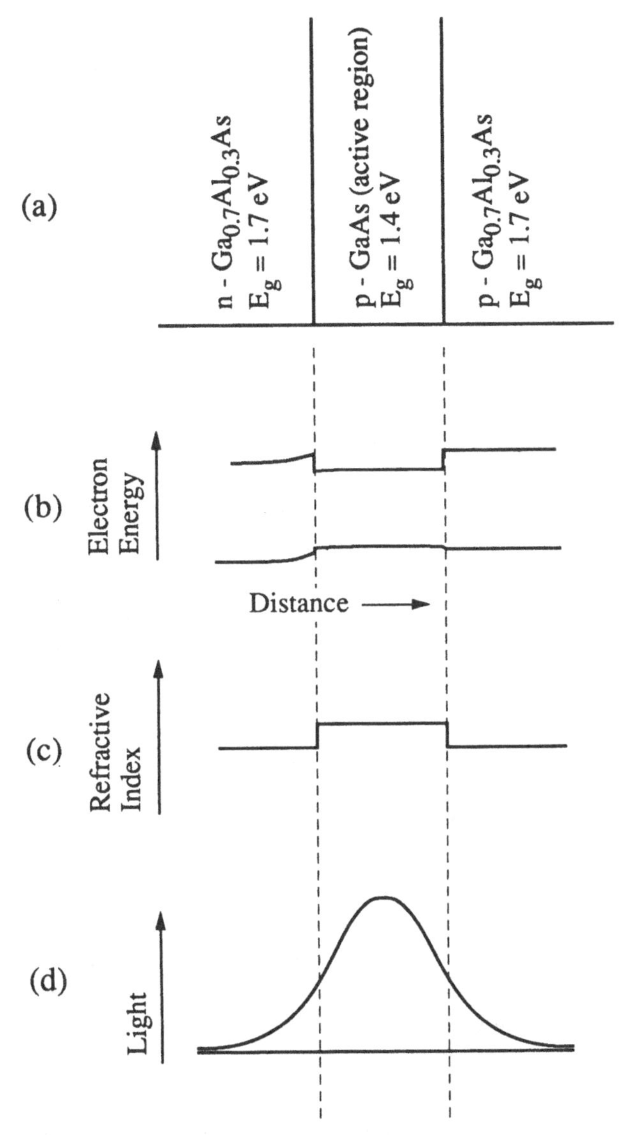

Figure 7.1. (*a*) Schematic diagram of a double heterostructure GaAs-$Ga_{0.7}Al_{0.3}As$ laser. (*b*) Energy diagram under forward bias. (*c*) Refractive index profile. (*d*) Optical-field intensity.

on either side by wide bandgap 'cladding layers'. The cladding layers shown are $Ga_{0.7}Al_{0.3}As$ with an energy gap of ≈ 1.7 eV. Upon forward bias of the p–n junction in the double heterostructure device, virtually all the injected charge can be confined by the $Ga_{0.7}Al_{0.3}As$ cladding layers to the GaAs active layer, even if the active region is quite thin (1000 A or less). Implementation of a thin lasing region allows high gain to be achieved at low pump current levels. The efficient use of electrical charge allows threshold current density levels of 500–1000 A/cm^2 to be achieved at room temperature[15].

Second, the wide bandgap $Ga_{0.7}Al_{0.3}As$ cladding layers also have a lower index of refraction ($\approx$3.4) than does the GaAs active layer ($\approx$3.6 at the 1.4 eV lasing wavelength). This refractive index change allows waveguiding and thus high overlap of the lasing optical wave with the electrical charge in the active region.

These two phenomena, carrier confinement to very thin active layers and good overlap of the optical wave with the injected carrier, still continue to be the key fundamental requirements for the design of high-power diode lasers and diode laser arrays. By 1976, 390 mW cw room temperature operation was demonstrated by utilizing an 80 μm-wide broad area emitter, a thin (for that time) 600 A-thick active layer, and a diamond heat sink[16].

Further significant improvements were achieved when, by 1983, 2.6 W of cw output power was demonstrated from one mirror of a multiple stripe laser array[17]. Three improvements incorporated in this 2.6 W cw laser, shown in Figure 7.2, were primarily responsible for the higher-power output of this device. First, the laser employed a 'quantum well' active region[18]. Quantum wells are very thin (<300 A-thick) low-bandgap layers sandwiched between layers of higher-energy gap material. These quantum wells allow higher gain to be achieved than in conventional double heterostructure lasers. Because of the modified band structure allowing higher gain and because the volume of material into which charge is injected is small, very low-threshold current densities ($\leq$300 A/cm^2) can be achieved. Recent improvements using thin 'strained layer' quantum wells have achieved threshold current densities as low as $\approx$50 A/cm^2 [19]. A low-threshold current density and low total current level is important in achieving high-power output from a diode laser since output power is often limited by how much waste heat is generated in the laser[20].

Second, the multiple quantum well laser shown in Figure 7.2 utilized an array of 40 emitters on 10 μm center-to-center spacings to spread the optical power over a large emitting area. This was in contrast to the early cw lasers developed in the 1970s which emitted from an aperture size of $\approx$10 μm[13,14]. Spreading the optical power over a large area is important for high-power cw operation since catastrophic facet damage caused by

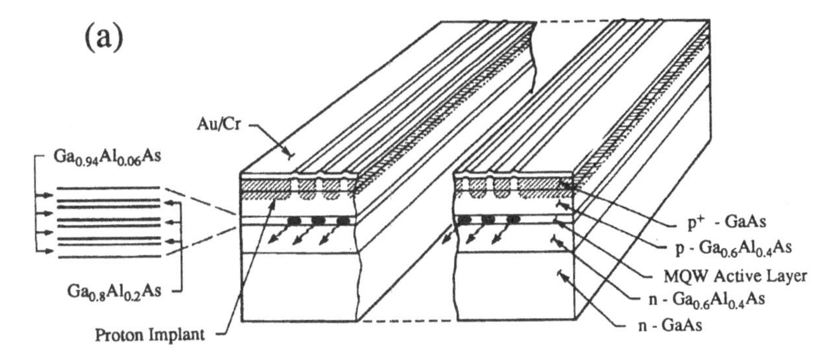

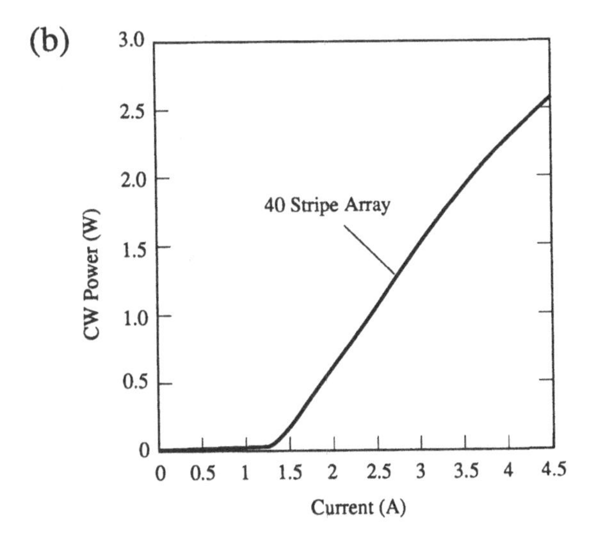

Figure 7.2. (*a*) Schematic diagram of a 40 emitter phase locked GaAlAs laser array. (*b*) cw front facet power from the array as a function of input current.

thermal runaway and subsequent melting of the cleaved laser mirror is another limiting factor in the high-power operation of GaAlAs lasers[21]. The use of multiple emitters also allowed phase-locked operation of the laser[22–26] (also see Chapter 1), thereby allowing higher laser brightness, as well as the potential for electronic beam steering and wavelength tuning[23].

Finally, for the laser in Figure 7.2, the cleaved laser mirror facets were overcoated with a relatively low-reflectivity (12%) front facet and high-reflectivity (95%) rear facet coating. This facet coating prevents rapid attack and erosion of the cleaved laser facet by water vapor in the air, thus helping to prevent the occurrence of catastrophic facet damage. Also, it has been found that the catastrophic facet damage level for high-power lasers is increased by using a low-reflectivity front facet[27]. By the mid-1980s, diode laser arrays with output power levels of 0.2 W–1 W cw and reliability of 10 000 hours or longer became commercially available.

Today, further significant increases in output power levels and laser utility have been achieved with other array structures. These will be described in the following section.

7.1.2 Laser array structures and high-power operating parameters

In the preceding section, we discussed the importance of several improvements which aided the achievement of high-power operation of diode lasers significantly. These included utilization of quantum well heterostructure active regions to achieve low-threshold, high-efficiency operation, the use of mirror coatings to reduce rapid mirror degradation and the use of efficient heat removal techniques. These parameters, when used in conjunction with a variety of laser array structures, have led to reliable, commercially available laser arrays for a variety of applications. This section will discuss the state-of-the-art performance at high power levels of these recent laser array structures.

7.1.2.1 Separately addressable single mode array

The simplest, and perhaps one of the most useful, array structures is the separately addressed single transverse mode array[28–33]. Data from a particular nine-element array is shown in Table 7.1[34]. In this particular array device, single transverse mode diode lasers are fabricated on 150 μm spacings. They are mounted junction-side up on a copper heat sink to facilitate the ability to provide individual wire bonds to each laser in the array. This allows the nine lasers in the array to be addressed separately. Each laser can be operated at an output power of 150 mW cw. A key to the reliable operation of such an array is the uniformity obtained in device properties. The uniformity of threshold current, operating current and wavelength is shown in Table 7.1. Such uniformity is currently obtained by MOCVD wafer growth and subsequent careful wafer processing. Large area arrays must be highly uniform in both material properties and thermal properties in order to avoid current channeling effects and thermal hot spots.

The nine-element, separately addressed, array shown in Table 7.1 has threshold current levels varying by less than 1 mA and wavelengths

Table 7.1. *Nine-element separately addressed array uniformity*

Device	1	2	3	4	5	6	7	8	9	Maximum deviation
I_{th} (mA)	11.0	11.5	11.5	12.0	11.5	11.0	11.0	12.0	12.0	1 mA
I (mA) at 150 mW cw	168	167	167	174	166	170	167	166	174	8 mA
λ (nm) at 10 mW cw	854.6	853.9	854.0	854.5	854.8	854.7	854.0	854.9	854.5	1 nm

varying by less than 1 nm. At 150 mW cw per element, the operating currents vary by less than 5%. Single element lifetests on similar single mode lasers operated at 100 mW cw indicate lifetimes greater than 100 000 hours at 25 °C[35]. Such separately addressed single transverse mode arrays are useful for multitrack optical recording and printing systems, among other applications.

7.1.2.2 High-power laser chips

If higher power is required from a laser diode for applications such as pumping solid state lasers or laser surgery, a laser diode quantum well array geometry consisting of a single die with an emitting aperture of ≈50 μm to ≈500 μm is generally used. Reliable commercial output power levels can range from 0.5 W cw from a 50 μm aperture laser to 4 W cw from a 500 μm aperture[36,37]. Such lasers may use either the multistripe array structure shown in Figure 7.2, or a broad area emitter[38].

The maximum output power levels reported for such lasers are presently 6 W cw for a 100 μm-aperture laser[39], 8 W cw for a 200 μm-aperture laser[39], and 11.5 W for a 370 μm-aperture laser[37]. Despite the optical facet power density of 60 mW/μm for the 100 μm-wide emitting device, the output power level in this device was limited by thermal rollover rather than by catastrophic facet damage. This high catastrophic facet damage threshold was a consequence of spreading the laser beam in the vertical direction (perpendicular to the p–n junction), as well as along the width of the junction. The application of a high-reflectivity rear facet coating (95%) and a low-reflectivity front facet coating (5%) also raised the catastrophic facet damage level of the laser[27]. As the aperture size was increased to 370 μm, the overall power conversion efficiency was decreased and thermal rollover dominated at still lower facet flux densities.

7.1.2.3 High-power cw bars

Still higher cw output power levels have recently been obtained from monolithic diode array bar structures. A monolithic laser diode bar array

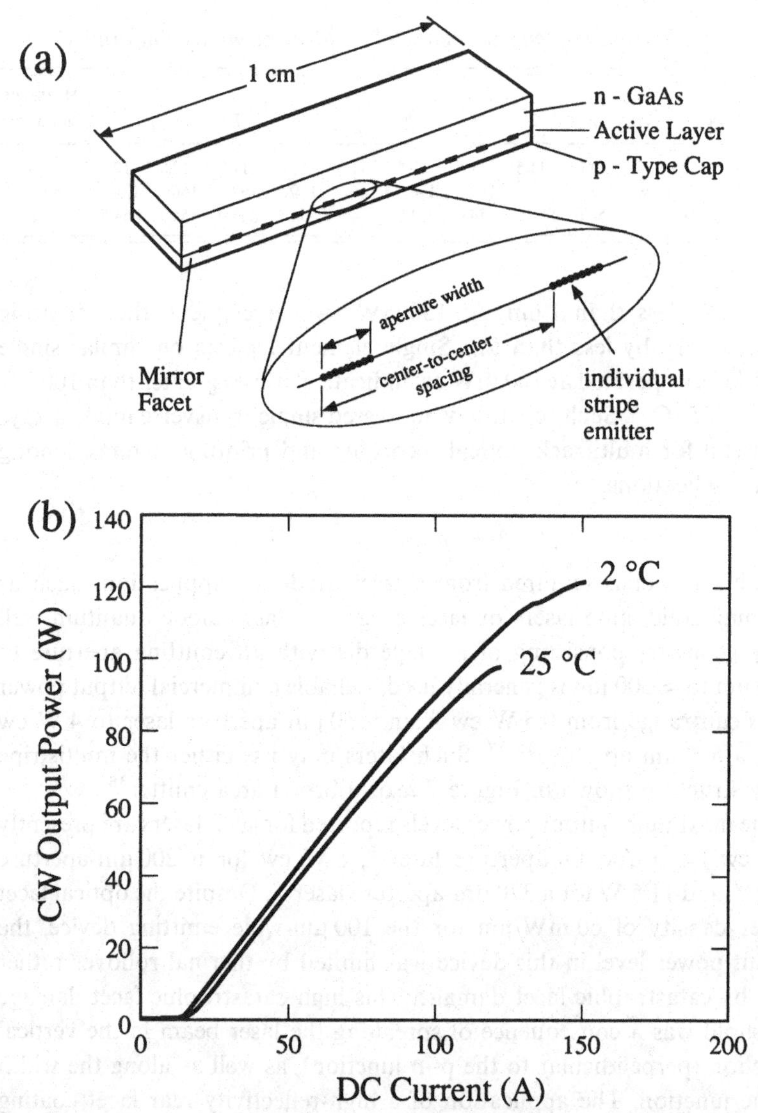

Figure 7.3. (*a*) Schematic diagram of a 1 cm monolithic laser bar. (*b*) cw light output power vs input current characteristics for such a bar at 2 °C and 25 °C heat sink temperatures.

has achieved 100 W cw operation at room temperature[7]. When the heat sink is cooled to 2 °C, over 120 W cw output is achieved. This 1 cm-long monolithic bar structure and its output power as a function of DC current at 25 °C and 2 °C is shown in Figure 7.3. Primary improvements allowing this further output power increase include the following: (1) the use of a 1 cm-wide monolithic laser bar which emits from 80 separate lasers, each having a 96 μm-wide aperture. (This geometry lases over a width of 7200 μm. At 100 W cw, this corresponds to an optical power density at the facet of $\approx$14 mW/μm. Thus the laser is far from the point of facet failure.) (2) The use of a diamond heatsink to remove waste heat efficiently from the laser array. The laser also exhibits a low operating threshold current level (17 A) and a high laser slope efficiency 1.05 W/A allowed by a single quantum well active region design. The front facet is coated with a $\approx$5% reflectivity mirror, while the rear facet is made nearly 100% reflecting.

7.1.2.4 High-power quasi-cw bars

Because of thermal limitations seen under cw operation, such 1 cm and larger diode laser bar arrays have also been operated in a so-called 'quasi-cw mode'[40]. This nomenclature describes diode lasers operating with pulse widths in excess of $\approx$1 μs and arose due to the fact that the thermal conductivity and typical thicknesses of a diode laser mounted with its junction directly on to a heat sink allows substantial heat transfer out of the laser chip for time periods longer than 1 μs. Quasi-cw operation thus differentiates the behavior from that of short pulse (<1 μs pulse width) devices which can operate to significantly increased peak power levels without incurring catastrophic facet failures[41]. Thus, low duty factor quasi-cw operation of devices often allows the removal of the thermal limitations in output power while maximizing the energy per pulse from a diode laser array. Quasi-cw diode laser bar arrays are thus most often used for optical excitation of solid state lasers where maximizing energy per pulse is the key operating specification[42–44].

Although early versions of quasi-cw bars had low fill factors, today's geometries are generally 1 cm or greater in length and 90–95% of the laser bar emits light. Such 1 cm quasi-cw bar arrays have been operated with 200 μs pulse widths at a 2% duty factor to power levels of greater than 200 W for both GaAlAs (208 W/cm)[45] and strained layer InGaAs (247 W/cm)[46] active layer material. Recently, it has been demonstrated that an 820 nm quasi-cw bar operated with 100 μs pulse width has a peak power exceeding 300 W/cm[47]. GaAlAs quasi-cw bars have also been reliably operated with 1 ms pulse width at power levels of 60 W/pulse, thereby achieving 60 mJ/pulse of optical energy[46]. Such lasers do not require diamond heat sinks or special cooling to operate at these power levels.

Quasi-cw arrays can be operated at high duty factor with proper heat sink configurations. Single 1 cm bars operating at 60 W peak output power levels at a 20% duty factor with constant current drive exhibit only an average decrease in output power of 1.6% over one billion 200 μs pulses[46]. Another report of 2 cm-long bars operating at a 25% duty factor at 80 W showed an average 5% power drop in one billion shots[48]. Thus, these quasi-cw lasers have reliability levels in excess of 1000 times that of the flash lamp technology which they are designed to replace.

7.1.2.5 Two-dimensional stacked arrays

Two-dimensional stacked arrays have also been demonstrated to emit high average and peak output power levels under quasi-cw operation[46–48]. This two-dimensional array structure is shown schematically in Figure 7.4(*a*). By changing the thickness of the bar bond plate, the number of diodes per cm^2 can be changed to provide either high peak power or high duty factor as required by different applications. Peak power output levels of 2.5 kW/cm^2 (5 kW total power) which will operate at up to 2% duty factor on the 50 bar 1 cm × 2 cm liquid cooled array shown in Figure 7.4(*b*) are now commercially available. The maximum pulse width for the 50 bar stack is 400 ms allowing an output power of 2 J per pulse. This type of array is useful for pumping slab-type solid state lasers. Also shown in Figure 7.4(*b*) is a 4.0 × 0.5 cm package which gives the same performance but is designed for pumping rod-type solid state lasers. The 4.0 × 0.5 cm package pictured in Figure 7.4(*b*) contains 16 laser bars and operates at a peak power level of 960 watts at up to 20% duty factor. A more complete description of two-dimensional quasi-cw operating parameters and structures is presented in Chapter 6 of this book.

7.1.2.6 Monolithic two-dimensional arrays

A further potential improvement in high-power arrays is the development of *monolithic* two-dimensional arrays. Three versions of such arrays are being actively studied. These are (1) the vertical cavity surface emitting laser[49–53], (2) the horizontal cavity angled facet laser[54–62], and (3) the grating coupled surface emitting laser[63–72]. Each of these structures allows the optical wave to be emitted through the surface of the wafer. To date, only the angled facet laser has shown operation at average power levels above 5 W cw. Nam *et al.*[62] achieved 50 W cw operation from such a monolithic two-dimensional angled facet surface emitting laser. The device shown in Figure 7.5 contains ≈1500 single mode lasers which emit at a wavelength of 944 nm. The two-dimensional laser array uses a single strained layer InGaAs quantum well[73]. Because the 944 nm laser light

(*a*)

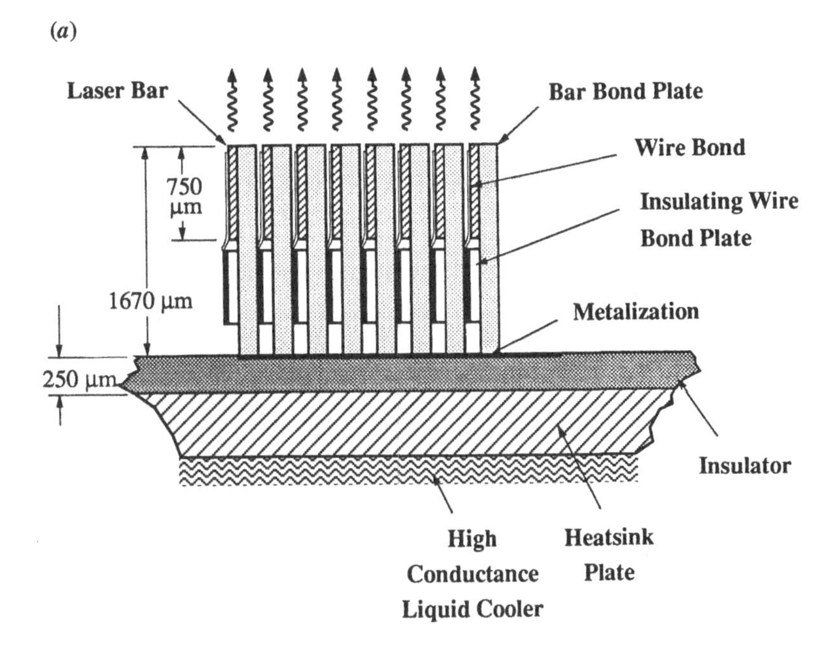

(*b*)

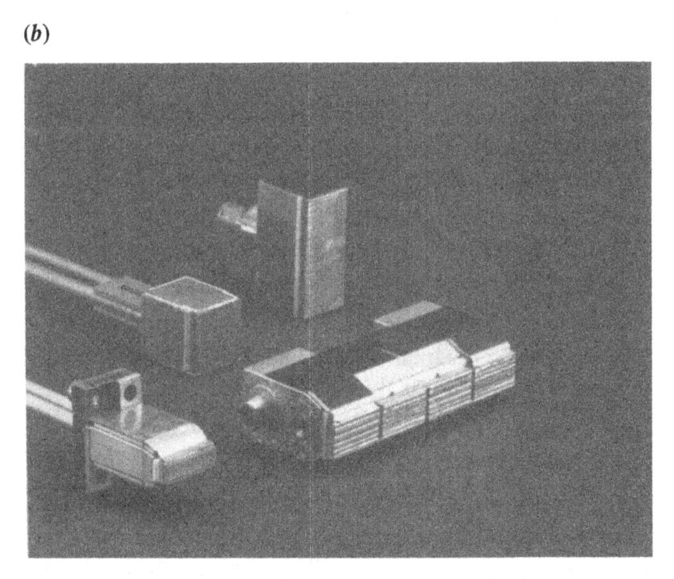

Figure 7.4. (*a*) Schematic diagram of stack array assembly. (*b*) Photos of various two-dimensional stack arrays. The 1 cm × 2 cm array emits 5 kW peak power while the 0.5 × 4.0 cm array emits ≈1 kW peak power and 200 W average power.

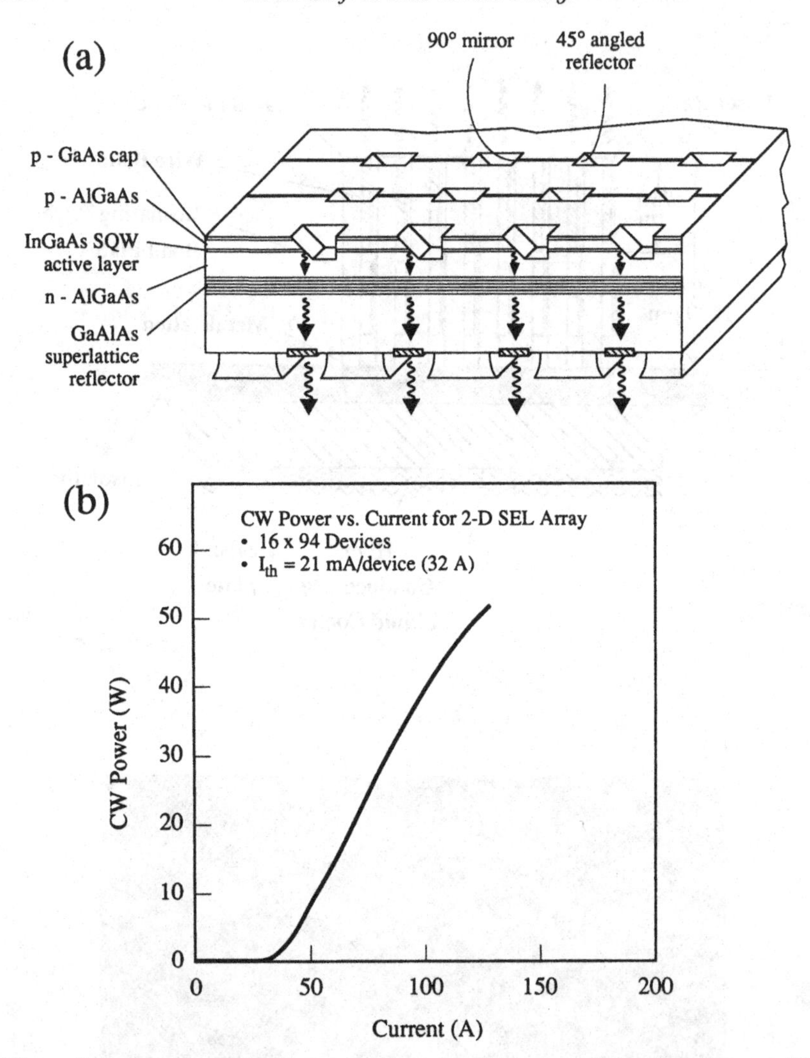

Figure 7.5. (*a*) Schematic diagram of a monolithic two-dimensional array. (*b*) Light output power vs input current characteristics of such an array. Maximum power is 50 W cw.

is not attenuated significantly by the GaAs substrate, the light can be taken out of the top of the wafer while the 1 cm^2 wafer is bonded junction-side down to a liquid cooled heat sink for efficient heat removal.

Monolithic two-dimensional arrays have several potential advantages over linear bar arrays since (1) catastrophic facet damage should be eliminated, (2) thermal heat removal can be optimized, and (3) the fabrication costs may ultimately be reduced. To date, however, none of these potential advantages have been fully realized for two-dimensional monolithic structures. Thus, today, linear bar arrays remain the best practical solution to achieving high cw or quasi-cw power.

7.1.3 Materials systems for high-power diode laser arrays

High-power diode laser arrays can be fabricated in a variety of semiconductor material systems. To date, most of the work reported in the literature and most of the commercially available high-power laser arrays are fabricated in the GaAlAs material system which emits in the wavelength range of $\approx$0.75–0.9 μm. However, high-power diode laser arrays and broad area diodes have now been demonstrated in the wavelength range from 0.6 μm to 2.0 μm. This section will describe some of these results.

7.1.3.1 Visible high-power lasers

At short wavelengths, MOCVD growth of $(Al_xGa_{1-x})_{0.5}In_{0.5}P$ lattice matched to GaAs substrates yielded demonstrations of 660 nm high-power visible diode laser arrays[74,75]. Recently, 633 nm lasers with power levels of 0.9 W cw and 3 W cw from 1 cm bar arrays with 20 separate 100 μm-wide broad area emitters have been demonstrated[76]. These 633 nm lasers are provided with tensile strain in the quantum well active region in order to reduce the threshold current density to $\approx$400 A/cm^2, thereby allowing high-power cw operation. However, the reliability of lasers with tensile strain is still questionable. At 680 nm, 20 W of cw output power[77] and 60 W[78] quasi-cw power levels have been achieved using compressively strained $Ga_{0.4}In_{0.6}P$ quantum wells surrounded by $(Al_{1-x}Ga_x)_{0.5}In_{0.5}P$ confining and cladding layers.

Such 680 nm broad area devices have been lifetested using 0.5 W cw lasers with 250 μm apertures exhibiting projected mean time to failure (MTTF) times of greater than 4000 hours[79].

One of the problems with the $(Al_xGa_{1-x})_{0.5}In_{0.5}P$ material system when used to fabricate high-power diode lasers is the rather high sensitivity of laser threshold current temperature on the junction temperature. This property can be modeled by a characteristic temperature, T_0, where at two temperatures, T_1 and T_2, the threshold current levels $I(T_1)$ and $I(T_2)$ are expressed as:

$$I(T_1)/I(T_2) = \exp((T_1 - T_2)/T_0). \tag{7.1}$$

For devices fabricated to date with the InGaAlP material system, a characteristic temperature (T_0) of 70–80 K is typical[75,80]. This is in contrast to the results for the GaAlAs material system where the T_0 values generally range from 150 to 200 K[81]. Material systems with a high T_0 value are far superior for the fabrication of high-power laser arrays since they are less susceptible to saturation of output power due to thermal runaway, or rollover.

7.1.3.2 High-power near IR lasers

High-power arrays have also been fabricated using InGaAs/AlGaAs strained quantum well layers[82–90]. Strain in a quantum well structure produces a reduction in the required threshold current density due to enhanced gain caused by a reduction in the effective mass of the valence band. In addition, when the strained layer is approximately equal to or less than the critical thickness defined by Matthews and Blakeslee[91], a reduction in the degradation associated with dark line defects has been reported[92]. Furthermore, this material has a high T_0 value in the range 150–250 K. Thus, InGaAs seems to be well suited to fabrication of high-power laser diode arrays. Output power levels of up to 4.9 W cw per facet from a 500 μm-wide laser operating with a heat sink temperature of 10 °C have been achieved[6]. One-hundred-micron-aperture devices have operated to 3 W cw power levels at 25 °C[84], while 0.5 W cw lasers have been obtained at 125 °C from a 500 μm-aperture device[6]. It should be noted that if the quantum well layer exceeds the critical thickness, misfit dislocations are present[91], leading to rapid device degradation.

The standard wavelength range for the InGaAs/GaAlAs system grown on a GaAs substrate is 0.8 μm to ≈1.1 μm. However, recent results using InGaAs/InGaAsP grown on InP substrates have demonstrated 10 °C output power levels of 1.6 W cw at a 2.0 μm[93] wavelength. This result is obtained for a 200 μm-aperture double quantum well device. The laser structure and power versus current curves are shown in Figure 7.6. The output power on this device is limited by thermal rollover owing to a value of 53 K for T_0, the characteristic temperature.

A further high-power result at a ≈2.1 μm wavelength has been reported for the GaInAsSb/GaAlAsSb material system. Specifically, 0.2 W cw per facet was achieved for a 100 μm-aperture laser at 20 °C[94]. It is predicted that high-power cw room temperature operation in this material system can be achieved out to 3–4 μm wavelengths.

Another material system which has demonstrated high cw output power levels in the 0.8 μm range suitable for pumping Nd:YAG lasers is $In_{1-x}Ga_xAs_{1-y}P_y$ lattice matched to a GaAs substrate. Garbuzov and

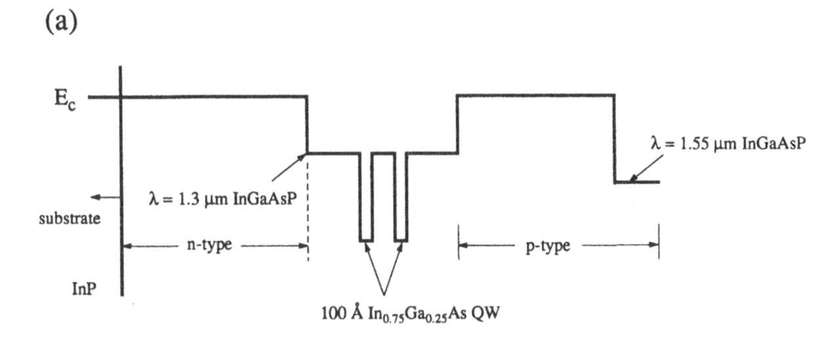

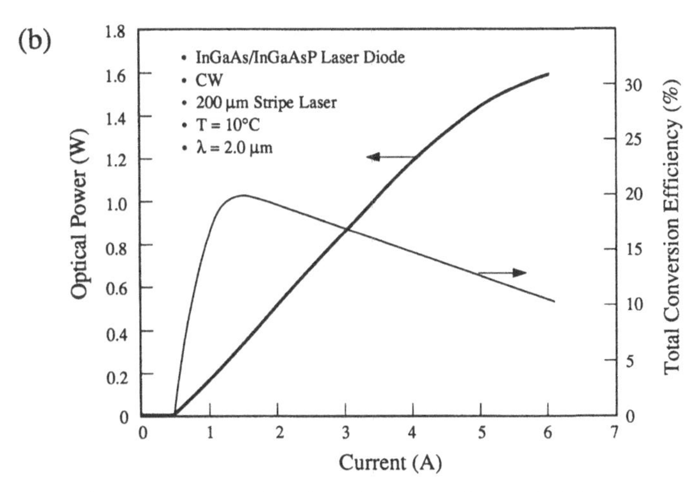

Figure 7.6. (*a*) Structure of InGaAs/InGaAsP laser. (*b*) Light output power and total conversion efficiency vs input current characteristics of such a laser emitting at a wavelength of ≈2.0 μm.

co-workers have demonstrated 5.3 W cw per facet from 100 μm-aperture lasers *without* the use of any facet coating[95]. This high output power level (53 mW/μm) in the absence of facet coatings is believed to arise from the fact that this material does not contain aluminum. It is believed that the reaction of aluminum with water or oxygen is the primary cause of facet related degradation in GaAlAs lasers[81,96,97]. Thus, these aluminum-free devices should be limited primarily by thermal rollover at high power levels.

The $In_{1-m}Ga_mAs_{1-n}P_n$ material system can also be lattice matched to InP substrates for emission at 1.3 μm. Such lasers emit up to 380 mW cw at −40 °C from apertures several microns in width in lasers with high-reflectivity rear facet coatings[98]. InGaAs/InGaAsP single mode devices have also been reported to emit output power levels in the range 200–300 mW for 1.3–1.48 μm[99,100]. InGaAsP devices operating at 1.3 μm and beyond are generally limited by thermal rollover because of a low T_0 and low efficiency caused primarily by Auger recombination. Thus, to date, they have not generally been used for higher-power laser array fabrication in the 1.3–1.5 μm range.

The results reported in this section have described the maximum power output from a variety of diode laser array structures and materials. It should be noted that there is always a trade-off between diode laser array output power and array reliability. This trade-off generally takes on two aspects for GaAlAs lasers in particular. First, if the array is operated at a power level which is greater than 50% of its catastrophic facet failure point, it can be expected to degrade rather rapidly and be susceptible to random sudden failures[101,102]. Second, if the array is operating at a high junction temperature, a gradual temperature activated degradation occurs[81]. These and other reliability mechanisms will be reported in detail in Section 7.2.

7.2 Failure mechanisms of high-power laser arrays

7.2.1 Introduction

For high-power diode lasers, we wish to have as high a power as possible yet the active region temperature should be as low as possible. Let us look at the parameters which are important in this regard. The relationship between operating power (P_{op}) and operating current (I_{op}) is expressed as

$$P_{op} = (I_{op} - I_{th})\eta_d, \tag{7.2}$$

where I_{th} is the threshold current and η_d is the slope or differential efficiency. Under the conditions that the gain has a linear dependence on current, these parameters are related to laser geometry factors, intrinsic material parameters and facet reflectivities as[16]:

$$I_{th} = J_0 w \cdot L \cdot \frac{d}{\eta_d} + \left(w \cdot L \cdot \frac{d}{\eta_d \beta}\right)\left\{\frac{1}{\Gamma}\left[\alpha + \frac{1}{2L}\ln\frac{1}{R_f \cdot R_r}\right]\right\} \tag{7.3}$$

and

$$\eta_d = \frac{\eta_i}{1 + [(2\alpha \cdot L)/\ln(R_f \cdot R_r)]}, \tag{7.4}$$

where L = cavity length, w = width of emitting region, d = active layer thickness, J_0 = transparency current density/μm, Γ = optical confinement factor, α = internal absorption, β = gain constant, R_f = front facet reflectivity, R_r = rear facet reflectivity, and η_i = internal quantum efficiency.

Optimizing the design of a semiconductor laser array for high-power operation involves minimizing the waste heat generated. Factors controlling this waste heat generation are illustrated in Equations (7.2)–(7.4). First, the differential quantum efficiency (η_d), which is a measure of the number of photons generated above lasing threshold relative to the carriers injected into the active region, must be high. The differential efficiency, which has been reported to be as high as 80–90% for GaAlAs lasers[103–105], is controlled by doping level which affects α[104], facet reflectivities and cavity length[105,106], and internal efficiency η_i[103]. A low threshold current density, low series resistance, and operation at current levels well above threshold also allows for a high total power conversion efficiency, η_T. Values of η_T have been reported to be greater than 60% for a single mode narrow stripe laser[107] and as high as 57% for a wide stripe laser[103] in the GaAlAs material system. Even higher levels of efficiency ($\eta_T = 66\%$) have been reported for InGaAsP/GaAs lasers at 0.86 μm[108].

When laser array products are in operation, they exhibit several aspects of degradation with time. These aspects generally depend on the type of application of the laser in the system. A main consideration is always power degradation, i.e. during constant current operation, the power decreases with time. With constant power operation (using an optical detector and electrical feedback control of the laser current), the current to maintain a constant power increases with time.

In addition to the power degradation, other parameters such as dynamic behavior, peak wavelength, mode stability, and the ratio of front to back facet operating power could change with time of operation. For some applications, the change of such parameters can also cause system failure.

For laser diode products stored at room temperature in a controlled environment, no degradation is expected. However, laser products are sometimes required to be exposed to hazardous environments such as high temperature, rapid temperature variation, high humidity, vibration or to a special environment such as that encountered in space. Consequently, reliability testing also requires one to determine the storage conditions under which laser products can survive. In various military applications, MIL specifications and detailed testing procedures for various tests are required.

Generally, there are three basic operational failure modes: rapid

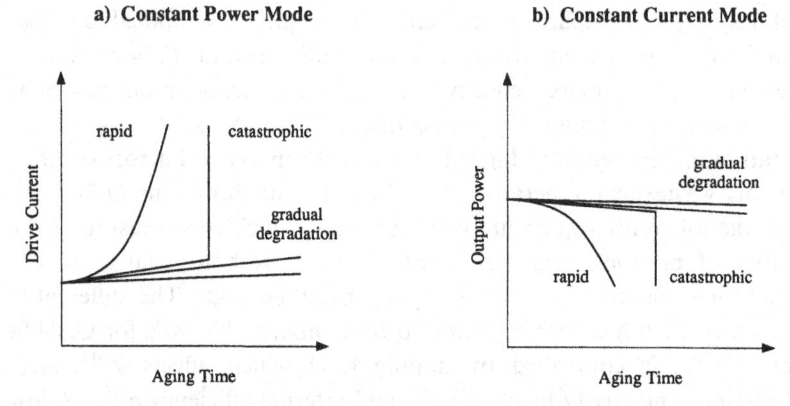

Figure 7.7. Typical degradation modes for laser arrays: (*a*) constant power mode; and (*b*) constant current mode.

degradation, gradual degradation and catastrophic failure. These are shown in Figure 7.7. A burn-in of each laser can readily eliminate the rapid failure mode from the population. As demonstrated in Figure 7.7(*a*) (constant power mode), the gradual degradation failure causes the operating current to gradually increase with time and finally reach the current compliance limit to cause system failure. The failed lasers often exhibit lasing characteristics but with higher threshold current and lower efficiency. This is because the intrinsic parameters such as optical absorption within the laser increase with aging. During catastrophic failure, the laser exhibits a sudden increase in current and finally ceases to lase over a short period of time, usually several hours or less. Figure 7.7(*b*) illustrates degradation modes under a constant current operating mode.

Fundamentally, the origin of the degradation is defects (or imperfections) in lasers. Such defects can be generated during fabrication processes, handling or operation. Through different mechanisms, such defects eventually cause degradation.

The lasers can be divided into three regions and the degradation associated with these regions are:

(a) *Bulk degradation*: this degradation is mainly associated with the property of the GaAlAs crystal which includes the epitaxially grown layers and the substrate.
(b) *Facet degradation*: this degradation is mainly associated with the two cleaved facets. Particularly for high-power lasers, extreme high power densities exist at the facet region.

(c) *Solder related degradation*: for high-power laser arrays, dissipation of several watts heat from the laser chip is required. Therefore, the lasers are generally mounted junction-side down to a heat sink with solder. This degradation is associated with the soldered junction and the metal interfaces involved.

7.2.2 ***Bulk degradation***

Although we have described several material systems in which high-power laser arrays are fabricated, we will primarily concern ourselves with degradation in the GaAlAs material system since most of the high-power lasers commercially available to date are made of GaAlAs.

GaAlAs diode lasers consist of multiple GaAlAs layers grown on a GaAs substrate with metal–organic–chemical-vapor deposition (MOCVD) processes[109] or molecular beam epitaxy (MBE)[110] or liquid phase epitaxy[111]. The p–n junction is formed during the growth. The laser arrays are fabricated on these grown layers with photolithography, ion implantation, thinning and metallization processes. Then, lasers are cleaved into bars and scribed into die.

The GaAlAs layers have a zinc blend crystal structure. For an ideal crystal, lattice constants for different layers are matched and group III atoms and group V atoms are arranged in a regular and repetitive pattern. However, in a real crystal, imperfections exist which disturb the regular arrangements of atoms. Such imperfections include point defects, line defects and plane defects[81,112]. The typical point defects as shown in Figure 7.8(*a*), are vacancies, interstitial atoms and locations where the lattice is severely distorted as a result of a lattice slip or the boundary between slipped and unslipped parts in the crystal. There are two types of dislocations, edge and screw dislocations, as illustrated in Fig. 7.8(*b*). In edge dislocation, the dislocation line is perpendicular to the slip direction, while in screw dislocation, the dislocation line is parallel to the slip direction. The typical plane defect, as shown in Figure 7.8(*c*), is a stacking fault in which the regular sequence of a stack of identical atom layers is locally interrupted in the crystal. In addition to crystal imperfections, defects can be generated by external means such as mechanical damage or metal diffusion. The deformation or breakage of the crystal also creates localized defects.

All defects form nonradiative recombination centers and increase absorption. As shown in Equations (7.3) and (7.4), the operating current increases and the quantum efficiency decreases as a result of an absorption

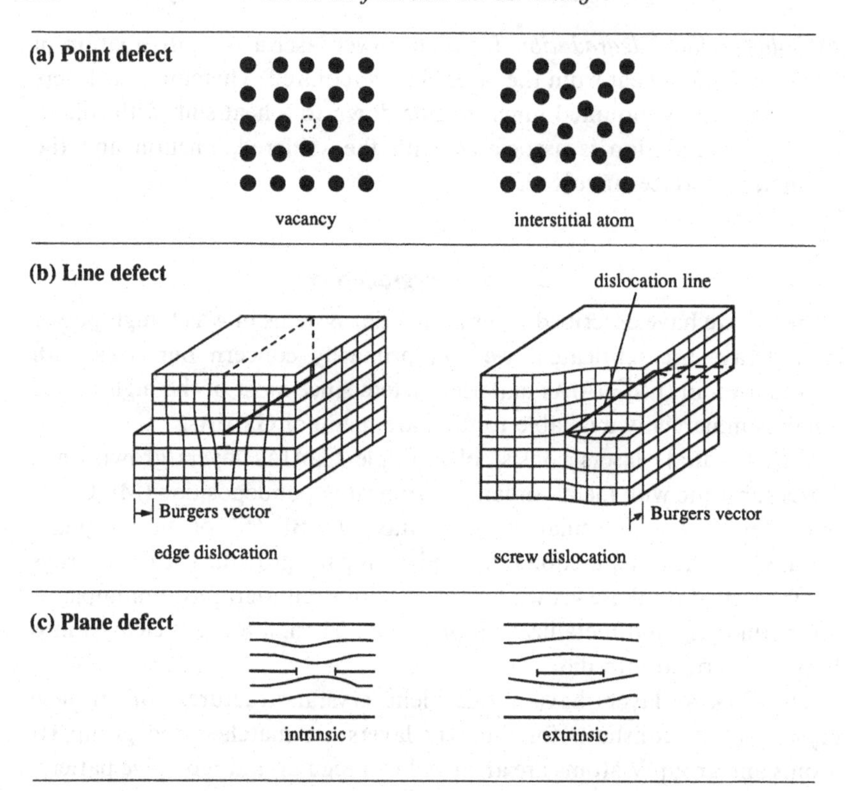

Figure 7.8. Basic concept of (*a*) point defects, (*b*) line defects, and (*c*) plane defects. (After Mitsuo Fukuda, Ref. 81.)

increase. Furthermore, the optical energy lost by the nonradiative recombination and by light absorption at defects is transformed into lattice vibrations through multiple phonon emission and can give rise to low-temperature defect motion. This is called recombination enhanced defect motion. There are two main defect motion mechanisms: climbing motion and gliding motion[113–115]. A review of such mechanisms can be found in several publications[81,116–119]. Defects can propagate due to these mechanisms to form dislocation networks. In such an area, the light is completely absorbed and the area becomes a dark area. Different types of defects can be identified by their shape as seen in an electroluminescent (EL) topograph. Such defects are called dark line defects (DLD) or dark spot defects (DSD).

The growth of defects, especially dark line defects, is a serious problem for GaAlAs/GaAs laser diodes. Such lasers exhibit a DLD growth rate which is reported to be more than two orders of magnitude higher than that in InGaAsP/InP laser diodes[119]. Dow and Allen[120] proposed that the defect growth proceeds through self-reproducing dangling-bond deep traps generated as a result of nonradiative recombination in GaAlAs crystals. Lack of deep traps in the InGaAsP crystal inhibits defect growth. Further understanding of this mechanism is required. To suppress such growth, early work by Kirkby[121] suggested that dislocation motion can be drastically reduced by adding In. Recently, Waters[122] *et al.* reported that in contrast to GaAlAs, the DLD growth is suppressed in the structure with a strained InGaAs quantum well. This is possibly due to dislocation pinning by the strained layer.

There are at least four causes which create dislocations in epitaxial layers: (1) propagation of defects from the substrate crystal; (2) defects introduced as a result of contamination in growth; (3) defects introduced as a result of lattice mismatch and stress; and (4) defects introduced as a result of mechanical damage. To eliminate defect propagation from the substrate, a superlattice layer may be grown to disturb the propagation path[91,123]. Also, strained InGaAs layers have been proposed to quench defect motion[92,122]. The defects induced by contamination in the GaAlAs epitaxial layers are controlled by the purity of the sources, the reactor condition, and the growth parameters such as growth temperature and III–V ratio. Although it is not clear how the dislocations are created due to contamination, impurities such as carbon, and especially oxygen, in epitaxial layers can degrade performance such as threshold current and quantum efficiency. Early work on DH lasers suggested that adding Al or Mg in the active layer as an oxygen getter can improve laser quality[124]. In recent years, due to improvement in source purity and epitaxial reactor design, reliable lasers can be achieved with GaAs active layers and the threshold and efficiency have been improved. In order to reduce the effects of the small lattice mismatch between GaAs and AlAs, compositional grading has been employed[125]. In this method, regions between layers are graded continuously or in small steps to relax the stress due to lattice mismatch.

7.2.3 Facet degradation

Outside the bulk region, two ends of the laser chips are normally cleaved on $\langle 110 \rangle$ crystal planes and are used as light reflectors. For high-power laser arrays, the optical power density at the cleaved facet region is extremely high, of the order of 1–10 MW/cm^2. A strong nonlinear temperature versus output power dependence is observed for uncoated

GaAlAs laser facets[126]. Due to this heating, oxidation and light absorption can rapidly occur at facets. As the lasers age, the reflectivity of the facet may change and the optical absorption in the facet region increases. This leads ultimately to a thermal runaway in the facet region and the crystal melts[127,128]. Consequently, GaAlAs lasers operated at high power or with uncoated facets exhibit very short lives[129]. Early work on laser reliability demonstrated that GaAlAs laser facets must be coated with dielectric layers in order to improve laser reliability[129–131]. The coated facets minimize the facet erosion caused by oxygen or water vapor, thereby reducing the temperature rise at the laser facet and increasing diode life[132]. A recent work by Moser *et al.*[102,133] concluded that the COD level is reduced with the time of the stress.

The rate of reduction of the COD level depends on the quantum well structure, reflectivity of coating and the surface treatment. As the COD decreases below the rated power level with time, catastrophic failure occurs. This type of failure can usually occur randomly after long time aging, even with coated laser facets.

It has been found that several mechanisms can increase the COD level for GaAlAs lasers. First, if the optical power density of the facet is reduced, sudden failures can be eliminated. Thus, spreading out the guided wave along or perpendicular to the junction plane increases the laser reliability.

Second, the fundamental mechanism causing catastrophic optical damage at facets is the strong absorption of the mirror region. Therefore, a laser structure with a nonabsorbing mirror (NAM)[134–146] could reduce mirror related failure. This type of laser is also called a window laser. It is reported that the COD level increases from 160 mW to over 300 mW for a BTRS laser[139] and generally increases by a factor of 2–4 for various NAM structures[140–144]. Furthermore, with the NAM structure, both gradual degradation and catastrophic damage of facets are reported to be significantly improved.

Third, workers at IBM have shown that laser lifetime was increased when GaAlAs laser mirrors were coated with $\lambda/4$-thick proprietary coatings[147]. A Raman study confirmed that facet heating was reduced by such a coating[102]. The results of the IBM study are shown in Figure 7.9. As shown, the time to catastrophic optical damage (COD) is inversely proportional to the facet linear power density and $\lambda/4$ coatings produce long-lived lasers at linear power densities of 40 mW/μm. An effective surface coating thus serves to minimize facet related degradation.

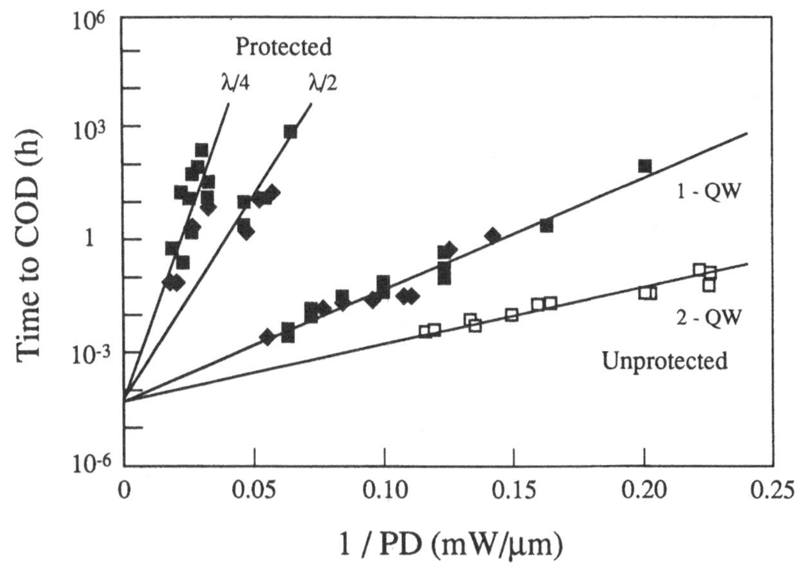

Figure 7.9. Time to catastrophic optical damage (COD) of the laser facet as a function of the inverse optical power per micron of emitter width. (After A. Moser, Ref. 147.)

7.2.4 Solder-related degradation

High-power laser arrays are normally mounted junction-side down to a heat sink with solder. During the mounting process, the solder and die metallization are melted together to form an alloyed joint. Several watts of heat are required to be dissipated from the laser chip through this solder joint to the heat sink. In addition, the heat must be dissipated uniformly without hot spots, which can cause local heating. Four degradation mechanisms are associated with this solder joint.

1. *Solder induced stress.* When a laser diode is soldered junction-side down at elevated temperatures and then cooled to room temperature, there is a built-in stress owing to the expansion mismatch between the laser die, the solder and the heat sink. In GaAlAs lasers, this bonding stress can cause the formation of dark line defects. Owing to this, soft solders such as In are generally used in the fabrication of high-power diode laser arrays which must be mounted on high thermal conductivity heat sinks sucn as copper or diamond, since neither diamond nor copper have a thermal expansion coefficient close to that of GaAs.
2. *Metal interdiffusion.* During the laser operation, the metal on the die

or the solder can diffuse through the GaAlAs crystal. In the case of deep diffusion, the metal penetrates through the p–n junction. As a result, the junction is shorted and the current no longer flows through the junction to emit light. In the case of a shallow diffusion, the metal diffuses to the cap layer to generate defects and such defects can propagate to the active layer to form dark line defects. Both mechanisms cause the laser to exhibit catastrophic failure.

3. *Metal electromigration.* During high-current and/or high-temperature stress, electromigration of the chip metallization or solder can occur. When a relatively high current flows in the metal, the thermally activated metal ion gains momentum in the direction of the electron flow in collision with electrons. Then, the metal ions move toward the positive potential via vacancies and form crystals, hillocks and whiskers, while the condensation of vacancies forms a void at the negative potential. Here, a whisker is a needle-like pure crystal of metal such as In or Sn[148]. Whisker growth depends on the solder material and is enhanced by temperature, humidity and mechanical surface stress.
4. *Increase in thermal resistance.* This problem is caused by the deterioration of, and formation of, voids in the solder[149]. During high-temperature storage, it is found that the Au atoms, for example, can diffuse into an In solder layer and intermix causing void formation and a subsequent increase in thermal resistance. When the laser is in operation, the local heating accelerates the deterioration of In solder and electromigration of Au atoms.

7.3 Lifetests of high-power diode arrays

7.3.1 Lifetests

To establish the reliability of diode lasers, a sample quantiy of diode layers or laser products is selected and operated for a long period of time (e.g. a few hundred hours to several years). This so-called lifetest can be performed with either a constant current or a constant power mode of operation. For high-power laser arrays, the constant power mode is generally used because some failure mechanisms are related to high optical power, as discussed in the preceding section. During a constant power lifetest, the operating current versus time is monitored. The degradation rate is defined as the slope of this curve. The end of life is generally defined as the time at which the current exceeds a certain value or exceeds a certain percentage of initial value.

The purpose of the lifetest is two-fold. The first is to demonstrate the lifetime of diode lasers. This is important because laser users require such a demonstration to ascertain the laser reliability in their application. The second is to reveal the failure mechanisms through such lifetests, and then to improve the product design or process to eliminate such failures. In order to shorten the test time, three accelerating factors are generally used. These are (1) power, (2) current, and (3) temperature. Empirically, the degradation rate r_{deg} is found to be related to power or current

$$r_{deg} \propto (P)^n (I)^m, \tag{7.5}$$

where n and m are accelerating factors for power (P) and current (I), respectively, and to be related to temperature (T) as:

$$r_{deg} \propto \exp\left(\frac{E_a}{kT}\right), \tag{7.6}$$

where E_a is an activation energy and k is Boltzmann's constant. However, it is very difficult to determine accelerating factors accurately. This is due to two reasons: (1) it is difficult to isolate, or even identify, failure modes which are due to different degradation mechanisms where each degradation mechanism may exhibit a different activation energy; and (2) determining acceleration factors requires a large quantity of lasers randomly selected from a uniform population. Therefore, acceleration lifetests are very effective for understanding failure mechanisms quickly. However, in order to determine life expectancy in a system, it is almost imperative to perform lifetests at system operating conditions and accumulate device-hours at least of the same order of magnitude as the lifetime requirement.

7.3.2 Reliability statistics

For semiconductor devices, failure rates which generally change with operating time exhibit a bathtub curve pattern (as shown in Figure 7.10)[150]. The curve consists of three periods: an infant failure period; a random failure period; and a wearout failure period. Diode laser array products exhibit similar behavior. However, the random failure period is not distinctive because the infant failure period and the wearout period overlap each other significantly. Therefore, the investigation of diode laser array product reliability has focused on only two failure modes. The infant failure mode generally occurs within several hundred hours of initial operation. Consequently, a proper burn-in process can eliminate this failure mode as far as the user is concerned[151,152]. The wearout failure mode occurs from several thousand hours to, perhaps, one million hours.

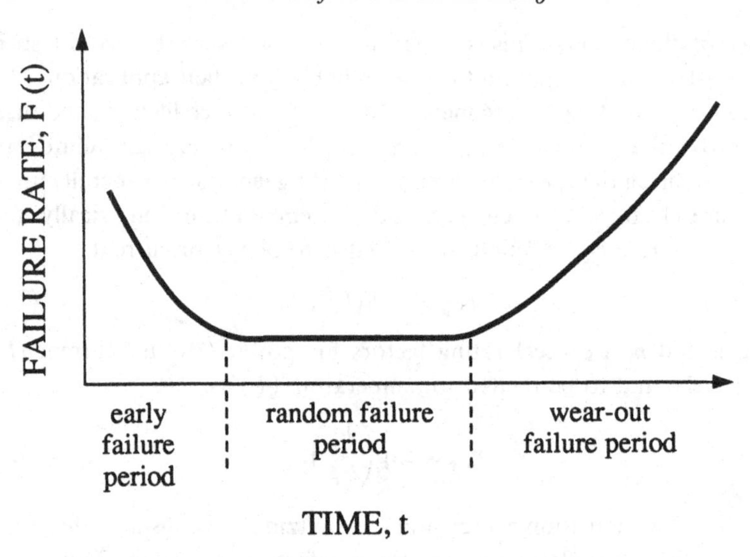

Figure 7.10. Failure rate vs time (bathtub curve).

The extended accelerated lifetest is used to characterize such a gradual degradation mode.

To analyze lifetest data in the wearout failure period statistically, the log-normal distribution is generally used as the cumulative failure distribution function $F(t)$, which is expressed as[150]

$$F(t) = \left\{1 + \operatorname{erf}\left[(2\sigma)^{-\frac{1}{2}} \ln\left(\frac{t}{t_m}\right)\right]\right\} \Big/ 2, \tag{7.7}$$

where t is operating time, t_m is median life, and σ is the standard deviation in a log-normal distribution. The MTTF, which is defined as the mean value of the time to failure, is calculated by the following equation:

$$\text{MTTF} = t_m \exp\left(\frac{\sigma^2}{2}\right). \tag{7.8}$$

For rapid degradation or a sudden failure, the time at which the failure occurs is the end of the life of the diode. For gradual degradation, since the failure may not occur during the time of the test, a projected life is extrapolated from the degradation rate and is defined as the end of the life of the diode. For example, if we define the end of life to be the time at which the current increases by 50% of its initial value, the life of the laser is defined as 0.5 times the initial current divided by the degradation rate. To determine the reliability, the cumulative failure and life of all

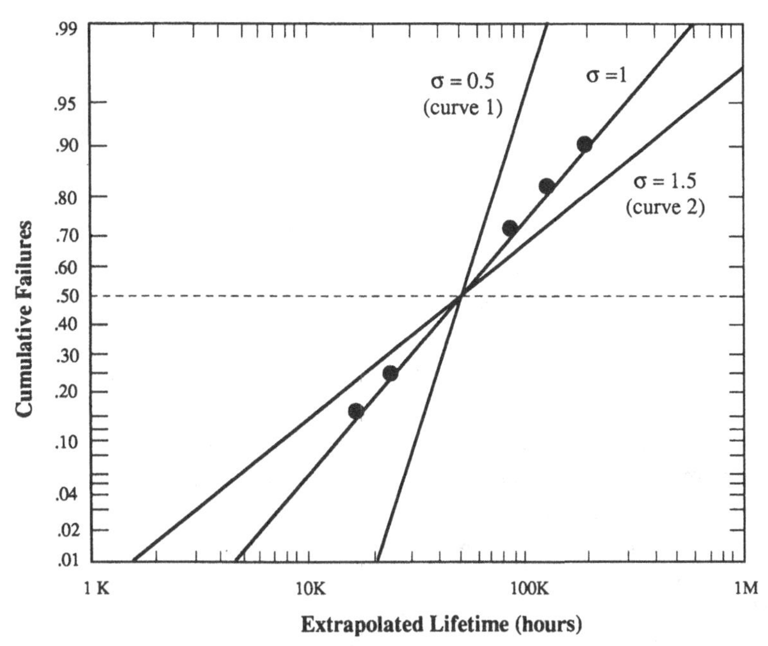

Figure 7.11. Illumination of log-normal distribution plots with standard deviations of 0.5, 1.0 and 1.5.

tested diodes are plotted on a log-normal distribution probability plot, as shown in Figure 7.11. From the linear regression fit of all data points, the median life t_m can be obtained from the time at which the cumulative failure $F(t)$ equals 0.5. Then the mean time to failure can be calculated according to Equation (7.8).

The median life, or MTTF, has commonly been used to specify laser diode reliability, but this single parameter sometimes does not specify the reliability adequately. This is because in a practical application, the user generally requires the laser to have less than a certain percentage of failures at a given operation time. Consequently, it is important to include a second parameter, standard deviation, σ, to specify the reliability. To illustrate this point, we use the following example. As shown in Figure 7.11, two products described by curves 1 and 2 have the same 50 000 h median life. Curve 2 with $\sigma = 1.5$ has $>25\%$ failure rate at 20 000 h, while curve 1 with $\sigma = 0.5$ has $<1\%$ failure rate at 20 000 h. Therefore, it is clearly shown that in order to specify the reliability of a diode laser product fully, median life or MTTF and standard deviation must be specified. Then, either from the plot or from Equations (7.7) and (7.8), the percentage failure rate at a given operation time can be derived.

7.3.3 Lifetests for high-power cw laser arrays

In this section, we report on lifetest results obtained for a variety of types of laser array operating under various power and ambient temperature conditions.

7.3.3.1 25 mW fiber coupled low-power arrays[153]

Sixty micron core 0.3 NA fibers were butt coupled to a six stripe GaAlAs multiple quantum well MOCVD grown diode laser array of the same geometry as in Figure 7.2. The total lasing emission aperture is 60 μm and the laser cavity length is 250 μm. Lasers are coated to produce mirror reflectivities of 30% and 90% on the front and rear facet respectively. The catastrophic optical damage level for this chip is typically 200 mW cw.

The complete fiber coupled diode laser is hermetically sealed and tested. In the lifetest, lasers are operated at 15 °C and at 14 mW cw out of fiber, which is equivalent to ≈25 mW from lasers. Two groups of lasers, a group of nine and a group of eight were selected separately from two years of production. The tested hours to date are greater than 31 000 h for the first group and 24 000 h for the second group. The plot of operating current versus stress time is shown in Figure 7.12. These lasers are operated at a facet COD derating factor of 8. Under such conditions, all diode laser arrays exhibited gradual degradation with no occurrence of catastrophic failure. The degradation rate ranged from 0.02%/kh to 0.44%/kh with a median rate of 0.11%/kh. If the failure is defined as the operating current

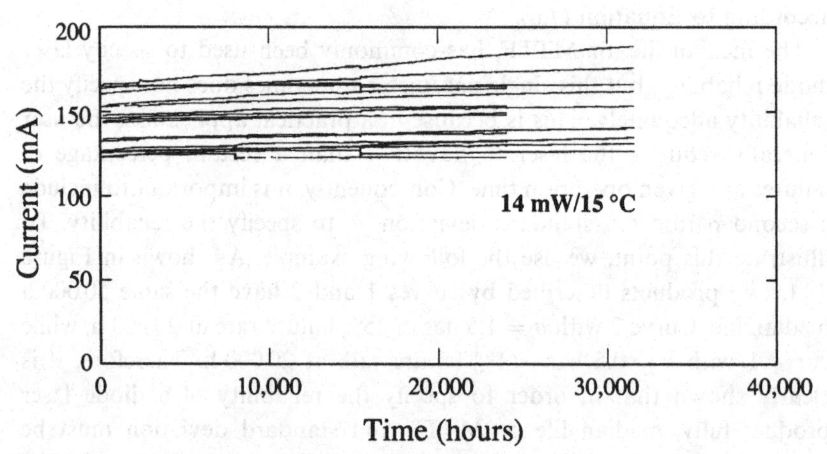

Figure 7.12. Constant power lifetest of fiber coupled 50 μm aperture low power GaAlAs arrays.

increasing 1.5 times the initial value, the median life of this group of lasers is 453 000 h at 15 °C. No failures have been observed in this test after 471 000 actual device hours. Thus, at low power levels of ≈0.5 mW/μm, high reliability is readily obtained.

7.3.3.2 High brightness diode lasers (10 mW/μm)

7.3.3.2.1 500 mW submounted lasers[154]. Five stripe 50 μm-aperture single quantum well (≈100 Å-thick $Ga_{0.93}Al_{0.07}As$ quantum well) laser arrays were operated at 0.5 W cw, 50 °C for 3000 h. The data is shown in Figure 7.13. The front and rear mirror reflectivities were 5% and 97% respectively. Twenty lasers from four MOCVD grown wafers were used in the test. The lasers were mounted on copper heat sinks with In solder. The catastrophic damage levels of the laser facets is typically greater than 1.5 W cw. Thus, the derating factor is ≈3.

As shown in Figure 7.13, there have been no catastrophic failures in a total of 60 000 device hours. However, one laser exhibited a high gradual degradation rate of 6% increase in operating current per kilohour. The degradation rate of the remaining group ranged from not measurable to 1.2%/kh. The median degradation rate was 0.17%/kh, which implies that for a 50% increase in operating current the median life of this group of lasers is 287 000 h at 50 °C. Assuming an activation energy of 0.3 eV for the gradual degradation mode, the median life would be 708 000 h at 25 °C. However, such an assumption may not be justifiable owing to the variation in COD with time[102]. Therefore, it is useful to look at the projected life prior to COD based on Figure 7.9. At a power level of

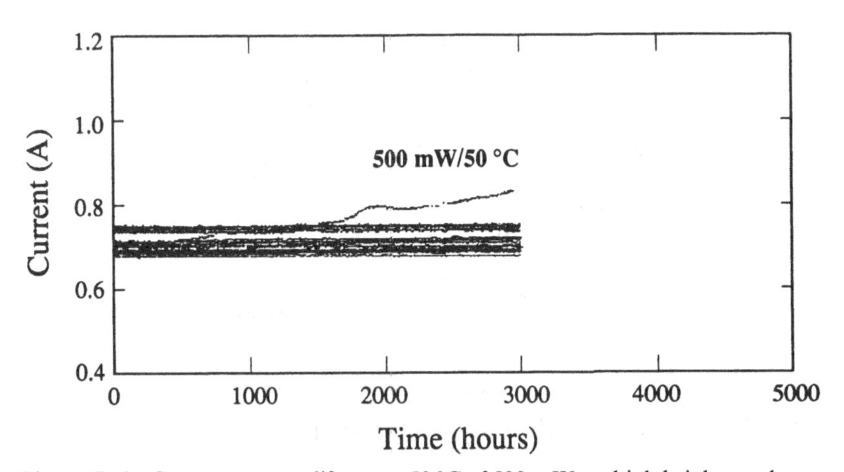

Figure 7.13. Constant power lifetest at 50 °C of 500 mW cw high brightness lasers.

10 mW/μm and with a properly prepared λ/4-thick coating, the COD life projected is well in excess of 10^9 h. Thus, at this power level (10 mW/μm) the COD process should not affect laser life for lasers with properly coated facets.

7.3.3.2.2 1 W submounted lasers[45,154]. These lasers have a similar design to the 50 μm-aperture 500 mW cw lasers except that the emission aperture is 200 μm and the pattern is a broad area without stripes. The typical COD limit is in excess of 6 W. In the lifetest, lasers were operated at an accelerated 50 °C temperature and at a power level of 1 W. This corresponds to a derating factor of 6. The test data plotted in a log-normal distribution is shown in Figure 7.14. No catastrophic failure occurred during the 1 W test. The projected life of the units is calculated from the

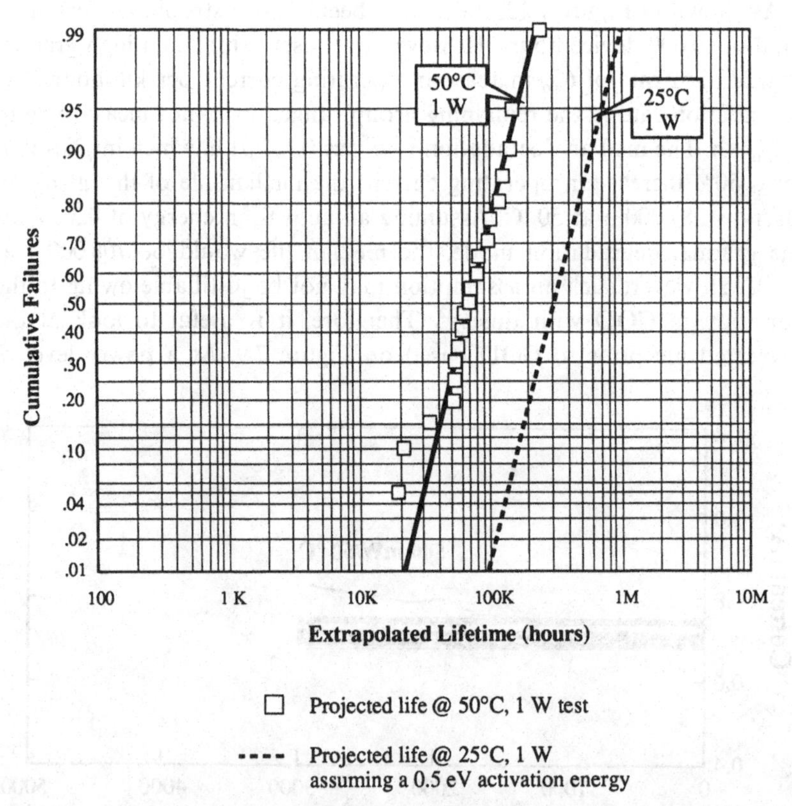

Figure 7.14. Log-normal reliability distribution plot of 1 W high-brightness lasers at 50 °C and projected reliability at 25 °C based on a 0.5 eV activation energy.

degradation rate based on a 50% increase in operating current, and data points of projected life are fitted with a least square fit. Extrapolated from the fitted curves, the median life at 50 °C for a 1 W laser is 72 600 h, and the σ is approximately 0.50. The median degradation rate is 15 mA/kh for the 1 W test. Also shown is the projected log-normal plot for these lasers operating at a 25 °C heat sink temperature and assuming a typical activation energy of 0.5 eV (see Equation (7.6)). As shown, the projected median life is greater than 300 000 h at 25 °C.

7.3.3.3 cw laser bars

The laser structure (shown schematically in Figure 7.3(*a*)) is a 1 cm laser bar which was grown by MOCVD and employed single quantum well separate confinement $Ga_{1-x}Al_xAs$ heterostructures (SQW–SCH)[155]. The wafers were processed to form gain-guided 6 μm-wide stripes on 10 μm centers by proton bombardment. After contact metallization, the wafers were cleaved into 1 cm-long bars which were mirror coated. Then they were bonded p-side down to a heat sink. Sakamoto *et al.* reported lifetimes for 10 W bars (20% packing density)[156], 15 W bars (35% packing density)[157], and 20 W bars (48% packing density)[158]. The lifetimes ranged from 4000 h to 20 000 h operating at 25 °C heat sink temperature. Recently a lifetest of 1 cm cw bars at 10 W, shown in Figure 7.15, has demonstrated very stable operation at 30 °C[159]. The packing density of such bars is 18%. As shown in Figure 7.15, after 700–1100 h test, the change in operating currents of all 20 bars are within the range caused by temperature fluctuations.

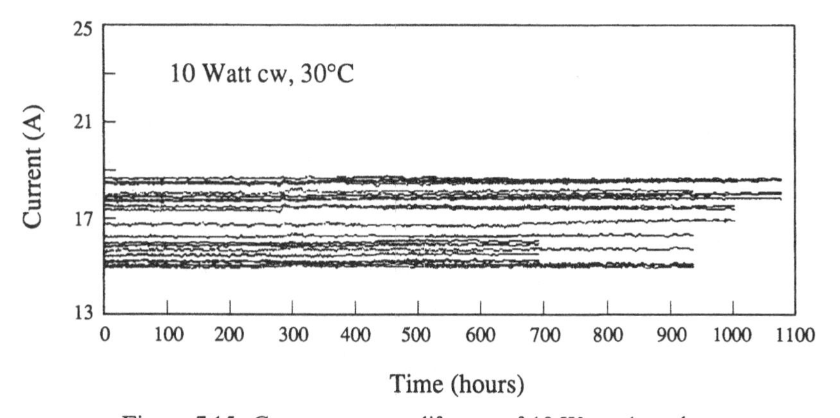

Figure 7.15. Constant power lifetests of 10 W cw 1 cm bars.

7.3.4 Lifetests for high-power laser arrays: quasi-cw mode

7.3.4.1 Quasi-cw laser bars

For solid state laser pumping, high-power laser arrays can be operated at higher peak power in the so-called quasi-cw operating mode[160,161]. The peak power that a 1 cm bar can reliably emit is 100 W, approximately a factor of 5 higher than cw laser bars.

Shown in Figure 7.16 is a lifetest for 100 W quasi-cw bars[162]. Pulse width is 200 μs, the duty factor is 2%, and the heat sink operating temperature is 30 °C. Based on this data, lifetimes of the order of 10^{10} shots are projected for these bars. For higher duty cycle tests, a group of four bars running at 20% duty factor at 50 W shows that the projected lifetime exceeds 10^{10} shots.

7.3.4.2 Quasi-cw bar stacks

Quasi-cw bars can also be stacked and mounted on a number of cooling subsystems to form a two-dimensional array. A schematic diagram is shown in Figure 7.4. Table 7.2 presents lifetests of five two-dimensional arrays, each containing sixteen 1 cm bars[163]. Each bar is operating at a 200 μs pulse width, 1 kHz repetition rate (20% duty factor). The peak output power of each two-dimensional array is 800 W and the average power is 160 W. A total of $>10^{11}$ device pulses has been recorded with no failures from the eighty 1 cm quasi-cw bars in the five two-dimensional arrays. The projected life per bar is 4.5×10^9 pulses for a 20% increase in initial operating current.

Table 7.2. *Stack lifetests at* 50 W/bar 20% *duty cycle*

Stack	No. of bars	Peak power	Avg. power	Hours (20%)	Stack pulses	Device pulses	Degradation (%)
1	16	800 W	160 W	356	1.3×10^9	2.0×10^{10}	15.6
2	16	800 W	160 W	405	1.5×10^9	2.3×10^{10}	11.9
3	16	800 W	160 W	383	1.4×10^9	2.2×10^{10}	12.6
4	16	800 W	160 W	383	1.4×10^9	2.2×10^{10}	14.6
5	16	800 W	160 W	243	0.9×10^9	1.4×10^{10}	7.1

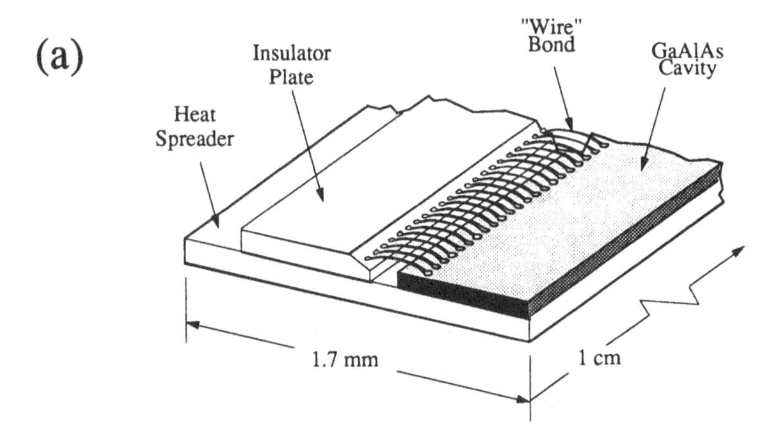

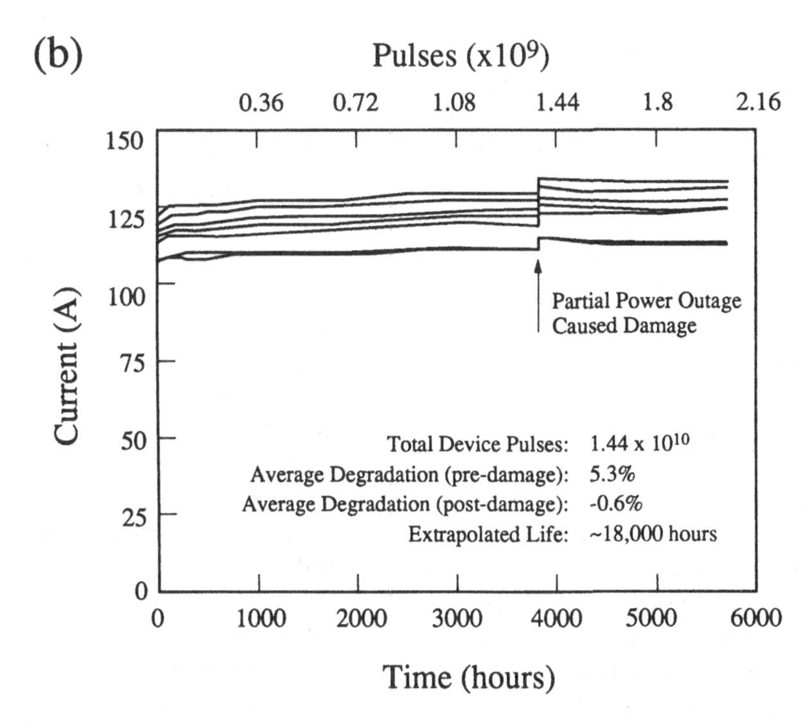

Figure 7.16. (*a*) Schematic diagram of a linear subassembly of a 1 cm Q-cw bar. (*b*) Lifetest of 100 W Q-cw bars at 30 °C.

7.4 Environmental tests of packaged laser diode arrays

In many applications, laser arrays are stored or operated under special environmental conditions. Therefore, environmental tests are performed to evaluate tolerance to environmental and mechanical stresses. Most of the tests are performed according to a military specification (MIL spec. test condition) which is specifically defined.

There are several MIL standards which can apply to packaged laser diodes. Among them, MIL-STD-883, which applies to microelectronic devices, is often used for high-power array products. With regard to testing methods, other standards such as MIL-STD-750, which applies to discrete semiconductor devices, and MIL-STD-810, which applies to systems, have similar testing methods.

7.4.1 Hermeticity tests

The purpose of this test is to determine the effectiveness of the seal of a laser package. The seal prevents moisture from penetrating into the package under a high humidity environment, and ensures a stable inert environment for laser operation. The test is designed to measure the leak rate, which is defined as that quantity of dry air at 25 °C in cubic centimeters flowing through leak paths when the high-pressure side is at 1 atmosphere (760 mm Hg absolute) and the low-pressure side is at a pressure not greater than 1 mm Hg absolute. The unit is atmosphere cubic centimeters per second (atm cc/s).

The technique commonly used to test leak rates of packaged laser diodes is the tracer gas helium method. Packaged laser diodes are stored under 60 psi pressure of helium gas for at least two hours. The process is called bombing. Then, a mass spectrometer leak detector which is calibrated for a certain helium leak rate is used to read the leak rate of the individual package. Sensitivity is generally required to read 10^9 atm cc/s. The leak rate is used to determine pass or fail and the reject limit is 5×10^{-8} atm cc/s. This test is called a fine leak test.

A large gross leak cannot be detected by this method since all helium will easily leak out of the package immediately after removing the laser from the helium bomb. Consequently, a gross leak test is often used. The test commonly used for laser diodes is a fluorocarbon gross leak test. Again, lasers are bombed under pressure up to 75 psi for up to ten hours. The lasers are then immersed in a 125 °C FC-72 fluid. If the laser package has a gross leak, bubbles will be observed.

7.4.2 Temperature cycle test

The temperature cycle test is conducted for the purpose of determining the resistance of a part to exposures at extremes of high and low temperatures, and to determine the effect of alternate exposure to these environmental extremes. For laser products, the extreme temperatures are generally −55 °C and +85 °C. During the test, parts are transferred between cold chambers and hot chambers. In each chamber, parts are stored for at least 10 minutes. In between transfer, parts are at room temperature for less than five minutes. After a temperature cycle test, the hermeticity test and the electrical test are performed for comparison with those readings taken before the test.

In this temperature range, the diode lasers should show no detectable change in electro-optical performance after 10–100 cycles. However, one major concern in design is the thermal expansion match of all the parts in the package. Particularly for a fiber coupled package, thermal expansion mismatch can place high stress on a fiber during the temperature cycle test. Consequently, this results in fiber misplacement or breakage. A package design requires one to consider thermal expansion when the material for individual parts is selected.

7.4.3 Neutron irradiation test[164]

Tests are performed to determine the susceptibility of packaged diode lasers to neutron irradiation. This is a destructive test. Generally, diode laser arrays are characterized before radiation; then they are irradiated at various neutron fluence levels. After each dose, diode laser arrays are characterized again for comparison. The dose varies from 10^{12} to 10^{15} n/cm^2.

When neutrons interact with semiconductor materials, they can cause point defects in the crystal. These defects occur as atoms are displaced from the lattice and with sufficient induced velocity. A neutron may displace other atoms to form a disordered region. These regions contain high densities of nonradiative recombination sites, and, consequently, neutron damage shortens the nonradiative carrier lifetime.

Figure 7.17 shows a typical test result of a neutron irradiated 100 μm diode laser array[164]. The *L–I* curves of the 100 μm laser array before and after a series of neutron irradiation doses from 9.9×10^{12} n/cm^2 to 1.2×10^{15} n/cm^2 are measured. Up to 10^{13} n/cm^2, there were no significant changes in threshold current. At 10^{14} n/cm^2, the threshold current increased by 35% and the slope efficiency dropped by 7%. With a fluence

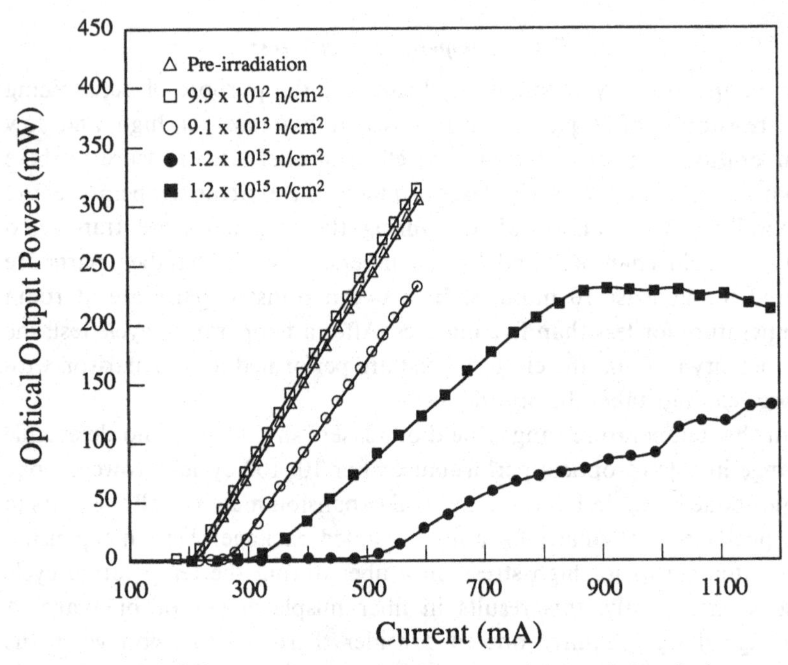

Figure 7.17. CW light output power vs DC current characteristics of a laser array after neutron irradiation at varying neutron dosage levels. (After Richard Carson and Stephen Fries, Ref. 164.)

of 10^{15} n/cm^2, the threshold increased by 150% to 480 mA and the slope efficiency dropped by 75% to 0.2 W/A. After several minutes of cw operation, the threshold current recovered to 310 mA and the efficiency recovered to 0.4 W/A. Because of this significant reduction in overall efficiency, the output power was thermally limited as the curves bend over at 900 mA. The recovery is due to that portion of radiation damaged defects which are annealed out during cw operation.

7.4.4 Random vibration test

This test is conducted for the purpose of determining the ability of a product to withstand the dynamic stress exerted by random vibration. Random vibration occurs in the type of environment in which missiles and rocket engines operate.

The vibration system generates a random vibration for which the magnitude has a normal amplitude distribution at a given frequency. The test requires the user to specify the rms value of acceleration (Grms), the vibration time (t), the axes (x, y, z) and the test method defined in the MIL spec.

The device is rigidly fastened on the vibration platform and the leads are adequately secured. For a fiber coupled package, the fiber needs to be secured. After the vibration test, the tested parts are electrically and optically characterized and an external visual inspection is performed.

7.4.5 Particle impact noise detection (PIND) test

This test provides a nondestructive means of identifying the parts containing loose particles. The tested parts are mounted on a fixture which is attached to a vibration shaker. The driver provides a sinusoidal motion to the shaker at specified conditions. An acoustic transducer is attached to the tested parts. The output of the transducer can detect the noise generated by the particle vibration. The PIND test is very sensitive to metal particles.

7.5 Future directions for high-power diode laser array research

This chapter has described the results to date in achieving high-power, high-reliability diode laser arrays. A number of highlights have been listed, including (a) 120 W cw monolithic 1 cm laser bars[7], (b) 66% total power conversion from an InGaAsP 810 nm array[108], and (c) cw laser arrays operating from 630 nm[76] to 2.1 μm[94], among others.

The question arises as to what directions will future research take and what might be the advances seen over the next ten years. It is clear that the search for higher power will require further research into a number of the following areas. These are (a) improved materials such as strained layer materials which allow lower thresholds, higher efficiency, higher T_0 values and reduction of defect propagation; (b) improved heat removal techniques such as active cooling and diamond heat sinks; (c) improved optics and fiber coupling of diode arrays so that the delivered optical power density may be high while the power is generated by spatially separated sources, thus reducing heat removal problems; (d) new materials for an even wider range of available emission wavelengths; and (e) high-power coherent diode arrays for achieving optical power densities comparable to, for example, CO_2 lasers, among others.

Furthermore, as the capabilities of these devices expand, the increased production volume will be likely to drive prices down dramatically over this time period. Such a price drop should further stimulate new demand and new applications. The future looks very bright for high-power laser diode arrays.

Acknowledgments

The authors wish to acknowledge Drs Richard Carson and Stephen Fries at Sandia National Labs for allowing them to use data from a Sandia report. The authors also wish to acknowledge their colleagues at SDL who have discussed and shared unpublished data, and Sean Ogarrio and Viola daSilva who assisted in the preparation of the manuscript.

References

1. R. N. Hall, G. E. Fenner, J. D. Kingsley, T. J. Soltys and R. O. Carlson, *Phys. Rev. Lett.*, **9**, 366 (1962).
2. M. I. Nathan, W. Dumke, G. Burns, F. H. Dill, Jr and G. Lasher, *Appl. Phys. Lett.*, **1**, 62 (1962).
3. N. Holonyak, Jr and S. F. Bevacqua, *Appl. Phys. Lett.*, **1**, 82 (1962).
4. T. M. Quist, R. H. Rediker, R. J. Keyes, W. E. Krag, B. Lax, A. L. McWhorter and H. J. Zeigler, *Appl. Phys. Lett.*, **1**, 91 (1962).
5. D. R. Scifres, R. D. Burnham and W. Streifer, *Appl. Phys. Lett.*, **41**, 1030 (1982).
6. H. K. Choi, C. A. Wang, D. F. Kolesar, R. L. Aggarwal and J. N. Walpole, *IEEE J. Quantum Electron.*, **QE-3**, 857 (1991).
7. M. Sakamoto, J. G. Endriz and D. R. Scifres, *Electron. Lett.*, **28**, 197 (1992).
8. W. Streifer, D. R. Scifres, G. L. Harnagel, D. F. Welch, J. Berger and M. Sakamoto, *IEEE J. Quantum Electron.*, **QE-24**, 883 (1988).
9. I. Hayashi, M. B. Panish and P. W. Foy, *IEEE J. Quantum Electron.*, **QE-5**, 211 (1969).
10. M. B. Panish, I. Hayashi and S. Sumski, *IEEE J. Quantum Electron.*, **QE-5**, 210 (1969).
11. H. Kressel and H. Nelson, *RCA Rev.*, **30**, 106 (1969).
12. Zh. I. Alferov, V. M. Andreev, E. L. Portnoi and M. K. Trukan, *Sov. Phys. Semicond.*, **3**, 1107 (1970). (Translated from *Fiz. Tekh. Poluprovodn.*, **2**, 1016, 1968).
13. I. Hayashi, M. B. Panish, W. Foy and S. Sumski, *Appl. Phys. Lett.*, **17**, 109 (1970).
14. Zh. I. Alferov, V. M. Andreev, D. Z. Garbuzov, Yu. V. Zhilyaev, E. P. Morozov, E. L. Portnoi and V. G. Trofim. *Sov. Phys. Semicond.*, **4**, 1573 (1971). (Translated from *Fiz. Tekh. Poluprovodn.*, **4**, 1826, 1970.)
15. H. Kressel and J. K. Butler, *Semiconductor Lasers and Heterojunction LEDs*, Figure 1.2, p. 4, Academic Press (1977).
16. N. Chinone, R. Ito and O. Nakada, *J. Appl. Phys.*, **47**, 785 (1976).
17. D. R. Scifres, C. Lindstrom, R. D. Burnham, W. Streifer and T. L. Paoli, *Electron. Lett.*, **19**, 169 (1983).
18. For a review of quantum-well semiconductor lasers and their properties, see G. Agrawal and N. K. Dutta, *Long-wavelength Semiconductor Lasers*, Chapter 9, p. 372, Van Norstrand Reinhold Company (1986).
19. N. Chand, E. E. Becker, J. P. van der Ziel, S. N. G. Chu and N. K. Dutta, *Appl. Phys. Lett.*, **58**, 1704 (1991).
20. J. Katz, *IEEE J. Quantum Electron.*, **QE-21**, 1854 (1985).
21. See for example, H. Yonezu, K. Endo, T. Kamejima, T. Tokikai, T. Yuasa and T. Furuse, *J. Appl. Phys.*, **50**, 5150 (1979).

22. J. E. Ripper and T. L. Paoli, *Appl. Phys. Lett.*, **17**, 371 (1970).
23. D. R. Scifres, W. Streifer and R. D. Burnham, *Appl. Phys. Lett.*, **33**, 616 (1978).
24. D. R. Scifres, R. D. Burnham and W. Streifer, *Appl. Phys. Lett.*, **33**, 1015 (1978).
25. D. R. Scifres, W. Streifer and R. D. Burnham, *IEEE J. Quantum Electron.*, **QE-15**, 917 (1979).
26. D. R. Scifres, R. A. Sprague, W. Streifer and R. D. Burnham, *Appl. Phys. Lett.*, **41**, 1121 (1982).
27. M. Ettenberg, H. S. Sommers, Jr, H. Kressel and H. F. Lockwood, *Appl. Phys. Lett.*, **18**, 571 (1971).
28. J. van der Ziel, R. A. Logan and R. M. Milulyuk, *Appl. Phys. Lett.*, **41**, 9 (1982).
29. D. Botez, J. C. Connolly, D. B. Gilbert, M. G. Harvey and M. Ettenberg, *Appl. Phys. Lett.*, **41**, 1040 (1982).
30. D. B. Carlin, B. Goldstein, J. B. Bednarz, M. G. Harvey and N. A. Dinkel, *IEEE J. Quantum Electron.*, **QE-23**, 476 (1987).
31. M. Tsunekane, K. Endo, M. Nido, I. Komazaki, R. Katayama, K. Yoshihara, Y. Yamanaka and T. Yuasa, *Electron. Lett.*, **25**, 1091 (1989).
32. R. L. Thornton, W. J. Mosby, R. L. Donaldson and T. L. Paoli, *Appl. Phys. Lett.*, **56**, 1623 (1990).
33. D. G. Mehuys, D. F. Welch, D. W. Nam and D. R. Scifres, *Electron. Lett.*, **26**, 1955 (1990).
34. R. Craig, D. Mehuys, D. Nam, E. Zucker, B. Chan and D. Welch, *Optical Data Storage Technical Conference Digest*, p. 178 (1991).
35. B. Gignac, E. Zucker and R. Craig, Spectra Diode Labs – private communication.
36. M. Sakamoto, D. F. Welch, H. Yao, J. G. Endriz and D. R. Scifres, *Electron. Lett.*, **26**, 729 (1990).
37. E. Wolak, M. Sakamoto, J. Endriz and D. R. Scifres, *LEOS '92 Digest*, Paper No. DLTA 5.1, p. 175 (1992).
38. D. K. Wagner, R. G. Waters, P. L. Tihanyi, D. S. Hill, A. J. Roza, Jr, H. J. Vollmer and M. M. Leopold, *IEEE J. Quantum Electron.*, **QE-24**, 1258 (1988).
39. D. F. Welch, B. Chan, W. Streifer and D. R. Scifres, *Electron. Lett.*, **24**, 113 (1988).
40. G. L. Harnagel, P. S. Cross, D. R. Scifres and D. Worland, *Electron. Lett.*, **22**, 231 (1986).
41. See, for example, H. Kressel and J. K. Butler, p. 535, *Semiconductor Lasers and Heterojunction LEDs*, Y. Pao and P. Kelley (eds.), Academic Press (1977).
42. D. Caffey, R. A. Utano and T. H. Allik, *Appl. Phys. Lett.*, **56**, 808 (1990).
43. R. C. Stoneman and L. Esterowitz, *Optics and Phot. News*, p. 10 (August 1990).
44. R. L. Burnham, *Optics and Phot. News*, p. 4 (August 1990).
45. D. R. Scifres, D. F. Welch, R. R. Craig, E. Zucker, J. S. Major, G. L. Harnagel, M. Sakamoto, J. M. Haden, J. G. Endriz and H. Kung, *SPIE '92*, vol. 1634, *Laser Diode Technology and Applications*, IV, p. 192 (1992).
46. G. L. Harnagel, M. Vakili, K. R. Anderson, D. P. Worland, J. G. Endriz and D. R. Scifres, *Electron. Lett.*, **28**, 1702 (1992).
47. D. Mundinger and J. Haden, Spectra Diode Labs – private communication.
48. R. Solarz, Lawrence Livermore National Lab – private communication. See Chapter 6.
49. R. D. Burnham, D. R. Scifres and W. Streifer, U.S. Patent No. 4,309,670, issued 5 January (1982).
50. K. Iga, F. Koyama and S. Kinoshita, *IEEE J. Quantum Electron.*, **QE-24**, 1845 (1988).

51. Y. H, Lee, J. L. Jewell, A. Scherer, S. L. McCall, J. Harbison and L. T. Florez, *Electron. Lett.*, **25**, 1377 (1989).
52. L. M. Zinkiewicz, T. J. Roth, L. J. Mawst, D. Tran and D. Botez, *Appl. Phys. Lett.*, **54**, 1959 (1989).
53. C. J. Chang-Hasnain, J. R. Wullert, J. P. Harbison, L. T. Florez, N. G. Stoffel and M. W. Maeda, *Appl. Phys. Lett.*, **58**, 31 (1991).
54. A. J. Springthorpe, *Appl. Phys. Lett.*, **31**, 524 (1977).
55. Z. L. Liau and J. N. Walpole, *Appl. Phys. Lett.*, **50**, 529 (1987).
56. J. P. Donnelly, W. D. Goodhue, T. H. Windhorn, R. J. Bailey and S. A. Lambert, *Appl. Phys. Lett.*, **51**, 1138 (1987).
57. J. Puretz, R. D. DeFreez, R. A. Elliott, J. Orloff and T. L. Paoli, *Electron. Lett.*, **23**, 130 (1987).
58. J. Kim, R. J. Lang, A. Larsson, L. Lee and A. A. Narayanan, *Appl. Phys. Lett.*, **57**, 2048 (1990).
59. J. J. Wang, M. Sergant, M. Jansen, S. S. Ou, L. Eaton and W. W. Simmons, *Appl. Phys. Lett.*, **49**, 1138 (1986).
60. S. S. Ou, M. Jansen, J. J. Yang and M. Sergant, *Appl. Phys. Lett.*, **59**, 1037 (1991).
61. M. Jansen, J. J. Yang, S. S. Ou, M. Sergant, L. Mawst, J. Rozenbergs, J. Wilcox and D. Botez, *Appl. Phys. Lett.*, **59**, 2663 (1991).
62. D. W. Nam, R. G. Waarts, D. F. Welch, J. S. Major, Jr and D. R. Scifres, *IEEE Phot. Tech. Lett.*, **5**, 281 (1993).
63. R. F. Kazarinov and R. A. Suris, *Sov. Phys. Semicond.*, **6**, 1184 (1973).
64. Zh. I. Alferov, S. A. Gurevich, R. F. Kazarinov, M. N. Mizerov, E. L. Portnoi, R. P. Seisyan and R. A. Suris, *Sov. Phys. Semicond.*, **8**, 541 (1974).
65. D. R. Scifres, R. D. Burnham and W. Streifer, *Appl. Phys. Lett.*, **27**, 295 (1975).
66. S. H. Macomber, J. S. Mott, R. J. Noll, G. M. Gallatin, E. J. Gratrix, S. L. O'Dwyer and S. A. Lambert, *Appl. Phys. Lett.*, **51**, 472 (1987).
67. N. W. Carlson, G. A. Evans, J. M. Hammer, M. Lurie, L. A. Carr, F. Z. Nawrylo, E. A. James, C. J. Kaiser, J. B. Kirk, W. F. Reichert, D. A. Truxal, J. R. Shealy, S. R. Chinn and P. S. Zory, *Appl. Phys. Lett.*, **52**, 939 (1988).
68. K. Kojima, M. Kameya, S. Noda and K. Kyuma, *Electron. Lett.*, **24**, 283 (1988).
69. S. Noda, K. Kojima and K. Kyuma, *Electron. Lett.*, **24**, 277 (1988).
70. D. F. Welch, R. Parke, A. Hardy, W. Streifer and D. R. Scifres, *Appl. Phys. Lett.*, **54** (1989).
71. J. S. Mott and S. H. Macomber, *IEEE Photonics Tech. Lett.*, **1**, 202 (1989).
72. N. W. Carlson, J. H. Abeles, R. Amantea, J. K. Butler, G. A. Evans and S. K. Liew, *SPIE 1634 Laser Diode Technology and Applications*, **IV**, 30 (1992).
73. W. D. Laidig, P. J. Caldwell, Y. F. Lin and C. K. Peng, *Appl. Phys. Lett.*, **44**, 653 (1984).
74. D. P. Bour and J. R. Shealy, *Appl. Phys. Lett.*, **51**, 1658 (1987).
75. H. B. Serreze, Y. C. Chen and R. G. Waters, *Appl. Phys. Lett.*, **58**, 2464 (1991).
76. R. S. Geels, D. Bour, D. W. Treat, R. D. Bringans, D. F. Welch and D. R. Scifres, *Electron. Lett.*, **28**, 1043 (1992).
77. R. S. Geels, D. F. Welch, D. R. Scifres, D. P. Bour, D. W. Treat and R. D. Bringans, CLEO '93, Paper CThQ3, Technical Digest Series, **11**, 478 (1993).
78. J. M. Haden, D. W. Nam, D. F. Welch, J. G. Endriz and D. R. Scifres, *Electron. Lett.*, **28**, 451 (1992).
79. R. Geels, Spectra Diode Labs – private communication.

80. P. J. A. Thijs, *13th IEEE International Semiconductor Laser Conference Digest*, p. 2 (1992).
81. See, for example, M. Fukuda, *Reliability and Degradation for Semiconductor Lasers and LEDs*, Artech House (1991).
82. D. P. Bour, D. B. Gilbert, L. Elbaum and M. G. Harvey, *Appl. Phys. Lett.*, **53**, 2371 (1988).
83. W. Stutius, Gavrilovic, J. E. Williams, K. Meehan and J. H. Zarrabi, *Electron. Lett.*, **24**, 1493 (1988).
84. D. F. Welch, W. Streifer, C. F. Schaus, S. Sun and P. L. Gourley, *Appl. Phys. Lett.*, **56**, 10 (1990).
85. W. D. Goodhue, J. P. Donnelly, C. A. Wang, G. A. Lincoln, K. Rauschenback, R. J. Bailey and G. D. Johnson, *Appl. Phys. Lett.*, **59**, 632 (1991).
86. H. K. Choi and C. A. Wang, *Appl. Phys. Lett.*, **57**, 321 (1990).
87. P. K. York, K. J. Beernink, G. E. Fernandez and J. J. Coleman, *Appl. Phys. Lett.*, **54**, 499 (1989).
88. J. S. Major, Jr, W. E. Plano, A. R. Sugg, D. C. Hall, N. Holonyak, Jr and K. C. Hsieh, *Appl. Phys. Lett.*, **56**, 105 (1990).
89. J. S. Major, Jr, D. C. Hall, L. J. Guido, N. Holonyak, Jr, Gavrilovic, K. Meehan, J. E. Williams and W. Stutuis, *Appl. Phys. Lett.*, **55**, 271 (1989).
90. J. N. Baillargeon, P. K. York, C. A. Zmudzinski, G. E. Fernandez, K. J. Beernink and J. J. Coleman, *Appl. Phys. Lett.*, **53**, 457 (1988).
91. J. W. Mathews and A. E. Blakeslee, *J. Crystal Growth*, **27**, 118 (1974).
92. R. G. Waters, R. J. Dalbey, J. A. Baumman, J. L. DeSanctis and A. H. Shepard, *IEEE Photonics Tech. Lett.*, **3**, 409 (1991).
93. J. S. Major, Jr, D. W. Nam, J. S. Osinski and D. F. Welch, *IEEE Photon. Technol. Lett.*, **5**, 594 (1993).
94. H. K. Choi and S. Eglash, *Appl. Phys. Lett.*, **61**, 1154 (1992).
95. D. Z. Garbuzov, N. Y. Antonishkis, A. D. Bondarev, A. B. Gulakov, S. N. Zhigulin, A. V. Kochergin and E. V. Rafailov, *IEEE J. Quantum Electron.*, **QE-27**, 1531 (1991).
96. T. Yuasa, M. Ogawa, K. Endo and H. Yonezu, *Appl. Phys. Lett.*, **32**, 119 (1978).
97. F. R. Nash and R. L. Hartman, *J. Appl. Phys.*, **50**, 3133 (1979).
98. T. Yamada and Y. Kawai, *Electron. Lett.*, **26**, 52 (1990).
99. H. Asano, S. Takano, M. Kawaradani, M. Kitamura and I. Mito, *IEEE Photonics Tech. Lett.*, **3**, 415 (1991).
100. S. Oshiba and Y. Tamura, *J. Lightwave Tech.*, **8**, 1350 (1990).
101. F. R. Gfeller and D. J. Webb, *J. Appl. Phys.*, **68**, 14 (1990).
102. A. Moser, E. E. Latta and D. J. Webb, *Appl. Phys. Lett.*, **55**, 1152 (1989).
103. R. G. Waters, D. K. Wagner, D. S. Hill, L. Tihanyi and B. J. Vollmer, *Appl. Phys. Lett.*, **51**, 1318 (1987).
104. R. G. Waters, D. S. Hill and S. L. Yellen, *Appl. Phys. Lett.*, **52**, 2917 (1988).
105. J. R. Shealy, *Appl. Phys. Lett.*, **50**, 1634 (1987).
106. J. Z. Wilcox, S. Ou, J. J. Yang, M. Jansen and G. L. Peterson, *Appl. Phys. Lett.*, **55**, 825 (1989).
107. D. Welch, R. Craig, W. Streifer, D. Scifres, *Electron. Lett.*, **26**, 1481 (1990).
108. N. Yu, Antonishkis, I. N. Arsent'ev, D. Z. Garbuzov, V. I. Kolyshkin, A. B. Komissarov, A. V. Kochergin, T. A. Nalet and N. A. Strugov, *Sov. Tech. Phys. Lett.*, **14**, 310 (1988).
109. For a review, see R. D. Dupuis and P. D. Dapkus, *IEEE J. Quantum Electron.*, **QE-15**, 128 (1979).

110. For a review, see R. D. Burnham and D. R. Scifres, Integrated optical devices fabricated by MBE, *Molecular Beam Epitaxy*, Brian R. Pamplin (ed.), Pergamon Press (1980).
111. H. Kressel, *J. Elec. Materials*, **3**, 747 (1974).
112. D. Hull, *Introductions to Dislocations*, 2nd edn, Pergamon Press (1975).
113. P. M. Petroff and R. L. Hartman, *J. Appl. Phys.*, **45**, 3899 (1974).
114. J. Matsui, K. Ishida and Y. Nannichi, *Jpn J. Appl. Phys.*, **14**, 1555 (1975).
115. P. M. Petroff and L. C. Kimerling, *Appl. Phys. Lett.*, **29**, 461 (1976).
116. See, for example, H. C. Casey, Jr and M. B. Panish, *Heterostructure Lasers Part B: Materials and operating characteristics*, Chapter 8, p. 277, Academic Press (1978).
117. See, for example, H. Kressel and J. K. Butler, *Semiconductor Lasers and Heterojunction LEDs*, Chapter 16, p. 533, Academic Press (1977).
118. See, for example, G. Agrawal and N. K. Dutta, *Long-Wavelength Semiconductor Lasers*, Chapter 10, p. 410, Van Nostrand Reinhold Company, New York (1986).
119. See, for example, O. Ueda, *J. Electrochem. Soc. Reviews and News*, **135**, 11C (1988).
120. J. D. Dow and R. E. Allen, *Appl. Phys. Lett.*, **41**, 672 (1982).
121. P. A. Kirkby, *IEEE J. Quantum Electron.*, **QE-11**, 562 (1975).
122. R. G. Waters, D. P. Bour, S. L. Yellen and N. F. Rugefieri, *IEEE J. Photon. Tech. Lett.*, **2**, 531 (1990).
123. M. A. Tischler, T. Katsuyama, N. A. El-Masry and S. U. Bedair, *Appl. Phys. Lett.*, **46**, 294 (1985).
124. H. Yonezu, T. Kamejima, M. Ueno and I. Sakuma, *Jpn J. Appl. Phys.*, **13**, 1679 (1974).
125. G. H. Olsen and M. Ettenberg, Growth effects in the heteroepitaxy of III–V compounds, in *Crystal Growth Theory and Techniques*, Volume 2, p. 1, C. H. L. Goodman (ed.), Plenum Press (1978).
126. H. Brugger and P. W. Epperlein, *Appl. Phys. Lett.*, **56**, 1049 (1990).
127. C. H. Henry, P. M. Petroff, R. A. Logan and F. R. Merritt, *J. Appl. Phys.*, **50**, 3721 (1979).
128. R. W. H. Engelmann and D. Kerps, *8th IEEE International Semiconductor Laser Conference Proceedings*, p. 26, J. C. Dyment (ed.) (1982).
129. M. Ettenberg, *Appl. Phys. Lett.*, **32**, 724 (1978).
130. I. Ladany, M. Ettenberg, H. F. Lockwood and H. Kressel, *Appl. Phys. Lett.*, **30**, 87 (1977).
131. Y. Shima, N. Chinone and R. Ito, *Appl. Phys. Lett.*, **31**, 625 (1977).
132. G. D. Henshall, *Solid State Elec.*, **20**, 595 (1977).
133. A. Moser, A. Oosenbrug, E. E. Latta, Th. Forster and M. Gasser, *Appl. Phys. Lett.*, **59**, 2642 (1991).
134. S. Takahashi, T. Kobayashi, H. Saito and Y. Furukawa, *Jpn J. Appl. Phys.*, **17**, 865 (1978).
135. K. Itoh, K. Asahi, M. Inoue and I. Teramoto, *IEEE J. Quantum Electron.*, **QE-13**, 623 (1977).
136. H. Yonezu, I. Sakuma, T. Kamejima, M. Ueno, K. Iwamoto, I. Hino and I. Hayashi, *Appl. Phys. Lett.*, **34**, 637 (1979).
137. J. Ungar, N. Bar-Chaim and I. Ury, *Electron. Lett.*, **22**, 279 (1986).
138. K. Y. Lau, N. Bar-Chaim, I. Ury and A. Yariv, *Appl. Phys. Lett.*, **45**, 316 (1984).
139. H. Naito, M. Kume, K. Hamada, H. Shimizu and G. Kano, *J. Quant. Elec.*, **25**, 1495 (1989).

140. H. O. Yonezu, M. Ueno, T. Kamejimi and I. Hayashi, *IEEE J. Quantum Electron.*, **QE-15**, 775 (1979).
141. H. Nakashima, S. Semura, T. Ohta and T. Kuroda, *Jpn J. Appl. Phys.*, **24**, L647 (1985).
142. S. Semura, T. Ohta, T. Kuroda and H. Nakashima, *Jpn J. Appl. Phys.*, **24**, L463 (1985).
143. Y. Kadota, K. Chino, Y. Onodera, H. Namizaki and S. Takamiya, *IEEE J. Quantum Electron.*, **QE-20**, 1247 (1984).
144. S. Yamamoto, H. Hayashi, T. Hayakawa, N. Miyauchi, S. Yano and T. Hijikata, *Appl. Phys. Lett.*, **42**, 406 (1983).
145. D. Botez and J. C. Connolly, *Electron Lett.*, **20**, 530 (1984), erratum **20**, 710 (1984).
146. Z. L. Liau and J. N. Walpole, *Appl. Phys. Lett.*, **50**, 528 (1987).
147. A. Moser, *Appl. Phys. Lett.*, **59**, 522 (1991).
148. K. Mizuishi, *J. Appl. Phys.*, **55**, 289 (1984).
149. K. Fujiwara, *Appl. Phys. Lett.*, **35**, 861 (1979).
150. For a review of reliability statistics, see Mitsuo Fukuda, *Reliability and Degradation of Semiconductor Lasers and LEDs*, Chapter 3, Artech House (1991).
151. E. I. Gordon, F. R. Nash and R. L. Hartman, *IEEE Elec. Dev. Lett.*, **4**, 465 (1983).
152. F. R. Nash, W. B. Joyce, R. L. Hartman, E. I. Gordon and R. W. Dixon, *AT&T Tech. J.*, **64**, 671 (1985).
153. H. Kung, D. P. Worland, H. Nguyen, W. Streifer, D. R. Scifres, D. F. Welch, L. Wood and S. Daudt, *SPIE*, **1044**, 2 (1989).
154. H. Kung, R. R. Craig, E. Zucker, B. Li and D. R. Scifres, *SPIE*, **1813**, 319 (1992).
155. M. Sakamoto, D. F. Welch, J. G. Endriz, E. P. Zucker and D. R. Scifres, *Electron. Lett.*, **25**, 972 (1989).
156. M. Sakamoto, M. R. Cardinal, J. G. Endriz, D. F. Welch and D. R. Scifres, *Electron. Lett.*, **26**, 422 (1990).
157. M. Sakamoto, D. F. Welch, H. Yao, J. G. Endriz and D. R. Scifres, *Electron. Lett.*, **27**, 902 (1991).
158. M. Sakamoto, J. G. Endriz and D. R. Scifres, *Electron. Lett.*, **QE-28**, 178 (1992).
159. M. Sakamoto, Spectra Diode Labs – private communication.
160. G. L. Harnagel, J. M. Haden, G. S. Browder, Jr, M. Cardinal, J. G. Endriz and D. R. Scifres, *SPIE*, **1219**, 19 (1990).
161. J. G. Endriz, M. Vakili, G. S. Browder, M. DeVito, J. M. Haden, G. L. Harnagel, W. E. Plano, M. Sakamoto, D. F. Welch, S. Willing, D. P. Worland and H. C. Yao, *IEEE J. Quantum Electron.*, **QE-28**, 952 (1992).
162. D. Mundinger, J. G. Endriz, J. Haden, G. Harnagel and D. Dawson, *SPIE*, vol. 1865, paper 15 (1993).
163. J. M. Haden, Spectra Diode Labs – unpublished data.
164. R. F. Carson and S. K. Fries, A radiation tolerant laser diode component for explosive and pyrotechnic ignition, Sandia Report, SAND88-2959.OC-2 (1989).

8

Strained layer quantum well heterostructure laser arrays

JAMES J. COLEMAN

8.1 Introduction

The oldest, best developed, and still most sophisticated heterostructure materials system for semiconductor diode laser arrays is the AlGaAs–GaAs system. This system has a unique combination of practical physical, electrical, optical and chemical features that has withstood the test of time. These features include a tractable chemical system that lends itself equally well to basic epitaxial growth methods, such as liquid phase epitaxy (LPE), and to more sophisticated thin layer epitaxial growth methods, such as metalorganic chemical vapor deposition (MOCVD) and molecular beam epitaxy (MBE). Controlled thin layer epitaxy has, in turn, allowed the development of AlGaAs–GaAs quantum well heterostructure lasers. Perhaps the single most important attractive feature of the AlGaAs–GaAs heterostructure system is the fortunate coincidence that AlAs and GaAs have, for all practical purposes, the same lattice constant ($\Delta a/a < 0.12\%$). This allows the design of any structure, with any combination of layer compositions, without regard to lattice mismatch or the associated dislocation formation.

The range of wavelengths available from lattice matched AlGaAs–GaAs conventional double-heterostructure lasers and quantum well heterostructure lasers is from $\lambda \approx 0.88$–0.65 μm. The long wavelength limit is defined by the band edge for GaAs, and AlGaAs active layers or the quantum size effect, or both, are used to shift the emission to shorter wavelengths. The short wavelength limit is not well defined, but arises from the intrusion of the large mass indirect conduction band minima at X and L on direct recombination processes. This wavelength range is sufficient to cover many important applications such as the use of high power AlGaAs–GaAs laser arrays at $\lambda \approx 0.81$ μm for diode pumped Nd:YAG solid state lasers. This is shown schematically in Figure 8.1,

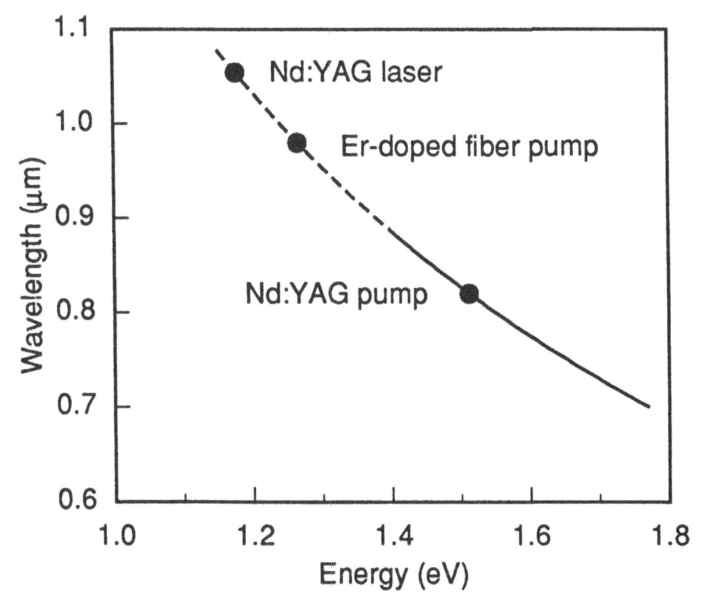

Figure 8.1. Conversion of wavelength versus energy, with the wavelength range supported by the AlGaAs–GaAs laser system shown as a solid line. The wavelengths of several important applications are shown as solid points.

which is simply the conversion of wavelength versus energy given by

$$\lambda = \frac{hc}{E}, \tag{8.1}$$

where h is Planck's constant and c is the speed of light. In Figure 8.1, the wavelength range supported by the AlGaAs–GaAs laser system is shown as a solid line. There are, however, a number of important applications that require laser emission at somewhat longer wavelengths, in the range of $\lambda \approx 0.88$–1.1 μm shown as a dashed line, which is unavailable in any well developed III–V lattice matched heterostructure laser materials system. These include frequency doubling of wavelengths near the 1.06 μm emission wavelength of Nd:YAG solid state lasers, and pumping the upper states of rare earth doped silica fiber amplifiers.

A simple examination of the available direct gap III–V compound semiconductor alloys suggests that perhaps the best material system for obtaining emission wavelengths in the range $\lambda \approx 0.88$–1.1 μm is $In_xGa_{1-x}As$. There is, however, no suitable binary substrate material that allows lattice matched compositions of $In_xGa_{1-x}As$ in the wavelength range of interest here. Thus, only a heterostructure materials system in which a very large mismatch must be accommodated ($\Delta a/a$ as great as 3%) may be considered, with the attendant possibility that strained layer materials may be unsuitable in terms of device reliability.

The metallurgical implications of accommodating strain elastically with layers thinner than some misfit-dependent critical thickness were described in the seminal work of Matthews and Blakeslee (1974). In the early 1980s, Tsang (1981) extended the lasing wavelength for $Al_xGa_{1-x}As$–GaAs double heterostructures by incorporating a small amount of In, and workers at the Sandia National Laboratories (Osbourn *et al.*, 1987) considered some of the effects of biaxial strain on the energy band structure, optical properties, and electrical transport in these structures. An important key in the continued development of strained layer InGaAs–GaAs heterostructure lasers was the work by Laidig and coworkers on injection laser diodes (Laidig *et al.*, 1984, 1985). This work showed the suitability of InGaAs–GaAs strained layer heterostructures for diode lasers at $\lambda \approx 1$ μm and were the first reports of reliable laser operation from strained layer lasers.

In recent years, the number of research laboratories studying various attributes of strained layer InGaAs quantum well heterostructure lasers has risen dramatically. In this chapter, we will consider the metallurgical aspects of critical thickness in stained layer heterostructure systems and the effects of strain and quantum size effect on emission wavelength. We will outline considerations of optical gain, threshold current density, and carrier-induced antiguiding in strained layer lasers and discuss the reliability of these laser structures. We will briefly describe a number of laser array structures for which the use of strained layer quantum well heterostructures, although not central to the fundamental operation of the laser arrays, provides significant advantage in terms of increased wavelength range, reduced threshold current density, or enhanced reliability. Finally, we will describe a laser array structure in which a specific property of strained layer heterostructures, carrier-induced antiguiding, plays a key role in the operation of the device.

8.2 Strained layer metallurgy and critical thickness

In a strained layer lattice mismatched system the elastic accommodation of the strain energy associated with the mismatch, without the formation of misfit dislocations, must be considered (Frank and van der Merwe, 1949; Matthews and Blakeslee, 1974). The unit cell of $In_xGa_{1-x}As$ can

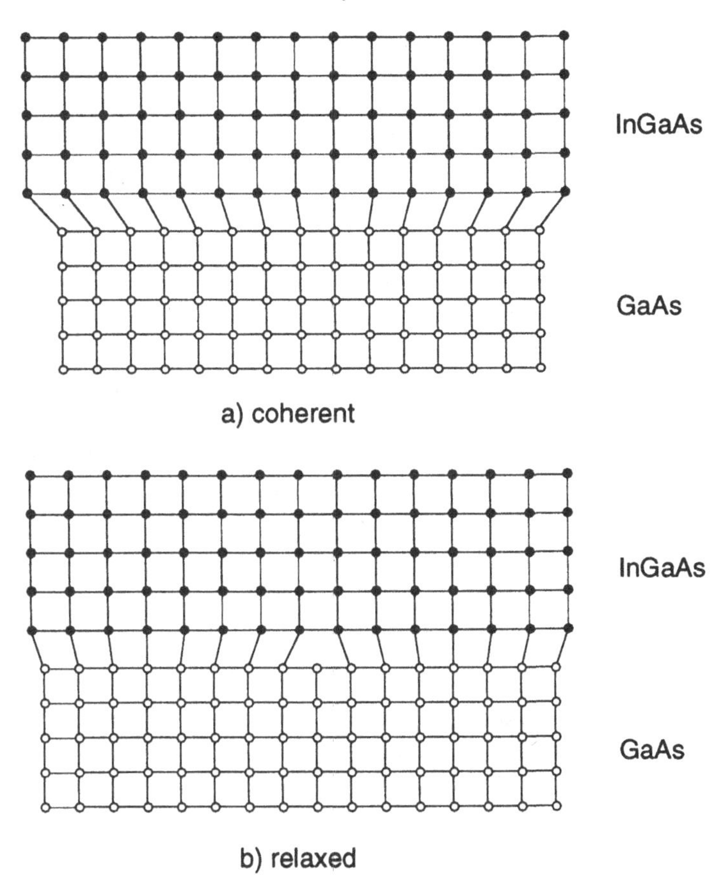

Figure 8.2. Interface of an $In_xGa_{1-x}As$–GaAs strained layer heterostructure shown schematically. When the strain is accommodated elastically, the layer is coherent (*a*). Otherwise, large numbers of misfit dislocations form, and the strained layer relaxes (*b*).

be much larger than GaAs (up to 3.6%), in contrast to $Al_xGa_{1-x}As$, which has a unit cell that is never more than about 0.13% larger than the GaAs unit cell. At the interface of an $In_xGa_{1-x}As$–GaAs heterostructure, shown schematically in Figure 8.2, the $In_xGa_{1-x}As$ cell is shortened in both directions parallel to the interface (biaxial compression), and elongated in the direction normal to the interface (uniaxial tension). The strain that results is approximately equal to the misfit and produces a force F_ε at the interface. If this force is sufficiently small, the strain is accommodated elastically, and the layer is coherent, as shown in Figure 8.2(*a*). If this force exceeds the tension F_l in a dislocation line, migration of a threading dislocation results in formation of misfit dislocations (Matthews and Blakeslee, 1974), and the strained layer relaxes (Figure 8.2(*b*)).

If the $In_xGa_{1-x}As$ cell is inserted between layers of the host as in a quantum well, assuming that the host is relatively thick on both top and bottom, both interfaces are under biaxial compression. The strain that results is again approximately equal to the misfit and, if the force F_ε exceeds twice the tension F_l in a dislocation line, migration of a threading dislocation results in the formation of two misfit dislocations (Matthews and Blakeslee, 1974). In the case of a superlattice, where the concept of a host does not apply and alternating layers are under either biaxial compression or biaxial tension, the misfit strain is distributed among all the layers and the strain is equal to approximately half of the misfit.

For each of these cases, a critical layer thickness h_c, below which the misfit strain is accommodated without formation of misfit dislocations, can be defined in terms of the elastic constants of the materials. Matthews and Blakeslee (1974) calculate the critical thickness for layers having equal elastic constants as

$$h_c = \frac{a}{\kappa\sqrt{(2)}\pi f}\,\frac{(1-0.25\nu)}{(1+\nu)}\left(\ln\frac{h_c\sqrt{(2)}}{a}+1\right), \tag{8.2}$$

where h_c is the critical thickness and a is the lattice constant of the strained layer. The misfit f is defined simply as

$$f = \frac{\Delta a}{a} \tag{8.3}$$

and ν is Poisson's ratio, which is defined as

$$\nu = \frac{C_{12}}{C_{11}+C_{12}}. \tag{8.4}$$

The coefficient κ has a value of 1 for a strained layer superlattice, 2 for a single quantum well, and 4 for a single strained layer. The critical thickness of Equation (8.2) for $In_xGa_{1-x}As$–GaAs as a function of indium composition in the range $x < 0.50$ can be solved simply by numerical methods and is shown in Figure 8.3 for (*a*) superlattice, (*b*) quantum well, and (*c*) single strained layer structures.

Others have reported variations of these models for critical thickness. People and Bean (1985) have proposed an alternative mathematical description for the critical thickness of a single layer by defining a balance between the strain energy in a layer and the energy required for formation of an isolated dislocation. Dodson and Tsao (1987) applied a model for

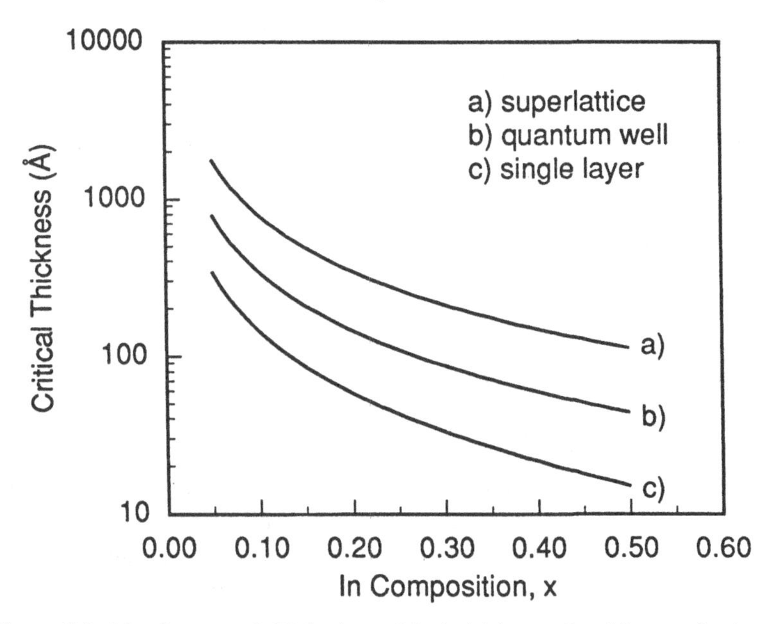

Figure 8.3. Matthews and Blakeslee critical thickness for (*a*) superlattice, (*b*) quantum well, and (*c*) single strained layer structures.

dislocation dynamics and plastic flow to account for the experimentally observed behavior in SiGe–Si strained layers. Vawter and Myers (1989) have described an equivalent strained layer model for design purposes which utilizes a reduced effective strain and a reduced effective thickness and applied this model to single- and multiple-quantum wells and strained layer superlattices. Most workers report reasonable agreement between experimentally determined critical thickness data for $In_xGa_{1-x}As$–GaAs quantum well structures with the Matthews and Blakeslee model of Equation (8.2) using $\kappa = 2$ (Fritz *et al.*, 1985; Andersson *et al.*, 1987; Yao *et al.*, 1988; Weng, 1989; Bertolet *et al.*, 1990).

8.3 The effects of strain on emission wavelength

The presence of biaxial strain results in changes in the effective masses of the conduction and valence bands of strained layer lattice mismatched systems and shifts in energy of the band edges relative to each other. In the unstrained compound semiconductor quantum well heterostructure laser, the band edges of interest are the conduction band edge and the degenerate light-hole and heavy-hole valence band edges which are shown in the schematic E–k diagram of Figure 8.4(a). In the strained AlGaAs–GaAs heterostructure system, biaxial compression defeats the cubic symmetry of the semiconductor (Olsen *et al.*, 1978). This results in removal of the degeneracy in the valence band edge because, although both heavy-hole and light-hole band edge energies increase in energy with respect to the conduction band edge with increasing strain, the increase in the light-hole band edge energy is greater, as shown schematically in Figure 8.4(b) (Asai and Oe, 1983; Schirber *et al.*, 1985; Jones *et al.*, 1985; Anderson *et al.*, 1986; O'Reilly, 1989). An orbital strain Hamiltonian for a given band at $k = 0$ can be written (Pikus and Bir, 1959, 1960) in terms of the components of the strain tensor ε_{ij}, the angular momentum operator **L**, the hydrostatic deformation potential Ξ, and the tetragonal and rhombohedral shear deformation potentials (b and d respectively). Assuming biaxial stress in the growth plane, the Hamiltonian can be simplified and eigenvalues calculated (Gavini and Cardona, 1970). This results in energy differences between the conduction band valence bands at $k = 0$ given to first order by (negative for compression, positive for tension)

$$\Delta E_{\mathrm{hh}} = \pm\left[2\Xi\varepsilon\left(\frac{C_{11} - C_{12}}{C_{11}}\right) - b\varepsilon\left(\frac{C_{11} + 2C_{12}}{C_{11}}\right)\right], \tag{8.5}$$

$$\Delta E_{\mathrm{lh}} = \pm\left[2\Xi\varepsilon\left(\frac{C_{11} - C_{12}}{C_{11}}\right) + b\varepsilon\left(\frac{C_{11} + 2C_{12}}{C_{11}}\right)\right], \tag{8.6}$$

where ΔE_{hh} is the shift in the heavy-hole valence band edge with respect to the conduction band edge, ΔE_{lh} is the shift in the light-hole valence band edge with respect to the conduction band edge, C_{ij} are elastic stiffness coefficients, and the strain ε is given by

$$\varepsilon = \frac{\Delta a}{a}. \tag{8.7}$$

The parameters necessary to calculate Equations (8.5) and (8.6) for $In_xGa_{1-x}As$ strained layers on GaAs can be interpolated from values (Adachi, 1982; Huang *et al.*, 1989; Madelung, 1982; Niki *et al.*, 1989; Yablonovitch and Kane, 1988) for the endpoint binary semiconductors, InAs and GaAs. Since the shift in the light-hole valence band edge with composition is much greater than the shift in the heavy-hole valence band edge, recombination in bulk strained layer $In_xGa_{1-x}As$ is dominated by transitions from the conduction band to the heavy-hole valence band. Since both band edges shift away from the conduction band edge, there

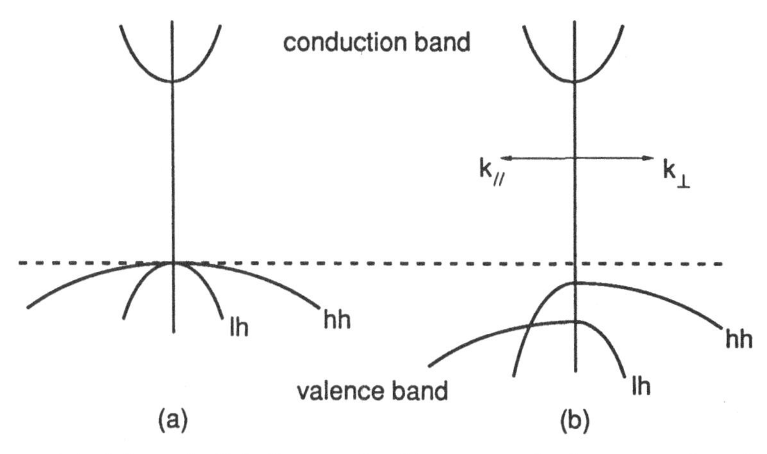

Figure 8.4. Schematic E–k diagram for (*a*) an unstrained compound semiconductor quantum well heterostructure laser and (*b*) the strained AlGaAs–GaAs heterostructure system. Biaxial compression defeats the cubic symmetry of the semiconductor and results in removal of the degeneracy in the valence band edge and reduction in hole effective mass in the direction parallel to the growth interfaces.

is a strain induced increase in the bulk band gap energy with indium composition that partially offsets the decrease in band gap energy with composition in unstrained $In_xGa_{1-x}As$.

An additional correction is necessary to account for energy shifts in thin layers resulting from the quantum size effect, because of the limits on strained layer thickness imposed by the critical thickness, and because of the advantages associated with the use of quantum wells in semiconductor lasers in general. The shifts in energy associated with the quantum size effect can be determined from solutions to the finite square potential well problem for both confined electrons in the conduction band and confined holes in the heavy-hole valence band, given the heights of the potential barriers and the effective masses of the particles. Here, we take the effective masses for electrons and heavy holes for $In_xGa_{1-x}As$ to be interpolated from the bulk values for the endpoint binary semiconductors, InAs and GaAs. The energy of the electron potential barrier is determined from the difference ΔE_g between the bulk band gap energy of the barrier layers and the strained band gap energy for $In_xGa_{1-x}As$, multiplied by a dimensionless conduction band heterostructure discontinuity fraction $\Delta E_c/\Delta E_g$. The energy of the heavy-hole potential barrier is determined from the same bulk energy difference multiplied by $1 - \Delta E_c/\Delta E_g$. A value $\Delta E_c/\Delta E_g = 0.65$ is used here (Y. Zou *et al.*, 1991). In any case, the overall transition energy from a bound state in the conduction band to a bound heavy-hole valence band state for a quantum well heterostructure laser is not particularly sensitive to the choice of $\Delta E_c/\Delta E_g$.

The range of sizes for a quantum well in an $In_xGa_{1-x}As$–GaAs laser structure is limited on the upper end by the critical thickness for elastic accommodation of strain, as described previously. In order to illustrate how the quantum size effect impacts the transition energy for a laser diode, we have chosen a value for the well thickness equal to the Matthews and Blakeslee critical thickness at the composition of interest. These values have appropriate bound states for compositions in the range of interest, are sufficiently thin that the quantum size effect is significant, and are comfortably in the range of practical thicknesses for modern epitaxial methods. The total transition energy for recombination from an $n = 1$ electron state to an $n = 1$ heavy-hole state in a strained layer quantum well heterostructure then consists of the bulk unconstrained energy gap for the well composition of interest, a correction for the shift in the heavy-hole valence band edge with strain, and corrections for shifts in both the conduction band and heavy-hole valence band associated with the quantum size effect. Thus, for $In_xGa_{1-x}As$

$$E(x, L_z) = E_g(x) + \Delta E_{hh}(x)|_{strain} + \Delta E_c(x)|_{L_z} + \Delta E_{hh}(x)|_{L_z}, \quad (8.8)$$

and the wavelength for this transition energy is given, of course, by

$$\lambda(x, L_z) = \frac{hc}{E(x, L_z)}, \quad (8.9)$$

where h is Planck's constant and c is the speed of light. Shown in Figure 8.5 is the transition wavelength as a function of composition for a strained layer in an $In_xGa_{1-x}As$–GaAs quantum well heterostructure laser having a well thickness equal to the Matthews and Blakeslee critical thickness, and for bulk (unstrained) and strained in $In_xGa_{1-x}As$. From Figure 8.5 it is clear that, even with corrections for strain and quantum size effect, the emission wavelength range of $\lambda = 0.9$–1.1 μm is adequately covered by $In_xGa_{1-x}As$–GaAs quantum well heterostructure lasers in the composition range of $x = 0.10$–0.40.

8.4 Threshold current density in strained layer structures

Perhaps more important for many laser array applications is that the changes in the valence band structure of strained layer $In_xGa_{1-x}As$–GaAs quantum well heterostructure lasers result in the expectation of reduced

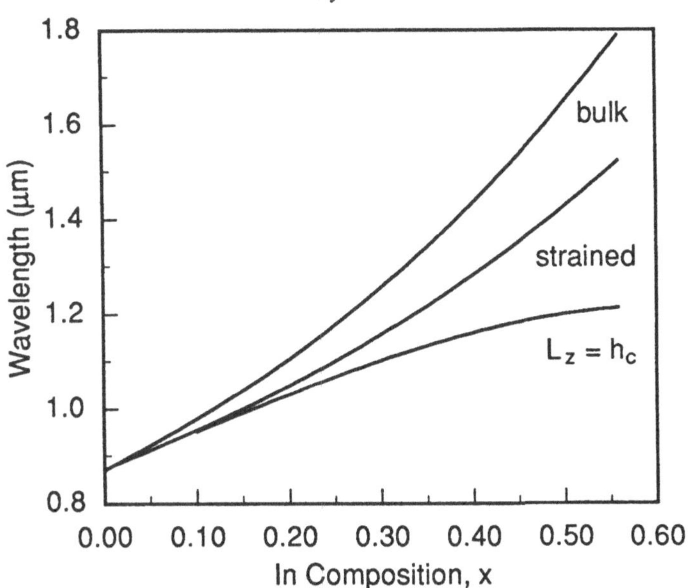

Figure 8.5. Transition wavelength as a function of composition for a strained layer $In_xGa_{1-x}As$–GaAs quantum well heterostructure laser having a well thickness equal to the Matthews and Blakeslee critical thickness, and for bulk (unstrained) and strained $In_xGa_{1-x}As$.

laser threshold current density. Yablonovitch and Kane (1986; 1988) and Adams (1986) have argued that the large asymmetry between the light conduction band effective mass and the much heavier valence band effective mass is a fundamental limitation on low threshold laser performance in conventional double heterostructure or quantum well heterostructure lasers. At threshold the quasi-Fermi level for electrons and holes must satisfy the condition (Bernard and Duraffourg, 1961) that

$$E_{F_n} - E_{F_p} > \hbar\omega, \tag{8.10}$$

where $\hbar\omega$ is greater than the band gap energy, or the transition energy in the case of a quantum well heterostructure laser. Assuming lightly doped active regions, detailed balance requires that the injected electron density δn and the injected hole density δp are equal. Of course, δn and the injected hole density δp are given by the integral of the product of the quantum well density of states and the Fermi function occupation probability. Thus, satisfying simultaneously Equation (8.10) and the equation

$$\delta n = \delta p \tag{8.11}$$

defines the values of E_{F_n} and E_{F_p}.

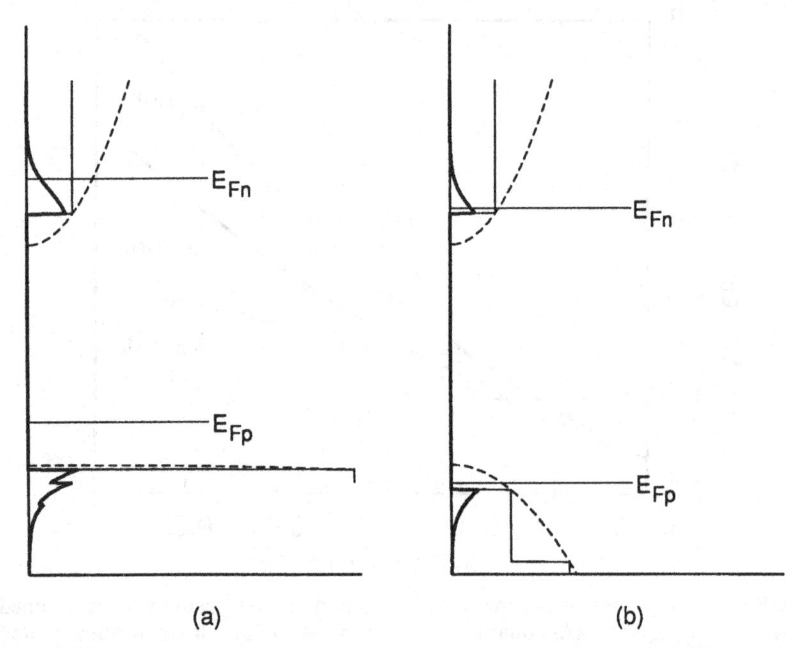

Figure 8.6. Density of states versus energy for (*a*) unstrained and (*b*) strained layer quantum well heterostructure lasers. The bulk parabolic density of states is shown as a dashed line for reference and the product of the quantum well density of states and the Fermi function is shown in bold.

In unstrained quantum well heterostructure lasers, the hole effective mass is so much greater than the electron effective mass that the electron quasi-Fermi level is degenerate while the hole quasi-Fermi level is far from degenerate. This is shown schematically in Figure 8.6(*a*), which is the density of states versus energy for an unstrained quantum well heterostructure laser. The bulk parabolic density of states is shown as a dashed line for reference and the product of the quantum well density of states and the Fermi function is shown in bold. The area under the bold curves corresponds to δn and δp which, when they are equal, yield the quasi-Fermi levels shown.

The reduction in hole effective mass in the direction parallel to the growth interfaces associated with strain results in a corresponding reduction in the injected carrier density required to reach threshold. This is illustrated in Figure 8.6(*b*), which is the density of states versus energy for a strained layer quantum well heterostructure laser. The lighter valence band effective mass results in a reduced density of states, and a correspondingly smaller value for injected carrier density. The lowest transparency carrier densities are obtained (Yablonovitch and Kane, 1986; 1988) when the electron and hole effective masses are equal and as small as possible. Even for unstrained $In_xGa_{1-x}As$ ($m_h^* \approx 0.45$), the transparency carrier density is smaller, simply because of the greater GaAs electron effective mass. Given a strain induced reduction of the hole effective mass to 0.1 (Adams, 1986), the transparency carrier densities in the strained layer wells are 1.5–2.4 times smaller than for comparable GaAs quantum wells. The relationship between transparency carrier density n_{tr}, and transparency current density J_0, is given by the well thickness, L_z, the fundamental charge q, and the spontaneous lifetime of the carriers in the well τ_{sp}, in the form

$$J_0 = \frac{q n_{tr} L_z}{\tau_{sp}}. \tag{8.12}$$

The spontaneous lifetime in the wells is a function of composition and increases rapidly with decreasing carrier densities below 1.5×10^{18} cm^{-3} (Olshansky *et al.*, 1984; Arakawa *et al.*, 1985). Shown in Figure 8.7 is the infinite barrier transparency current density J_0, versus hole effective mass.

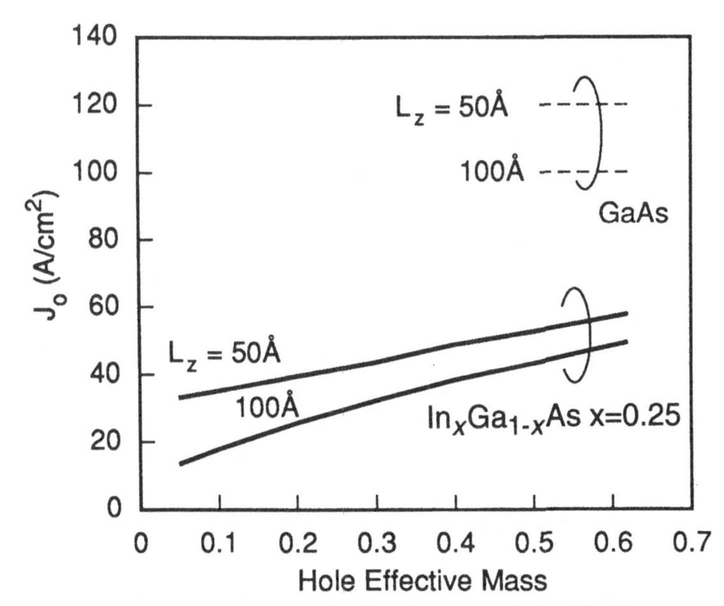

Figure 8.7. Transparency current density J_0, versus hole effective mass for an infinitely deep $In_xGa_{1-x}As$ strained layer single quantum well. Shown as dashed lines are the ideal transparency current densities for infinitely deep GaAs quantum wells of 50 and 100 Å.

Shown also for reference as dashed lines in Figure 8.7 are the ideal transparency current densities for infinitely deep GaAs quantum wells of 50 and 100 Å. From Figure 8.7, it can be seen that the modest reductions in transparency carrier density associated with strain are amplified by the effects of the associated increases in the spontaneous lifetime, and result in decreases in the transparency current density of from 3.5 to 6 times compared to similar GaAs quantum well structures.

The first report of injection laser diodes from strained layer AlGaAs–GaAs quantum well heterostructures (Laidig *et al.*, 1984) were for MBE-grown triple quantum well devices emitting at ≈ 1.0 μm. These early lasers were characterized by somewhat high threshold current densities (≈ 1.2 kA/cm^2) but were quickly followed by reports of lower threshold current densities (465 A/cm^2, Laidig *et al.*, 1985). After 1986, research activity on this materials system for broad area lasers increased rapidly, with other groups reporting low broad area threshold current densities at $\lambda \approx 1$ μm (Feketa *et al.*, 1986; Choi and Wang, 1990), high power cw operation at $\lambda \approx 0.93$ μm (Bour *et al.*, 1988), high power conversion efficiency at $\lambda \approx 0.93$ μm (Bour *et al.*, 1989a), and details of the operating characteristics, such as threshold current density J_{th}, differential quantum efficiency η, and characteristic temperature of threshold current T_0 in the range $\lambda \approx 0.93$–1.0 μm (Bour *et al.*, 1989b).

A series of studies has been reported (Beernink *et al.*, 1989a, b; York *et al.*, 1990a) in which different structural parameters of strained layer $In_xGa_{1-x}As$–GaAs–$Al_yGa_{1-y}As$ quantum well heterostructure lasers have been varied over a wide range and the resulting laser characteristics compared. The results for a series of broad area (150 μm) oxide defined stripe $Al_{0.20}Ga_{0.80}As$–GaAs–$In_xGa_{1-x}As$ single quantum well separate confinement laser structures having a fixed well size (70 Å) and varying well indium composition are summarized in Figure 8.8 (Beernink *et al.*, 1989a). The cavity length for these structures is 815 μm. Figure 8.8(*a*) shows the thickness as a function of composition for each of the laser structures with reference to the Matthews and Blakeslee critical thickness (solid line) for a single quantum well. Shown in Figure 8.8(*b*) is the measured emission wavelength for each laser composition along with the emission wavelength (solid line) expected from analysis of the quantum size effect, including the strain related shifts in the valence band edges.

Figure 8.8(*c*) is the laser threshold current density as a function of indium composition in the well. A minimum in J_{th} versus composition approaching 100 A/cm^2 is observed for indium fractions in the range $0.20 \leq x \leq 0.30$. Higher threshold current densities are observed for

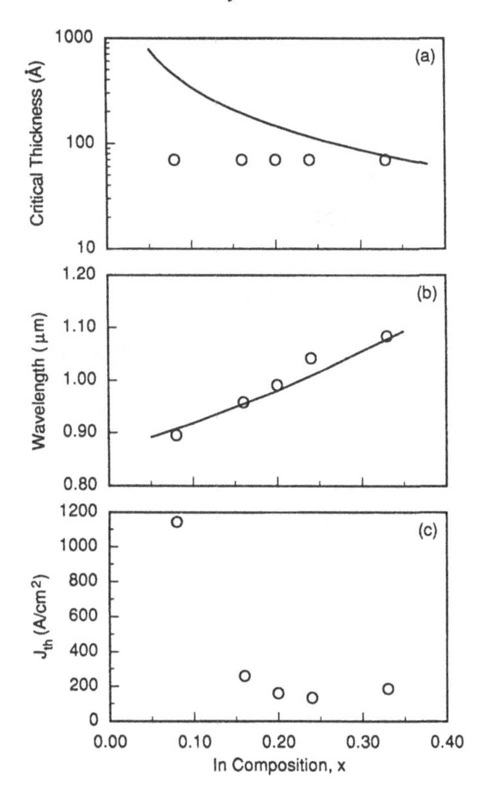

Figure 8.8. Laser data versus well indium composition including (*a*) thickness with reference to the Matthews and Blakeslee critical thickness (solid line), (*b*) the measured and expected (solid line) emission wavelength, and (*c*) laser threshold current density, for broad area (150 μm) oxide defined strip $Al_{0.20}Ga_{0.80}As$–GaAs–$In_xGa_{1-x}As$ single quantum well separate confinement lasers having a fixed well size (70 Å).

compositions below $x \approx 0.20$ because of insufficient carrier confinement in the well. Smaller confining energies result in a significant portion of the electron population in the active region having energies greater than the barrier height and, thus, unconfined in terms of the quantum well. As the indium fraction in the wells is increased and the well is made deeper, the state is better confined, resulting in decreased J_{th}. A similar decrease in J_{th} might be expected if the barrier layers were $Al_yGa_{1-y}As$ instead of GaAs and the aluminum composition y was increased.

The expectation for very low laser threshold current densities $In_xGa_{1-x}As$–GaAs quantum well heterostructure lasers has been realized experimentally in many laboratories (Beernink *et al.*, 1989a, b; Bour *et al.*,

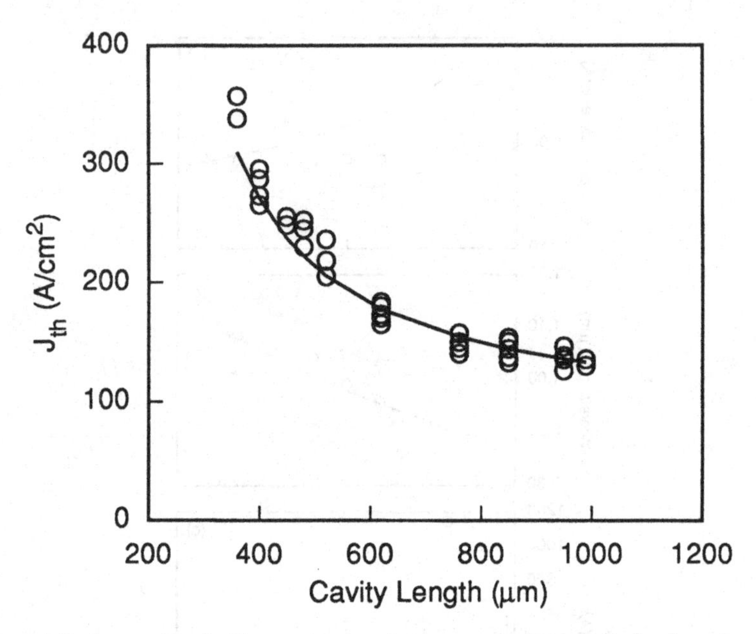

Figure 8.9. Laser threshold current density versus cavity length for broad area $Al_{0.20}Ga_{0.80}As$–GaAs–$In_{0.25}Ga_{0.75}As$ (L_z = 100 Å) quantum well heterostructure lasers.

1989a; Choi and Wang, 1990; Eng *et al.*, 1989; Feketa *et al.*, 1986; Larsson *et al.*, 1989; Offsey *et al.*, 1990; Saint-Cricq *et al.*, 1991; Stutius *et al.*, 1988; Wang and Choi, 1991a; Welch *et al.*, 1990; Williams *et al.*, 1991; York *et al.*, 1989a, b, 1990a, b, 1991). Shown in Figure 8.9 are laser threshold current density versus cavity length data for a broad area $Al_{0.20}Ga_{0.80}As$–GaAs–$In_{0.25}Ga_{0.75}As$ (L_z = 100 Å) quantum well heterostructure laser (York *et al.*, 1989, 1990a, b, 1991). At short cavity lengths, the lasers exhibit the usual upturn in threshold current density associated with gain saturation in short cavity single quantum well structures (Zory *et al.*, 1986). At longer cavity lengths, the threshold current density approaches 100 A/cm^2. For thin single quantum well lasers, the threshold current density J_{th} depends on cavity length L, facet reflectivity R, gain coefficient β, distributed losses α, internal quantum efficiency η, and the optical waveguide confinement factor Γ. The functional dependence of laser threshold current density on these parameters is given by

$$J = \frac{J_0}{\eta} \exp \frac{\alpha + (1/L) \ln(1/R)}{\Gamma \beta J_0}, \tag{8.13}$$

where the dependence of gain γ on current density (McIlroy *et al.*, 1985; Zory *et al.*, 1986; Wilcox *et al.*, 1988) is taken from $\gamma = \beta J_0 \ln(J/J_0)$. An analysis of the data in Figure 8.9 (shown as a solid line) gives a value of 42.4 A/cm^2 for the transparency current density and a value for the gain coefficient β of 28 cm/A (Corzine *et al.*, 1990). It is difficult to make a direct comparison between published results from different laboratories because of widely varying device parameters, including well size and composition, waveguide structure and composition, stripe width, cavity length, and even growth method. In some cases, the reported data are not sufficiently complete for this detailed comparison, or the data has been analyzed assuming a linear model for threshold current density and must be reanalyzed. It is worth noting that in most cases internal quantum efficiencies are high ($>90\%$) and distributed losses, which are more a function of the $Al_xGa_{1-x}As$–GaAs waveguide than the strained layer quantum well, are low (2–10 cm^{-1}). A key point is that the transparency current densities are very low (40–50 A/cm^2) and gain coefficients are high (>25 cm/A). These values are a significant improvement over similar $Al_xGa_{1-x}As$–GaAs single quantum well lasers (W. X. Zou *et al.*, 1991).

8.5 Antiguiding in strained layer lasers

Oxide defined stripe strained layer $In_xGa_{1-x}As$–GaAs quantum well heterostructure lasers having stripe widths of less than 75 μm are characterized by very high apparent laser threshold current densities and operation from higher order quantum states or inner barrier band edges (Shieh *et al.*, 1989a, b; Beernink *et al.*, 1991). This is observed in the same laser material which exhibits high performance as broad area (>150 μm) devices. The explanation for this behavior is an unusually large index antiguide resulting from injected carriers in the strained layer $In_xGa_{1-x}As$ active region. These characteristics and the antiguiding parameters associated with them are much greater than for comparable AlGaAs–GaAs quantum well heterostructure diode lasers. The high diffraction loss associated with the strong carrier-induced antiguiding causes some devices to laser on higher transitions in the quantum well and in the GaAs barriers, and results in near field widths that are much larger than the actual stripe width and diverging double-lobed far field emission patterns.

Shown in Figure 8.10 is a schematic cross-section of an oxide defined stripe laser along with the lateral distribution of carriers, gain (or loss) and real index of refraction. The injected carriers spread under the insulating stripe in the cap layer, upper confining layer and in the active

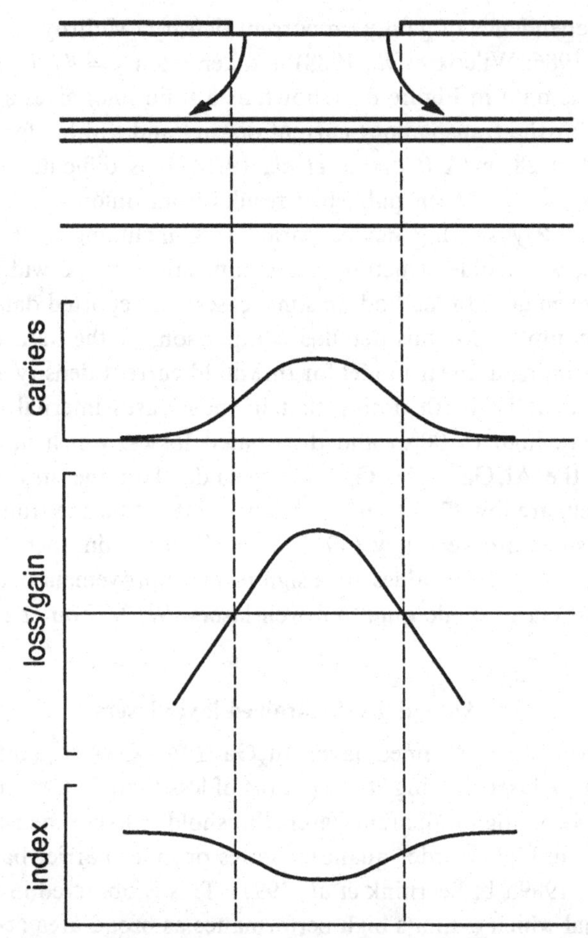

Figure 8.10. Schematic cross-section of an oxide defined stripe laser along with the lateral distribution of carriers, gain (or loss) and real index of refraction.

region. Given sufficient injected carrier density, there is a peak in the gain, and thus a peak in the imaginary refractive index, corresponding to the peak in the carrier distribution. Similarly, there is a maximum carrier-induced reduction in the real index of refraction at the same point. The formation of a gain guide or antiguide depends on the relative contributions of these two processes.

An antiguiding factor b (Streifer *et al.*, 1982) can be defined as the ratio of the decrease in the real refractive index Δn_r to the increase in the imaginary refractive index Δn_i,

$$b = -\frac{\Delta n_r}{\Delta n_i}. \tag{8.14}$$

The imaginary refractive index is related to the gain g by

$$\Delta n_i = \frac{g}{2k_0}, \tag{8.15}$$

where $k_0 = 2\pi/\lambda_0$ and λ_0 is the free space wavelength. The decrease in the real refractive index depends on the injected carrier distribution under the stripe, offset somewhat by an increase from the thermal contribution, and is proportional to (Paoli, 1977)

$$\Delta n_r \propto \frac{\lambda^2}{m_e}, \tag{8.16}$$

where m_e is the electron effective mass. Thus, b is a measure of the strength of the carrier-induced index depression relative to the gain that results from the same carrier density. From Equation (8.16), the carrier-induced reduction in the real refractive index in strained layer $In_xGa_{1-x}As$–GaAs lasers should be greater than in comparable AlGaAs–GaAs quantum well lasers. Negative values of b result in guiding, while positive values give antiguiding, with increased curvature of the wavefronts outside the stripe and a double-lobed far field emission pattern for narrow stripe widths. Far field emission patterns for $Al_{0.20}Ga_{0.80}As$–GaAs-$In_{0.25}Ga_{0.75}As$ single quantum well (70 Å) lasers just above threshold, for stripe widths of 50 and 4 μm are shown in Figure 8.11 (Beernink *et al.*, 1991). As the stripe width is decreased from 50 μm, the far field broadens and develops two lobes. The intensity at facet normal (0°) decreases relative to that in the lobes as the stripe width is further reduced. This is exactly the behavior expected from narrow stripe gain-guided devices with strong antiguiding (Streifer *et al.*, 1982). Beernink *et al.* (1991) have calculated a value of $b = 6.4$ for these devices, which is considerably larger than the values 0.5–3.0 reported for AlGaAs–GaAs lasers (Kirkby *et al.*, 1977). The antiguiding also results in a large diffraction loss due to light propagating out from under the stripe. The very high threshold current densities of the narrow stripe devices are a result of this high diffraction loss (Shieh *et al.*, 1980a, b; Beernink *et al.*, 1991).

Assuming a step index profile in the transverse direction, leaky waves within the device produce a lobe in the far field at an angle on either side

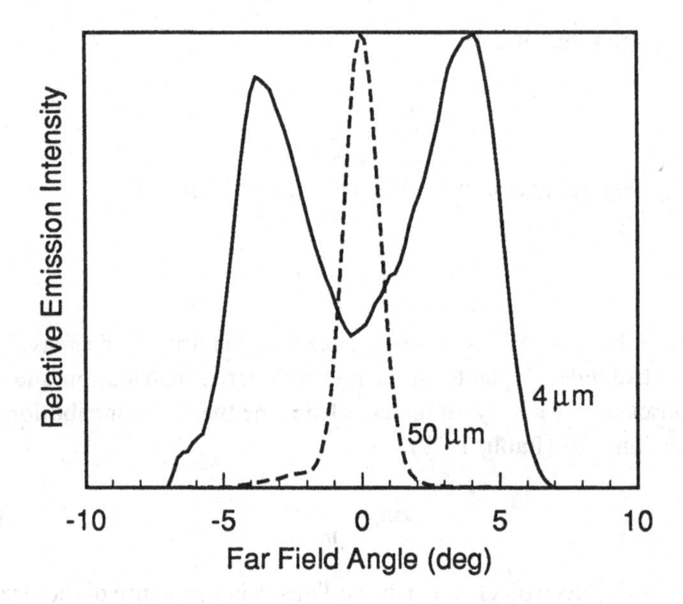

Figure 8.11. Far field emission patterns for oxide defined stripe $Al_{0.20}Ga_{0.80}As$–GaAs–$In_{0.25}Ga_{0.75}As$ single quantum well (70 Å) lasers just above threshold, for stripe widths of 50 (dashed line) and 4 μm (solid line).

of the facet normal given by (Ackley and Engelmann, 1980):

$$\sin\theta \approx (2\,\Delta n\, n_{\text{eff}})^{1/2}, \tag{8.17}$$

where θ is the far field angle, Δn is the magnitude of the index depression, and n_{eff} is the effective index of the region outside the stripe. The angular position relative to facet normal of one side-lobe in the far field is plotted as a function of drive current in Figure 8.12 for a 4 μm stripe width $Al_{0.20}Ga_{0.80}As$–GaAs–$In_{0.25}Ga_{0.75}As$ single quantum well (70 Å) laser over a range of currents. Also shown on the plot is the index depression calculated from Equation (8.17). The optical confinement factor for these devices is 0.015, so that the approximate carrier-induced index depression in the well is calculated to range from 0.032 to 0.41 over this range of current. The lobe position of the spontaneous emission from the well begins to decrease slightly at a current corresponding to the onset of lasing in the GaAs barrier layers. As the stimulated emission in the barrier layers grows, fewer carriers are available to be captured by the InGaAs as well, and the carrier density in the well, and the magnitude of the carrier-induced index depression, decrease.

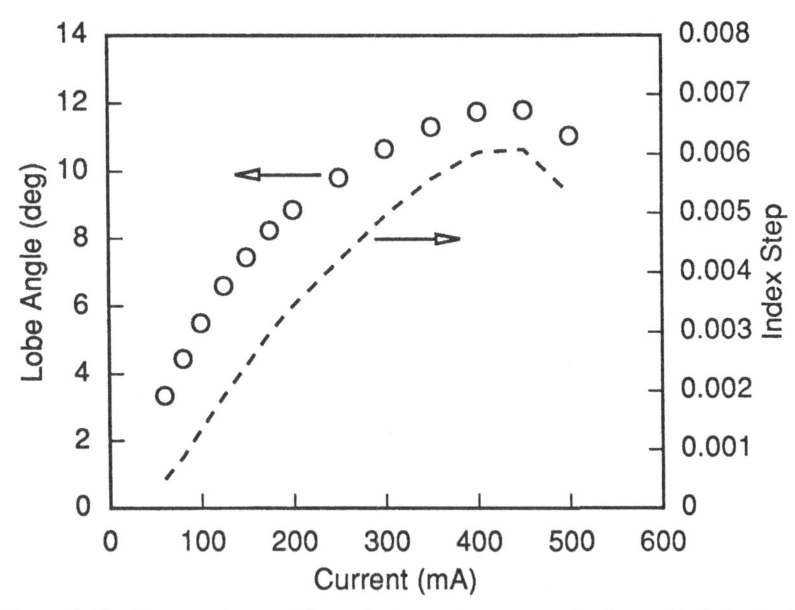

Figure 8.12. The angular position relative to facet normal of one side-lobe in the far field (open circles) and the calculated index depression (dashed line) versus drive current for a 4 μm stripe width $Al_{0.20}Ga_{0.80}As$–GaAs–$In_{0.25}Ga_{0.75}As$ single quantum well (70 Å) laser.

8.6 Reliability

Strained layer InGaAs–GaAs quantum well heterostructure lasers are of little practical consequence if the devices are unreliable. Early studies of photopumped InGaAs–GaAs strained layer superlattices (Camras *et al.*, 1983; Ludowise *et al.*, 1983) demonstrated cw 300 K laser operation but suggested that these strained materials are inherently unreliable. The first limited reports of reliable laser operation from strained layer InGaAs–GaAs heterostructure lasers appeared in 1985 (Laidig *et al.*, 1985), followed by reports of cw operation for a few hundred hours in 1988 (Stutius *et al.*, 1988). Fischer *et al.* (1989) reported 5000 h cw operation of strained layer InGaAs–GaAs lasers with 1.8% per kh degradation rates and 100% survival. Bour *et al.* (1990) reported 10 000 h operation with a degradation rate under 1% per kh. Both of these degradation rates are significantly better than for comparable AlGaAs–GaAs quantum well heterostructure laser diodes (Harnagel *et al.*, 1985; Waters and Bertaska, 1988).

The thickness of the InGaAs quantum well with respect to the critical thickness plays an important role (Beernink *et al.*, 1989b). The results of reliability testing suggest that the maximum practical thickness for $In_{0.25}Ga_{0.75}As$ quantum well active regions is less than the critical thickness predicted by the mechanical equilibrium model of Matthews and Blakeslee. The characteristics of strained layer InGaAs–GaAs quantum well lasers include very low gradual degradation rates and the absence of sudden failures, even for devices that have no facet coatings and were not screened or burned-in, with lifetimes exceeding 20 000 h (Fischer *et al.*, 1989; Waters *et al.*, 1990a; Yellen *et al.*, 1990). Waters *et al.* (1990b) have also reported on the reliability of InGaAs–GaAs strained layer quantum well heterostructure lasers having a step graded strained layer active region and emitting at 1.1 μm, which have operated for more than 20 000 h with very low gradual degradation (Yellen *et al.*, 1991). Researchers at NTT (Okayasu *et al.*, 1990; Takeshita *et al.*, 1990) have reported excellent reliability and high power results for 0.98 μm ridge waveguide strained layer InGaAs–GaAs quantum well lasers. Scifres *et al.* (1992) have compared the catastrophic damage limits and long term reliability of high power laser diodes from three materials systems, InGaP–InGaAlP, AlGaAs–GaAs and InGaAs–AlGaAs, covering the wavelength range of 0.69–1 μm. Strained layer lasers have also been shown to be capable of operation (Fu *et al.*, 1991; van der Ziel and Chand, 1991) at very high ambient temperatures ($>180\,°C$).

8.7 Conventional edge emitter and surface emitter laser arrays

Semiconductor diode laser arrays formed from strained layer quantum well heterostructure materials can be broadly characterized in two categories. In the first, and by far the largest, category are those strained layer laser arrays in which the unique properties of strained layers are not fundamental to the operation of the array. These laser arrays are generally structures that have been developed in lattice matched materials systems, usually AlGaAs–GaAs. They have been adapted to strained layer quantum well systems in order to obtain significant advantage in laser threshold current density, emission wavelength, or device reliability. In this section, we will outline highlights of a number of these structures, keeping in mind that many are discussed in much greater detail elsewhere in this volume.

A number of groups have reported conventional multiple-element coupled stripe laser arrays. Baillargeon *et al.* (1988) reported a high power phase locked non-planar corrugated substrate $Al_{0.20}Ga_{0.80}As$–GaAs–

$In_{0.25}Ga_{0.75}As$ ($L_z = 120$ Å) quantum well heterostructure array. These index guided arrays consist of 4 μm mesas, grown over a substrate etched 0.5 μm deep, with 8 μm center-to-center spacing. Two hundred element arrays yield more than 2.5 W of pulsed power per uncoated facet at an emission wavelength of 1.03 μm. The far field emission patterns of these laser arrays indicate the presence of phase locked operation with a 180° phase shift between neighboring elements.

Stutius *et al.* (1988) have reported ten stripe InGaAs–GaAs quantum well laser arrays formed by proton isolation, having 5 μm stripe widths and 9 μm center-to-center spacing. These arrays operate at a wavelength of 0.95 μm and yield up to 200 mW of cw power per uncoated facet. Welch *et al.* (1990) have reported ten stripe proton implanted InGaAs–GaAs quantum well laser arrays having several different In compositions ($x = 0.0$–0.20), and emission wavelengths, to 0.96 μm. These facet coated arrays had low threshold current densities (115 A/cm^2) and yielded up to 3 W cw for 100 μm apertures. Major *et al.* (1990) reported strained layer InGaAs–GaAs quantum well heterostructure laser arrays formed by hydrogenation. These ten-element arrays, formed from 4 μm stripes on 12 μm centers, emitted at 1.06 μm with a double-lobed 180° out-of-phase locked far field emission pattern. This same group, in collaboration with researchers at Polaroid (Major *et al.*, 1989), reported index-guided ten stripe InGaAs–GaAs quantum well heterostructure laser arrays formed by silicon-impurity-induced layer disordering. These arrays have 0.89 μm wavelength emission and operated phase locked in the highest order array mode. Richard *et al.* (1991) utilized native oxidation to form ten stripe arrays (5 μm elements on 10 μm centers) emitting near 0.91 μm from InGaAs–GaAs quantum well heterostructure material. Depending on the extent of the oxide isolation between elements, these arrays operated as either phase locked (180° out-of-phase) or as uncoupled elements).

Two-dimensional arrays of multiple element strained layer InGaAs–GaAs quantum well heterostructure lasers have been reported by several groups. The 'rack and stack' approach has been used (Bour *et al.*, 1989c) to form an oxide defined stripe 536 element uncoupled array of $\lambda = 0.95$ μm strained layer InGaAs–GaAs quantum well emitters. Almost 200 W of pulsed power has been obtained from these arrays. Coherent monolithic two dimensional strained layer InGaAs–GaAs quantum well laser arrays utilizing grating surface emission have been reported by Evans *et al.* (1989, 1990). These are arrays of distributed Bragg reflectors that emit normal to the junction plane and make use of the fact that the GaAs substrate is transparent to the strained layer emission wavelength (≈ 1 μm) to enable junction down mounting. Up to 14 W of peak pulsed power has been obtained from a 140-element array (Evans *et al.*, 1990). Goodhue *et al.* (1991) have made two dimensional InGaAs strained layer arrays using a horizontal folded cavity structure with two dry etched 45° internal reflectors and two top surface partially transmitting reflectors. Sixteen-element arrays have yielded more than 13 W of pulsed power at an emission wavelength of 1.03 μm. Nam *et al.* (1992) have obtained 50 W of cw power (≈ 940 nm) from a similar 1504-element array fabricated with 45° and 90° ion milled facets.

Vertical cavity surface emitting lasers, although presently limited in terms of output power, have a great deal to offer in terms of integrability and beam shaping. Since the gain path has been oriented normal to the epitaxial growth planes, the limitations of conventional asymmetric laser geometries are avoided, and it is possible to consider densely packed arrays of symmetric circular beams. There are several advantages, in addition to higher gain and lower threshold current densities, to using strained layer InGaAs–GaAs quantum well heterostructures for vertical cavity surface emitting laser arrays. As mentioned above for the grating coupled surface emitters, the GaAs substrate is transparent to the InGaAs emission wavelengths, allowing for higher reflectivity top surface reflectors and extraction of the beam through the substrate. The use of InGaAs quantum wells results in a greater range of $Al_xGa_{1-x}As$ compositions (essentially $x \approx 0$–1) for the layers that comprise the lower multi-layer reflector (Lee *et al.*, 1989; Geels *et al.*, 1990; Lei *et al.*, 1991). It is also possible to place high gain, efficient InGaAs strained layer quantum wells only at the peaks of the standing wave within the vertical cavity (Corzine *et al.*, 1989; Raja *et al.*, 1989), increasing the effective modal gain of the laser and avoiding wasted pumping of the volume near the standing wave nulls. There have been a few reports of vertical cavity surface emitting strained layer quantum well heterostructure two dimensional laser arrays. Yoo *et al.* (1990) reported a phase locked 160-element two dimensional array of 1.3 μm-diameter vertical cavity InGaAs strained layer lasers emitting near 1 μm. A two dimensional vertical cavity InGaAs strained layer laser array, in which the array elements are defined by a modulated reflector formed by ion milling part of the top surface metal reflector, has been reported (Orenstein *et al.*, 1991). Maeda *et al.* (1991) have operated a 16-element two dimensional vertical cavity InGaAs strained layer laser array with 16 independent wavelengths. Each laser was shown to be capable of 5 Gb/s operation.

8.8 Leaky mode and antiguided strained layer laser arrays

If there is coupling in the conventional edge emitting index guided strained layer laser arrays described above, the natural array mode is the highest order mode, in which adjacent elements are coupled 180° out-of-phase. This takes place because the presence of loss in the regions between the elements favors the highest order mode. It is also possible to introduce a real index antiguide into an array structure (Ackley and Engelmann, 1980) and obtain high power in-phase array mode operation if the device is laterally resonant (Mawst *et al.*, 1989). Shiau *et al.* (1990) have extended this concept to form InGaAs–GaAs strained layer quantum well heterostructure leaky mode laser arrays. They used thin patterned transparent GaAs:n^+ (greater index) embedded in the upper AlGaAs:p^+ (lower index) confining layer of an InGaAs–GaAs strained layer quantum well heterostructure to separate the gain and waveguide regions. Far field emission patterns for the structure show a major on-axis (0°) lobe that has a FWHM of 1.5° at 1.5 times threshold, and 2.4° at five times threshold. Major *et al.* (1991) have designed a similar structure that operates in a mode where the stripes are not resonantly coupled. These devices operate to power levels of more than 500 mW (cw) with a FWHM of ≈ 1.5 times the diffraction limit. Zmudzinski *et al.* (1992) achieved 1 W (pulsed) diffraction-limited operation from a resonant strained-layer array.

Beernink *et al.* (1990) have made use of the natural carrier-induced antiguiding in strained layer InGaAs quantum well heterostructure lasers described earlier to form five-element oxide defined arrays with 8 μm stripe width and center-to-center spacings from 14 μm to 30 μm and emitting at $\lambda \approx 1$ μm. Arrays with large interelement spacing exhibit a twin-lobed far field pattern corresponding to the highest order supermode with 180° phase shift between adjacent elements. Devices with smaller interelement spacing exhibit a single lobe in the far field corresponding to in-phase operation. The far field pattern of a device with 14 μm center-to-center spacing is shown in Figure 8.13 at (*a*) 1.25 I_{th} and (*b*) 3 I_{th}. The single lobe in the far field is typical of array operation with all emitters in phase. The half-widths of 1.8° at 1.25 I_{th}, and 2.3° at 3 I_{th} correspond to 2.0 and 2.6 times the diffraction limit respectively. Arrays on 15 μm centers also exhibit a narrow, single lobe in the far field. For comparison, the far field pattern at 2 I_{th} of a single stripe device with a stripe width of 50 μm is shown in Figure 8.13 as a dashed line. The transverse modes of such broad area devices are unstable with drive current and the far field has spread to $\approx 10°$ FWHM.

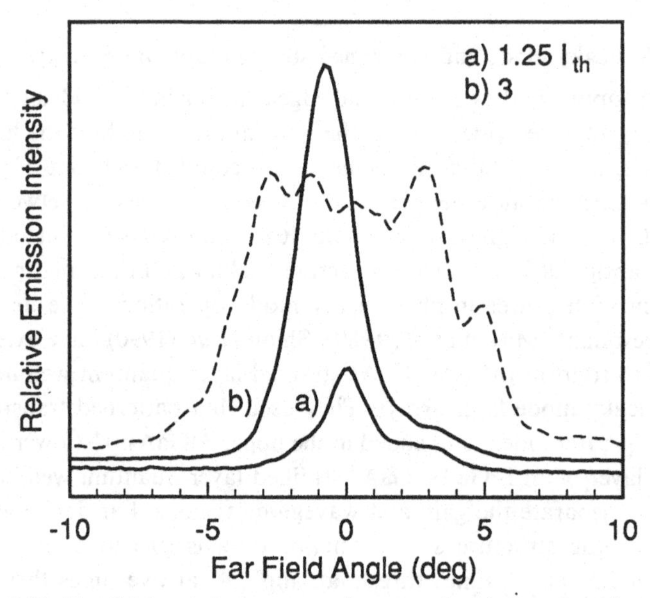

Figure 8.13. Far field emission patterns for an antiguided strained layer InGaAs quantum well heterostructure laser array with 14 μm center-to-center spacing at (*a*) 1.25 I_{th} and (*b*) 3 I_{th}. The far field pattern at 2 I_{th} of a single stripe device with a stripe width of 50 μm is also shown (dashed lines).

The *L–I* characteristics for a single stripe 50 μm device and for an array on 15 μm centers are shown in Figure 8.14. The array operated with a kink-free characteristic up to the limit of the current source, emitting 730 mW per uncoated facet at 1.8 A. In contrast, the single stripe device characteristic is non-linear with the appearance of several kinks. Although the arrays of closely spaced stripes with leaky modes seemingly approach broad area device behavior, the improved stability of the far field and the linearity of the *L–I* characteristic are evidence of the superior performance of the arrays.

8.9 Other strained layer laser materials systems

As many of the attractive features of strained layer quantum well heterostructure lasers have developed, it is not surprising that strained layers are also being incorporated into otherwise lattice matched quantum well heterostructure laser materials systems. Although the incorporation of other strained layer materials into laser arrays is just beginning, a significant effort is being directed toward developing these materials into

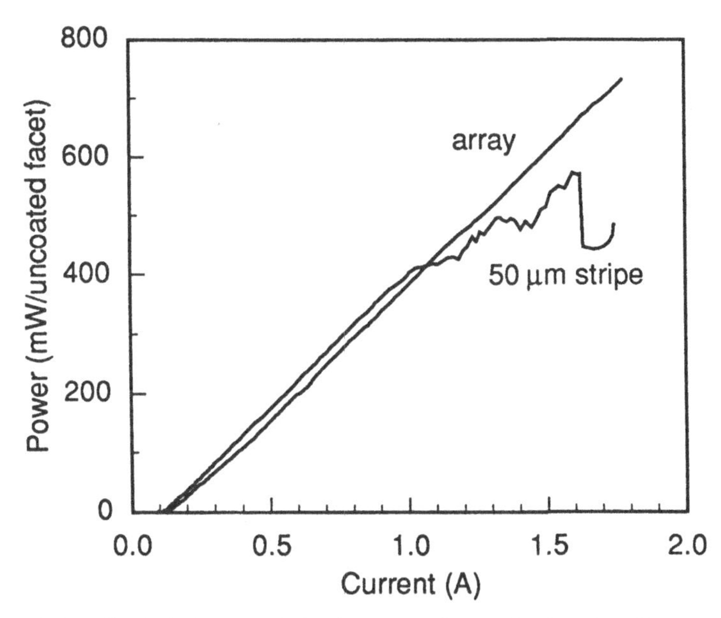

Figure 8.14. The *L–I* characteristics for a broad area single stripe 50 μm strained layer InGaAs quantum well heterostructure laser and for an antiguided strained layer InGaAs quantum well heterostructure laser array on 15 μm centers.

single-element devices. It is logical to expect that the extension of this work toward laser arrays will take place rapidly in the near future.

In addition to lower threshold currents and higher differential gain, the likelihood that Auger recombination and intervalence band absorption can be greatly diminished in strained layer systems provides strong incentive to apply strain to InGaAsP–InP quantum well heterostructure lasers, which have thus far been disappointingly slow in developing. Of course, the applications requiring 1.3–1.5 μm laser emission are an important factor. Melman *et al.* (1989) reported photoluminesence emission to 1.3 μm from strained layer InGaAs–GaAs heterostructures. More recently, high efficiency, high power and low threshold currents have been obtained in $In_xGa_{1-x}As$–InP with x adjusted for biaxial compressive strain. Although Tothill *et al.* (1989) saw no improvement in the threshold current density or T_0 of strained layer InGaAs–InP lasers emitting at 1.5 μm, others have reported substantial improvements in the laser characteristics of these devices compared to lattice matched structures, including lower threshold, higher efficiency, higher output power, higher T_0 and other interesting optical properties (Thijs and van Dongen, 1989; Temkin *et al.*, 1990; Koren *et al.*, 1990; Gershoni and Temkin, 1989;

Tanbun-Ek *et al.*, 1990; Coblentz *et al.*, 1991; Osinski *et al.*. 1991; Zah *et al.*, 1991a; Osinski *et al.*, 1992). A number of workers have constrasted otherwise similar devices with either compressive or tensile strain in the InGaAs–InP materials system. Corzine and Coldren (1991) predict much larger material and modal gain in structures under biaxial compression. Threshold currents have been observed experimentally by Temkin *et al.* (1991) to be lower in strained layer lasers under compression than under tension. O'Reilly *et al.* (1991), however, argues that biaxial tension will significantly enhance TM gain compared to TE gain and, by suppressing spontaneous emission, reduce the threshold current density. Several groups (Zah *et al.*, 1991b; Thijs *et al.*, 1991; Tiemeijer *et al.*, 1991) have reported experimental results that indicate that the threshold current densities are lower in strained layer lasers under tension than under compression. It is not clear yet how this discrepancy will be resolved but it is worth noting that the lowest threshold current density reported thus far (Osinski *et al.*, 1992) for any InGaAs–InP strained layer laser is $\approx 140\ A/cm^2$. These results are for lasers under 1.8% compressive strain.

There are several applications for which the attractive features of strained layer InGaAs–GaAs lasers, reliability in particular, are desirable, but the range of emission wavelengths is unsuitable. For example, strained layer lasers with emission wavelengths near 0.8 μm would be attractive as pumps for Nd:YAG solid state lasers. One approach to this problem is the use of strained layer quaternary AlInGaAs as the active region in an otherwise conventional AlGaAs quantum well heterostructure. High power, lower threshold current density, reliable operation of strained layer AlInGaAs–AlGaAs lasers has been reported by several groups (Wang *et al.*. 1991b; Buydens *et al.*, 1991; Choi *et al.*, 1991; Hughes *et al.*, 1992). There is an increasing interest in short wavelength lasers for applications requiring visible laser emission. The use of strained layers in separate confinement AlGaInP–GaInP quantum well heterostructure lasers has resulted in low threshold currents, high output powers and evidence of long term reliability at visible ($\lambda \approx 630$–690 nm) emission wavelengths (Katsuyama *et al.*, 1990; Hashimoto *et al.*, 1991; Serreze and Chen, 1991; Nitta *et al.*, 1991; Chang-Hasnain *et al.*, 1991; Welch and Scifres *et al.*, 1991; Scifres *et al.*, 1992).

Acknowledgments

It is a pleasure to acknowledge the many helpful discussions and the contributions to this work of D. E. Ackley, A. R. Adams, K. J. Beernink,

D. Botez, D. P. Bour, P. D. Dapkus, G. A. Evans, M. E. Givens, E. S. Koteles, J. LaCourse, L. M. Miller, R. Murison, T. L. Paoli, D. R. Scifres, W. Streifer, W. T. Tsang, R. G. Waters, P. K. York and P. S. Zory. This work was supported by the National Science Foundation (DMR 89-20538 and ECD 89-43166), the Joint Services Electronics Program (N00014-90-J-1270) and SDIO/IST (DAAL03-89-K-0080).

References

Ackley, D. E. and Engelmann, R. W. H. (1980). *Appl. Phys. Lett.*, **37**, 866.
Adachi, S. (1982). *J. Appl. Phys.*, **53**, 8775.
Adams, A. R. (1986). *Electron. Lett.*, **22**, 249.
Anderson, N. G., Laidig, W. D., Kolbas, R. M. and Lo, Y. C. (1986). *J. Appl. Phys.*, **60**, 2361.
Andersson, T. G., Chen, Z. G., Kulakovskii, V. D., Uddin, A. and Vallin, J. R. (1987). *Appl. Phys. Lett.*, **51**, 752.
Arakawa, Y., Sakaki, H., Nishioka, M. and Yoshino, Y. (1985). *Appl. Phys. Lett.*, **46**, 519.
Asai, H. and Oe, K. (1983). *J. Appl. Phys.*, **54**, 2052.
Baillargeon, J. N., York, P. K., Zmudzinski, C. A., Fernandez, G. E., Beernink, K. J. and Coleman, J. J. (1988). *Appl. Phys. Lett.*, **53**, 457.
Beernink, K. J., York, P. K. and Coleman, J. J. (1989a). *Appl. Phys. Lett.*, **55**, 2585.
Beernink, K. J., York, P. K., Coleman, J. J., Waters, R. G., Kim, J. and Wayman, C. M. (1989b). *Appl. Phys. Lett.*, **55**, 2167.
Beernink, K. J., Alwan, J. J. and Coleman, J. J. (1990). *Appl. Phys. Lett.*, **57**, 2764.
Beernink, K. J., Alwan, J. J. and Coleman, J. J. (1991). *J. Appl. Phys.*, **69**, 56.
Bernard, M. G. A. and Duraffourg, G. (1961). *Phys. Status. Solidi*, **1**, 699.
Bertolet, D. C., Hsu, J. K., Agahi, F. and Lau, K. M. (1990). *J. Electron. Mater.*, **19**, 967.
Bour, D. P., Gilbert, D. B., Elbaum, Gilbert, L. and Harvey, M. G. (1988). *Appl. Phys. Lett.*, **53**, 2371.
Bour, D. P., Ramon, U., Martinelli, D. B., Elbaum, Gilbert, L. and Harvey, M. G. (1989a). *Appl. Phys. Lett.*, **55**, 1501.
Bour, D. P., Evans, G. A. and Gilbert, D. B. (1989b). *J. Appl. Phys.*, **65**, 3340.
Bour, D. P., Stabile, P., Rosen, A., Janton, W., Elbaum, L. and Holmes, D. J. (1989c). *Appl. Phys. Lett.*, **54**, 2637.
Bour, D. P., Gilbert, D. B., Fabian, K. D., Bednarz, J. P. and Ettenberg, M. (1990). *IEEE Photon. Tech. Lett.*, **2**, 173.
Buydens, L., Demeester, P., van Ackere, M. and van Daele, P. (1991). *Electron. Lett.*, **27**, 618.
Camras, M. D., Brown, J. M., Holonyak, N. Jr, Nixon, M. A., Kaliski, R. W., Ludowise, M. J., Dietze, W. T. and Lewis, C. R. (1983). *J. Appl. Phys.*, **54**, 6183.
Chang-Hasnain, C. J., Bhat, R. and Koza, M. A. (1991). *Electron. Lett.*, **27**, 1553.
Choi, H. K. and Wang, C. A. (1990). *Appl. Phys. Lett.*, **57**, 321.
Choi, H. K., Wang, C. A., Kolesar, D. F., Aggarwal, R. L. and Walpole, J. N. (1991). *Photon. Tech. Lett.*, **3**, 857.
Coblentz, D., Tanbun-Ek, T., Logan, R. A., Sergent, A. M., Chu, S. N. G. and Davisson, P. S. (1991). *Appl. Phys. Lett.*, **59**, 405.
Corzine, S., Geels, R. S., Yan, R. H., Scott, J. W. and Coldren, L. A. (1989). *IEEE Photon. Tech. Lett.*, **1**, 52.

Corzine, S. W., Yan, R. H. and Coldren, L. A. (1990). *Appl. Phys. Lett.*, **57**, 2835.
Corzine, S. W. and Coldren, L. A. (1991). *Appl. Phys. Lett.*, **59**, 588.
Dodson, B. W. and Tsao, J. Y. (1987). *Appl. Phys. Lett.*, **51**, 1325.
Eng, L. E., Chen, T. R., Sanders, S., Zhuang, Y. H., Zhao, B. and Yariv, A. (1989). *Appl. Phys. Lett.*, **55**, 1378.
Evans, G. A., Bour, D. P., Carlson, N. W., Hammer, J. M., Lurie, M., Butler, J. K., Palfrey, S. L., Amantea, R., Carr, L. A., Hawrylo, F. Z., James, E. A., Kirk, J. B., Liew, S. K. and Reichert, W. F. (1989). *Appl. Phys. Lett.*, **55**, 2721.
Evans, G. A., Carlson, N. W., Bour, D. P., Luie, M., Defreez, R. K. and Bossert, D. J. (1990). *Electron. Lett.*, **26**, 1380.
Feketa, D., Chan, K. T., Ballantyne, J. M. and Eastman, L. F. (1986). *Appl. Phys. Lett.*, **49**, 1659.
Fischer, S. E., Waters, R. G., Fekete, D. and Ballantyne, J. M. (1989). *Appl. Phys. Lett.*, **54**, 1861.
Frank, F. C. and van der Merwe, J. H. (1949). *Proc. Roy. Soc.* (London) **A198**, 216.
Fritz, I. J., Picraux, S. T., Dawson, L. R., Drummond, T. J., Laidig, W. D. and Anderson, N. G. (1985). *Appl. Phys. Lett.*, **46**, 967.
Fu, R. J., Hong, C. S., Chan, E. Y., Booher, D. J. and Figueroa, L. (1991). *Photon. Tech. Lett.*, **3**, 308.
Gavini, A. and Cardona, M. (1970). *Phys. Rev. B*, **1**, 672.
Geels, R. S., Corzine, S. W., Scott, J. W., Young, D. B. and Coldren, L. A. (1990). *IEEE Photon. Tech. Lett.*, **2**, 234.
Gershoni, D. and Temkin, H. (1989). *J. Luminescence*, **44**, 381.
Goodhue, W. D., Donnelly, J. P., Wang, C. A., Lincoln, G. A., Rauschenbach, K., Bailey, R. J. and Johnson, G. D. (1991). *Appl. Phys. Lett.*, **59**, 632.
Harnagel, G. L., Paoli, T. L., Thornton, R. L., Burnham, R. D. and Smith, D. L. (1985). *Appl. Phys. Lett.*, **46**, 118.
Hashimoto, J., Katsuyama, T., Shinkai, J., Yoshida, I. and Hayashi, H. (1991). *Appl. Phys. Lett.*, **58**, 879.
Huang, K. F., Tai, K., Chu, S. N. G. and Cho, A. Y. (1989). *Appl. Phys. Lett.*, **54**, 2026.
Hughes, N. A., Connolly, J. C., Gilbert, D. B. and Murphy, K. B. (1992). *Photon. Tech. Lett.*, **4**, 113.
Jones, E. D., Ackermann, H., Schirber, J. E. and Drummond, T. J. (1985). *Solid State Comm.*, **55**, 525.
Katsuyama, T., Yoshida, I., Shinkai, J., Hashimoto, J. and Hayashi, H. (1990). *Electron. Lett.*, **26**, 1375.
Kirkby, P. A., Goodwin, A. R., Thompson, G. H. B. and Selway, P. R. (1977). *IEEE J. Quantum Electron.*, **QE-13**, 705.
Koren, U., Oron, M., Young, M. G., Miller, B. I., DeMiguel, J. L., Raybon, G. and Chien, M. (1990). *Electron. Lett.*, **26**, 465.
Laidig, W. D., Caldwell, P. J., Lin, Y. F. and Peng, C. K. (1984). *Appl. Phys. Lett.*, **44**, 653.
Laidig, W. D., Lin, Y. F. and Caldwell, P. J. (1985). *J. Appl. Phys.*, **57**, 33.
Larsson, A., Cody, J. and Lang, R. J. (1989). *Appl. Phys. Lett.*, **55**, 2268.
Lee, Y. H., Jewell, J. L., Scherer, A., McCall, S. L., Walker, S. J., Harbison, J. P. and Florez, L. T. (1989). *Electron. Lett.*, **25**, 1377.
Lei, C., Rogers, T. J., Deppe, D. G. and Streetman, B. G. (1991). *Appl. Phys. Lett.*, **58**, 1122.
Ludowise, M. J., Dietze, W. T., Lewis, C. R., Camras, M. D., Holonyak, N. Jr, Fuller, B. K. and Nixon, M. A. (1983). *Appl. Phys. Lett.*, **42**, 487.

Madelung, O. (ed.) (1982). *Numerical Data and Functional Relationships in Science and Technology*, Group III, vol. 17, Berlin: Springer-Verlag.
Maeda, M. W., Chang-Hasnain, C., Von Lehmen, A., Izadpanah, H., Lin, C., Iqbal, M. Z., Florez, L. and Harbison, J. (1991). *IEEE Photon. Tech. Lett.*, **3**, 863.
Major, J. S. Jr, Guido, L. J., Hsieh, K. C., Holonyak, N. Jr, Stutius, W., Gavrilovic, P. and Williams, J. E. (1989). *Appl. Phys. Lett.*, **54**, 913.
Major, J. S. Jr, Guido, L. J., Holonyak, N. Jr, Hsieh, K. C., Vesely, E. J., Nam, D. W., Hall, D. C. and Baker, J. E. (1990). *J. Electronic Materials*, **19**, 59.
Major, J. S. Jr, Mehuys, D., Welch, D. F. and Scifres, D. R. (1991). *Appl. Phys. Lett.*, **59**, 2210.
Matthews, J. W. and Blakeslee, A. E. (1974). *J. Crystal Growth*, **27**, 118.
Mawst, L. J., Botez, D., Roth, T. J. and Peterson, G. (1989). *Appl. Phys. Lett.*, **55**, 10.
McIlroy, P. W. A., Kurobe, A. and Uematsu, Y. (1985). *J. Quant. Electron.*, **21**, 1958.
Melman, P., Elman, B., Jagannath, C., Koteles, E. S., Silletti, A. and Dugger, D. (1989). *Appl. Phys. Lett.*, **55**, 1436.
Nam, D. W., Waarts, R. G., Welch, D. F., Major, J. S. Jr and Scifres, D. R. (1992). Unpublished data.
Niki, S., Lin, C. L., Chang, W. S. C. and Wieder, H. H. (1989). *Appl. Phys. Lett.*, **55**, 1339.
Nitta, K., Itaya, K., Nishikawa, Y., Ishikawa, M., Okajima, M. and Hatakoshi, G. (1991). *Appl. Phys. Lett.*, **59**, 149.
O'Reilly, E. P. (1989). *Semicond. Sci. Technol.*, **4**, 121.
O'Reilly, E. P., Jones, G., Ghiti, A. and Adams, A. R. (1991). *Electron. Lett.*, **27**, 1417.
Offsey, S. D., Schaff, W. J., Tasker, P. J. and Eastman, L. F. (1990). *IEEE Photon. Tech. Lett.*, **2**, 9.
Okayasu, M., Fukuda, M., Takeshita, T. and Uehara, S. (1990). *Photon. Tech. Lett.*, **2**, 689.
Olsen, G. H., Nuese, C. J. and Smith, R. T. (1978). *J. Appl. Phys.*, **49**, 5523.
Olshansky, R., Su, C. B., Manning, J. and Powazinik, W. (1984). *J. Quantum Electron.*, **QE-20**, 838.
Orenstein, M., Kapon, E., Stoffel, N. G., Harbison, J. P., Florez, L. T. and Wullert, J. (1991). *Appl. Phys. Lett.*, **58**, 804.
Osbourn, G. C., Gourley, P. L., Fritz, I. J., Biefeld, R. M., Dawson, L. R. and Zipperian, T. E. *Semiconductor and Semimetals*, R. K. Willardson and A. C. Beer (eds.) (Academic Press, New York, 1987), vol. 24, 459.
Osinski, J. S., Grodzinski, P., Zou, Y. and Dapkus, P. D. (1991). *Electron. Lett.*, **27**, 469.
Osinski, J. S., Zou, Y., Grodzinski, P., Mathur, A. and Dapkus, P. D. (1992). *Photon. Tech. Lett.*, **4**, 10.
Paoli, T. L. (1977). *IEEE J. Quantum Electron.*, **QE-13**, 662.
People, R. and Bean, J. C. (1985). *Appl. Phys. Lett.*, **47**, 322.
Pikus, G. E. and Bir, G. L. (1959). *Sov. Phys. Solid State*, **1**, 136.
Pikus, G. E. and Bir, G. L. (1960). *Sov. Phys. Solid State*, **1**, 1502.
Raja, M. Y. A., Brueck, S. R. J., Osinski, M., Schaus, C. F., McInerney, J. G., Brennan, T. M. and Hammons, B. E. (1989). *IEEE J. Quantum Electron.*, **25**, 1500.

Richard, T. A., Kish, F. A., Holonyak, N. Jr, Dallesasse, J. M., Hsieh, K. C., Ries, M. J., Gavrilovic, P., Meehan, K. and Williams, J. E. (1991). *Appl. Phys. Lett.*, **58**, 2390.
Saint-Cricq, B., Bonnefont, S., Azoulay, R. and Dugrand, L. (1991). *Electron. Lett.*, **27**, 865.
Schirber, J. E., Fritz, I. J. and Dawson, L. R. (1985). *Appl. Phys. Lett.*, **46**, 187.
Scifres, D. R., Welch, D. F., Craig, R. R., Zucker, E., Major, J. S., Harnagel, G. L., Sakamoto, M., Haden, J. M., Endriz, J. G. and Kung, H. (1992). *Proc. SPIE–Int. Soc. Opt. Eng.*, **1634**, 192.
Serreze, H. B. and Chen, Y. C. (1991). *Photon. Tech. Lett.*, **3**, 397.
Shiau, T. H., Sun, S., Schaus, C. F., Zheng, K. and Hadley, G. R. (1990). *IEEE Photon. Tech. Lett.*, **2**, 534.
Shieh, C., Mantz, J., Lee, H., Ackley, D. and Engelmann, R. (1989a). *Appl. Phys. Lett.*, **54**, 2521.
Shieh, C., Lee, H., Mantz, J., Ackley, D. and Engelmann, R. (1989b). *Electron. Lett.*, **25**, 1226.
Streifer, W., Burnham, R. D. and Scifres, D. R. (1982). *IEEE J. Quantum Electron.*, **QE-18**, 856.
Stutius, W., Gavrilovic, P., Williams, J. E., Meehan, K. and Zarrabi, J. H. (1988). *Electron. Lett.*, **24**, 1494.
Takeshita, T., Okayasu, O. and Uehara, S. (1990). *Photon. Tech. Lett.*, **2**, 849.
Tanbun-Ek, T., Logan, R. A., Olsson, N. A., Temkin, H., Sergent, A. M. and Wecht, K. W. (1990). *Appl. Phys. Lett.*, **57**, 224.
Temkin, H., Tanbun-Ek, T. and Logan, R. A. (1990). *Appl. Phys. Lett.*, **56**, 1210.
Temkin, H., Tanbun-Ek, T., Logan, R. A., Cebula, D. A. and Sergent, A. M. (1991). *Photon. Tech. Lett.*, **3**, 100.
Thijs, P. J. A. and van Dongen, T. (1989). *Electron. Lett.*, **25**, 1735.
Thijs, P. J. A., Tiemeijer, L. F., Kuindersma, P. I., Binsma, J. J. M. and van Dongen, T. (1991). *J. Quant. Elect.* **27**, 1426.
Tiemeijer, L. F., Thijs, P. J. A., de Waard, P. J., Binsma, J. J. M. and van Dongen, T. (1991). *Appl. Phys. Lett.*, **58**, 2738.
Tothill, J. N., Westbrook, L., Hatch, C. B. and Wilkie, J. H. (1989). *Electron. Lett.*, **25**, 580.
Tsang, W. T. (1981). *Appl. Phys. Lett.*, **38**, 661.
van der Ziel, J. P. and Chand, N. (1991). *Appl. Phys. Lett.*, **58**, 1437.
Vawter, G. A. and Meyers, D. R. (1989). *J. Appl. Phys.*, **65**, 4769.
Wang, C. A. and H. K. Choi (1991a). *J. Quant. Elect.* **27**, 681.
Wang, C. A., Walpole, J. N., Choi, H. K. and Missaggia, L. J. (1991b). *Photon. Tech. Lett.*, **3**, 4.
Waters, R. G. and Bertaska, R. K. (1988). *Appl. Phys. Lett.*, **52**, 179.
Waters, R. G., Bour, D. P., Yellen, S. L. and Ruggieri, N. F. (1990a). *IEEE Photon. Tech. Lett.*, **2**, 531.
Waters, R. G., York, P. K., Beernink, K. J. and Coleman, J. J. (1990b). *J. Appl. Phys.*, **67**, 1132.
Welch, D. F., Steifer, W., Schaus, C. F., Sun, S. and Gourley, P. L. (1990). *Appl. Phys. Lett.*, **56**, 10.
Welch, D. F. and Scifres, D. R. (1991). *Electron. Lett.*, **27**, 1915.
Weng, S. L. (1989). *J. Appl. Phys.*, **66**, 2217.
Wilcox, J. Z., Peterson, G. L., Ou, S., Yang, J. J., Jansen, M. and Schechter, D. (1988). *J. Appl. Phys.*, **64**, 6564.
Williams, R. L., Dion, M., Chatenoud, F. and Dzurko, K. (1991). *Appl. Phys. Lett.*, **58**, 1816.

Yablonovitch, E. and Kane, E. O. (1986). *J. Lightwave Technol.*, **LT-4**, 504.
Yablonovitch, E. and Kane, E. O. (1988). *J. Lightwave Technol.*, **LT-6**, 1292.
Yao, J. Y., Andersson, T. G. and Dunlop, G. L. (1988). *Appl. Phys. Lett.*, **53**, 1420.
Yellen, S. L., Waters, R. G., Chen, Y. C., Soltz, B. A., Fischer, S. E., Fekete, D. and Ballantyne, J. M. (1990). *Electron. Lett.*, **26**, 2083.
Yellen, S. L., Waters, R. G., York, P. K., Beernink, K. J. and Coleman, J. J. (1991). *Electron. Lett.*, **27**, 552.
Yoo, H. J., Scherr, A., Harbison, J. P., Florez, L. T., Paek, E. G., van der Gaag, B. P., Hayes, J. R., Von Lehmen, A., Kapon, E. and Kwon, Y. S. (1990). *Appl. Phys. Lett.*, **56**, 1198.
York, P. K., Beernink, K. J., Fernandez, G. E. and Coleman, J. J. (1989a). *Appl. Phys. Lett.*, **54**, 499.
York, P. K., Beernink, K. J., Kim, J., Coleman, J. J., Fernandez, G. E. and Wayman, C. M. (1989b). *Appl. Phys. Lett.*, **55**, 2476.
York, P. K., Langsjoen, S. M., Miller, L. M., Beernink, K. J., Alwan, J. J. and Coleman, J. J. (1990a). *Appl. Phys. Lett.*, **57**, 843.
York, P. K., Beernink, K. J., Fernandez, G. E. and Coleman, J. J. (1990b). *Semiconductor Sci. and Tech.*, **5**, 508.
York, P. K., Beernink, K. J., Kim, J., Alwan, J. J., Coleman, J. J. and Wayman, C. M. (1991). *J. Crystal Growth*, **107**, 741.
Zah, C. E., Bhat, R., Favire, F. J. Jr, Menocal, S. G., Andreadakis, N. C., Cheung, K. W., Hwang, D. M. D., Koza, M. A. and Lee, T. P. (1991). *J. Quant. Elect.*, **27**, 1440.
Zah, C. E., Bhat, R., Pathak, B., Caneau, C., Favire, F. J., Andreadakis, N. C., Hwang, D. M., Koza, M. A., Chen, C. Y. and Lee, T. P. (1991b). *Electron. Lett.*, **27**, 1414.
Zmudzinski, C., Mawst, L. J., Botez, D., Tu, C. and Wang, C. A. (1992). *Electron. Lett.*, **28**, 1543.
Zory, P. S., Reisinger, A. R., Mawst, L. J., Costrini, G., Zmudzinski, C. A., Emanuel, M. A., Givens, M. E. and Coleman, J. J. (1986). *Electron. Lett.*, **22**, 475.
Zou, W. X., Chuang, Z. M., Law, K-K., Dagli, N., Coldren, L. A. and Merz, J. L. (1991). *J. Appl. Phys.*, **69**, 2857.
Zou, Y., Grodzinski, P., Menu, E. P., Jeong, W. G., Dapkus, P. D., Alwan, J. J. and Coleman, J. J. (1991). *Appl. Phys. Lett.*, **58**, 601.

9

Vertical cavity surface-emitting laser arrays

CONNIE J. CHANG-HASNAIN

9.1 Introduction

Semiconductor diode lasers emitting normal to the substrate plane, known as surface-emitting lasers, are extremely promising for addressing a range of applications from optical interconnects, optical communications and optical recording to remote sensing. The most promising aspect perhaps lies in the prospect of eliminating low yield laser fabrication steps, i.e. laser packaging processing including wafer lapping, cleaving and dicing, facet coatings and diode bonding. The possibility of being able to make any number of lasers anywhere on a wafer is also an increasingly important factor for applications such as optical interconnects. At present two completely different approaches are aimed at realizing surface-emitting lasers. The first represents an extension of the existing technology for semiconductor edge-emitting lasers that uses a 45° slanted mirror[1] or a second-order grating[2] to vertically couple the light out (Figure 9.1(1). (2)). The second, pioneered by K. Iga in 1979[3], uses highly reflective mirrors to clad the active region, resulting in a vertical cavity that produces an output beam propagating normal to the substrate surface (Figure 9.1(3)).

The vertical cavity design offers important advantages over other surface-emitting laser designs. The unique topology of a vertical cavity facilitates large-scale processing, on-wafer testing and pre-process screening. The small lateral dimensions allow for fabrication of large 2-D arrays with high packing density and integration with other optical and electronic devices. The vertical cavity lasers emit circular low divergence output beams, resulting in simpler and more efficient optical coupling with bulk optics and optical fibers. In addition, the active volume can potentially be scaled down to very small size (10^{-13}–10^{-14} cm^{-3}) such that extremely low threshold current (μA or lower) is potentially obtainable.

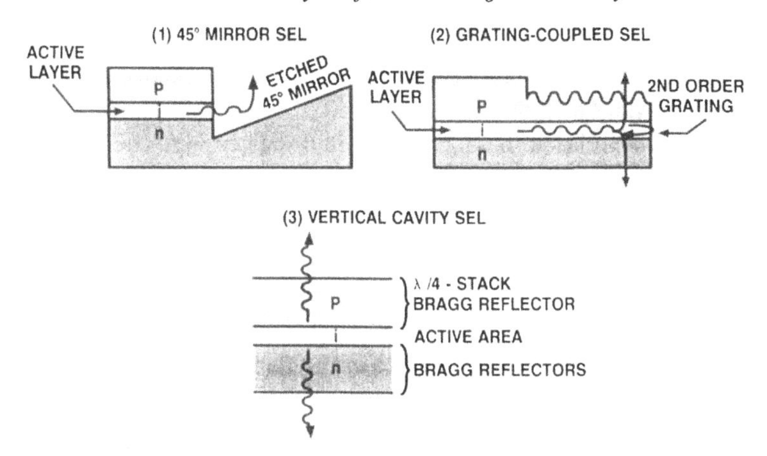

Figure 9.1. Three major approaches used to obtain laser emission in the direction normal to the substrate: (1) using a 45° slanted mirror; (2) a second-order grating; (3) a vertical cavity configuration.

Earlier work on vertical cavity lasers[4–6] used a combination of dielectric mirrors and epitaxially grown semiconductor mirrors. Most of the lasers had bulk active regions. The thick active region which provides higher one-pass gain was necessary because the mirror reflectivity achievable then was not sufficiently high. Recently, with the advances of epitaxy technologies, particularly in the control of thickness and composition of epitaxy, semiconductor Bragg reflectors with extremely high reflectivity can be obtained. This has resulted in the realization and the first demonstration of a continuous wave (cw), room-temperature operated, low-threshold-current vertical cavity laser[7]. The laser structure involves a quantum well active region sandwiched inbetween n- and p-doped semiconductor Bragg reflectors. With this VCSEL configuration, the laser mirrors are epitaxially grown and, thus, have neither defects nor dangling bonds as regular cleaved facets typically do. Thus, a much reduced mirror degradation rate at higher optical output, and hence longer laser lifetime, is expected. Triggered by this demonstration, rapid and exciting progress has been made in threshold current reduction[8], wavelength tuning[9], monolithic integration with a photodetector[10] and with a thyristor[11]. In the meantime, many advances have been reported on 2-D vertical cavity laser arrays including a novel multiple wavelength array[12,13], phase-locked 2-D arrays[14,15] and large addressable arrays[16–18]. Demonstrations in systems experiments using VCSEL arrays have also been performed for wavelength-division-multiplexed (WDM) optical communications[19,20] and board-to-board optical interconnections[18].

In this chapter, we will focus mostly on the recent advances in 2-D VCSEL arrays. The chapter is organized as follows. The general guideline of a vertical cavity laser design is discussed in Section 9.2. The fabrication process and typical laser characteristics are discussed in Sections 9.3 and 9.4. Novel and highly functional VCSEL arrays are described in Sections 9.5 and 9.6. Section 9.7 discusses the laser array addressing and packaging issues. Finally, future prospects are discussed.

9.2 Vertical cavity surface-emitting laser design

A generic VCSEL heterostructure is illustrated in Figure 9.2. The structure can be divided into three components: the top distributed Bragg reflector (DBR), the bottom DBR, and the center spacer.

The upper and lower DBRs differ only in the last layers due to different output boundary conditions. The Bragg reflectors consist of alternating layers of quarter-wave-thick high and low refractive index epitaxial materials, e.g. (Al)GaAs and AlAs. Figure 9.3 shows a calculated reflectivity spectrum for a plane wave ($\lambda = 0.98\ \mu m$) entering from a semi-infinite GaAs material into a 20-pair AlAs/GaAs DBR and exiting into air. The reflectivity at the designed wavelength, 0.98 μm, is as high as 0.99 with a broad bandwidth of 100 nm due to the relatively large refractive index difference. The peak reflectivity as a function of the number of DBR pairs is depicted by Figure 9.4 with air and GaAs, respectively, as exiting media.

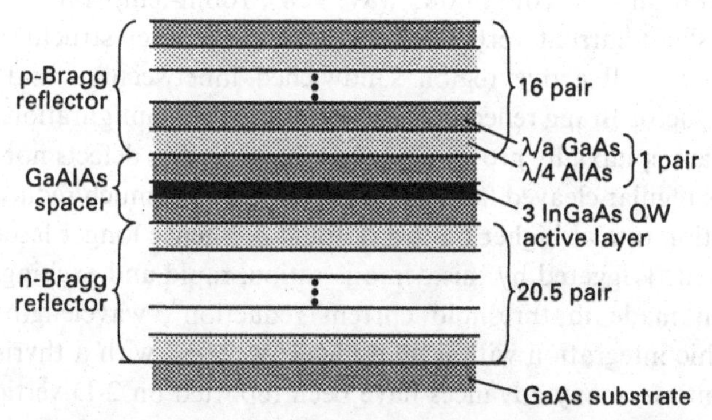

Figure 9.2. The schematic of a typical bottom-emitting VCSEL design consisting of a 20.5-pair n-doped quarter-wave GaAs/AlAs stack as the lower DBR, three 80 Å $In_{0.2}Ga_{0.8}As$/GaAs strained quantum wells in the center of a one-wave $Ga_{0.5}Al_{0.5}As$ spacer, and a 16-pair p-doped upper GaAs/AlAs DBR, grown on an n^+-GaAs substrate. The topmost layer is a half-wave-thick GaAs layer to provide phase matching for the metal contact. (Ref. 33, © 1991 IEEE.)

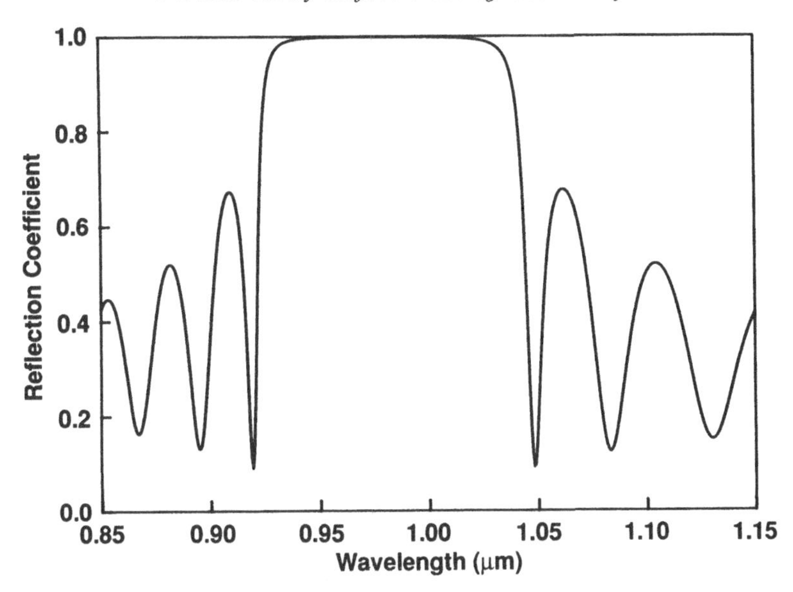

Figure 9.3. Calculated reflectivity spectrum for a plane wave ($\lambda = 0.98$ μm) entering from a semi-finite GaAs material into a 20-pair AlAs/GaAs DBR and exiting into air.

For the same reflectivity, the former requires many fewer pairs due to the large refractive index difference between GaAs and air.

A VCSEL emitting from the epitaxy side (top-emitting) has air and GaAs as the boundaries for its upper and lower DBRs, respectively, whereas a VCSEL with its emission through the back of the substrate (bottom-emitting) may have metal and GaAs, respectively, as the boundaries. An appropriate phase matching layer or a half-wave layer may be used to match the boundary condition with metal. The bottom DBRs for both types of VCSEL usually have an integer number of pairs plus half a pair.

The optical thickness of the center spacer is an integer multiple of one wavelength. Thus, the maximum of the optical field of the Fabry–Perot mode is located at the center where the quantum well active layers are, and the gain in the quantum wells can be used more efficiently due to a larger confinement factor (overlap of optical intensity profile and gain). Figure 9.5 shows the optical intensity of the Fabry–Perot mode and the refractive index as a function of distance in a typical VCSEL. The effective cavity length taken at $1/e$ points of the optical intensity profile is as short as 1 μm. The center spacer typically includes an AlGaAs region with an

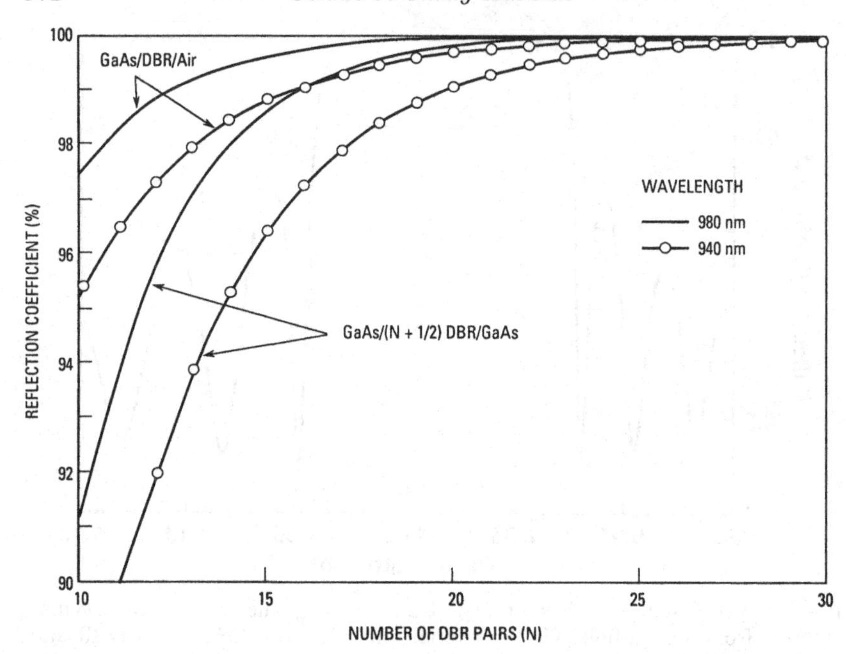

Figure 9.4. Peak reflectivity as a function of the number of DBR pairs with air and GaAs as exiting media.

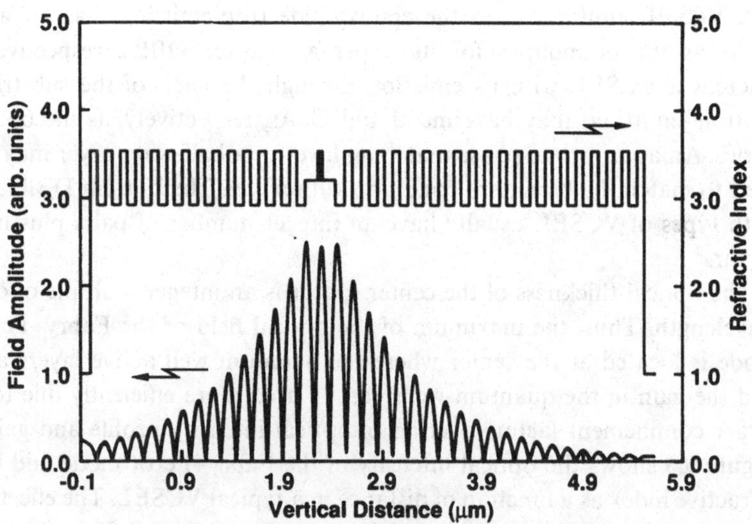

Figure 9.5. Optical intensity of the Fabry–Perot mode and the refractive index as a function of distance in a typical VCSEL.

active region consisting of quantum wells at its center. A graded refractive index separate confinement heterostructure (GRIN–SCH) around the quantum well region is often used for better carrier confinement but at the expense of a lower optical confinement factor.

Although the parameters of the VCSELs are at an extreme, e.g. ultra-short cavity length and ultra-high mirror reflectivity, the design criteria are no different from those of edge-emitting lasers. Hence, the threshold conditions and general characteristics are, as for edge-emitting lasers, obtained from the wave equation and rate equations.

The barriers at the heterojunctions of the Bragg reflectors contribute to significant increase in drive voltage and series resistance, which create a serious heating problem. To reduce the voltage and resistance, step or graded layers are placed at the heterojunctions[8]. Although effective, this measure has not eliminated the problem completely.

9.3 VCSEL array fabrication

There are a number of variations on the techniques used to fabricate VCSELs[7–10,21–24]. The transverse optical guiding mechanisms are either gain- or index-guided, identical to those for edge-emitting layers. The index-guided structure typically involves etching laser-posts (Figure 9.6). Such nonplanar structures are more difficult for array fabrication. The highly index-guided post-like VCSELs also tend to emit multiple transverse modes[25]. The gain-guided VCSELs, on the other hand, are made by providing current isolation to the laser surroundings. The isolation can

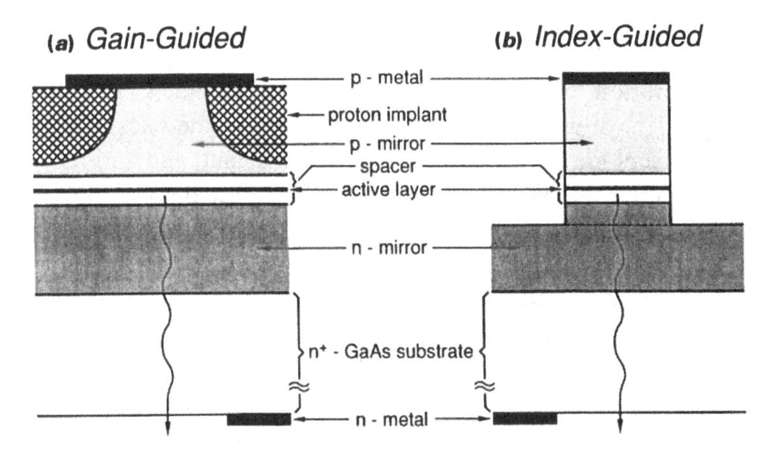

Figure 9.6. Device schematic of a (*a*) gain- and (*b*) index-guided VCSEL. (Ref. 25.)

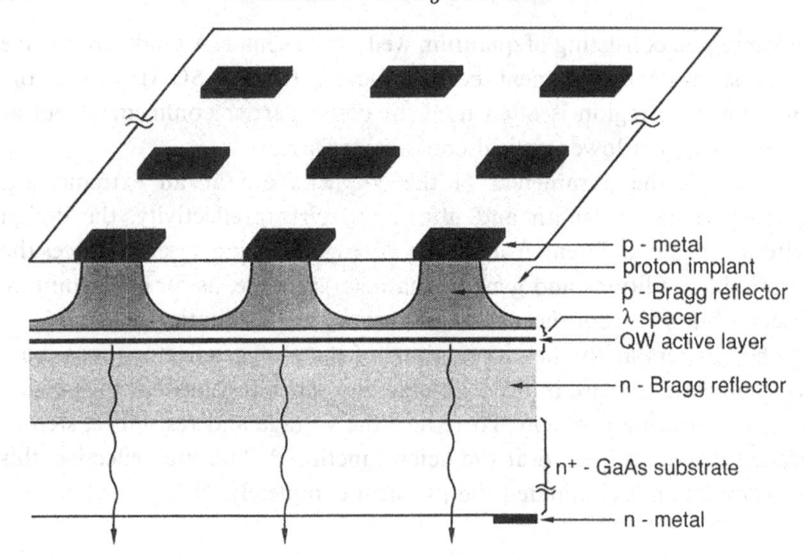

Figure 9.7. The schematic of a 2-D gain-guided VCSEL array. (Ref. 33, © 1991 IEEE.)

easily be provided by damaging the material with ion implantation, such as oxygen[21] or proton[22] implants. These lasers are easy to fabricate with high yield and are therefore suitable for arrays. In addition, at low current levels, a gain-guided VCSEL emits a single pure transverse mode. The modal evolution is well behaved, as will be discussed in Section 9.4.1. This also makes them more preferable for systems prototype experiments.

Figure 9.7 shows the schematic of a 2-D gain-guided, bottom-emitting VCSEL array. The processing steps[12] include evaporation of the p-contacts, photolithography to mask over the p-contacts, proton implantation, evaporation of contact pads over the p-contacts, polishing of the back of the substrate, n-metal evaporation on the back side of the substrate except for the area where the VCSELs emit, and anti-reflection (AR) coating on the back side of the substrate over the windows where the VCSELs emit. The proton implantation energies and dosages vary depending on the specific VCSEL design and performance requirement. The VCSELs fabricated with deeper implant at higher energies exhibit lower threshold currents and voltages[24]. At this stage, the lasers can be probed and tested independently. An additional mask such as that shown in Figure 9.8 can be evaporated on the wafer to provide contact pads for bonding. Figure 9.8 is a photograph of a fabricated 8 × 8 independently addressable VCSEL array[17].

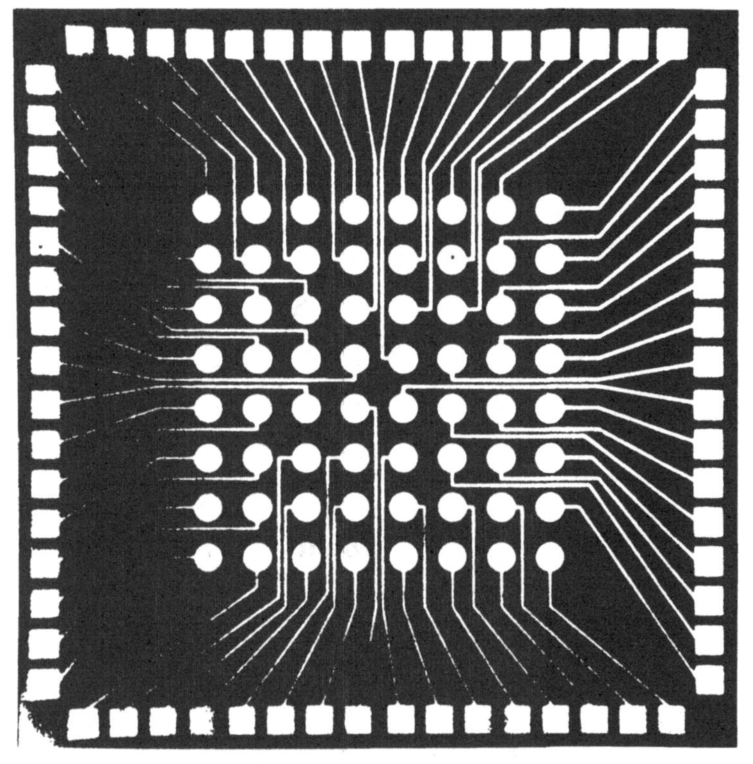

Figure 9.8. Photograph of a fabricated 8 × 8 independently addressable VCSEL array. (Ref. 17.)

For high speed or high power operation, the VCSELs can be mounted junction-side down on a metallized BeO substrate with In solder bumps, which is subsequently put into an appropriate package. The flip-chip bonding requires much thicker p-side metal, which is typically done by Au plating, and a better electrical isolation between the devices, which is done by deep etching between the lasers. Figure 9.9 shows the schematic of a flip-chip mounted 2 × 8 VCSEL array and a photograph of such an array in a high speed package[26]. The performance of these packaged lasers will be discussed in Section 9.7.3.

For larger arrays, independently addressing becomes difficult. Using a matrix (also known as row–column) addressing scheme, N^2 contacts can be reduced to $2N$ contacts for an $N \times N$ array. The fabrication of a matrix addressable array[16], shown in Figure 9.10, requires the epitaxial material to be grown on a semi-insulating substrate, and some additional processing steps such as: etching grooves between the rows of lasers to isolate the

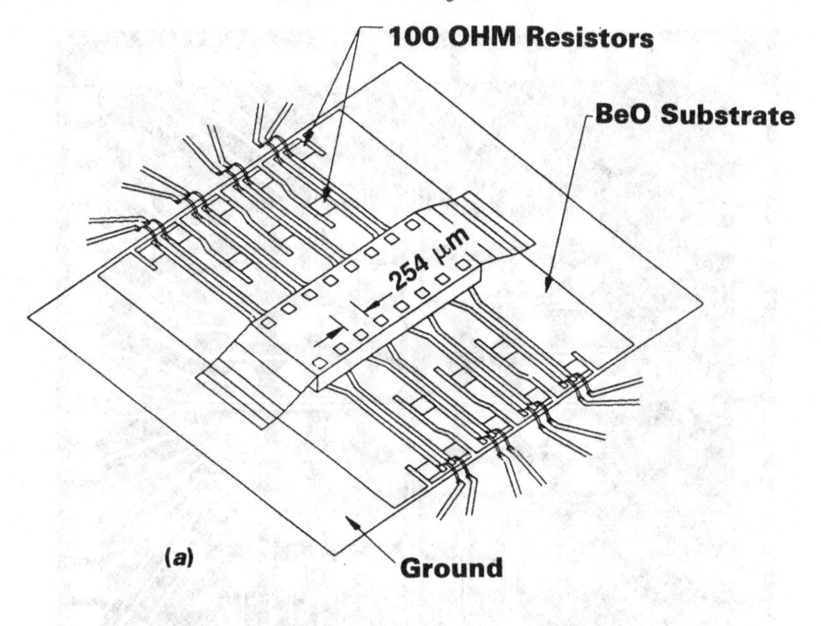

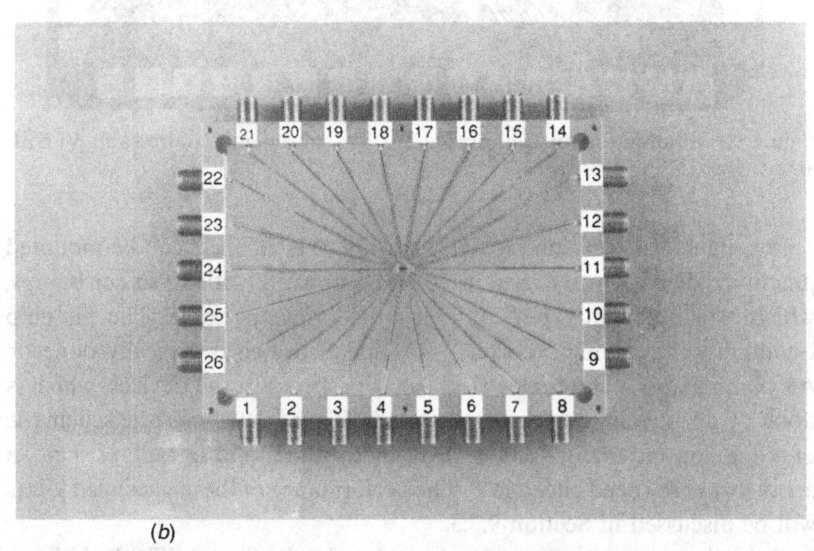

Figure 9.9. (a) Schematic of a 2×8 VCSEL array flip-chip mounted on to a metallized BeO substrate. The metallized pattern on the BeO substrate is matched to the spacings of the VCSEL array. Additional metallized lines are for wire bonding. The resistors are fabricated on the BeO substrate to provide impedance matching for the VCSELs. (Ref. 26, © 1991 IEEE.) (*b*) Photograph of a flip-chip 2×8 VCSEL array mounted in a high speed package. The laser spacings are 254 and 508 μm for the lasers in one row and one column respectively.

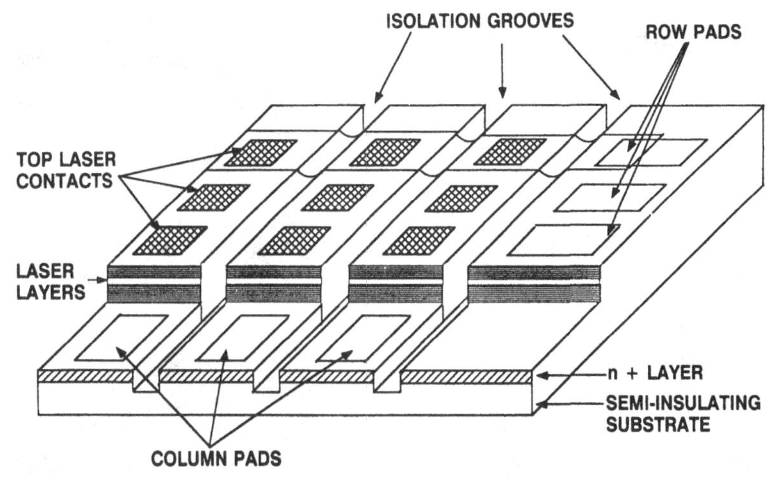

Figure 9.10. Schematic of a matrix addressable VCSEL array. The p-contacts of all the lasers in a row are connected, whereas the n-contacts of lasers in a column are connected. Thus, by applying bias to the ith row and jth column, the ijth laser is turned on. (Ref. 16a.)

p-doped layers between columns, etching the edges of each row down to the n-doped layers and evaporating the n-contacts for the rows, planarizing the wafer by filling the etched grooves with polyimide, and then evaporating p-metal to connect the p-contacts of the columns.

9.4 VCSEL characteristics

9.4.1 Transverse modes

The transverse modes of a gain-guided VCSEL are TEM modes. This can be understood because its transverse dimension is typically significantly larger than its effective cavity length. Thus, the complete bases of transverse modes are the same as for lasers irrespective of the shapes of their contacts. However, since the current/carrier distribution determines the gain profile, whose overlap with the TEM modes determines the modal gain for each transverse mode, the specific onset of a particular higher order mode may vary with the VCSEL shape.

The near-field patterns of a 15 μm square VCSEL at various cw current levels are shown in Figure 9.11. The laser starts to lase in the fundamental TEM_{00} mode, as shown in Figure 9.11(a). At higher currents, higher order modes successively appear due to gain spatial hole burning. The TEM_{01*} and TEM_{10} modes dominate for only a small range of drive currents (Figure 9.11(b, c)). At still higher cw current, the laser emits both TEM_{00}

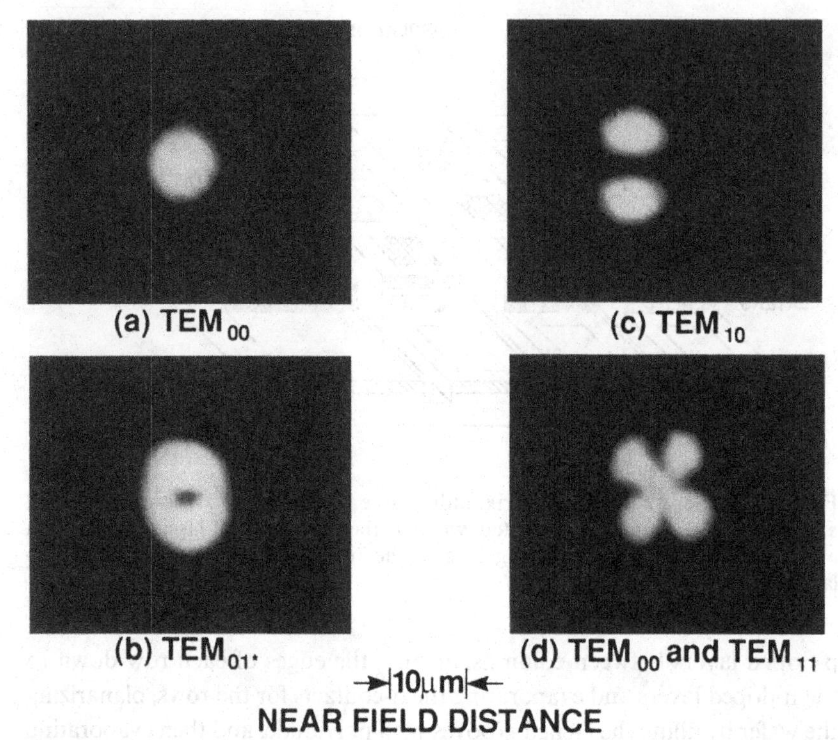

Figure 9.11. CW near-field patterns of a typical 15 μm square gain-guided VCSEL emitting a single (*a*) TEM_{00} mode; (*b*) TEM_{01*} mode; (*c*) TEM_{10} mode; and (*d*) TEM_{00} and TEM_{11} mode. (Ref. 25.)

and TEM_{11} (Figure 9.11(*d*)). In addition, the larger the VCSEL, the more modes it emits. A 5 μm square VCSEL emits only the TEM_{00} mode and a 10 μm × 5 μm rectangular VCSEL emits two modes, TEM_{00} and TEM_{10}. On the other hand, a 20 μm square laser supports transverse modes up to TEM_{22}. These observations are consistent with those found in well-behaved gain-guided edge-emitting lasers[27].

The optical spectrum of a VCSEL emitting a single TEM_{00} mode typically exhibits a sidemode suppression larger than 45 dB and a linewidth of about 100 MHz[28]. The wavelength separations between adjacent transverse modes are about 2–3 Å. Since the wavelength difference between TEM_{00} and TEM_{11} modes is larger, they can be separated using the spectrally resolved near-field measurements shown in Figure 9.12.

The transverse modes of VCSEL are linearly polarized in uncorrelated directions[29]. When two modes are excited simultaneously, they often exhibit orthogonal polarization directions due to unintentional inhomo

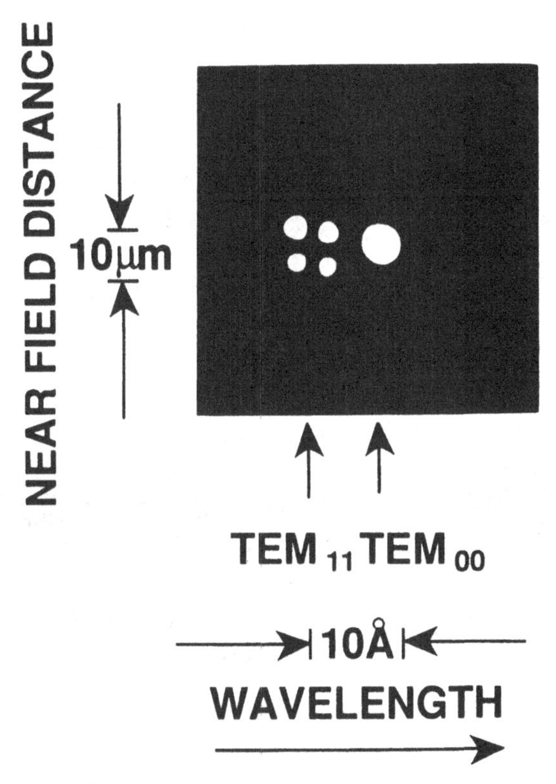

Figure 9.12. Spectrally resolved near-field image corresponding to Figure 9.11(*d*). A large input spectrometer slit was used in this case to allow the entire 2-D near-field image to pass through. The two modes are spatially separated due to their large wavelength difference. (Ref. 25.)

geneity in the gain or thermal distribution. Figure 9.13 shows the spectrally and polarization-resolved near-field patterns of a VCSEL at various cw current levels. Near threshold, the laser emits a single highly linearized polarized TEM_{00} mode. At higher current, when two transverse modes are excited simultaneously, the two modes tend to oscillate at two orthogonal polarizations. Not only the polarization for a particular mode is observed to vary randomly from one laser to the next, it also varies with drive current. At 1.8 I_{th}, the laser emits the donut-shaped mode TEM_{01*} (Figure 9.11(*b*)), whose spectrally resolved near field shows a single mode-like pattern (Figure 9.13(*e*)) limited by spectral resolution. However, with the use of a polarizer, this mode can be clearly separated into the TEM_{01} mode at 84° and the TE_{01} mode at 4° (Figure 9.13(*f*)). This demonstrates that the ring-shaped TEM_{01*} mode is indeed a superposition of TEM_{10} and TEM_{01} modes.

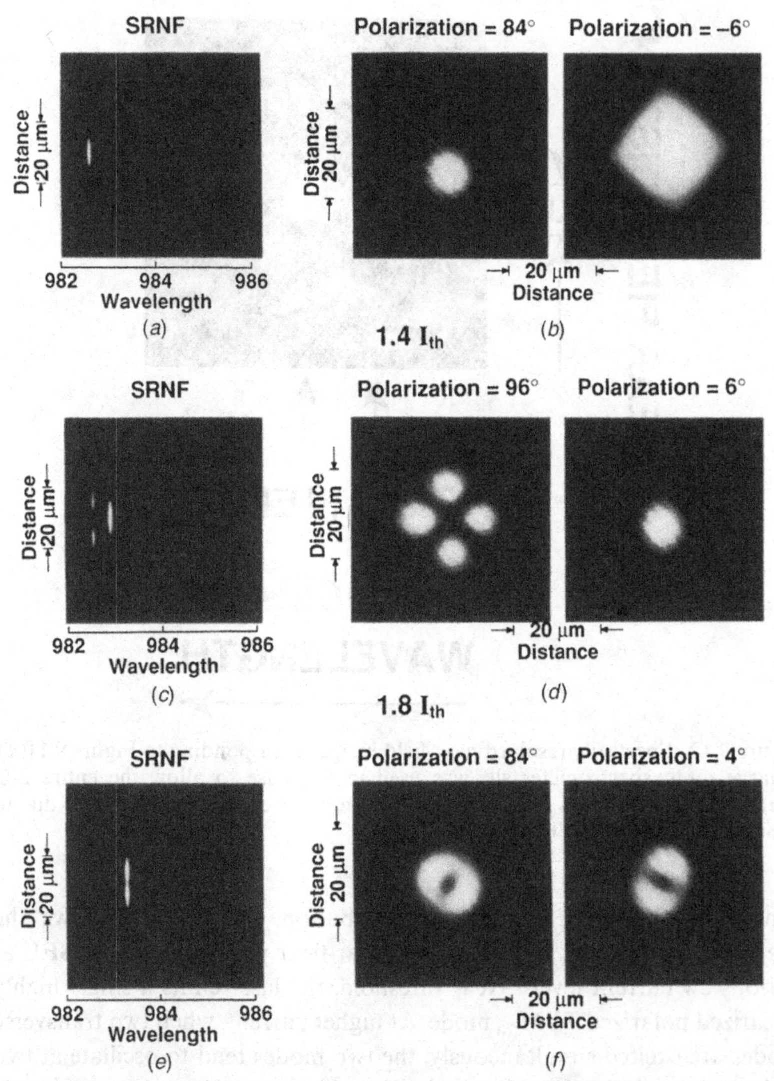

Figure 9.13. The spectrally and polarization-resolved near-field patterns for the same VCSEL at various cw current levels. (*a*) and (*b*) 1.05 I_{th}: the laser emits a single TEM_{00} mode. The near-field images are filtered through a polarizer with its angle set at 84° and −6° with respect to the [011] crystal axis. The two pictures in (*b*) are not aligned with each other. The intensity of the image at −6° is about 1000 times weaker than that at 84°. (*c*) and (*d*) 1.4 I_{th}: the near-field images filtered through a polarizer with angles set at 96° and 6°. The two transverse modes exhibit orthogonal polarizations. The TEM_{00} mode changed its polarization orientation with current from (*b*) to (*d*). (*e*) and (*f*) 1.8 I_{th}: near-field images filtered through a polarizer with angles set at 84° and 4°. (Ref. 29.)

9.4.2 Light and voltage vs current characteristics

The optical light output vs current (*L–I*) and voltage vs current (*V–I*) characteristics[30] of a top-emitting 10 μm diameter VCSEL under cw operation is shown by Figure 9.14. The drive current and voltage at the laser threshold for this particular laser are 4.3 mA and 4 V. This laser exhibits a fairly high electric power to optical power conversion efficiency (wall-plug efficiency) of 7.5%. It emits a single TEM_{00} mode up to ≈ 1 mW output power at 5 mA bias current, above which the second transverse mode is excited. This causes a mild kink to appear on the *L–I* curve. The *L–I* curve saturates due to heating[31], which is typically observed for VCSELs.

The *L–I* and *V–I* characteristics for a bottom-emitting VCSEL are ideally not very different from those of a top-emitting laser. The best reported wall-plug efficiency for this type of VCSEL is 17.3% for a 30 μm-diameter multimode VCSEL[32]. However, highly nonlinear *L–I* curves are often observed with these lasers due to the external feedback provided by polished substrate, which acts as the third mirror to the original optical cavity[33]. This additional feedback can be minimized by

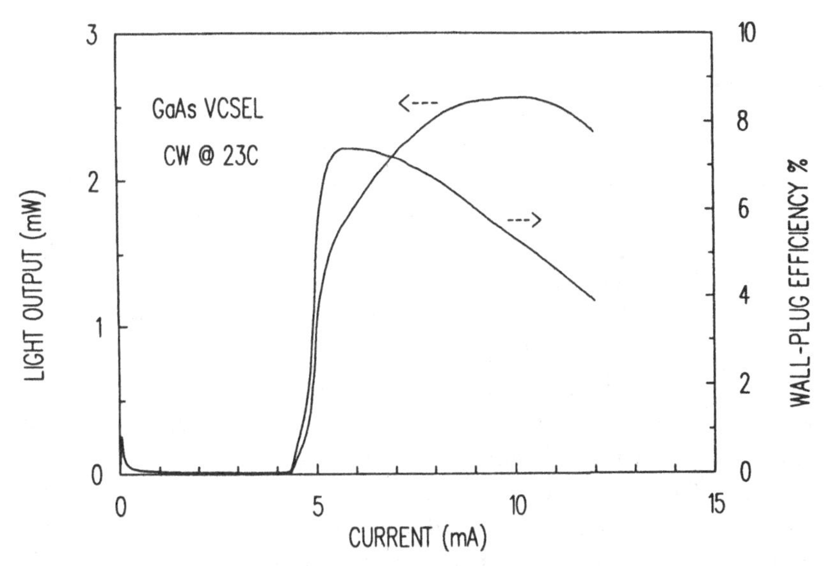

Figure 9.14. Light intensity versus current and voltage versus current characteristics for a top-emitting VCSEL with high wall-plug efficiency under cw operation. (Ref. 30.).

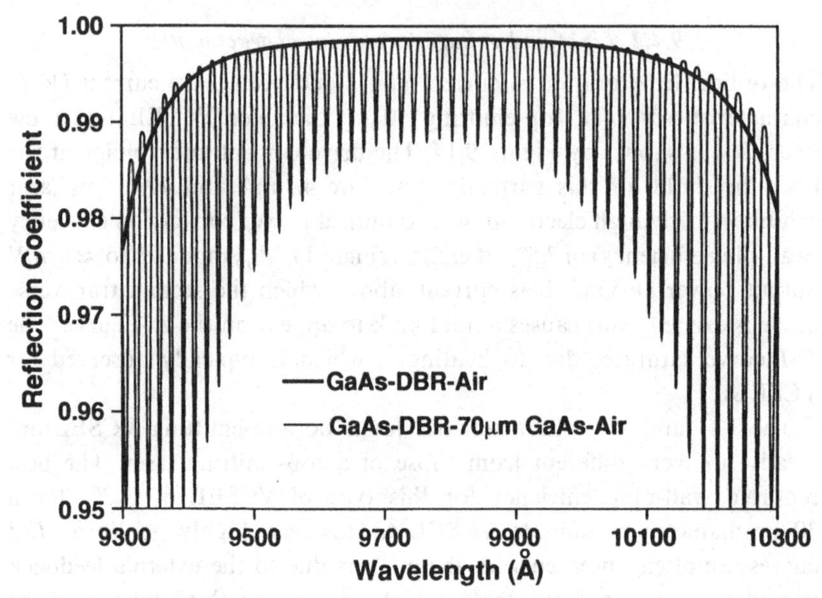

Figure 9.15. Calculated reflectivity spectra of a plane wave incident from GaAs on to two media. The dark line represents the case when the medium is a 20-pair DBR exiting into air, whereas the lighter line represents the case when the last quarter-wave GaAs layer before air is replaced by a 70 μm-thick GaAs substrate. (Ref. 33, © 1991 IEEE.)

using an antireflection coating on the back of the substrate. Conversely, this feedback can be enhanced to create a coupled-cavity tunable laser.

Figure 9.15 shows the calculated reflectivity spectra for a plane wave in GaAs (the λ-spacer) incident on to a simple 20.5 pair DBR (shown by the dark curve) as opposed to that incident on to a combination of the DBR, a 70 μm-thick GaAs substrate and a polished GaAs/air interface (shown by the ligher curve). A spectrum with periodic resonance is obtained for the latter case, instead of the smooth reflectivity spectrum obtained for the former. These resonances are caused by the reflection at the GaAs/air interface (M_3 in Figure 9.16). The spacing between the resonances is 19.4 Å for this case. It equals $\lambda^2/2nL$, where λ is the laser wavelength, n is the refractive index of GaAs and L is the substrate thickness. The amplitude of the resonances depends on the reflectivities of the DBR (M_2 in Figure 9.16) and M_3. The resonances can be enhanced or smoothed out with high or low reflectivity coating respectively. The mode selection mechanism (Figure 9.16) in a three-mirror coupled-cavity laser can be explained simply as follows: when the FP mode is at a

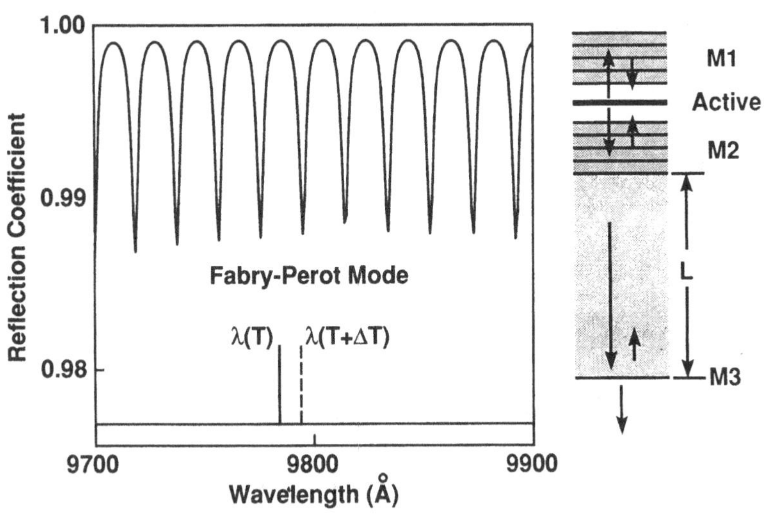

Figure 9.16. Fabry–Perot mode selection mechanism in a three-mirror VCSEL: M_1 and M_2 denote the top and bottom DBR respectively; M_3 indicates the substrate/air interface.

wavelength corresponding to high reflectivity regions of the combined mirror, the device will lase. As the drive current increases, the FP mode red shifts due to heating, whereas the reflectivity spectrum of M_2 is not shifted since the substrate is much cooler. When the FP mode red shifts into a wavelength corresponding to low reflectivity regions of M_2, since a VCSEL requires very high mirror reflectivity to lase, the mirror losses cannot be compensated by the gain, and the device stops lasing. Increasing the current further, the FP mode red shifts to the next higher reflectivity resonance, and the device lases again. The laser emission wavelength can therefore be tuned *discretely* with the drive current[33]. A unique and interesting on–off-type *L–I* characteristic is obtained[33] which may have other potential applications.

9.4.3 Time-resolved spectra

The cw and pulsed (300 ns at 5 kHz) spectra for a VCSEL fabricated with shallow implant and thus having high threshold voltage of 6.1 V and 100 Ω series resistance are shown by Figure 9.17[24]. The cw spectrum consists of a clean single mode with linewidth being resolution limited to 0.08 Å. The pulsed spectrum, on the other hand, is very broad with a 2 Å-wide full width half-maximum and is 28 Å shorter than the cw

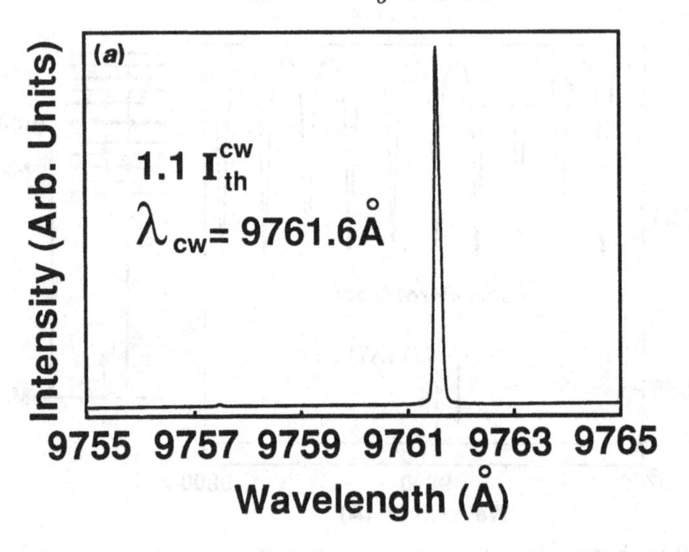

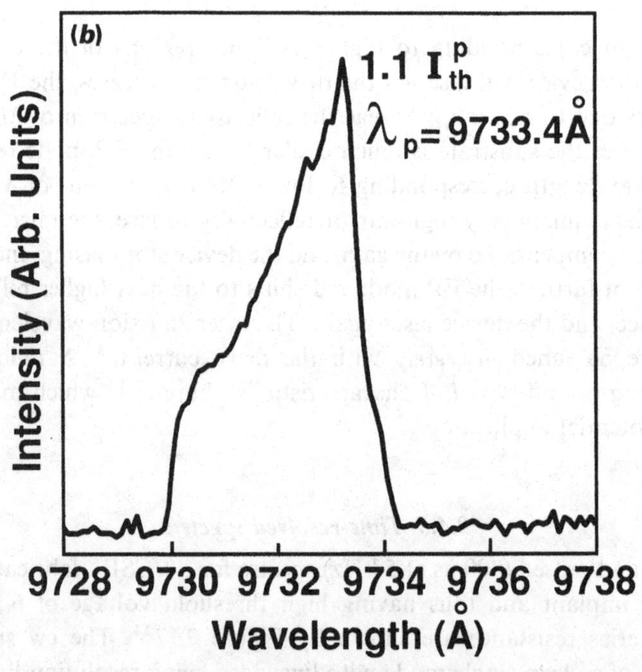

Figure 9.17. The (*a*) cw and (*b*) pulsed (300 ns at 5 kHz) lasing spectrum for a VCSEL with high voltage and series resistance at 1.1 times cw and pulse threshold current respectively. (Ref. 24.)

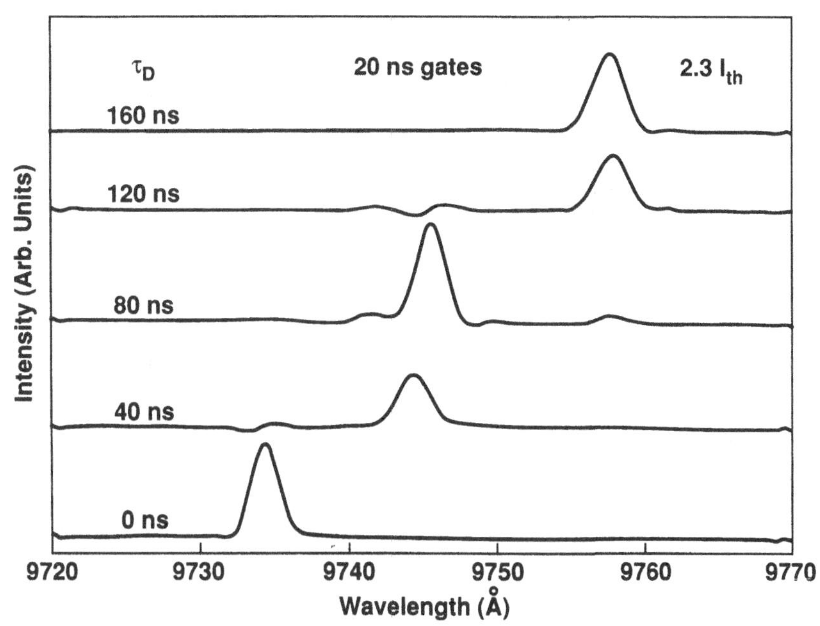

Figure 9.18. The time-resolved spectra of a VCSEL with high threshold voltage and resistance under pulse operation at 2.3 I_{th}. The spectra were measured with 20 ns gates positioned at the beginning of the light pulse, and with a delay τ_D of 40, 80, 120 and 160 ns. (Ref. 24.)

wavelength. The broadening of the pulsed spectrum and the large red shift for the laser under cw operation are caused by the high resistive heating close to the laser junction.

Time-resolved spectra of such a VCSEL under pulsed operation at 2.3 I_{th} is shown in Figure 9.18. The spectra were measured with 20 ns gates positioned at the beginning of the light pulse and with a delay τ_D of 40, 80, 120 and 160 ns. At $\tau_D = 0$, the laser spectrum is single-lobed and the peak wavelength is at $\approx$9733 Å. At $\tau_D = 40$ ns, the laser emission red shifts. The discontinuous red shift is due to the reflection from the substrate/air interface introducing destructive interference at some periodically separated wavelengths at which the laser stops lasing, as mentioned in the preceding section. The laser continues to red shift at $\tau_D = 120$ and 160 ns. The total red shift in 200 ns is as much as 23 Å. The total chirp depends considerably and monotonically on the currrent level and series resistance of the device.

For a VCSEL fabricated with deep implant and thus having reduced voltage of 2.6 V and resistance 70 Ω, its pulsed spectrum exhibits a clean

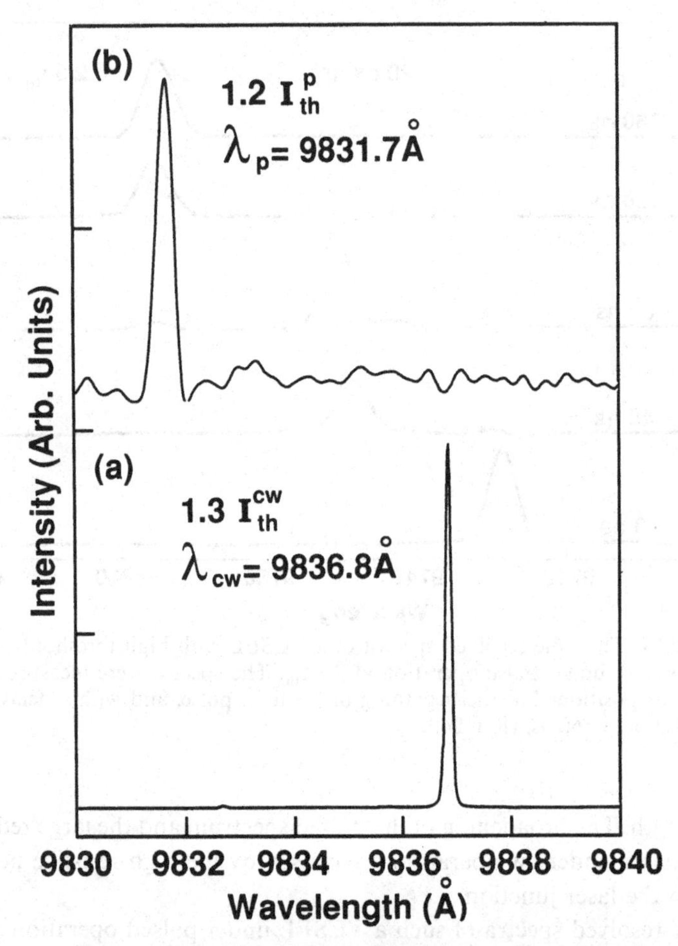

Figure 9.19. The (*a*) cw and (*b*) pulsed lasing spectrum for a VCSEL with low threshold voltage and series resistance. (Ref. 24.)

single-lobed signal similar to that of its cw spectrum. The lasing wavelength in this case is much closer to the cw wavelength, only 5 Å shorter, with a linewidth of 0.30 Å (Figure 9.19). The time-resolved spectra for this laser under pulsed operation at 2.5 I_{th} shows that the total red shift in 200 ns is much reduced to only 0.5 Å (Figure 9.20).

From these measurements, it is evident that high operating voltage and series resistance are detrimental for high speed operation. Furthermore, the resistive heating also limits the density of VCSELs that can be operated simultaneously in a 2-D array. Therefore, optimizing these parameters is extremely important for VCSELs.

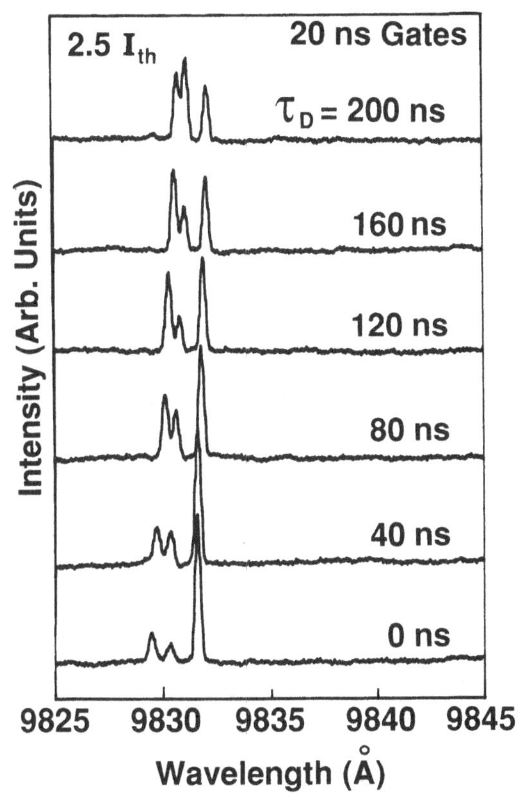

Figure 9.20. The time-resolved spectra for a VCSEL with low threshold voltage and resistance under pulse operation at 2.5 I_{th}. The spectra were measured with 20 ns gates at the beginning of the pulse, and with delay τ_D equal to 40, 80, 120, 160 and 200 ns. (Ref. 24.)

9.5 Two-dimensional multiple wavelength VCSEL array

9.5.1 Introduction

Optical sources capable of Tb/s (10^{12} bits/s) data rates are essential for applications in multi-media optical fiber communications, interconnection of large computers, and real-time optical signal and image processing. The more conventional approach to achieving high data rates is to use a single high speed or mode-locked diode laser. However, their bandwidths are limited from a few tens to a hundred GHz[34]. Moreover, the laser, as well as its driver and packaging, becomes very expensive as its speed increases. A highly promising approach which has attracted considerable research effort is to use a system which utilizes the laser wavelength as an *additional* parameter for multiplexing and coding, also known as wavelength-division-multiplexing (WDM).

An important optical source for WDM systems is a monolithic array of single-wavelength lasers emitting distinct wavelengths with uniform wavelength spacings. Fabricating such an array using the conventional distributed feedback (DFB) edge-emitting lasers[35,36] has been very difficult and expensive because highly precise variation of the grating periods is required for the entire laser array. Moreover, the number of laser wavelengths and the wavelength spacing are both limited.

In this section, we describe a VCSEL array which can provide hundreds of independent wavelengths through control during wafer epitaxial growth[33]. The wavelength spacing demonstrated is uniform and as small as 0.3 nm. The spacing can, in principle, be made smaller than 0.1 nm or as large as several nanometers. Using this novel design, not only can a larger number of controllable wavelengths be obtained, but the processing is as simple as that of a single VCSEL.

A WDM-based system experiment using part of the laser array is also described here, which represents the first system experiment using a VCSEL array[19,20]. The results indicate negligible optical or electrical crosstalk between the lasers.

9.5.2 Multiple-wavelength array design

The calculated reflectivity spectra for a complete VCSEL with ideal designed thickness and with two pairs of Bragg reflectors being 10% thicker than designed are shown by Figure 9.21. The calculation shows that only one Fabry–Perot mode exists within the mirror bandwidth due to the VCSEL's ultra-short cavity. This mode, which is the VCSEL lasing wavelength, depends critically on the total cavity length. In fact, the FP mode not only depends on cavity thickness variation but also on the position and the number of layers that are varied, as shown in Figure 9.22. The wavelength varies nearly linearly and monotonically with the layer thickness variation. The peak reflectivity value, on the other hand, is not very sensitive to small thickness variation in the Bragg reflectors. Shown in Figure 9.23 is the reflectivity spectrum for a 20-pair DBR with two pairs having 10% thickness variation. It is to be compared with that in Figure 9.3 for DBR with ideal thickness. The required high reflectivity in the mirrors for a VCSEL is not compromised with small variations in Bragg reflector thickness. Thus, we show that the VCSEL lasing wavelength can be tailored to either longer or shorter wavelengths with a small

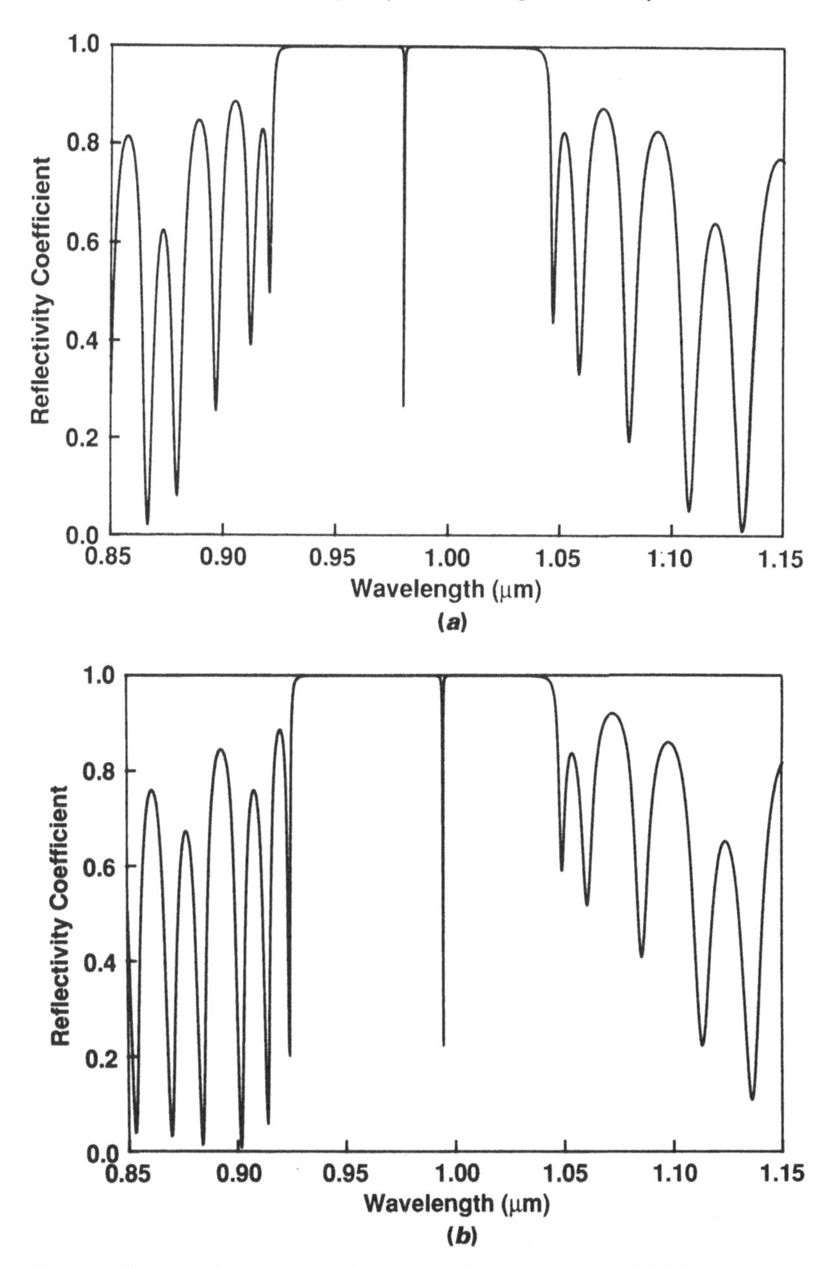

Figure 9.21. Calculated reflectivity spectra for a complete VCSEL with (*a*) ideal designed thickness and (*b*) two pairs of Bragg reflectors being 10% thicker than designed. The calculation shows that only one Fabry–Perot mode exists within the mirror bandwidth due to the VCSEL's ultra-short cavity. This mode depends critically on the total cavity length.

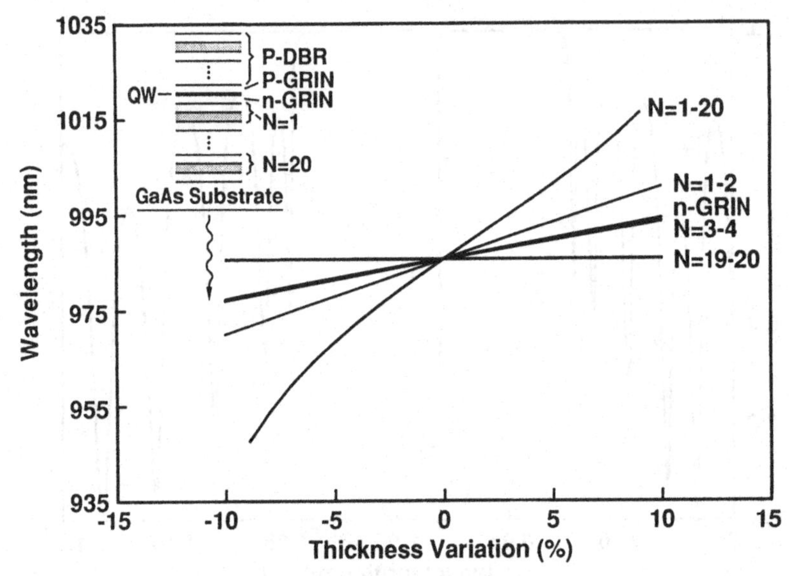

Figure 9.22. The Fabry–Perot mode wavelength as a function of thickness variation for different combinations of layers being varied. It varies nearly linearly and monotonically with the layer thickness variation. The closer the variation is to the active layer the greater its effect; similarly, the more layers are varied, the more wavelength shift can be obtained.

variation in cavity thickness, without significant compromise in the lasing characteristics.

Since the VCSEL wavelength depends sensitively on cavity length, a thickness gradient across the wafer can translate into a wavelength gradient of the lasers fabricated across the wafer. Figure 9.24 shows the schematic of a three-element laser array based on this idea. With spatial thickness gradient being created in two layers closest to the central spacer (the amount of thickness variation shown here is highly exaggerated), the lasing wavelengths for these three lasers are thus tailored according to the thickness variation.

One way to create such a thickness gradient is to simply keep the wafer stationary during part of the molecular beam epitaxy (MBE) growth. The thickness variation originates from the fact that the atomic sources in an MBE system are incident to the wafer at an angle off normal ($\approx 33°$ for the Varian Gen II system we used) and hence the number of atoms arriving at the wafer varies monotonically in the direction parallel to the planes of incidence of the sources. Figure 9.25 shows schematically the arrangement of the atomic sources in an MBE system. Since the MBE

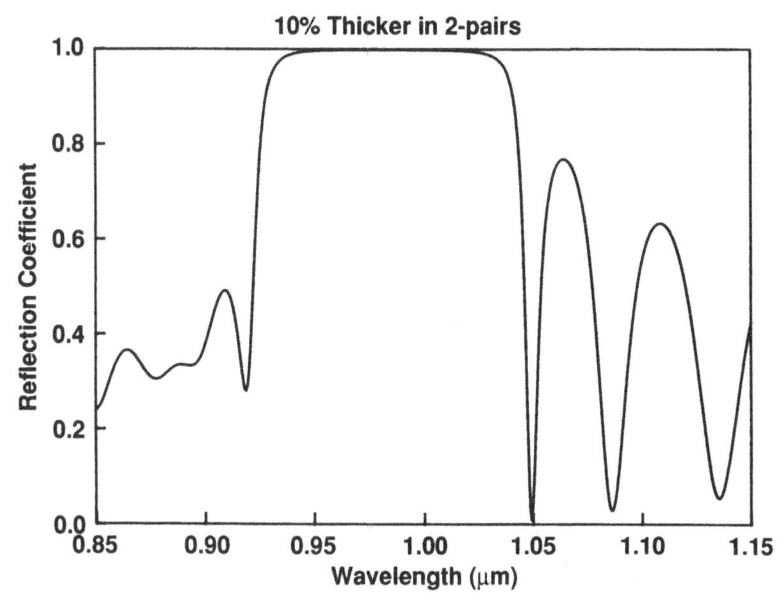

Figure 9.23. The reflectivity spectrum for a 20-pair DBR with two pairs having 10% thickness variation. It is to be compared with that in Figure 9.3 for DBR with ideal thickness. The peak reflectivity is not affected significantly by small variations in Bragg-reflector thickness.

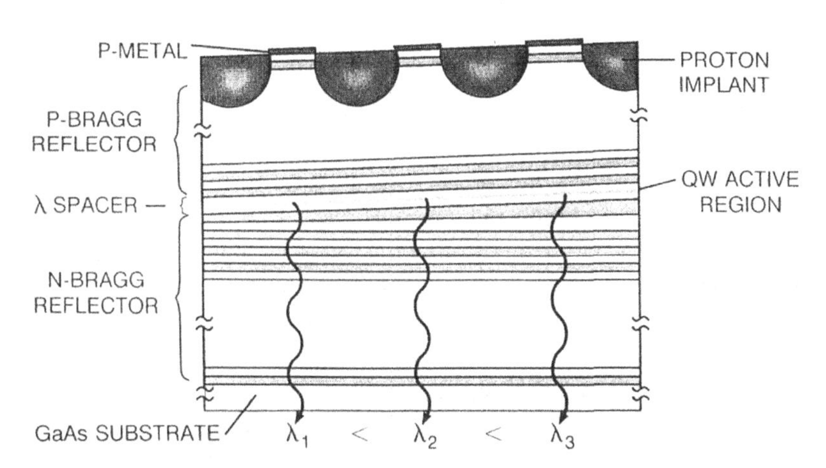

Figure 9.24. Schematic of an array consisting of three VCSELs as defined by proton implant. One pair of Bragg reflectors near the active region is shown to have a thickness gradient, which is used to alter the Fabry–Perot mode wavelength of the adjacent lasers. The amount of thickness variation in the spatially chirped layers is highly exaggerated.

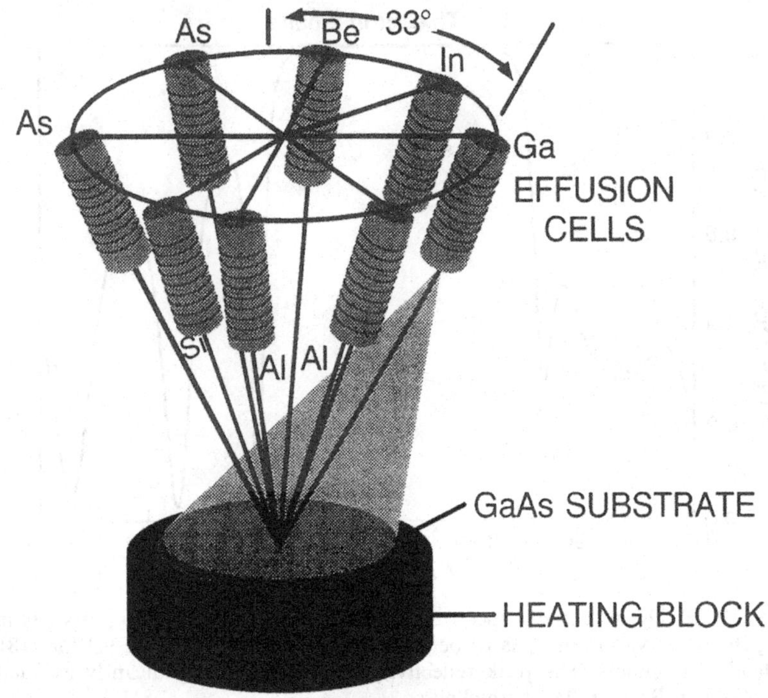

Figure 9.25. Schematic of the arrangement of atomic sources in a molecular beam epitaxy (MBE) system. All the sources are placed in a cone off normal from the substrate. (Ref. 33, © 1991 IEEE.)

material is grown in an As rich environment, the thickness of the MBE growth is determined by the number of group III sources that arrive at the wafer. Thus, the directions of thickness variation for GaAs and AlAs layers are parallel to the directions of the Ga and Al sources respectively. Therefore, we can obtain the desired small, but definite, thickness variation across the wafer by rotating the wafer for uniformity during the MBE growth of the VCSEL structure *except* for two pairs of AlAs and GaAs DBR layers during whose growth the wafer is kept stationary. Since the Ga and Al sources are placed next to each other in our MBE system, the direction of the resulting cavity thickness variation runs parallel to the line intersecting the directions of the two sources.

Although the wafer used here is grown with thickness variation in only one direction, $\hat{t}$, nevertheless, a 2-D laser array having no redundant wavelengths and a well-defined wavelength variation among the lasers can be obtained if the direction $\hat{t}$ does not coincide with the array axes. We have shown that the laser wavelength depends linearly on the cavity

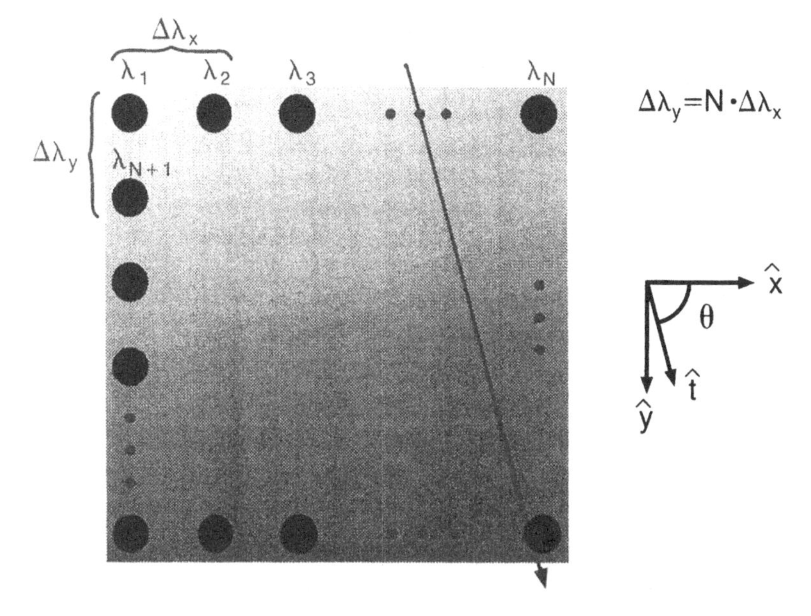

Figure 9.26. Schematic of the 2-D rastered multiple wavelength (RMW) VCSEL array. By aligning the array obliquely with respect to the direction of thickness variation, a 2-D laser array with no redundant wavelengths is obtained. (Ref. 33, © 1991 IEEE.)

thickness; in addition, the thickness gradient implemented by keeping the substrate stationary is linear over a small distance (small compared to the distance between the atomic sources and the wafer, which is 5–10 cm). Hence, the wavelength separation $\Delta\lambda$ between any two lasers is proportional to the distance between the lasers projected on to $\hat{t}$. For a 2-D array with its x-axis making an oblique angle θ_x with $\hat{t}$, as shown in Figure 9.26, we obtain

$$\Delta X \approx d \cdot \cos\theta_x \qquad \text{and} \qquad \Delta Y \approx d \cdot \sin\theta_x, \tag{9.1}$$

where d is the spacing between neighboring lasers and $\Delta\lambda_x$ and $\Delta\lambda_y$ are the wavelength separation between neighboring lasers in the x and y directions, respectively. Therefore,

$$\Delta\lambda_y \approx \tan\theta_x \cdot \Delta\lambda_x = N\,\Delta\lambda_x,$$
$$N \equiv \tan\theta_x.$$

Hence, the wavelength of the $(N + 1)$th laser on a row is the same as that of the first laser on the next row, provided that $\tan\theta_x$ is close to an integer. Restricting the array to N columns, a 2-D array with no redundant wavelengths is obtained. To obtain a large N, θ_x must be large and close to 90°. The number of rows in such an array, M, can be made very large. Ideally, M is limited only by the total gain bandwidth ≈ 100 nm[37], which has to be larger or equal to the total wavelength span

$$\lambda_{\text{span}} = (NM - 1) \cdot \Delta\lambda_x.$$

However, the actual physical dimension may impose a limit if uniform wavelength separation throughout the array is required. The direction of increasing wavelength rasters through the elements of the array. Such an array is a rastered multiple wavelength or RMW laser array.

In an experimental demonstration, the angle θ_x was chosen to be $\approx 82°$, and then $\Delta\lambda_y$ is seven times $\Delta\lambda_x$. Although larger θ_x would give a larger N, alignment becomes progressively more difficult.

9.5.3 Laser array characteristics

The wavelength distribution of a 7 × 20 RMW array of 20 μm square lasers is shown in Figure 9.27. The total wavelength range is as large as

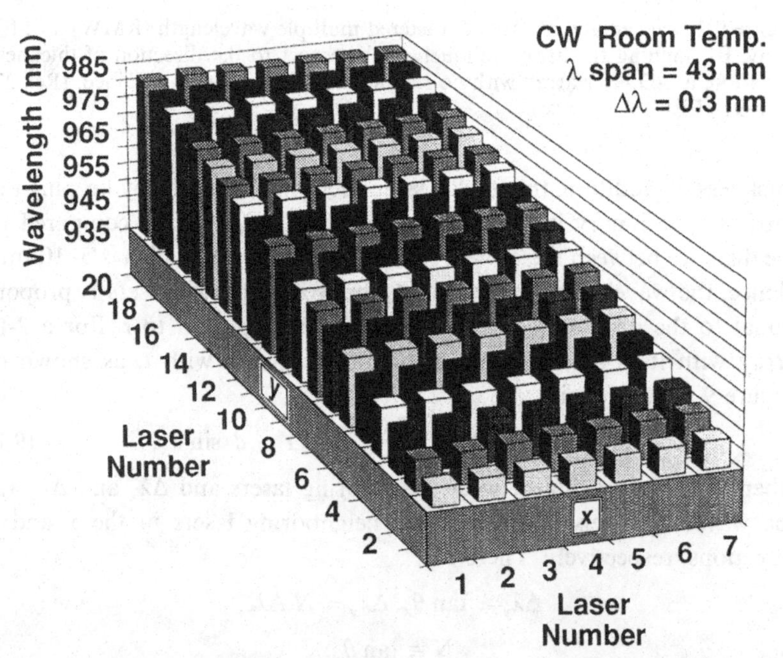

Figure 9.27. Experimentally measured cw wavelength distribution of the 7 × 20 RMW VCSEL array. Each laser emits a unique wavelength. The direction of increasing wavelength rasters through the 2-D array. (Ref. 33, © 1991 IEEE.)

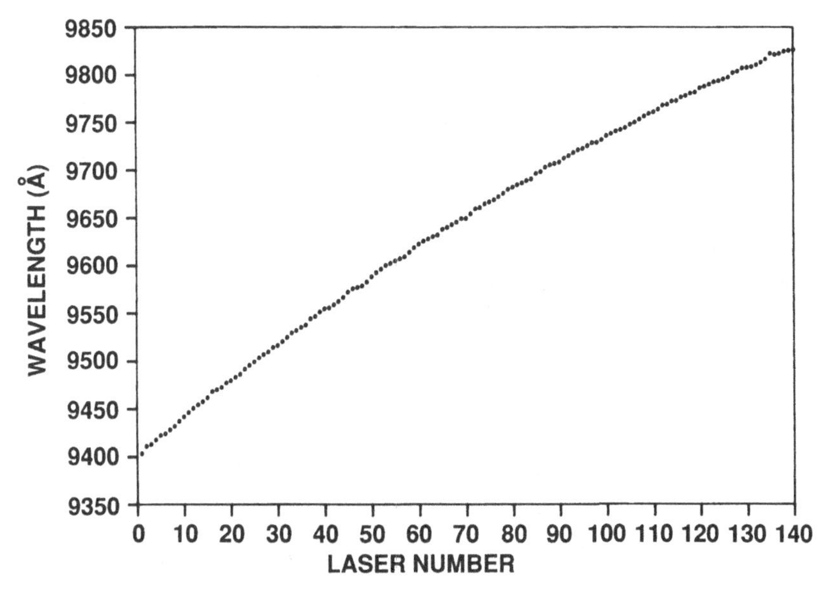

Figure 9.28. The measured cw wavelength shown in Figure 9.27 is replotted with respect to a new laser number K to absorb the rastering information. K is defined as $(X-1)+7(Y-1)+1$, where $X=1,2,\ldots,7$ and $Y=1,2,\ldots,20$. The curve is nearly linear indicating nearly uniform wavelength separation throughout the 140 lasers.

43 nm, from 940 to 983 nm. The average wavelength separation between two neighboring lasers in a row and column are 0.3 and 2.1 nm respectively. Each physical position on the 2-D array is represented by a unique wavelength. The laser spectra are measured at ≈ 1.1 times the threshold currents under cw operation with the lasers emitting a single TEM_{00} mode each. The wavelength separation can easily be made smaller by reducing the laser spacing or the number of chirped layers. It can also easily be made larger by doing the opposite. Figure 9.28 shows the wavelength distribution replotted as a function of laser number K, where

$$K=(X-1)+7(Y-1)+1 \qquad X=1,2,\ldots,7 \quad \text{and} \quad Y=1,2,\ldots,20.$$

A nearly linear relation is obtained indicating uniform wavelength separation and rastering between the lasers.

Figure 9.29 shows the cw threshold current distribution of the 2-D RMW laser array. The average cw threshold current is 9.5 mA. The 2-D laser array exhibits good uniformity with 85% of the lasers having threshold currents within ± 2 mA of the average values. All lasers show comparable electrical and optical characteristics.

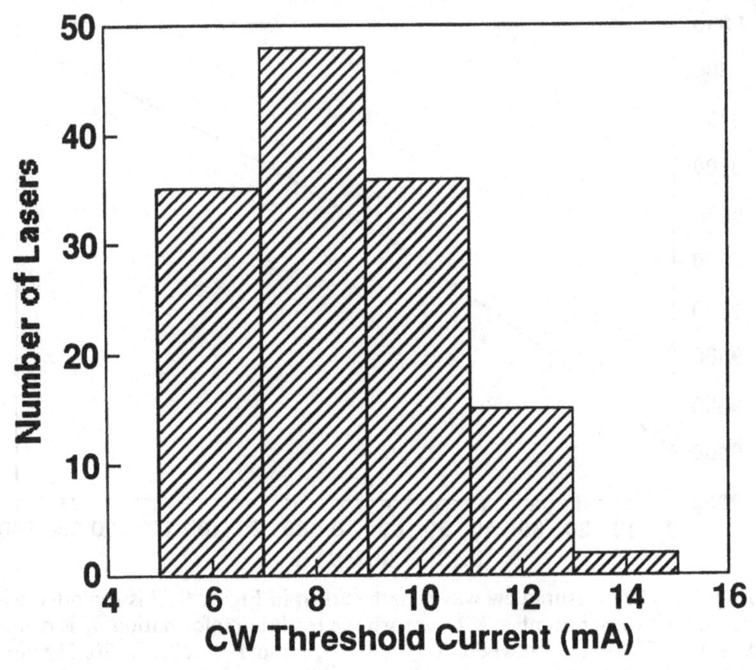

Figure 9.29. Distribution of measured cw threshold currents of the 7 × 20 RMW wavelength VCSEL array. (Ref. 33, © 1991 IEEE.)

The physical implementation of the chirped layers was achieved by keeping the wafer stationary during the MBE growth of two pairs of DBR layers. This technique offers simplicity and elegance. It is worth noting that many large 1-D arrays can be fabricated on the same wafer having the same central wavelength and wavelength separation. As for 2-D arrays, many with the same wavelength range can be obtained simply by stacking them as shown in Figure 9.30. None the less, the arrays made from the portion closer to Ga and Al sources will be different from the ones made from the portion away from the sources.

The 2-D array with unique wavelengths can also be obtained by variations in our approach[38]. The spacing between the lasers can be varied in both x and y directions independently. As indicated in Equation (9.1), the wavelength separation will be varied accordingly. Alternatively, 2-D chirped layers can be obtained by keeping the wafer stationary for one layer and then a few more layers with the wafer rotated by 90°.

The spatially tapered layers can also be made using a number of alternative techniques[38]. For example, by using growth on a nonplanar

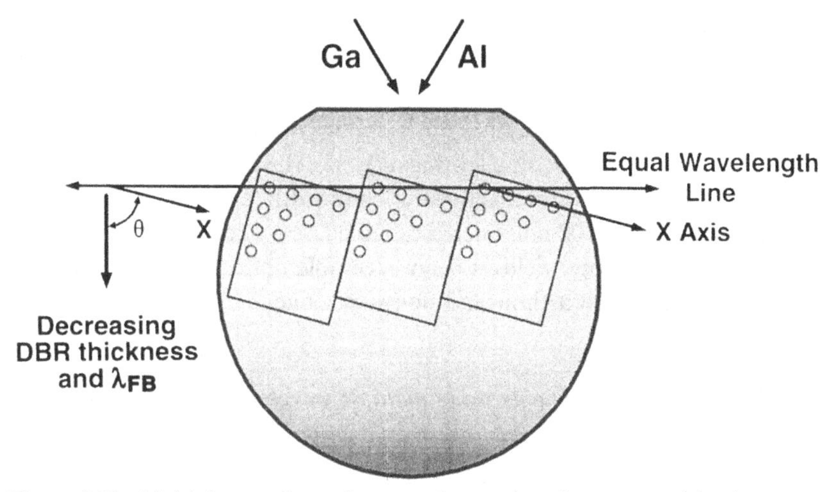

Figure 9.30. Multiple number of rastered wavelength arrays with the same wavelength spans can be made from the same wafer by stacking them next to each other with a distance offset.

substrate[39], VCSELs with different layer thicknesses can be obtained since the growth rate depends on the size of the pattern, e.g. mesas, on the substrate. Another approach is to pattern-etch either the substrate holder (the *Mo* block)[40] or the back side of the substrate to provide a temperature gradient, which then transfers into a growth gradient on the substrate. Using MOCVD (metal–organic chemical vapor deposition) as the growth technique, laser-induced desorption[41] or selective-area epitaxy may also be used to produce the thickness gradient.

These lasers emit in the 980 nm regime. The techniques described here, however, are readily extendable to both longer, 1.3–1.55 μm, and shorter, 0.6–0.7 μm, wavelength regimes.

9.5.4 Enabled new applications

Very high data rate throughput is expected when using the additional dimension of freedom – wavelength. Indeed, an $N \times M$ VCSEL array can be fabricated to have only M independent wavelengths but a large number, N, of redundancies for each wavelength, or to have hundreds, $N \times M$, of unique wavelengths. The former case increases the tolerance on the array reliability, while the latter enhances the system capabilities. The RMW laser array should also be useful for optical imaging and optical-signal processing using the unique one-to-one mapping between wavelengths and physical positions of the lasers.

Using the entire array as an ensemble of optical sources, a fast wavelength-tunable laser with a large tuning range is obtained. The tuning speed in this case can be much faster than 1 ns, since it is simply the speed of each laser. The wavelength step is the wavelength separation of the lasers, which can be made smaller than 0.1 nm. The total wavelength range is limited by the quantum-well gain bandwidth and the DBR-mirror bandwidth to about 100 nm. Such a tunable laser, when further combined with an optical grating, yields a beam-steerable optical source capable of resolving hundreds to a thousand points per line.

9.5.5 System experiments using multiple wavelength VCSEL array

A preliminary WDM system experiment was demonstrated[19,20] using a 4×1 linear VCSEL array with a wavelength and physical spacing of 1.5 nm and 508 μm respectively. This wavelength spacing was limited by the passband of the optical filter used at the receiving end. The lasers used in this experiment were not packaged or heat-sunk; the substrate was positioned on a sapphire window, through which the output beams were collected. None the less, simultaneous operation of four lasers was attained.

The lasers were modulated with $2^{15} - 1$ word-length non-return-to-zero (NRZ) pseudo-random patterns at 155 Mb/s and the output light was butt-coupled into four 50/125 multimode fibers (MMF) that were fixed in a silicon V-groove holder. The typical coupling efficiency was 3 dB, limited by the beam divergence through the substrate and a sapphire window (450 μm total thickness). The variation in the coupling efficiency for the four fibers was less than 1.5 dB. Higher coupling efficiency is expected with a more optimized laser design and the use of lensed fibers[42]. The cross-coupling from the adjacent 508 μm-spaced lasers into the optical fiber was less than -60 dB. The signals were combined into a single fiber with a star coupler and transmitted over approximately 20 m of fiber. The maximum output power from each laser was approximately -6 dBm with simultaneous cw operation of four lasers. The lasers were biased at $1.4\,I_{th}$, corresponding to ≈ 12 mA, under cw operation. A higher output power is expected with proper heat sinking of the lasers.

Figure 9.31(*a*) shows the optical spectrum at the output of the star coupler. Four uniformly separated peaks with nearly equal intensities were obtained. The optical filter used in the experiment was an electronically tunable liquid-crystal Fabry–Perot etalon[43] that can also be potentially constructed and operated as a 2-D filter array. The filtered output is

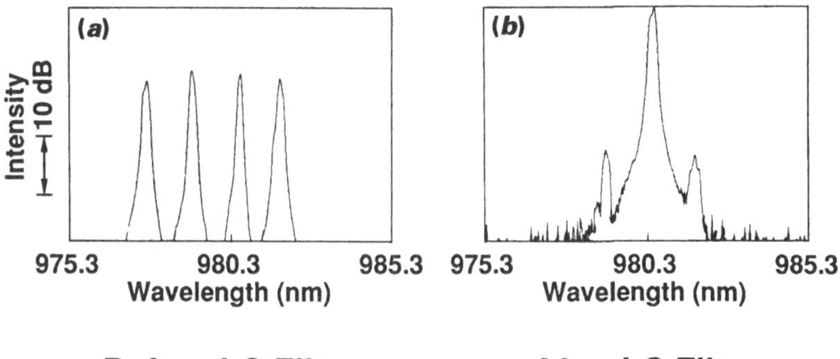

Figure 9.31. The optical spectrum of (*a*) a 1 × 4 multiple wavelength, simultaneously cw-operated VCSEL array (Ref. 20), and (*b*) the output filtered through a liquid-crystal etalon wavelength-tunable filter. (Ref. 19, © 1991 IEEE.)

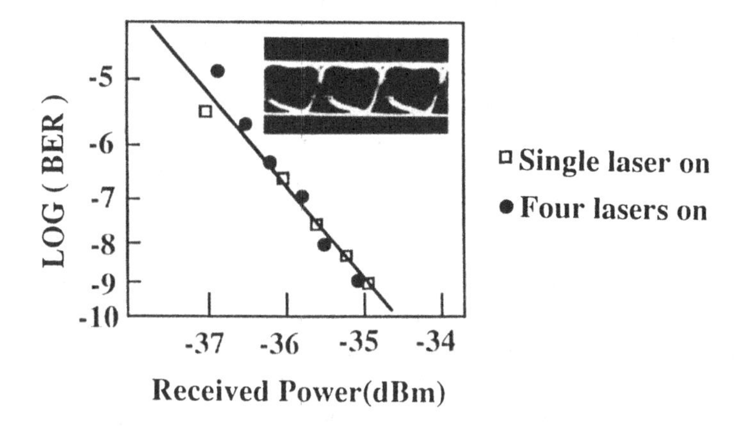

Figure 9.32. Bit error rate versus received power for the case when just one laser was operated as compared with all four lasers being on. Inset shows an open eye diagram at $<10^{-9}$ BER. (Ref. 19, © 1991 IEEE.)

shown in Figure 9.31(*b*). The filter had a free-spectral-range of 18 nm and a passband of approximately 0.2 nm. The power ratio of the filtered signal to the rejected channels is more than 25 dB. The filtered light was detected by an InGaAs receiver and the bit-error-rate (BER) was measured to study the crosstalk between the multiplexed signals.

Figure 9.32 shows the BER as a function of received power for the case when only one laser is modulated as compared with the case when all four operate simultaneously. The well-overlapped lines for both cases show that no sensitivity penalty was measured with simultaneous operation of the VCSELs. A BER of 10^{-9} was achieved with a received power of -35 dBm. An open eye diagram was obtained, as shown by the inset of Figure 9.32, for lasers modulated with $2^{15} - 1$ word-length NRZ pseudo-random code at 155 Mb/s with less than 10^{-9} BER. This bit rate was limited by the impedance mismatch between the lasers and the drive circuitry and the probes used to drive the lasers. Higher speed of multigigabit/s operation was obtained with proper packaging of the lasers.

9.6 Phase-locked VCSEL arrays

Phase-locked laser arrays, in which all the elements of the array are at a fixed phase relationship with each other, are of interest as a method of scaling the output power while maintaining a high-quality beam and a narrow spectrum. Although significant progress has been made in phase-locked horizontal-cavity lasers[44], studies on 2-D phase-locked VCSEL arrays have just commenced. In particular, VCSELs are uniquely interesting for research on phase-locked arrays because the array pattern can be almost arbitrarily defined by photolithographic processing to implement various coupling mechanisms for phase locking.

Using a structure similar to the early gain-guided, phase-locked, edge-emitting laser arrays[45], a 20×20 phase-locked VCSEL array emitting primarily into a single out-of-phase (highest order) array mode has been demonstrated[14]. The array elements are 3 μm by 3 μm square with a 4 μm center-to-center spacing, arranged in a square-grid format. The far-field intensity profile exhibits a 2×2-lobed pattern. Shown in Figure 9.33(*a*) and (*b*) are the experimentally measured and the calculated near-field and far-field patterns, respectively, of such an array. The lobewidth is 0.8° near threshold, and the separation between the nearest two lobes is 14°. Both agree reasonably well with the calculated values.

In an effort to obtain a single-lobed output beam from a 2-D VCSEL array, a binary phase-shifting mask has been used to compensate the phase relationship of the elements of the array[15]. The mask consists of transparent material of variable thickness to produce a 180° phase-shift between adjacent elements. This method resembles earlier publications on edge-emitting laser arrays[44,46], where phase-shifters were made by evaporation and preferential etching at the cleaved facet of the laser. The measured near-field and far-field patterns of a 2-D array with phase-shifting mask is

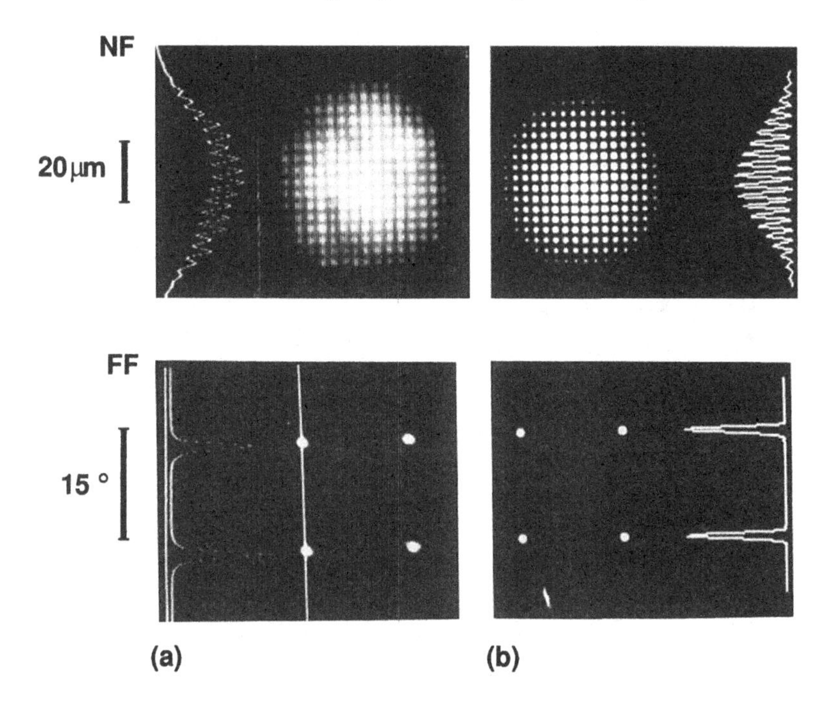

Figure 9.33. (*a*) Experimentally measured and (*b*) the calculated near-field and far-field patterns of an evanescent coupled 20 × 20 VCSEL array. (Ref. 14.)

shown in Figure 9.34(*a*). Most of the power is emitted into the central lobe. This is in contrast with an array without the mask (Figure 9.34(*b*)), which simply emits the 2 × 2-lobed out-of-phase array-mode pattern described above. The angular width of the central lobe shown in Figure 9.34(*a*) is 2° FWHM, approximately twice the calculated diffraction limit. However, at this stage, this structure still needs to be modified for electrical pumping.

Both of the above methods use the nearest-neighbor coupling mechanism (or 'series coupling') between the array elements, which has been demonstrated in 1-D linear arrays to be ineffective for phase-locking[44]. Similar to 1-D arrays, resonant leaky-wave coupled 2-D VCSEL arrays have been shown theoretically[47] to be promising for obtaining stable in-phase-mode operation. However, there have not been any experimental results yet. Due to the extremely short gain path, phase-locked VCSEL arrays may not necessarily lead to practical sources of high coherent power (1 W and upwards); they are, nevertheless, very attractive for controlled beam steering in two dimensions.

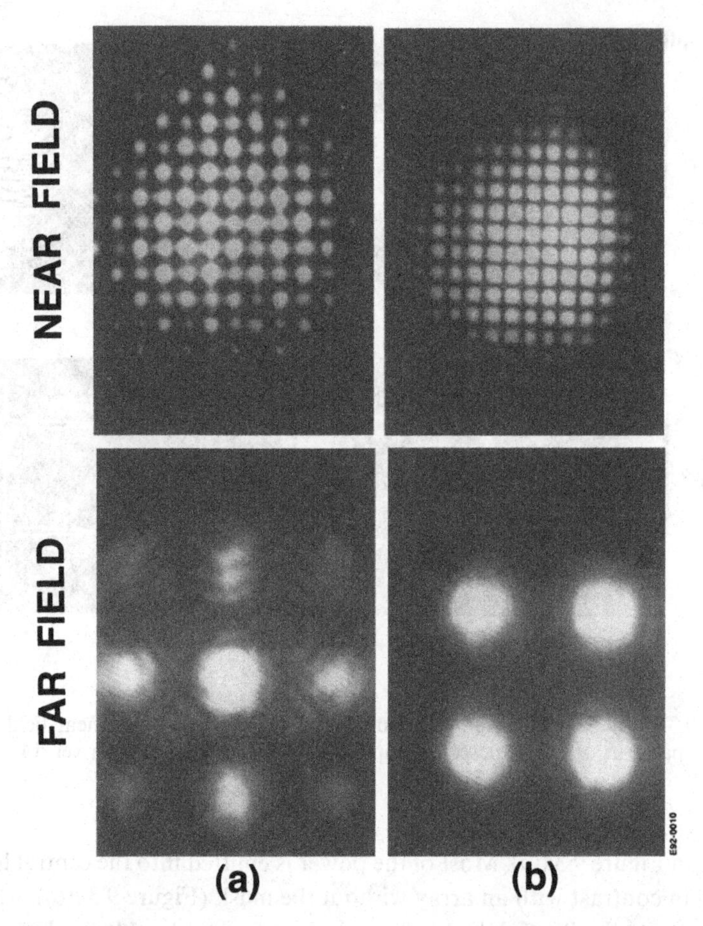

Figure 9.34. Near- and far-field emission from an optically pumped 2-D phased-locked VCSEL array with (*a*) and without (*b*) integrated phase-shift mask. (Ref. 15.)

9.7 Addressable VCSEL array

9.7.1 Independently addressable array

The yield and uniformity of independently addressable VCSEL arrays can be very high due to simplicity in fabrication. Typical threshold-current variation for an 8×8 VCSEL array is about 2 mA[17], although this is clearly wafer-dependent. The array size, however, reaches its limit at fairly small numbers (a few hundred) for most device packages.

Systems experiments using individually addressable VCSEL arrays have

Figure 9.35. Photograph of a 36-channel optical interconnection scheme using a 2×18 top-emitting VCSEL array, which is pigtailed into a 2×18 optical fiber bundle.

been demonstrated for optical data link applications[18]. Figure 9.35 shows a photograph of a complete 36-channel optical interconnection scheme using a 2×18 top-emitting VCSEL array, which is pigtailed into a 2×18 optical fiber bundle, and a 2×18 GaAs p–i–n detector array. The schematic of one such connection is shown in Figure 9.36. The VCSEL array exhibits high uniformity, as shown by the *L–I* characteristics in Figure 9.37. Data rates at higher than 622 Mb/s per channel were demonstrated, resulting in an aggregate bandwidth of 22 Gb/s.

9.7.2 Matrix-addressable array

The matrix addressing technique is more suitable for addressing large VCSEL arrays. A 32×32 matrix-addressable array was demonstrated[16], and it is not difficult to imagine still further expansion. The key limit of such arrays is that, at any time, only 32 lasers (M for $M \times M$ array) can be operated simultaneously. All the lasers in a column are connected electrically via a common n^+ epitaxial layer; whereas each column is electrically isolated from all the other columns. In each row, all the p contacts of the lasers are connected. By applying a voltage in between

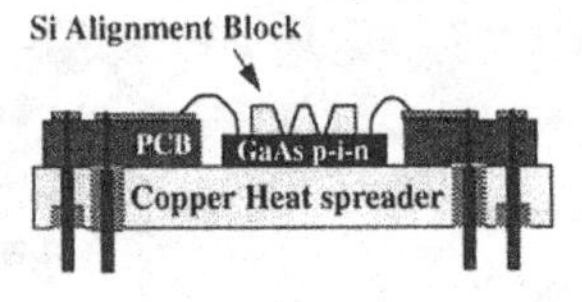

36 channel 2D-ODL prototype

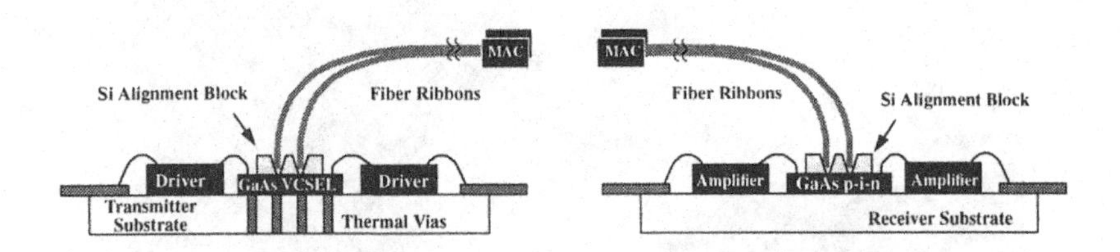

36 channel 2D-ODL concept

Figure 9.36. The schematic of one of the fiber connections. (Ref. 18.)

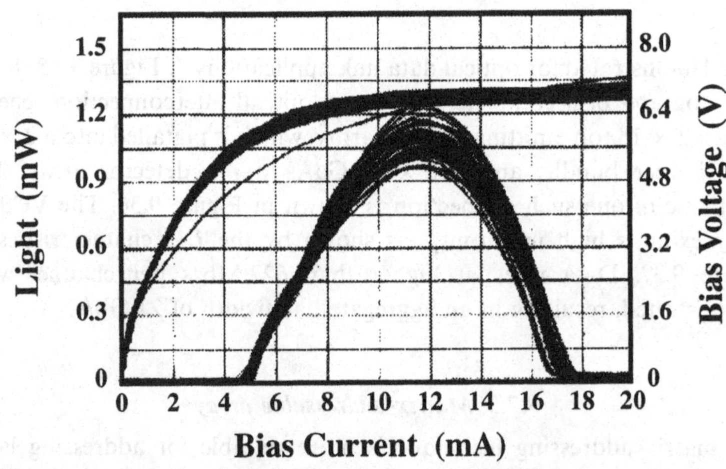

Figure 9.37. *L–I* and *I–V* characteristics of the 2 × 18 VCSEL array used in the 36-channel optical interconnection experiment. (Ref. 18.)

a specific row *i* and a specific column *j* while keeping all the other contacts open, the laser in the row–column intersection, laser *ij*, is thus turned on.

The measurements on this array were performed under pulsed operation (200 ns pulse width; 10 kHz repetition rate)[16]. The threshold current and

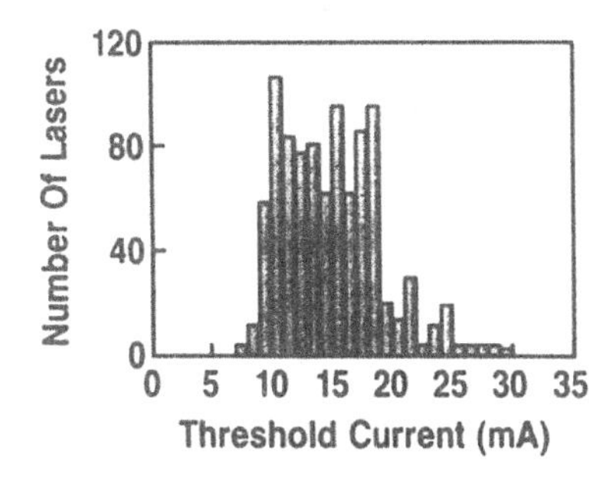

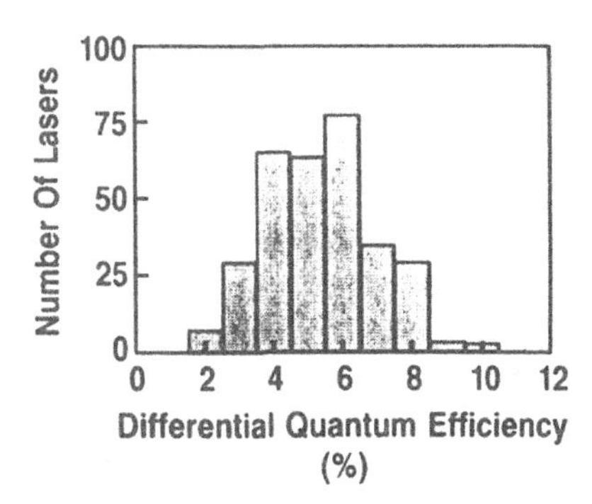

Figure 9.38. Pulsed threshold current and differential quantum efficiency distribution of a 32 × 32 matrix-addressable VCSEL array. (Ref. 16b.)

differential efficiency distribution for all the VCSELs of the array are shown in Figure 9.38. All 1024 lasers operated with a most probable (MP) threshold current of 6.8 mA and output differential quantum efficiency (DQE) of 8%. The spread over the entire array is about 18% for the threshold current and about 30% for the DQE. In addition, these lasers exhibit 10^{-9} BER under 622 Mb/s modulation rate.

9.7.3 High speed multiple wavelength VCSEL array

Flip-chip bonding is a common technique used for a variety of devices. One important step is the alignment of the substrate with active devices to a metallized substrate with In solder bumps. The most crucial process when mounting VCSELs is the prevention of the solder and Au in the contact from diffusing into the epitaxial layers, which alters the mirror reflectivity.

Figure 9.39 shows the *L–I* and *I–V* characteristics of one typical VCSEL under cw operation before and after the flip-chip bonding on to a BeO substrate. In both cases, the lasers are tested free-standing without heat sinks. The laser cw output power increases to about 6 mW with bonding, indicating the importance of an appropriate heat removal mechanism[48].

Figure 9.9(*a*) shows the schematic of a VCSEL flip-chip mounted on to a BeO substrate[13,26]. Such a BeO carrier is then mounted into a high speed package. Figure 9.9(*b*) shows a photograph of a flip-chip mounted 2 × 8 VCSEL array mounted in a high speed package. The laser spacings are 254 and 508 μm for the lasers in one row and one column respectively.

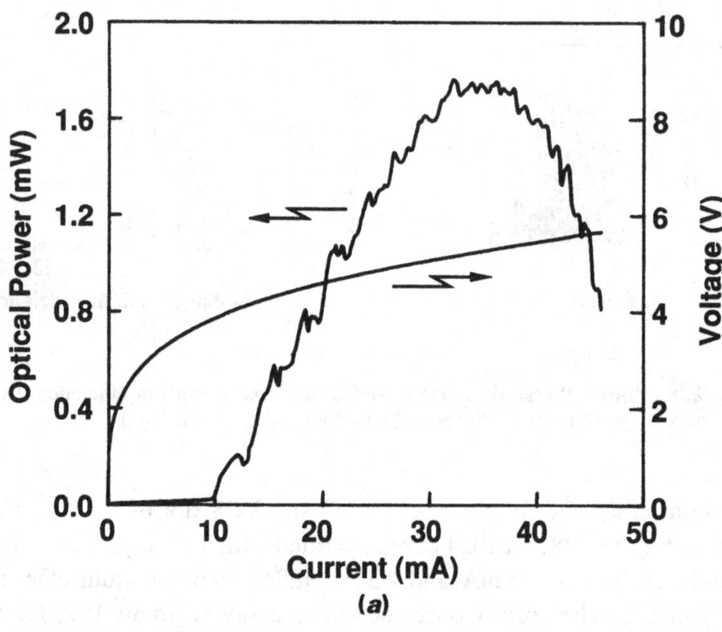

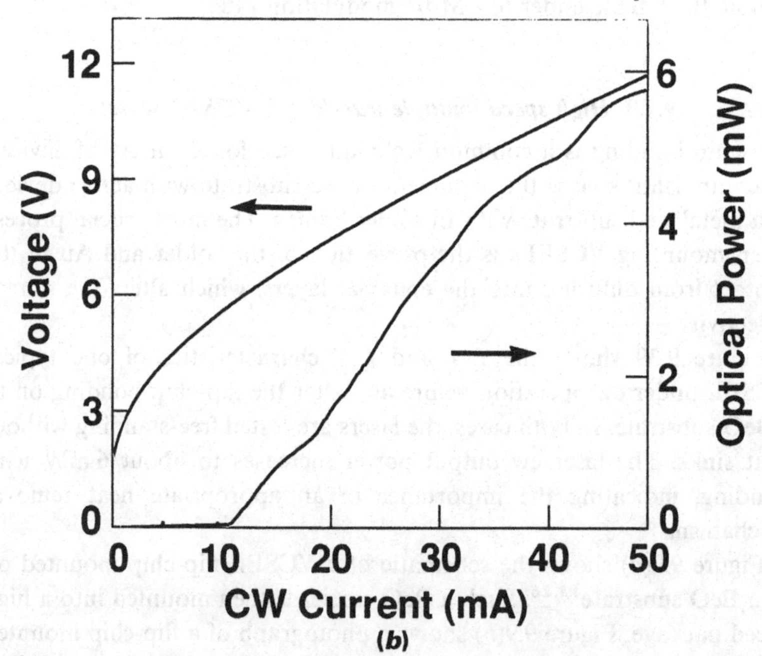

Figure 9.39. CW L–I and I–V characteristics of a typical VCSEL (*a*) before (Ref. 33, © 1991 IEEE) and (*b*) after flip-chip bonding, both without heat sinks.

Though comparable with those used for edge-emitting laser arrays, these spacings are designed to be fairly large such that thermal crosstalk can be minimized and most of the arrays can be operated simultaneously. Each laser is connected in parallel with a 100 Ω resistor fabricated on the BeO substrate to provide high speed impedance matching. Each microstrip line from the BeO substrate terminates at an SMA connector. The lasers share the common ground contact and the ground termination was fabricated via ribbon wires from the n-type substrate at the top of the bonded device. The light output from each laser is transmitted through the ≈200 μm-thick substrate and butt-coupled into a single-mode fiber. The output power from the laser is about 0.5 mW at 1.1 I_{th}. Up to eight lasers were biased simultaneously in our experiment. Optical isolators were not used in the measurements.

Figure 9.40 shows the output spectra of the 16 lasers. The sidemode

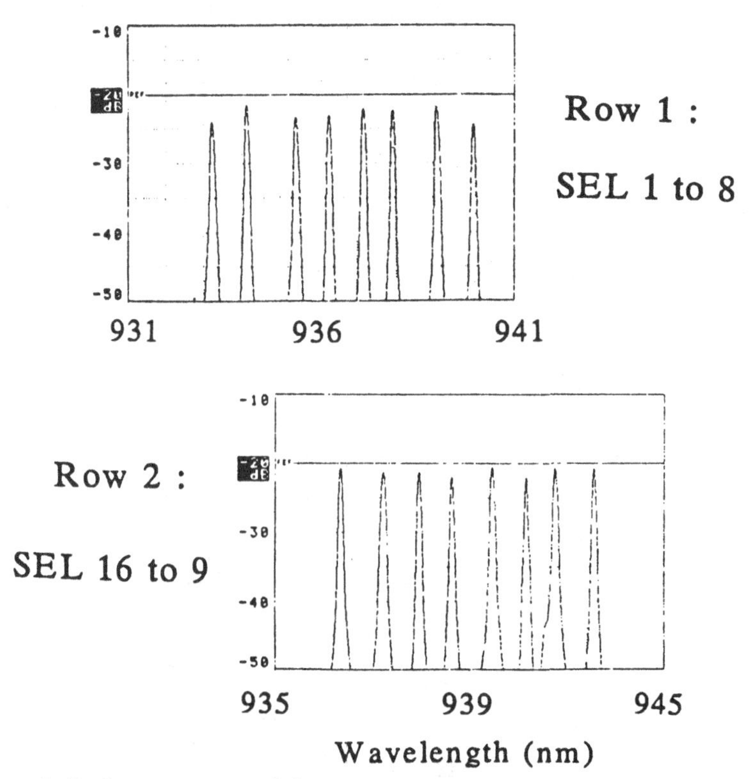

Figure 9.40. Output spectra of the 16 lasers. The wavelength spacing is approximately 0.9 nm between the row elements. There is a small overlap in the wavelength range covered by the lasers in the two rows; however, all 16 lasers can be used as independent wavelength sources. (Ref. 26, © 1991 IEEE.)

suppression for each laser is greater than 45 dB and the chirp-broadened 20 dB spectral width of the lasers under 5 Gb/s modulation is under 0.3 nm. The wavelength spacing is approximately 0.9 nm between the row elements. There is a small overlap in the wavelength range covered by the lasers in the two rows. However, due to the limited spectral width, all 16 lasers can be used as independent wavelength sources. The use of wavelength-rastered arrays can also remove wavelength duplication in larger arrays. Simultaneous operation was demonstrated on eight VCSELs of the same row, limited solely by the availability of equipment.

Figure 9.41 displays the typical small-signal intensity modulation response of the laser, showing the increasing resonance frequency as the bias current level is raised. The maximum speed of the device is limited to about 8 GHz due to the onset of multiple lasing modes.

In order to verify that the lasers are well-behaved under large-signal modulation, sensitivity measurements were carried out at 5 Gb/s and 2.5 Gb/s with a $2^7 - 1$ pseudorandom pattern. Longer word lengths were not used to the low frequency cutoff in the receiver response. Figure 9.42 shows bit error rate as a function of received power for VCSEL 12 modulated alone and with two nearest-neighbor lasers (254 μm) being modulated simultaneously by two independent sources. At 10^{-9} error rate, a sensitivity penalty of 0.7 dB was measured. No crosstalk effect

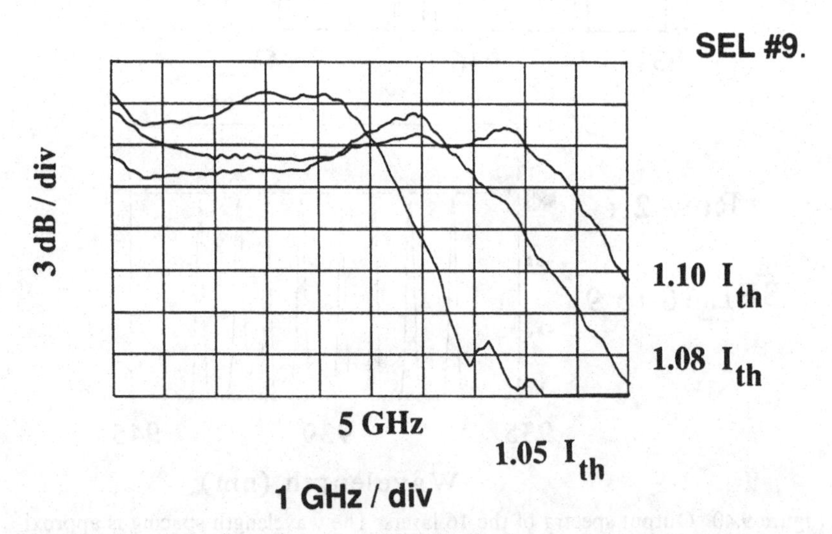

Figure 9.41. Typical small-signal intensity modulation response of the laser showing the increasing resonance frequency as the bias current level is raised. (Ref. 26, © 1991 IEEE.)

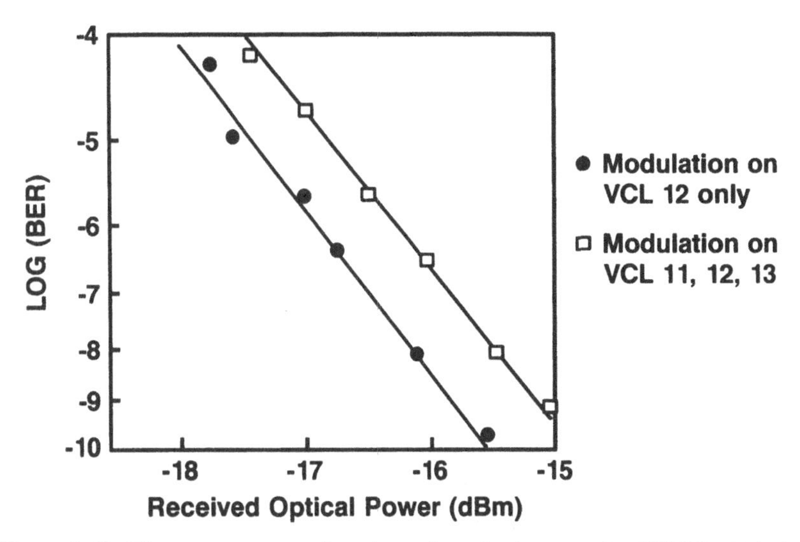

Figure 9.42. Bit error rate as a function of received power for VCSEL 12 being modulated at 5 Gb/s and with two nearest-neighbor lasers being modulated simultaneously by two independent sources also at 5 Gb/s. (Ref. 13.)

was observed when the modulation was applied to the next nearest (508 μm-spaced) neighbors. Thus, this 16-wavelength VCSEL array[13] promises a record-high aggregate bandwidth of 80 Gb/s.

These measurements were repeated at 2.5 Gb/s, shown in Figure 9.43. An error rate of 10^{-9} was attained at a receiver input power of -27.5 dBm. Negligible effect on the sensitivity curve (<0.2 dB) was observed with the two nearest-neighbor lasers being modulated simultaneously. Although the loss and dispersion in optical fibers were not optimum for 980 nm (≈ 1 dB/km and ≈ 40 ps/(nm km), respectively), the feasibility of fiber transmission using an 8 km span of fiber for these VCSELs at 2.5 Gb/s was demonstrated, as shown in Figure 9.43. The 1.4 dB penalty is attributed to the optical reflections in the systems and can be largely eliminated by an optical isolator.

There are three crosstalk mechanisms for laser arrays: optical, thermal and electrical. With the laser spacing of 254 μm and the substrate thickness of about 200 μm, the optical cross-coupling of adjacent lasers into the fiber is less than -50 dB. The thermal RF crosstalk, measured as a wavelength shift induced by the bias change of an adjacent laser, is between 0.02 and 0.05 Å/mA. The dominant component of crosstalk in this experiment is, thus, electrical crosstalk, which can be improved by better packaging design.

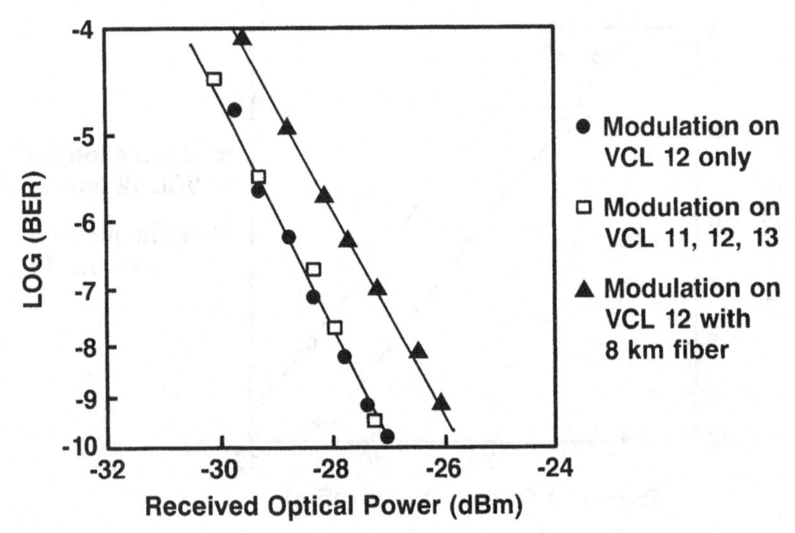

Figure 9.43. Bit error rate as a function of received power for VCSEL 12 being modulated at 2.5 Gb/s and with two nearest-neighbor lasers being modulated simultaneously by two independent sources also at 2.5 Gb/s. An error ratio of 10^{-9} was attained at a receiver input power of -27.5 dBm. (Ref. 13.)

9.8 Future prospects

The VCSELs offer the unique combination of easily fabricatable large 2-D arrays and novel wavelength tailorable functionality. The low-divergence circular output beam and potential for very low threshold are also important advantages. It is expected that these lasers should find important applications both as low cost, high performance lasers and as highly functional laser arrays. The two most important aspects of VCSELs that need to be improved are the wall-plug efficiency and stable single-transverse-mode operation.

On the issue of efficiency, it is the wall-plug efficiency that dictates the ultimate suitability, packing density and reliability of VCSELs in real-world applications. The wall-plug efficiency is dependent on the threshold current, threshold voltage, series resistance and the differential quantum efficiency. It is important to note that the combination of all four parameters determines the merit of a VCSEL; not just the drive voltage or the resistance alone. Therefore, an improvement on some of the parameters at the expense of the others *will not* lead to a more efficient VCSEL.

As far as stable single-mode operation is concerned, it is desirable to fabricate a buried heterostructure VCSEL with a small built-in index guide in the transverse direction, again similar to edge-emitting lasers. Recent results from a passive-antiguide-region (PAR) buried heterostructure VCSEL show promising results of stable, single-transverse-mode operation to high drive currents[49,50]. More research activities will no doubt be undertaken in this area.

VCSELs emitting at both visible and long-wavelength (1.3–1.5 μm) regimes are also of great interest. The former have a variety of applications ranging from compact disk (CD) players, to laser scanners, to optical interconnects. The latter offer numerous advantages for optical fiber communications including high coherence, potential for very low cost, excellent and simple fiber coupling, and facilitation of large multiple-wavelength arrays. Indeed, many advances on VCSELs at both wavelength regimes are continuously being made[51–55].

References

1. Z. L. Liau and J. N. Walpole, *Appl. Phys. Lett.*, **50**, 528–30, 1987.
2. N. W. Carlson, G. A. Evans, D. P. Bour and S. K. Liew, *Appl. Phys. Lett.*, **56**, 16–18, 1990.
3. H. Soda, K. Iga, C. Kitahara and Y. Suematsu, *Japan J. Appl. Phys.*, **18**, 2329–30, 1979.
4. K. Iga, F. Koyama and S. Kinoshita, *IEEE J. Quantum Electron.*, **QE-24**, 1845, 1988.
5. D. Botez, L. M. Zinkiewicz, T. J. Roth, L. J. Mawst and G. Peterson, *IEEE Photon. Tech. Lett.*, **1**, 205, 1989.
6. R. Burnham, D. R. Scifres and W. Streifer, US Patent No. 4309670, January 1982.
7. J. L. Jewell, A. Scherer, S. L. McCall, Y. H. Lee, S. Walker, J. P. Harbison and L. T. Florez, *Electron. Lett.*, **25**, 1123–4, 1989.
8. R. S. Geels, S. W. Corzine, J. W. Scott, D. B. Young and L. A. Coldren, *IEEE Photon. Tech. Lett.*, **2**, 234–6, 1990.
9. C. J. Chang-Hasnain, J. P. Harbison, C. E. Zah, L. T. Florez and N. C. Andreadakis, *Electron. Lett.*, **27**, (11), 1002–3, 1991.
10. G. Hasnain, K. Tai, Y. H. Wang, J. D. Wynn, K. D. Choquette, B. E. Weir, N. K. Dutta and A. Y. Cho, *Electron. Lett.*, **27**, (18), 1630–2, 1991.
11. Ping Zhou, Julian Cheng, C. F. Schaus, S. Z. Sun, C. Hains, K. Zheng, A. Torres, D. R. Myers and G. A. Vawter, *Appl. Phys. Lett.*, **59**, (20), 2504–6, 1991.
12. C. J. Chang-Hasnain, J. R. Wullert, J. P. Harbison, L. T. Florez, N. G. Stoffel and M. W. Maeda, *Appl. Phys. Lett.*, **58**, (1), 31–3, 1991.
13. C. J. Chang-Hasnain, M. W. Maeda, A. Von Lehman, Chinlon Lin and H. Izadpanah, *LEOS Annual Meeting*, San Jose, CA, November, 1991; M. W. Maeda, A. von Lehman, J. R. Wullert, M. Allersma, H. Izadpanah, C. J. Chang-Hasnain, M. Z. Iqbal and Chinlon Lin, *Optical Fiber Communications Conf.*, San Jose, CA, February 1992.
14. M. Orenstein, E. Kapon, J. P. Harbison, L. T. Florez and N. G. Stoffel, *Appl. Phys. Lett.*, **60**, 1535 (1992).

15. M. E. Warren, P. L. Gourley, G. R. Hadley, G. W. Vawter, Y. M. Brennan, B. E. Hammons and K. L. Lear, *Appl. Phys. Lett.*, **61**, 1484–6, 1992.
16. (a) M. Orenstein, A. von Lehman, C. J. Chang-Hasnain, N. G. Stoffel, L. T. Florez, J. P. Harbison, J. Wullert and A. Scherer, *Electron. Lett.*, **27**, (5), 437–8, 1991; (b) A. von Lehman, M. Orenstein, C. J. Chang-Hasnain, N. G. Stoffel, L. T. Florez and J. P. Harbison, *OSA Annual Meeting*, Boston, MA, November 1990.
17. A. von Lehman, C. J. Chang-Hasnain, J. Wullert, L. Carrion, N. G. Stoffel, L. T. Florez and J. P. Harbison, *Electron. Lett.*, **27**, (7), 583–5, 1991.
18. G. Hasnain, R. A. Novotny, J. D. Wynn and R. Leibenguth, *Conf. on Lasers and Electro-Optics*, Anaheim, CA, 1992.
19. M. W. Maeda, C. J. Chang-Hasnain, Chinlon Lin, J. S. Patel, H. A. Johnson and J. A. Walker, *Photon. Tech. Lett.*, **3**, 268–70, 1991.
20. C. J. Chang-Hasnain, M. W. Maeda, J. P. Harbison, L. T. Florez and Chinlon Lin, *J. Lightwave Technol.*, **9**, 1665–73, 1991.
21. K. Tai, R. J. Fischer, C. W. Seabury, N. A. Olson, T.-C. D. Huo, Y. Ota and A. Y. Cho, *Appl. Phys. Lett.*, **55**, 2473 (1989).
22. M. Orenstein, A. von Lehman, C. J. Cnang-Hasnain, N. G. Stoffel, J. P. Harbison and L. T. Florez, *Appl. Phys. Lett.*, **57**, (24), 2384–6, 1990.
23. Y. H. Lee, B. Tell, K. F. Brown-Goebeler, J. L. Jewell and J. V. Hove, *Electron. Lett.*, **26**, pp. 710–11, 1990.
24. C. J. Chang-Hasnain, L. T. Florez, G. Hasnain, C. E. Zah, J. P. Harbison, N. G. Stoffel and T. P. Lee, *Appl. Phys. Lett.*, **58**, (12), 1247–9, 1991.
25. C. J. Chang-Hasnain, M. Orenstein, A. von Lehman, L. T. Florez, J. P. Harbison and N. G. Stoffel, *Appl. Phys. Lett.*, **57**, 218–20, 1990.
26. M. W. Maeda, C. J. Chang-Hasnain, A. von Lehman, H. Izadpanah, Chinlon Lin, M. Z. Iqbal, L. T. Florez and H. P. Harbison, *Photon. Tech. Lett.*, **3**, 863–65, 1991.
27. C. J. Chang-Hasnain, E. Kapon and R. Bhat, *Appl. Phys. Lett.*, **54**, 205–7, 1989.
28. R. S. Geels, S. W. Corzine and L. A. Coldren, *IEEE J. Quantum Electron.*, **QE-27**, (6), 1359–67, 1991.
29. C. J. Chang-Hasnain, J. P. Harbison, L. T. Florez and N. G. Stoffel, *Electron. Lett.*, **27**, (2), 163–5, 1991.
30. G. Hasnain, J. D. Wynn, S. Gunapalo and R. E. Leibengnth, paper JTHA2, *Conference on Lasers and Electrooptics*, Anaheim, CA, 14 May, 1992.
31. G. Hasnain, K. Tai, L. Yang, Y. H. Wang, R. J. Fischer, J. D. Wynn, B. Weir, N. K. Dutta and A. Y. Cho, *IEEE J. Quantum Electron.*, **QE-27**, (6), 1377–85, 1991.
32. M. G. Peters, D. B. Young, F. H. Peters, B. Thibeault, J. Scott, M. L. Majewski and L. A. Coldren, *Tech. Dig. CLEO '93 Conference*, pp. 136–8, Baltimore, MD, May 1993.
33. C. J. Chang-Hasnain, J. P. Harbison, C.-E. Zah, M. W. Maeda, L. T. Florez, N. G. Stoffel and T.-P. Lee, *IEEE J. Quantum Electron.*, **QE-27**, (6), 1368–76, 1991.
34. K. Y. Lau and A. Yariv, *IEEE J. Quantum Electron.*, **QE-21**, 125 (1985).
35. M. Nakao, K. Sato, T. Nishida, T. Tamamura, A. Ozawa, Y. Saito, I. Okada and H. Yoshihara, *Electron. Lett.*, **25**, 148–50, 1989; see also, M. Nakao, K. Sato, T. Nishida and T. Tamamura, *IEEE J. Select. Areas Commun.*, **8**, 1178–82, 1990.
36. C. E Zah, K. W. Cheung, S. G. Menocal, R. Bhat, M. Z. Iqbal, F. Favire,

N. C. Andreadakis, P. S. D. Lin, A. S. Gozdz, M. A. Koza and T. P. Lee, Paper WB5, *Opt. Fiber Conf., San Diego*, CA, February 1991.
37. D. Mehuys, M. Mittelstein, A. Yariv, R. Sarfaty and J. E. Ungar, Paper FL4, presented at the *Conf. Lasers and Electrooptics*, Baltimore, MD, 1989.
38. C. J. Chang-Hasnain, US Patent No. 5029176, July 1991.
39. C. J. Chang-Hasnain, E. Kapon, J. P. Harbison and L. T. Florez, *Appl. Phys. Lett.*, **56**, (5), 429–31, 1990.
40. D. E. Bossi, W. D. Goodhue, M. C. Finn, K. Rauschenbach, J. W. Bales and R. H. Rediker, *Appl. Phys. Lett.*, **56**, 420–22, 1990.
41. T. L. Paoli, D. W. Treat, R. L. Thornton and R. D. Bringans, *Proc. Int. Semiconductor Laser Conf.*, Davos Switzerland, 146–7, 1990.
42. K. Tai, G. Hasnain, J. D. Wynn, R. J. Fischer, Y. H. Wang, B. E. Weir, J. Gamelin and A. Y. Cho, *Electron. Lett.*, **26**, (19), 1990.
43. J. S. Patel, M. A. Saifi, D. W. Berreman, C. Lin and N. Andreadakis and S. D. Lee, *Appl. Phys. Lett.*, **57**, 1718–20, 1990.
44. Chapter 1 by Dan Botez, this book.
45. D. R. Scifres and W. Streifer, *IEEE J. Quantum Electron.*, **QE-15**, 917, 1979.
46. M. Matsumoto, M. Taneya, S. Matsui, S. Yano and T. Hijikata, *Appl. Phys. Lett.*, **50**, 1541, 1987.
47. G. R. Hadley, *Optics Lett.*, **15**, 1215–17, 1990.
48. C. J. Chang-Hasnain, unpublished.
49. C. J. Chang-Husnain, Y. A. Wu, G. S. Li, G. Hasnain, K. D. Choquette, C. Caneau and L. T. Florez, *Appl. Phys. Lett.*, **63**, 10, 1993.
50. Y. A. Wu, C. J. Chang-Hasnain and R. Nabiev, *Electronics Lett.*, **29**(21), 1861, 1993.
51. H. Wada, D. I. Babic, D. L. Crawford, T. E. Reynolds, J. J. Dudley, J. E. Bowers, E. L. Hu, J. L. Merz, U. Koren and M. G. Young, *IEEE Photon. Tech. Lett.*, **3**, 977–9, 1991.
52. B. Tell, R. E. Leibenguth, K. F. Brown-Goebeler and G. Livescu, *IEEE Photon. Tech. Lett.*, **4**, 1195–6, 1992.
53. T. Baba, Y. Yogo, K. Suzuki, F. Koyama and K. Iga, *OSA Quantum Optoelectronics Conference*, paper PD2-2, Palm Springs, CA (1993).
54. J. J. Dudley, D. I. Babic, L. Yang, R. Mirin, B. I. Miller, R. J. Ram, T. Reynolds, E. L. Hu and J. E. Bowers, *LEOS Annual Meeting*, paper SCL 4.1, San Jose, CA, Nov. 1993.
55. R. Schneider and J. A. Lott, *Appl. Phys. Lett.*, **63**(7), 917, 1993.

10

Individually addressed arrays of diode lasers

DONALD B. CARLIN

10.1 Introduction

One method by which the rate of information transmission can be increased is by the use of multiple, independently addressable sources to form parallel data channels. In the optical domain, this may be implemented by semiconductor diode lasers, which can be fabricated into monolithic arrays of emitters. The purpose of this chapter is to describe the development of linear arrays of index-guided diode lasers, and indicate directions in which the technology may proceed. Note that all the discussions in this chapter will be confined to semiconductor diode lasers in which the light propagates parallel to the epitaxial planes.

A few words are in order to explain the motivations that have driven the development of individually addressed diode laser arrays. Early work included the fabrication of both AlGaAs[1] and InGaAsP[2] devices for parallel data transmission through bundles of optical fibers, and arrays of lasers have often been considered in optical printing or optical data processing[3,4] applications. The application that has been most responsible for the development of these devices, however, is optical data storage. This discussion will focus on the use of multielement arrays in optical recording systems, as a very useful example that has driven diode laser array technology towards certain performance capabilities.

The commercial availability of reliable AlGaAs semiconductor lasers emitting up to 50 mW cw in a single spatial mode has allowed the practical realization of several different types of optical data storage systems that are now consumer and commercial products. Examples include: the very popular compact disk format for high-quality audio reproduction; the CD-ROM, which is the digital data equivalent of the audio disk; video disks that were once offered and abandoned, but which are now being marketed again; write-once-read-many-times (WORM) disks for archival applications; and erasable disk systems. These types of commercial product are targeted to meet data transfer rate and density requirements well below those that are already technologically possible.

10.2 Multichannel optical recording

Many applications are envisioned in which massive amounts of data (10^{11}–10^{15} bits) must be both recorded and retrieved at rates of many hundreds of megabits per second, often with fast random access to any bit of data in storage. In some of these applications, the recording medium must be erasable for repeated use.

Prototype optical recording systems were developed as early as 1981, which record and play back multiple channels of independent data simultaneously allowing commensurately multiplied data rates. As much as the early single-channel optical recording systems were based on the use of argon ion lasers, so were the initial investigations of multiple channel recording.

The maximum rate at which data can be recorded on a spinning optical disk is determined by the optical power density at the spot focused on the disk, the optical sensitivity of the recording medium, the disk rotation velocity, and the response of the servomechanisms required to keep the laser focused on the moving target. For single-channel systems using an argon ion laser as the recording source, the practical continuous recording rate available to the system user is about 33 Mb/s, although higher performance has also been reported[5]. Systems based on AlGaAs diode laser sources have decreased capabilities due to the longer source wavelength ($\approx$830 nm versus 488 nm). Despite this, the AlGaAs diode laser has completely supplanted other types as the preferred source in recording systems because of its compactness, high electrical-to-optical power conversion efficiency, ease of modulation, low cost, and long life. Some diode laser optical memory systems are now available that operate at data rates up to 24 Mb/s.

Such data rates, while impressive, will not suffice for imaging instruments and supercomputers planned for deployment and installation in the next decade. For example, the NASA Space Station is intended to be the platform for several imagers, each transmitting enormous amounts of data ($\approx 10^{13}$ bits/day) at 300 Mb/s data rates, the data transfer rate of a single channel of the Tracking and Data Relay Satellite System (TDRSS). One response to the need for large libraries of stored digital data has been the development of 'Jukeboxes' of optical disks by RCA in the early 1980s[6].

Two such systems are located at NASA's Marshall Spaceflight Center and the Rome Laboratory of the Air Force, each storing a total of 10^{13} bits distributed on 128 disks, with random access to any bit of data on any disk within six seconds. Both units use an argon ion laser source. The beam is split into two channels, each externally modulated by the use of acousto-optic deflectors in the NASA machine to double the data transfer rates. RCA had, in fact, implemented nine-channel heads in developmental argon laser recorders during this era, attaining recording rates of 300 Mb/s. The direction of the industry, however, demanded that practical high-performance optical recorders be based on solid-state diode laser technology.

Furthermore, the use of erasable media is critical for many data storage applications. Some systems must retain data only for relatively short periods. The disks must be erased frequently for these buffer memories to perform their function. In other applications, the availability of reusable media greatly decreases the cost of operation.

The most available erasable media currently being manufactured is based on a magneto-optic effect, in which magnetic domains of the material change the direction of their magnetization due to local heating by the laser in the presence of a magnetic bias field. In order to erase the data in a single pass of the laser, as is desired in most applications, the laser must be operated continuously at the power level required for recording. This doubles the thermal load on the laser and necessitates the use of a diode with approximately twice the output power capability as is necessary for archival recordings. Improvements in the output power capabilities of AlGaAs lasers have made diode-laser-based erasable optical recorders possible.

10.3 Individually addressable arrays

A monolithic linear array of individually addressed diode lasers, each falling within the field of view of imaging optics, is an obviously appropriate configuration for the realization of compact, practical, high-performance optical recorders. A schematic drawing of a ten-element array, which was developed to demonstrate the feasibility of such a source for optical recording[7], is shown in Figure 10.1[8]. Each diode laser is independently, directly modulated by the injection current supplied to the lasers via individual contacts, and the separate data channels are recorded simultaneously on multiple parallel tracks on a moving optical recording medium. The data transfer rate of the system is the sum of the data

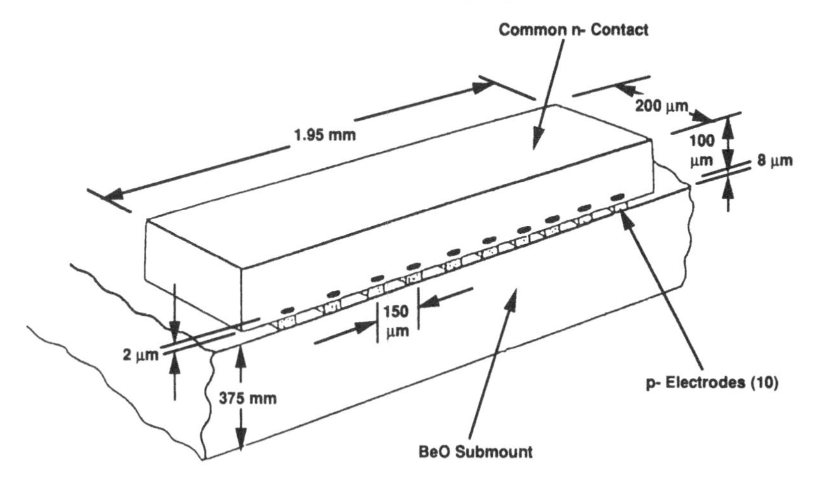

Figure 10.1. Schematic drawing of a ten-element array mounted junction-down on a BeO submount. (Ref. 8 – David Sarnoff Research Center/NASA.)

rates of each independent channel. Fabrication of the sources on a single semiconductor chip ensures that they will be in spatial and angular registry. All the sources lie on a straight line perpendicular to the optical axis, allowing the use of a single, wide field-of-view collimation lens to collect the light.

Because this chapter will use optical recording as an example of the incorporation of individually addressed arrays of semiconductor lasers in an optical system, it is important to understand the primary performance characteristics required of the lasers for them to be of practical use in this application. In virtually all practical optical recording systems, light is emitted from an index-guided semiconductor laser (that is, a laser in which the optical field is confined within a structure having a real index-of-refraction difference with respect to the surrounding material)[9]. The emitted radiation will be relatively free of optical aberrations and can, therefore, be focused to a diffraction-limited spot on the recording medium by the use of high-quality optics. The full-width-at-half-maximum (FWHM) of the diffraction pattern on the recording medium generally defines the recording area for most systems. The recorded spot size, d_0, is approximately:

$$d_0 = 0.56\lambda/NA, \tag{10.1}$$

where NA is the numerical aperture of the focusing objective and λ is the recording wavelength. For example, systems operating at 830 nm with a focusing objective of $NA = 0.65$ may record spots having diameters of

0.71 μm. Other critical output performance characteristics for arrays used in optical recording systems are summarized below.

10.3.1 Output power

Most erasable optical memories require the light source to be operated continuously (cw) at full power during the erase cycle; WORM systems may operate at an average duty-cycle of 50% or so, depending on the modulation code employed. In both cases, at least 30 mW from the emitting facet are required for the recording of data that can later be retrieved (with the laser operating cw at a few mW or less) with sufficient signal-to-noise and system margin for low bit-error-rates. Most systems manufacturers specify lasers emitting 40 or 50 mW cw for these reasons. Some systems require output powers of the order of 100 mW cw from the laser because of either the relative insensitivity (and, therefore, archival stability) of the medium or high media transport speeds for increased data transfer rates.

10.3.2 Transverse (to the junction plane) far-field emission profile

The light from the lasers must be collected and collimated by an optical system, the first element of which is basically a microscope objective having an infinite conjugate distance. The numerical aperture of this collimation objective should be at least equal to the full-width at $1/e^2$ of the maximum intensity of the emitted beam for optimized utilization of the optical power to form an intense spot at the recording medium[10]. This angle is roughly 1.7 times the full-width-at-half-maximum parameter that is most often quoted when discussing diode laser emission patterns. Thus, a diode laser having a transverse far field of 35° FWHM would require a collection objective of $NA = 0.5$. While this is not an excessively difficult NA to achieve for single on-axis emitters, it represents a reasonable limit for an objective that must also have a wide, flat field-of-view, as required by monolithic semiconductor array sources.

10.3.3 Parallel (to the junction plane) far-field emission profile

The far-field emission pattern of index-guided diode lasers parallel to the epitaxial layers is determined by the width and strength of the waveguide that forms the laser structure. The strength can be determined by the difference in effective index of refraction between the guide and the surrounding material. Typical parallel FWHMs for index-guided AlGaAs lasers are about 10°. After collimation, an anamorphic beam is formed. The anamorphic ratio is simply defined as the ratio of the transverse to parallel far-field beam dimensions. For efficient utilization, the beam cross-section should be made round to fill the final objective used to focus the light on to the medium. This anamorphic correction is usually effected by the use of one or two prisms, often in a Littrow arrangement[11], between the two objectives to expand or contract the beam in one dimension. The placement of anamorphic beam correction optics is often critical for multichannel array systems, in which the field of view of the system must be large. Anamorphic ratios of as high as five can be corrected by use of a high-index-of-refraction prism pair. A perfectly round emergent beam is most easily accommodated by the optical system. This is not usually the case for diode lasers optimized for high output power, but it is possible to use diode laser structures that have beam cross-sections that are not overly elliptical.

10.3.4 Astigmatism

Astigmatism is the primary Seidel aberration[12] of the emitted radiation pattern inherent in the semiconductor laser that can significantly degrade systems performance. Astigmatism can be defined as the positional difference between the apparent line sources from which light is emitted parallel and perpendicular to the junction. It arises in diode lasers because of the different optical confining mechanisms in the directions parallel and transverse to the epitaxial planes[13]. In the transverse direction, the light is confined by the real index-of-refraction difference between the active layer and the cladding layers of the semiconductor laser. This results in a well-defined transverse output profile that has a well-defined apparent line source that does not vary with injection current level. The guide parallel to the junction plane may be formed in many different ways. The apparent line source parallel to the junction is not generally at the same distance from the facet as the one transverse to the junction. In addition, the lateral waveguide often has an index-of-refractive difference that has an imaginary component that does vary slightly with output power. Thus, there may be both a static astigmatism and a component that varies with drive current. Most index-guided diode lasers have a static astigmatism of a few microns. For example, the channeled-substrate-planar (CSP) device was determined experimentally to exhibit approximately 2.5 μm of astigmatism by Hitachi, whereas gain-guided lasers had much larger values[14]. Note that an astigmatic focal difference of $d_0/2NA$ at the recording medium will result in the growth of the spot to $\sqrt{(2)}d_0$, an intolerable amount. The static astigmatism, however, can be canceled by proper adjustment of the anamorphic expansion elements of the optical system. It is the dynamic component that is uncorrectable by conventional optical systems. The effect of astigmatism at the source, however, is mitigated by the magnification effects of the optical system. If the magnification in the parallel dimension is denoted by M_p, then the astigmatism at the disk is decreased by a factor M_p^2 with respect to the source astigmatism. This can easily be a factor of ten or more in optical recording systems. Thus, even dynamic astigmatism is not a problem if index-guided structures are used.

10.3.5 Array size

The primary benefit of using arrays is the multiplicity of data channels allowed. Often, it is necessary that all operating laser elements of the array be contiguous; that is, dead, defective, or degraded emitters cannot be tolerated within the body of the array being used. It is possible to design and fabricate microscope objectives for collection and collimation of the laser light having relatively wide, flat fields-of-view. The cost of such optics grows rapidly, however, and the practical limit is a field-of-view of approximately 1.5 mm for objectives designed for optical recording. The physical dimensions of the emitting portion of the array must be contained completely within the field-of-view of the optical system, often with an additional margin for system error. An example of this would be an additional ±100 μm required in the field-of-view for operation of the tracking servomechanism in optical recorders.

10.3.6 Reliability

The operational lifetime requirements for field deployment of laser sources range from several thousand hours for some consumer and commercial products to over 10^5 hours for unattended systems in remote locations. An example of the latter might be an optical buffer memory for satellite communications[15]. These reliability requirements pertain to the entire source. Thus, if system functionality requires that all elements of an array operate to a certain specification, then the end-of-life of the source is determined by the first failure of any element of that array.

The reliability of diode lasers is often described in terms of log-normal statistics, in which the failure probability density function, $f_{\ln}(t)$, is written as

$$f_{\ln}(t) = (2\pi)^{-1/2} \exp(-\{[\ln(t/t_m)]^2/2\sigma^2\}), \tag{10.2}$$

where t_m is the median life and σ is the standard deviation of the distribution. The reliability function, $R(t)$, which is the percentage of devices still operating after time t is given by[16]

$$R(t) = \int_t^\infty f(t')\, dt'. \tag{10.3}$$

The probability that n emitters will still be functioning reliably after time t can then be expressed as

$$R_n(t) = \left[\int_t^\infty f(t')\, dt'\right]^n. \tag{10.4}$$

This assumes that the failures are independent. This analysis would, therefore, be appropriate for either an n-element array, in which the failures of the elements were uncorrelated, or for a sample of identical, individual lasers.

The mean-time-to-failure of an n-element array, τ_n, can then be written as

$$\tau_n = \int_0^\infty R_n(t)\, dt \tag{10.5}$$

or

$$\tau_n = \int_0^\infty \left[\int_t^\infty f(t')\, dt'\right]^n dt. \tag{10.6}$$

Equation (10.6) can be solved using the log-normal statistics of Equation (10.2). A numerical integration is shown in Table 10.1, which summarizes the expected mean-times-to-failure for two-, four-, six-, eight- and ten-element arrays as a function of σ. The expected life of an n-element array in terms of the expected life of a single laser is represented by τ_n/τ_1. Table 10.1 indicates that it is critical to have laser elements that are described by a failure probability density function having a low σ if reasonable life is to be expected of arrays of more than a few contiguous emitters.

Early investigations of AlGaAs laser failures[17] indicated that $\sigma \approx 1.45$ for a distribution of individual oxide-stripe devices that obeyed log-normal statistics. The diode laser structures that were incorporated in the first reported monolithic high-power arrays[7,18,19] were all double heterostructures grown by liquid phase epitaxy (LPE) and could, therefore, be expected to have relatively high standard deviations. It must be noted,

Table 10.1. *Ratio of the expected mean-time-to-failure of an n-element array to that of a single emitter*

σ	n: 2	4	6	8	10
0.25	0.86	0.76	0.72	0.69	0.67
0.5	0.72	0.56	0.49	0.45	0.43
1.0	0.48	0.27	0.21	0.17	0.15
1.5	0.29	0.12	0.07	0.06	0.05

however, that arrays of contiguous elements differ from populations of the same size of individual emitters in a very significant way. The array elements all come from a confined area of a single wafer. The growth and fabrication of the devices can be expected to be more uniform than for devices selected at random across a single wafer, and considerably more uniform than for devices selected from different wafers. In addition, defects that lead to element degradation are likely to be clustered rather than randomly distributed. It is not at all clear that log-normal failure probability distributions based on individual emitter failures are rigorously correct. It is more appropriate to develop empirical data on the failures of n-element arrays, and consider each array as the individual unit rather than the laser element that comprises the array. Nevertheless, the statistics shown above provide a useful framework for predicting the failure of arrays.

Other statistical distributions have also been applied to the failure of monolithic arrays. NEC[20], for example, has used Weibull statistics, in which the failure probability density function is expressed as[16]:

$$f_w(t) = m(t^{m-1}/t_0^m)\exp[-(t/t_0)^m], \tag{10.7}$$

where m is a shape parameter that determines the standard deviation of the failure distribution and t_0 is a characteristic time for failure. The mean-time-to-failure, τ_1, can then be calculated to be

$$\tau_1 = t_0\Gamma(1 + 1/m), \tag{10.8}$$

where Γ is the gamma function[21]. Generalizing this expression for an n-element array allows the mean-time-to-failure to be expressed as

$$\tau_n = n^{-1}/m\tau_1. \tag{10.9}$$

The parameters t_0 and m may be determined empirically. For example, NEC has estimated that the mean-time-to-failure of their single-element double-quantum-well laser is 2500 hours[22] and has also estimated[20] that $m \approx 5$. For this value of m, the ratio τ_n/τ_1 becomes 0.76 for a four-element array and 0.66 for an eight-element unit. This experimental data confirms the fact that the standard deviation of the failure probability density function is low; it would be in the neighborhood of 0.25 for log-normal statistics, as shown in Table 10.1. Thus, expected array reliability can approach that of single devices, at least for quantum-well structures grown by metal–organic chemical vapor deposition (MOCVD) techniques.

10.3.7 Wavelength stability and noise

The uniformity and stability of the emission wavelength across the array can be important because of dispersion in the optical system and/or signal-to-noise requirements for data retrieval. In the first case, some optical elements, such as prisms and gratings, are strongly wavelength dispersive. If the laser elements in an array oscillate over an appreciable wavelength range, then these dispersive effects will shift the direction of each beam. These effects can be appreciable, especially in LPE-grown structures in which variations in emission wavelength of ± 5 nm were often observed[18].

A typical index-guided AlGaAs laser may operate in one or several Fabry–Perot longitudinal modes (typically separated by 0.3 nm) and may vary with temperature at the rate of 0.3 nm/°C. While these effects seem small, they can shift the location of the spots focused on the recording medium by several tenths of a micron. The thermal drift effect is not normally of great concern for optical recording applications, as the temperature of each element of the array changes by roughly the same amount.

Mode hops, however, can occur independently at each array site. In addition, mode hops are a source of noise during data retrieval when the laser elements are operated cw at low power. This mode-hop noise can affect the bit-error-rate of the system, especially in magneto-optic data storage systems. These systems require linearly polarized light at the disk surface, which is rotated by the Kerr effect. The resultant elliptical polarization is detected as the signal. Mode-hop noise can be serious in magneto-optic recording systems because of the limited signal level available through the Kerr rotation. In most non-erasable systems, the problem is less serious because the data is recorded by thermal effects and is retrieved by either intensity contrasts or phase shifts. In either case, linear polarization is not necessary. Consequently, polarizing beam splitters and quarter-wave retardation plates may be positioned in the optical system for the dual purpose of easily extracting the readout signal and minimizing the light reflected from the recording medium directly back into the laser, a condition which exacerbates mode hops.

Because this is not possible in magneto-optic systems, other techniques, such as high-speed modulation of the laser during data retrieval[23,24], are used to reduce the effects of feedback. Another technique could be the use of a wavelength-stable laser structure, such as the distributed Bragg reflector (DBR) or distributed feedback (DFB) laser, which would minimize the systems effects of thermal drift and eliminate mode hops.

Relative intensity noise (RIN)[24] can be measured as a function of optical feedback to characterize the noise performance of lasers operating at low power for data retrieval in optical memory systems. Values of the order of -130 dB/Hz are often required for high-performance, high-data-rate systems.

10.3.8 Modulation rates

The data rates of individual array elements that are modulated are quite modest by communcations standards. The maximum rate typically used is 24 Mb/s, although higher rates have been demonstrated[18]. The noise-suppression technique of modulation of the lasers at hundreds of MHz in the readout mode[23] requires that the electrode pattern on the submount and the leads to the submount be designed for simultaneous operation at such rates.

10.4 Interelement isolation and array packaging

Three types of interactions must be considered as sources of crosstalk between array elements: thermal, optical and electrical. Thermal crosstalk can be a major limiting factor in the performance of individually addressed arrays and, consequently, is discussed in some detail in the following section.

Two other considerations are important in determining the interelement separation. These are the mechanical alignment tolerances for positioning the laser array chip on the individual addressing electrodes of the submount and the field-of-view of the optical system for which the laser array is intended.

For junction-down mounted arrays, alignment of the individual contacts on the laser elements to the contact electrodes can be difficult because the interface is not directly visible. The electrodes must be separated by an adequate distance to ensure that solder bridges will not form during the bonding operation. Such constraints generally impose a minimum distance allowed between the electrode fingers of the order of 10 μm.

As noted earlier, the flat field-of-view of collection objectives is limited to approximately 1.5 mm. Since many of the early systems were aimed at operating with eight or ten multiple channels, it was clear that an upper limit of approximately 150 μm between laser emitters was allowed.

10.4.1 Interelement interactions

Early work on the CDH–LOC arrays by Botez and Connolly[7] was aimed at determining the minimum separation allowed by optical interactions in this type of laser structure. This structure incorporates large-optical-cavity, which is a guide layer designed to leak light from the emitting region towards the adjacent elements. Arrays of such laser were grown with 64, 74, 84, 94, 150 and 200 μm interelement separations. The combined far field showed spatial modulation, at periods corresponding to the diffraction pattern to be expected from the emitter spacings, when optical coupling was present. No optical interactions were observed at device separations ≥ 94 μm. Laser structures that do not incorporate a guide layer should be expected to exhibit optical interaction effects over shorter distances.

Electrical isolation between elements is necessary because the heavily doped confining layers will conduct carriers laterally. Isolation is typically effected by either etching channels through the active layer of the epitaxial structure or by proton bombardment in the regions between laser elements. Both electrical isolation schemes have achieved good results and electrical crosstalk has generally not been significant, at least for recording applications. For example, Mehuys *et al.*[25] observed that electrical interactions between adjacent elements, separated by 50 μm, with each emitting 30 mW, were suppressed by 33 dB at frequencies near 1 MHz, as shown in Figure 10.2.

10.4.2 Thermal issues

For all diode layers, performance and life degrade at elevated temperatures. This problem is exacerbated for the case of arrays of closely spaced emitters. The junction temperature of each element is not only determined

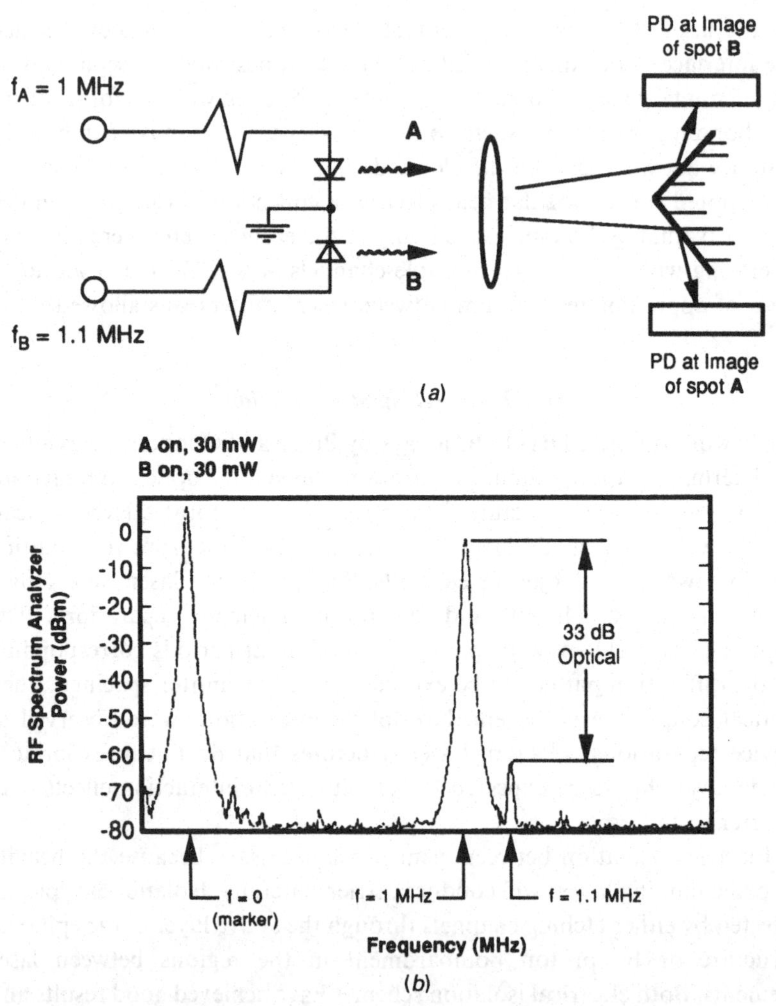

Figure 10.2. Electrical crosstalk between two adjacent array elements: (*a*) measurement technique; (*b*) spectrum. Laser A is modulated at 1 MHz while laser B is modulated at 1.1 MHz, with both lasers emitting 30 mW. The 1.1 MHz component in the output of B is reduced by 33 dB compared to the 1 MHz signal. (From Ref. 25 – SDL/*Electronics Letters*.)

by the spreading of its own heat through the body of the array and the heat sink but also by the heat generated by neighboring lasers. The waste heat generated by each diode laser (up to 250 mW) can be substantial, and could, therefore, easily perturb the operating conditions of adjacent devices in an array. Thermal interactions can be manifested as either decreased output power of adjacent operating elements or as emission wavelength variations. The magnitude of the thermal effects depends on the light-conversion efficiency of the laser elements, the thickness of the chip and the proximity of the heat sinks, as well as the choice of substrate materials.

The array is generally bonded to a submount, which provides individual electrical contact to each of the diodes. The design of the submount is also dictated by the need to minimize the thermal resistance between the heat sources (the diode laser junctions) and a heat sink. Beryllia (99.5% BeO) is a suitable material for this purpose as it has a relatively high thermal conductivity (2.2 W/(cm °C)) and a coefficient of linear thermal expansion which matches that of GaAs, minimizing thermally induced stress at the array–submount interface. Other materials have also been used successfully. Early work on the thermal performance of arrays at IBM[26] was based on the use of silicon as a submount. Sanyo has also used silicon. Both NEC and NTT have reported the use of array submounts made of silicon carbide. Other candidate materials include cubic boron nitride, aluminum nitride, Type IIa diamond, and, recently, CVD-deposited diamond. Table 10.2 lists submount materials that have been used or considered for high-power diode laser array applications.

Arrays may be mounted junction-side down or junction-side up on the submount. Junction-down mounted arrays require submount materials that are electrically insulating, whereas it is possible to use electrically conducting materials for junction-up arrays. Junction-down mounting reduces the temperature rise of the junction by eliminating the part of the thermal resistance attributed to most of the low thermal conductivity

Table 10.2. *Submount materials*

Material	Thermal conductivity (W/(cm °C))	Coefficient of thermal expansion ($\times 10^6$/°C)
GaAs	0.54	6.8
BeO	2.2	7.0
Al_2O_3	0.37	5.3
Si	1.4	3.0
Type IIA diamond	20.0	0.8
CVD diamond	13.0	2.0
cBN	6.0	3.7
SiC	2.7	3.4
Cu	4.0	16.8
Thermkon 62	2.1	6.8

($K \approx 0.54$ W/(cm °C)) GaAs substrate itself. This component of the thermal resistance can easily be in the range of 100 °C/W for the relatively short ($\approx$250 μm cavity length), narrow ($\approx$5 μm) devices that typify early LPE-grown structures. These devices were also relatively inefficient and could easily contribute over 100 mW/element to the thermal load, imposing an added temperature rise of the order of 10 °C. Junction-down mounting, however, places greater constraints on the accuracy with which the chip must be mounted on the metallized submount. The active layer is only $\approx$2 μm from the chip–submount interface. Therefore, the front facet of the array must be positioned to within several microns of the submount edge across the entire length of the emitting portion of the array. This is to avoid obscuring or reflecting part of the highly divergent output beam (transverse to the junction plane) if the array chip is too far back from the edge, and also to avoid the generation of hot spots if the chip overhangs the submount.

It is, therefore, much simpler to mount the arrays junction-side up mechanically. This was not possible for early high-power arrays. Many of the newer, more efficient devices, however, allow this geometry despite the inherently higher thermal resistance.

10.4.3 Thermal crosstalk

Because of the sensitivity of array operation and life, significant attention has been given to the calculation of the temperature rise of the individual elements, as well as to the additional temperature increases due to the operation of neighboring array elements. Laff *et al.*[26] modeled the heat flow for an array structure by modification of a thermal analysis for single diode lasers[27]. More recently, various finite element analyses have been used to model thermal effects in arrays.

A common measure of thermal crosstalk between array elements is the percentage decrease in output power of one element when a neighboring element is also excited. Such thermal interactions increase with decreasing interelement separation. This effect is larger at low output powers typical of data retrieval, when the lasers are operating just above threshold, than at the high powers used to record or erase data, although the total thermal load on the array obviously increases with increasing output power.

Yamaguchi *et al.*[28] measured thermal crosstalk for four-element arrays of LPE-grown channeled substrate inner-stripe devices on 100 μm centers. A novel feature of this work was the incorporation of a channeled silicon light guide, positioned at the rear facet of the array, to enable independent monitoring of the emission of the lasing elements by rear-mounted photodetectors. One purpose of this arrangement is to allow the use of photofeedback to minimize the effects of thermal interactions between the elements[29]. Crosstalk of 3.6% was observed for adjacent emitters operating at 30 mW. This decreased to 2.6% when the photodetectors were used in a power control servosystem, demonstrating the effectiveness of the technique. This level of crosstalk is still substantial, however. Such LPE-grown devices have relatively low electrical-to-optical power conversion efficiencies of ≈20% and, therefore, generate substantial heat (≈120 mW) during operation. These authors also calculated the temperature rise distribution and modeled the effects of thinning the substrate of this junction-up mounted array. The elements adjacent to a 30 mW emitter exhibited a temperature rise of 4 °C when the total array thickness was 120 μm with a cavity length of 250 μm. This was reduced to 2.5 °C when the substrate was thinned to 60 μm. In addition, the effect of cavity length on thermal interactions was evaluated. Because of the greater area over which heat can be dissipated, 120 μm-thick arrays having a 400 μm cavity length also showed a reduced temperature rise, in this case to 3 °C. This work has been extended to eight-element arrays on 50 μm centers[30]. A temperature rise of 2.4 °C for adjacent emitters was observed at 20 mW output.

Mehuys *et al.*[25] have described thermal crosstalk in quantum-well arrays fabricated by MOCVD techniques in the same terms. They note that for lasing elements that have relatively low threshold currents (determined by the output facet reflectivity), the crosstalk between adjacent lasers is relatively insensitive to peak output power. In this study, the array elements were separated by 50 μm. For low threshold current array elements (10 mA) and with each element operating at the same peak power, crosstalk ranged from about 1% at 5 mW to 0.6% at 40 mW. For higher threshold devices ($I_{th} \approx 22$ mA), the crosstalk was about 8% at the low power levels typically used for data retrieval, decreasing rapidly and leveling off to about 1.5% at output powers exceeding 20 mW. Even the latter results demonstrate that substantially decreased thermal crosstalk can be achieved due to the greater power conversion efficiency of quantum-well devices. The benefit of low crosstalk as a result of the use of low threshold structures has also been noted by Thornton *et al.*[31] for very closely spaced devices (10 μm) operating at lower output powers.

Other techniques have also been investigated for reducing the thermal resistance of arrays and minimizing thermal crosstalk. NTT[32,33] has shown that neither thinning the cap layer of a junction-down device from 7 μm to 3 μm, nor increasing the depth of interelement isolation channels improves the thermal resistance significantly. Murata and Nishimura[32,34] have also implemented a second heat sink, bonded to the top side of a junction-down mounted array. This second heat sink is made of copper and is indium-soldered to both the n-side of the arrays of window diffusion stripe lasers[35] and the submount to the sides of the array chip itself. The use of this secondary heat sink was reported to reduce the temperature rise of an eight-element array on 50 μm centers, operating at 100 mW cw, from 26 °C to 16 °C[34].

Thermal crosstalk can also cause a variation in the emission wavelengths of the lasing elements. As the data reported indicates, temperature variations of several degrees centigrade are predicted when neighboring elements are turned on and off. As noted earlier (in Section 10.3.7 'Wavelength stability and noise'), an emission wavelength thermal variation of about 0.3 nm/°C is typical of AlGaAs Fabry–Perot lasers. Thus, several nanometers of wavelength drift are expected. It is often important, therefore, to measure the emission spectra of neighboring array elements to determine the system effects of interelement thermal crosstalk.

10.5 Representative array structures and reported results

During the early 1980s, diode laser technology was driven by the need for a high-power (≈20 mW cw at that time) index-guided device capable of emitting laser light in a fundamental spatial mode for optical recording applications. The light output from index-guided diode lasers is characterized by having negligible astigmatism, and can therefore be focused to a diffraction-limited spot by the use of suitable optics to produce high-quality recordings. These efforts led to the development of the CDH–LOC (constricted double-heterojunction, large optical cavity)[36], which was the first index-guided diode laser used to write data in an optical recording system[37].

Soon afterwards, development of the CSP structure[38] resulted in useful, high-power, single spatial mode lasers. Both the CDH–LOC[7,18] and CSP[19] structures were used as the laser elements of early individually addressed arrays. These types of lasers are fabricated by liquid phase epitaxial growth. As a representative example, the CSP is grown over a single channel etched using photolithographic techniques into a GaAs substrate. The layers of the structure grow flat, as the word 'planar' in

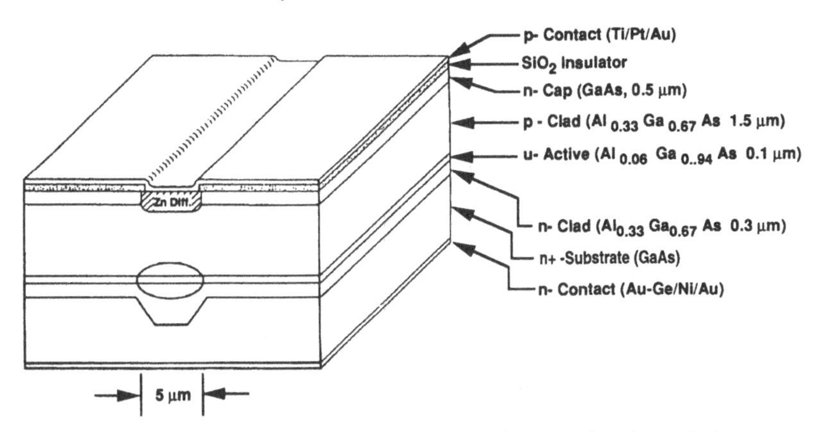

Figure 10.3. Schematic drawing of a CSP laser, showing the channeled substrate and Zn-diffusion for current injection.

the name of the device suggests; the laser growth is, therefore, relatively easy to reproduce. The structure of this device is depicted in Figure 10.3, with the compositions and thicknesses of the layers shown on the right of the diagram. Light is generated in a thin ($\approx$0.1 μm-thick) active layer of undoped $Al_{0.06}Ga_{0.94}As$, sandwiched between cladding layers of p-doped and n-doped AlGaAs (top and bottom of the diagram, respectively) of higher aluminum concentration. This stack of layers forms a real-index guide, which confines the laser light, in the transverse direction (perpendicular to the heterojunction plane). The injection current is confined both by a narrow contact stripe in the p-side metallization, defined by an SiO_2 insulation film deposited on the GaAs cap layer, and by the diffusion of a zinc finger partially through the p-cladding layer. The injection current would itself confine the lateral spatial mode (parallel to the junctions) by affecting the gain distribution, but such a gain-guided laser structure would emit an astigmatic beam of light. The lateral spatial mode is also well-confined by the shoulders of the channel. The laser light has a higher energy than the bandgap of the n–GaAs substrate material and is, therefore, heavily absorbed outside the channel region. In addition, the indices of refraction of the layers support radiation modes into the substrate which also increase the loss outside the channel.

Figure 10.4 depicts three elements of an array of CSP diode lasers. The lasers are separated by 150 μm, defined by the mask pattern used to etch the channels and contact stripes photolithographically. In this case, electrical crosstalk between elements is eliminated by the formation of

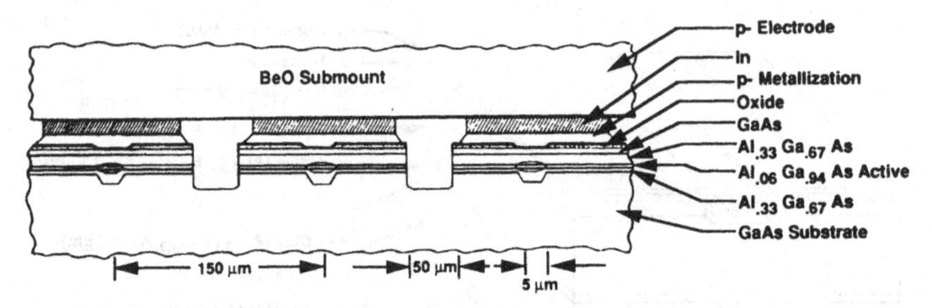

Figure 10.4. Details of a monolithic array on a BeO submount, showing three CSP elements on 150 μm centers. The etched 50 μm-wide isolation channels (not drawn to scale) are approximately 3 μm-deep. (Ref. 19 – David Sarnoff Research Center/*IEEE Journal of Quantum Electronics*.)

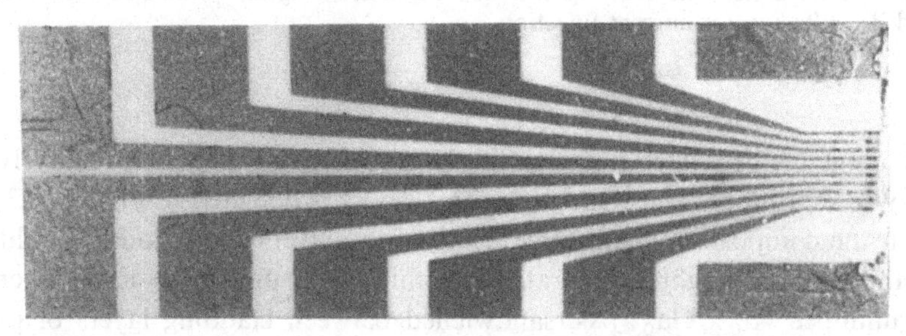

Figure 10.5. Photograph of a ten-element CSP array mounted junction-down on a BeO submount. Two large-cross-section gold ribbons attach the n-side of the array to two contact pads at the right of the submount in this picture. (Ref. 18 – David Sarnoff Research Center/*Applied Optics*.)

50 μm-wide channels between the elements. These channels are ion milled through the p-contact metallization (comprising Ti–Pt–Au), cap, p-confinement, and active layer of the CSP structure. Optical interactions between laser elements are insignificant at 150 μm separations in the CSP design.

A photograph of a ten-element array bonded to a submount is shown in Figure 10.5. This 5.3 mm × 16.3 mm × 180 μm-thick submount is made of beryllia (99.5% BeO). The p-side of the array is soldered directly to ten electrodes formed on the submount. Connections to the n-side of the arrays are made through two large cross-section gold ribbons to two electrode pads on either side of the array chip. The submount is attached to a thermoelectrically cooled copper heat sink.

Figure 10.6 shows superpositions of the power-versus-current characteristics, lateral far fields, and spectra of the ten emitters of this array after

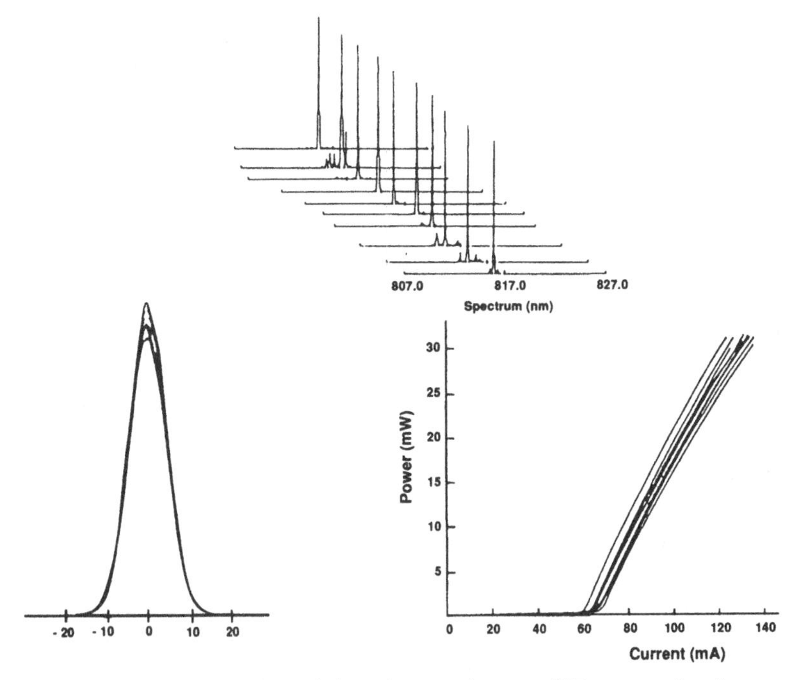

Figure 10.6. Output characteristics of a ten-element CSP array, showing superpositions of the ten lateral far-field intensity profiles and cw power-versus-current characteristics, as well as the spectra of the emitters. These data were taken at 23 °C after 100 hours of burn-in at 30 mW cw. (Ref. 40 – David Sarnoff Research Center/NASA.)

100 hours of burn-in at 30 mW cw[33]. The average FWHM of the lateral far fields is 10°. For this array, the emission wavelengths range from 815.1 to 818.1 nm. Some of the elements are not emitting in a single spectral mode at 30 mW cw, although in each case most of the emitted power is predominantly in a single mode. The threshold currents range from 58 to 68 mA. The uniformity of these characteristics is good for such a large array grown by liquid phase epitaxy. The relatively low power conversion efficiencies, around 15%, generate considerable waste heat that affects long-term performance and life negatively.

Over the past decade, metal–organic chemical vapor deposition techniques and source materials have developed rapidly. MOCVD offers benefits in uniformity of layer thicknesses and compositions over those possible by LPE growth. In addition, MOCVD growth allows the incorporation of quantum-well active regions, which allow greater efficiency

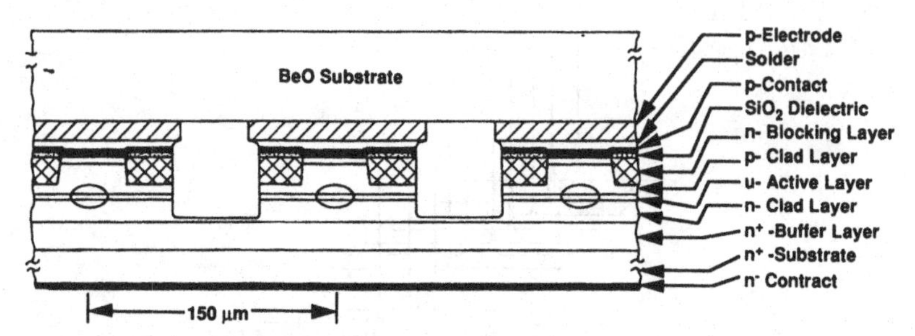

Figure 10.7. Schematic showing three elements of an individually addressed ICSP array. (Ref. 40 – David Sarnoff Research Center/NASA.)

incorporation of quantum-well active regions, which allow greater efficiency than do double heterostructures. MOCVD growth does not fill in steps in the same manner as does LPE and, therefore, different laser structures must be designed.

One such is the inverted channeled-substrate-planar (ICSP)[39], as depicted schematically in an array geometry in Figure 10.7. The device structure shown is a double heterostructure, but arrays have also been fabricated having separate confinement heterostructure quantum-well active regions. Such SCH–QW ICSP arrays have exhibited power-conversion efficiencies as high as 50%, with each element emitting more than 30 mW cw. In either case, the epitaxial layers are grown as for any simple planar laser. A ridge is then created by photolithography and etching through the p-cap and part of the p-confinement region. Guiding in the ICSP is also by absorption of light, as in the CSP. In the ICSP, however, the absorption is in the regrown GaAs, indicated by the cross-hatched portions of the structure in Figure 10.7. This layer is n-doped to also provide additional current blocking for lateral current confinement. Threshold currents of 12 mA were measured for laser arrays fabricated at the David Sarnoff Research Center[40].

NEC also began development of individually addressed arrays by using an LPE-grown laser structure[41], and demonstrated four-track optical recording with this device[42]. In this case, a new planoconvex waveguide (n–PCW) structure was grown over a channel in an n–GaAs substrate, similar to the growth method for CSP devices. This group then incorporated their version of an MOCVD-grown ICSP double-heterostructure into an eight-element array on 100 μm centers[43]. Each element emitted more than 60 mW cw in a fundamental transverse spatial mode. This laser structure was later modified to incorporate a double-quantum-well separate-confinement-heterostructure active region[22]. The authors compared double-quantum-well lasers to those having a single well. They found that the relative intensity noise was about 15 dB lower for the DQW than for the SQW lasers. The double-quantum-well devices maintained a RIN below −130 dB/Hz at 3 mW, modulated at 650 MHz, even in the presence of optical feedback levels as high as 10%. The output power from this type of array was reported to be more than 80 mW cw/element, with threshold currents of 20 ± 1 mA[44].

A window diffusion stripe structure[35] has been used as array elements by workers at the Mitsubishi Electric Corporation[45,46]. In this structure, planar epitaxial layers are grown by MOCVD. Zinc diffusion through an n-type active layer defines the narrow (≈2 μm) p-type active stripes. Termination of the Zn-diffusion 15 μm from the cleaved facets leaves a non-absorbing window region, which increases the output power at which catastrophic optical damage occurs. Each of the four lasers in an array on 100 μm centers, mounted junction-up on a cubic boron nitride submount, operated to 100 mW cw in a stable fundamental transverse spatial mode. At 50 mW cw at 25 °C, the emission wavelengths were 830.6 ± 0.6 nm, indicating good device uniformity. The lateral and transverse far fields measured 9–11° and 29–31° FWHM respectively. Astigmatism was also measured at 5 ± 1 μm for each of the lasers, and did not vary substantially or systematically as a function of output powers up to 50 mW cw.

Spectra Diode Laboratories has also reported on the performance characteristics of multielement arrays fabricated by MOCVD growth[47–50]. Nine-element arrays, on 150 μm centers, exceed 150 mW cw in a single fundamental spatial mode mounted junction-up on a submount. The threshold currents were 15.0–15.5 mA, and the emission wavelengths varied from 853.9 to 854.9 nm, again demonstrating the high degree of uniformity associated with MOCVD growth techniques. Eighteen-element arrays on 50 μm centers were also fabricated and mounted junction-up. Each element emits 80 mW cw, as shown in Figure 10.8, in fundamental spatial modes that are typically 10° × 30° FWHM.

10.6 Surface-emitting arrays

Thus far, progress in the development of individually addressed arrays of high-power index-guided lasers has been predominantly confined to linear, edge-emitting devices. Recent advances in high-power surface emitting, of both the deflecting mirror[51–53] and grating-surface-emitter (GSE)[54–59], types (also see Chapter 2 in this volume by Welch and

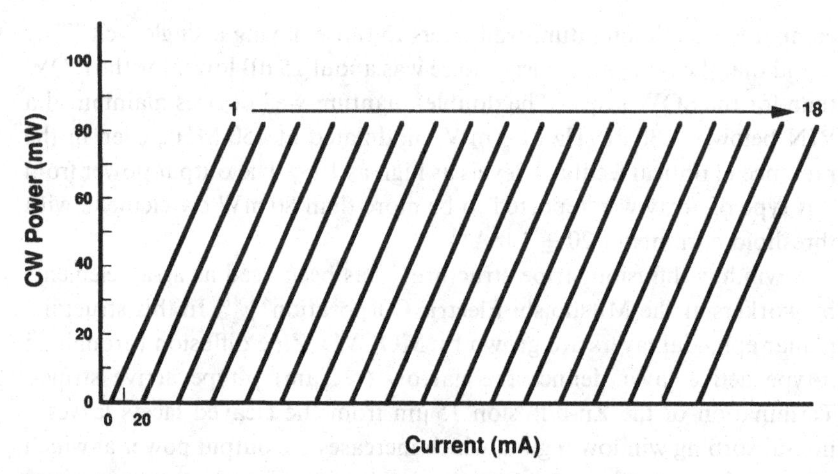

Figure 10.8. Power-versus-current characteristics of an 18-element array on 50 μm centers. (From Ref. 48 – SDL/*Electronics Letters.*)

Mehuys), can be configured into two-dimensional source geometries, with concomitant potential benefits for many applications. In both cases, light is generated and propagates along the direction of the epitaxial planes, as opposed to the vertical-cavity-surface-emitter laser (also see Chapter 9 in this volume by Chang-Hasnain). The emitted light is deflected out of the junction plane by either an etched mirror or a Bragg diffraction grating, as shown in Figure 10.9.

A variety of geometries can be implemented using either type of structure. An example of an individually addressed GSE–DBR array is shown in Figure 10.10. In this case, a second-order Bragg output grating diffracts light (in first order) normal to the wafer surface. Second-order diffraction provides feedback to the gain section. Either a first-order grating (depicted) or an etched facet can provide feedback on the other side of the gain region with no output emission from that end. The gratings also impart wavelength stability to the output. As shown in the diagram, a two-dimensional array may be configured as two columns of parallel emitters. For increased system life, the second column can provide redundancy for elements that might fail in the first column. This could be critical for operation in remote locations where the source is not accessible for replacement. Another function that this type of geometry affords is a second column of emitters for direct-read-after-write (DRAW) confirmation that data that were intended to be recorded were, in fact, recorded.

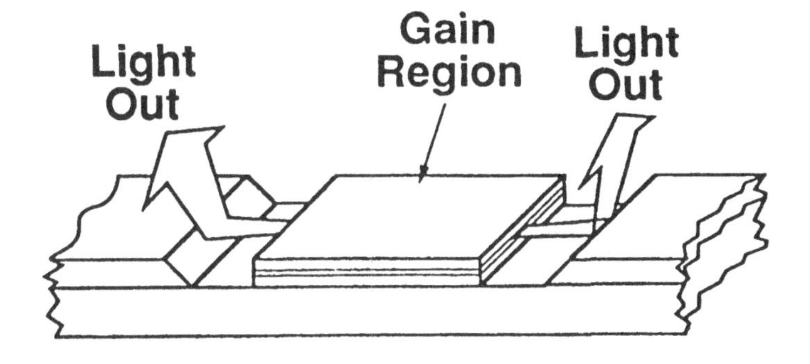

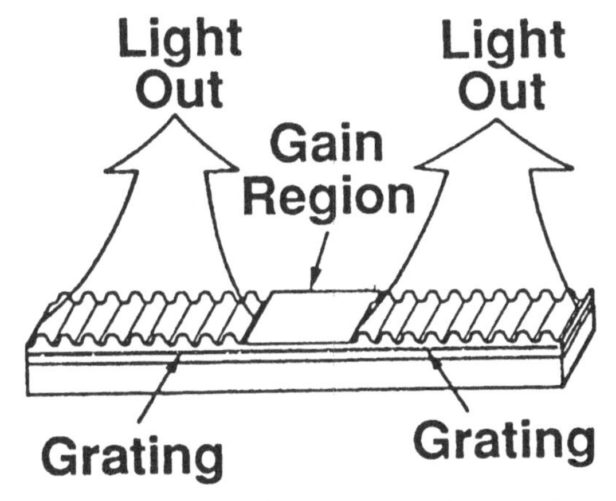

Figure 10.9. Etched-mirror-surface-emitter (top) and grating-surface-emitter (bottom) lasers.

Figure 10.11 shows an alternative geometry in which the emitting gratings (or mirrors) are interdigitated. This allows a factor of approximately two closer packing of the array than would be allowed by edge-emitting devices. This is because the heat generated by the array elements comes from the gain sections and, in the geometry shown, the gain sections for adjacent emitters can still be relatively widely separated.

Grating surface emitters have also shown the capability of emitting high temporally[54] and spatially[55,58] coherent power. High input power is critical for efficient frequency doubling of diode laser light in non-linear materials. Generation of short-wavelength light is an area of active development with important benefits for optical recording in particular. As can be seen from Equation (10.1), the data transfer rate and storage density increase as the recording wavelength and its square respectively.

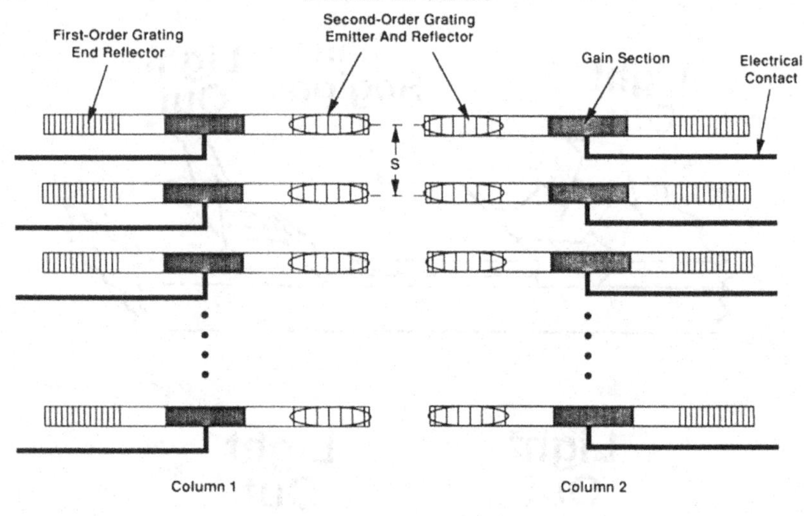

Figure 10.10. Two-dimensional grating-surface-emitting array suitable for optical recording applications. The structures shown have second-order emitting gratings and first-order gratings to provide feedback to the gain sections. The second column of emitters can be used as either redundant elements, for improved system reliability, or as sources for direct-read-after-write error detection.

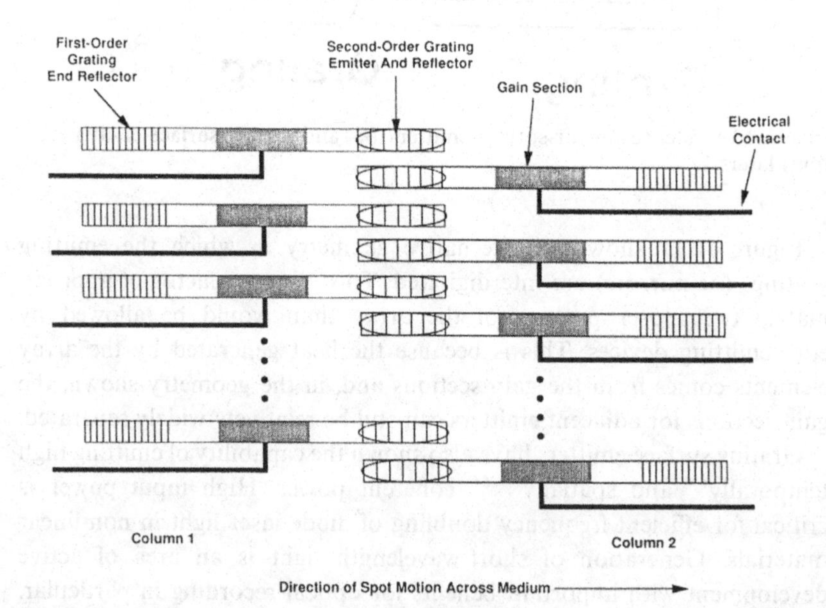

Figure 10.11. Interdigitated GSE array to pack the sources closer together while maintaining large separations between the sources of heat.

10.7 Summary

The advantages of multiple emitters in monolithic linear arrays as light sources for parallel, high rate data transfer have been proven, not only in optical recording applications, but also in optical communications through fiber bundles and in printing as well. Considerable advances have been made in fabrication technologies that have allowed continued improvements in the output power, power-conversion-efficiency, and uniformity of emission characteristics across long arrays. These improved arrays show strong evidence of having reasonably long life, consistent with their installation in practical commercial equipment. A variety of packaging geometries and materials have been developed to minimize the effects of dissipated heat while addressing each laser element independently. In fact, a number of semiconductor laser manufacturers offer array products aimed at recording applications having up to four elements, or have developed prototype units having greater numbers of lasers. In addition, there are many possibilities for lower-power arrays in communications, printing and scanning applications that may be realized with these same technologies using lower power, more closely spaced arrays having more contiguous elements.

Also, the development of surface emitters allows two-dimensional geometries for individually addressed arrays that can potentially improve the performance of many types of system. The high-power capabilities of these new devices may lead to arrays of multiple emitters in the blue region of the spectrum through frequency doubling.

Acknowledgments

I would like to thank many of my current and former colleagues at the David Sarnoff Research Center for their direct contributions to work in this field over the years. In particular, the efforts of John Connolly, Gerry Alphonse, Maria Harvey, Charlie Kaiser, Nancy Hughes, Bernard Goldstein, Dan Botez and Jim Bednarz were critical to early successes in the development of high-power index-guided arrays and their use in multichannel optical recording systems. In addition, I would like to acknowledge the help provided by Dietrich Meyerhofer, Mike Lurie, Bob Amantea, Paul Stabile, and Steve Miller in useful and informative discussions during the course of writing this chapter. Lastly, I would like to thank Bob Bartolini and Mike Ettenberg for their continued support of this and related work.

References

1. J. D. Crow, L. D. Comerford, J. S. Harper, M. J. Brady and R. A. Laff, Gallium arsenide laser-array-on-silicon package, *Appl. Opt.*, **17**, 479 (1978).
2. L. A. Koszi, B. P. Segner, H. Temkin, W. C. Dautremont-Smith and D. T. Huo, 1.5 μm InP/GaInAsP linear array with twelve individually addressable elements, *Electron. Lett.*, **24**, 217 (1988).
3. J. P. van der Ziel, R. A. Logan and R. M. Mikulyak, A closely spaced (50 μm) array of 16 individually addressable buried heterostructure GaAs lasers, *Appl. Phys. Lett.*, **41**, 9 (1982).
4. T. Kajimura, S. Yamashita, T. Kuroda, M. Nakamura and J. Umeda, GaAlAs visible laser arrays, *IEEE Trans. Electron Devices*, **ED-27**, 2181 (1980).
5. R. A. Bartolini, Optical recording: high-density information storage and retrieval, *IEEE Proc.*, **70**, 589 (1982).
6. O. Bessette and D. Thomas, 2500 gigabyte capacity 6 Mbyte/sec. data rate, dual port NASA optical disk jukebox, *Optical Data Storage Topical Meeting*, Vancouver, Canada, *IEEE Tech. Digest Paper 2.1* (1990).
7. D. Botez, J. C. Connolly, D. B. Gilbert, M. G. Harvey and M. Ettenberg, High-power individually addressable monolithic array of constricted double heterojunction large-optical-cavity lasers, *Appl. Phys. Lett.*, **41**, 1040 (1982).
8. B. Goldstein, G. N. Pultz, D. B. Carlin, S. E. Slavin and M. Ettenberg, AlGaAs heterojunction lasers, *NASA Contractor Report 4119*, Langley Research Center (1988).
9. See, for example, G. H. B. Thompson, *Physics of Semiconductor Laser Devices*, John Wiley & Sons, Chichester (1980).
10. D. B. Carlin, Spot-forming efficiency of optical recording systems using high-power diode lasers, *Proc. SPIE*, **382**, 211 (1983).
11. See, for example R. Kingslake, *Applied Optics and Optical Engineering*, Vol. V *Optical Instruments, Part II*, Academic Press, New York (1969).
12. See, for example, M. Born and E. Wolf, *Principles of Optics*, Pergamon Press, Oxford (1975).
13. D. D. Cook and F. R. Nash, Gain-induced guiding and astigmatic output beam of GaAs lasers, *J. Appl. Phys.*, **46**, 1660 (1975).
14. K. Tatsuno and A. Arimoto, Measurement and analysis of diode laser wave fronts, *Appl. Opt.*, **20**, 3520 (1981).
15. M. L. Levene, High performance optical disk recorder – preliminary test results and spaceflight model projections, *Optical Data Storage Topical Meeting*, Vancouver, Canada, *IEEE Tech. Digest Paper 2.2* (1990).
16. See, for example, M. Fukuda, *Reliability and Degradation of Semiconductor Lasers and LEDs*, Artech House, Norwood, MA (1991).
17. M. Ettenberg, A statistical study of the reliability of oxide-defined stripe cw lasers of (AlGa)As, *J. Appl. Phys.*, **50**, 1195 (1979).
18. D. B. Carlin, J. P. Bednarz, C. J. Kaiser, J. C. Connolly and M. G. Harvey, Multichannel optical recording using monolithic arrays of diode lasers, *Appl. Opt.*, **23**, 3994 (1984).
19. D. B. Carlin, B. Goldstein, J. P. Bednarz, M. G. Harvey and N. A. Dinkel, A ten-element array of individually addressable channeled-substrate-planar AlGaAs diode lasers, *IEEE J. Quantum Electron.*, **QE-23**, 476 (1987).
20. R. Katayama, private communication, 1 April 1991.
21. See, for example, *CRC Handbook of Chemistry and Physics*, David R. Lide (ed.), pp. A99–101, CRC Press, Boca Raton, FL (1991).

22. S. Ishikawa, M. Nido, K. Endo, I. Komazaki, K. Fukagai and T. Yuasa, 830 nm high-power low-noise self-aligned AlGaAs/GaAs double-quantum-well lasers, *Electron. Lett.*, **25**, 1398 (1989).
23. A. Arimoto, M. Ojima, N. Chinone, A. Oishi, T. Gotoh and N. Ohnuki, Optimum conditions for the high frequency noise reduction method in optical videodisc players, *Appl. Opt.*, **25**, 1398 (1986).
24. M. Ojima, A. Arimoto, N. Chinone, T. Gotoh and K. Aiki, Diode laser noise at video frequencies in optical videodisc players, *Appl. Opt.*, **25**, 1404 (1986).
25. D. Mehuys, D. F. Welch, D. W. Nam and D. R. Scifres, High power (>100 mW), low crosstalk (<0.4%), individually addressed laser diodes, *Electron. Lett.*, **26**, 1955 (1990).
26. R. A. Laff, L. D. Comerford, J. D. Crow and M. J. Brady, Thermal performance and limitations of silicon-substrate packaged GaAs laser arrays, *Appl. Opt.*, **17**, 778 (1978).
27. W. B. Joyce and R. W. Dixon, Thermal resistance of heterostructure lasers, *J. Appl. Phys.*, **46**, 855 (1975).
28. T. Yamaguchi, K. Yodoshi, K. Minakuchi, Y. Inoue, N. Tabuchi, K. Komeda, H. Hamada and T. Niina, Monolithic four-beam semiconductor laser array with built-in monitoring photodiodes, *SPIE*, **1043**; *Laser Diode Technology and Applications*, p. 17 (1989).
29. D. R. Scifres, F. A. Ponce and W. Stutius, Integrated output power detection for AlGaAs laser array, *IEEE J. Quantum Electron.*, **QE-16**, 502 (1980).
30. K. Minakuchi, Y. Bessho, Y. Inoue, K. Komeda, N. Tabuchi, K. Tominaga, A. Tajiri, K. Yodoshi and T. Yamaguchi, High-power, 790 nm, eight-beam AlGaAs laser array with a monitoring photodiode, *Technical Digest of the International Symposium on Optical Recording, 1991*, Sapporo, Japan, p. 21 (1991).
31. R. L. Thornton, W. J. Mosby, R. M. Donaldson and T. L. Paoli, Properties of closely spaced independently addressable lasers fabricated by impurity-induced disordering, *Appl. Phys. Lett.*, **56**, 1623 (1990).
32. S. Murata and K. Nishimura, Improvements in thermal properties of a multi-beam laser diode array, *Japan J. Appl. Phys.*, **28**, Supplement 28-3, p. 165 (1989).
33. S. Murata and K. Nishimura, A simple new laser diode array model for thermal interaction analysis, *J. Appl. Phys.*, **70**, 4715 (1991).
34. S. Murata and K. Nishimura, Thermal improvement of a 50 μm-spaced 8-beam laser diode array by a heat pass wire, *Tech. Digest CLEO '91, Paper CWF38*, Optical Society of America (1991).
35. K. Isshiki, T. Kamizato, A. Takami, A. Shima, S. Karakida, H. Matsubara and W. Susaki, High-power 780 nm window diffusion stripe laser diodes fabricated by an open-tube two-step diffusion technique, *IEEE J. Quantum Electron.*, **26**, 837 (1990).
36. D. Botez, cw high-power single-mode operations of constricted double-heterojunction AlGaAs lasers with a large optical cavity, *Appl. Phys. Lett.*, **36**, 190 (1980).
37. R. A. Bartolini, A. E. Bell and F. W. Spong, Diode laser optical recording using trilayer structures, *IEEE J. Quantum Electron.*, **QE-17**, 69 (1981).
38. K. Aiki, M. Nakamura, T. Kuroda, J. Umeda, R. Ito, N. Chinone and M. Maeda, Transverse mode stabilized $Al_xGa_{1-x}As$ injection lasers with

channeled-substrate-planar structure, *IEEE J. Quantum Electron.*, **QE-14**, 89 (1978).
39. J. J. Yang, C. S. Hong, J. Niesen and L. Figueroa, High-power operation of index-guided inverted channel substrate planar (ICSP) lasers, *Electron. Lett.*, **21**, 751 (1985).
40. G. Alphonse, D. B. Carlin and J. C. Connolly, Linear laser diode arrays for improvement in optical disk recording for space stations, *NASA Contractor Report 4322*, Langley Research Center (1990).
41. M. Tsunekane, K. Endo, S. Ishikawa, R. Katayama, K. Yoshihara, K. Kubota and T. Yuasa, Monolithic eight-channel high-power low-astigmatism AlGaAs laser diode array, *Japan. J. Appl. Phys.*, **28**, L468 (1989).
42. R. Katayama, K. Yoshihara, Y. Yamanaka, M. Tsunekane, K. Yoshida and K. Kubota, Multi-beam magneto-optical disk drive for parallel read/write operation, *SPIE*, **1078**, 98 (1989).
43. M. Tsunekane, K. Endo, M. Nido, I. Komazaki, R. Katayama and K. Yoshihara, High-power individually addressable monolithic laser diode array, *Electron. Lett.*, **25**, 1091 (1989).
44. M. Tsunekane, S. Ishikawa, K. Yoshihara, R. Katayama, K. Endo and T. Yuasa, High-power individually addressed monolithic AlGaAs laser diode arrays for optical disc memories, *IEEE LEOS Annual Meeting, Paper SDL5.3*, Boston, MA (1990).
45. K. Isshiki, T. Kadowaki, A. Takami, S. Karakida, T. Kamizato, Y. Kokubo and M. Aiga, High-power, low-threshold current, individually addressable monolithic four-beam array of GaAlAs window diffusion stripe lasers, *Tech. Digest CLEO '91, Paper CTuQ1*, Optical Society of American (1991).
46. K. Isshhiki, A. Takami, S. Karakida, T. Kamizato, S. Kakimoto and M. Aiga, High-power individually addressable monolithic four-beam array of GaAlAs window diffusion stripe lasers, *IEEE J. Quantum Electron.*, **QE-28**, 804–10 (1992).
47. D. G. Mehuys, D. F. Welch, R. A. Parke and D. R. Scifres, High-power, individually addressed, dual channel laser diodes, *Tech. Digest CLEO '90, Paper CFA7*, Optical Society of America (1990).
48. D. W. Nam, R. R. Craig, D. G. Mehuys and D. F. Welch, Uniform high power nine and 18 element individually addressable laser diode arrays, *Electron. Lett.*, **27**, 464 (1991).
49. D. W. Nam, R. R. Craig, D. G. Mehuys, D. F. Welch and D. Scifres, Uniform high power 9 and 18 element individually addressable laser diode array, *Tech. Digest CLEO '91, Paper CTuQ2*, Optical Society of America (1991).
50. R. Craig, D. Mehuys, D. Nam, E. Zucker, B. Chan and D. Welch, High power addressable laser arrays, *Optical Data Storage Topical Meeting, Technical Digest Series*, Volume 5, Colorado Springs, CA, p. 178 (1991).
51. J. H. Kim, R. J. Lang, L. P. Lee and A. A. Narayanan, High-power AlGaAs/GaAs single quantum well 45°-beam deflecting surface-emitting lasers, *LEOS '90 Conference Digest*, p. 208, Boston, MA (1990).
52. M. Jansen, J. J. Yang, S. S. Ou, M. Sergant, L. Mawst, J. Rosenbergs and J. Wilcox, Monolithic surface emitting devices with etched micromirrors, *LEOS '90 Conference Digest*, p. 210, Boston, MA (1990).
53. D. W. Nam, R. G. Waarts, D. F. Welch, J. S. Major, Jr and D. R. Scifres, Operating characteristics of high continuous power two-dimensional surface emitting laser arrays, *Photon. Tech. Lett.*, **5** (3), 281–4 (1993).

54. G. A. Evans, D. P. Bour, N. W. Carlson, R. Amantea, J. M. Hammer, H. Lee, M. Lurie, R. C. Lai, P. F. Pelka, R. E. Farkas, J. B. Kirk, S. K. Liew, W. Reichert, C. A. Wang, H. K. Choi, J. N. Walpole, J. K. Butler, W. F. Ferguson, Jr, R. K. DeFreez and M. Felisky, Characteristics of coherent two-dimensional grating surface emitting diode laser arrays during cw operation, *IEEE J. Quantum Electron.*, **QE-27**, 1594 (1991).
55. D. F. Welch, D. Mehuys, R. Parke, R. Waarts, D. Scifres and W. Streifer, Coherent operation of monolithically integrated master oscillator amplifiers, *Electron. Lett.*, **26**, 1327 (1990).
56. N. W. Carlson, J. H. Abeles, D. P. Bour, S. K. Liew, W. F. Reichert, P. S. D. Lin and A. S. Gozdz, Demonstration of a monolithic, grating-surface-emitting laser master-oscillator-cascaded power amplifier array, *IEEE Photon. Tech. Lett.*, **2**, 708 (1990).
57. N. W. Carlson, R. Amantea, G. A. Evans, D. P. Bour and S. K. Liew, Applications of surface emitting lasers to coherent communications systems, *LEOS '90 Conference Proceedings*, p. 406, Boston, MA (1990).
58. D. Mehuys, D. F. Welch, R. Parke, R. G. Waarts, A. Hardy and D. Scifres, High power, diffraction-limited emission from monolithically integrated active grating master oscillator power amplifier, *Electron. Lett.*, **27**, 492 (1991).
59. S. K. Liew, N. W. Carlson, R. Amantea, D. P. Bour, G. A. Evans and E. Vangieson, Operation of distributed-feedback grating surface emitting laser with a buried grating structure, *LEOS '90 Conference Proceedings*, Boston, MA, p. 410 (1990).

Index

active grating, 85
amplified spontaneous emission (ASE), 99, 114
amplifier arrays
  cascaded, 74
  two-dimensional, 82, 83
amplifiers, filamentation in, 99
analytic matching technique, 191, 197
anamorphic correction, 419
antiguide
  built-in, 22
  carrier-induced, 351
antiguided arrays, 22, 25, 26, 200, 237, 359
  stabilization of, 239, 248
antiguiding factor, b, 352
aperture filling, 144
  microlenses for, 144
  phase filtering for, 145
  Talbot effect for, 147
array modes, 5, 186
  evanescent-type, 7, 10, 14, 31
  leaky-type, 7, 14, 28, 37, 40
arrays of antiguides, 22
  diffraction-limited power, 53, 56
  distributed feedback analogy, 37
  dynamics of, 238, 246
  gain-guided structures, 25, 62, 219, 297, 357, 359, 400
  injection-locking of, 61
  intermodal discrimination, 35, 40
  modulation bandwidth, 61
  nonresonant, 56
  percentage of energy in main lobe, 55
  real-index-guided structures, 26, 219
  resonant, 33, 53
  rigorous modeling, 50, 197, 200
  single-frequency operation of, 61
  Talbot-type spatial filter, 47, 49
  threshold-current density, 49
  two-dimensional optical confinement factor, 30, 42
  two-dimensional surface emitting, 62, 218, 221, 401
arrays of positive-index guides, 7
  coupling coefficient, 11, 229
  diffraction-limited beamwidth, 12
  diffraction coupled, 2, 19
  dynamics of, 230, 239, 244, 248
  evanescent-wave coupled, 2, 8
  percentage of energy in main lobe, 12
  tree-type, 17
  two-dimensional surface emitting, 218, 400
  uniform-intensity profile, 21
  X-junction coupled, 20
  X–Y-junction coupled, 17
  Y-junction coupled, 2, 17, 239

beam-propagation method, 193
bit-error rate (BER), 399, 400
Bloch-function array analysis, 28, 46, 189
boundary conditions at dielectric interfaces, 185
broad area amplifiers
  non-tapered, 96–103
  tapered, 104–109
broad area lasers
  single-stripe, 211, 228, 296, 299, 305–7, 360
  arrays of, 305, 357

carrier diffusion, 207
catastrophic optical damage (COD), 288–90, 296, 298, 314, 322
catastrophic optical mirror damage, *see* catastrophic optical damage
CDH–LOC laser, 425, 430
chaotic pulsations, 235, 236
characteristic temperature, T_0, 208, 305
  AlGaAs/GaAs structures, 208, 256, 306
  InGaAs(P)/InP structures, 208, 306, 308, 361

InGaAlP/GaAs structures, 306
InGaAs/AlGaAs/GaAs structures, 306, 348
CHIC cooler, 263, 266–70, 274
coherence, effects of, 125
collimation lens, 417, 418
compressive strain, 339, 362
concentration limits, light, 163
concentration systems, light, 168
coordinate transforming optics, 168
coupled stripe arrays, *see* phase-locked laser arrays
coupled-mode theory, 10, 188, 228
coupling, *see* interelement coupling
coupling coefficient, 228
coupling-induced instabilities, 226, 243
critical thickness, 308, 338
crosstalk
edge-emitting arrays, 424–6, 428
VCSEL arrays, 388, 399, 407–9
crystal defects, 311
CSP laser, 430
current spreading, 205

degradation mechanisms, 288–90, 308–16
diffraction efficiency, microlens, 137
diffusion equation, 235
discrete-element MOPA, 96
double-pass, 97
single-pass, 97
filamentation in, 99
flared, 104
distributed Bragg reflector (DBR), 75, 109, 370, 371, 382, 388, 392, 396, 398
double-pass amplifiers, 97
duty cycle, 261, 168, 270, 277, 285, 301, 302, 324, 418
amplifiers, 81, 98, 100
duty factor, *see* duty cycle

effective index method, 185
electrical isolation, 425
environmental tests, 326
erasable disk systems, 415
etendue, 163

Fabry–Pérot mode, 371, 382, 383, 388, 398
facet coating,
amplifiers, 97, 104, 110
lasers, 55, 297
laser lifetime as a function of, 288, 314
far-field beam-pattern formulation
antiguided phase-locked arrays, 55
positive-index-guided phase-locked arrays, 11
fast-Fourier-transform (FFT) methods, 194
fiber lasers, pumping of, 171
fibers, illumination of, 164, 171
field-of-view, 418
fill factor, effects of, 128, 156
flared-amplifier MOPA
discrete, 104, 109
monolithic, 109
spectral stability of, 117
virtual-source displacement with drive level, 106, 114
Fox–Li iteration, 193
Fresnel zones, 136

gain coefficient, β, 49, 309, 351
gain-guided arrays, 3, 25, 62, 296, 357, 359, 374, 400
modeling of, 191, 193, 196, 235
gain saturation
in amplifiers, 77, 87, 110
in lasers, 207
in quantum-well structures, 350
gain spatial hole burning, 7, 207
antiguided arrays, 55, 56
positive-index-guided arrays, 10, 19
gain-sheet approximation, 194
grating-outcoupled surface-emitting laser arrays, 64, 357, 436
grating output coupler, 76, 436
grating surface emitter, 436
gratings, 141
surface relief, 141
volume, 143

hermeticity, 326
heterostructure-discontinuity fraction, 343
high-average-power operation of incoherent arrays, 280, 285, 287
hole effective mass, 346
holograms, beam combining with, 143
hydraulic power, 274, 275

ICSP laser, 434
illuminators, incoherent, 170
individually addressable arrays
collimation lens, 417, 418
edge-emitting, 298, 414
electrical isolation, 325
field of view, 418
horizontal-cavity surface emitting, 435
laser reliability, 420
optical interactions, 425
parallel data transmission, 414
spatially coherent, 62
submounts, 427
temporally coherent, 83, 250, 357, 436
thermal crosstalk, 428
vertical-cavity surface emitting, 374, 402
wavelength stability, 423
incoherent arrays, one-dimensional, 255, 294, 300, 301, 305, 357

$In_xGa_{1-x}As$, 337
injected signal, detuning of, in M-MOPA device, 75
interelement coupling, 2, 3
  diffraction, 2, 19
  evanescent-wave, 2, 8
  leaky-wave, 1, 37, 425
  resonant-leaky-wave, 33, 62
  types of overall, 3
  Y-junction, 2, 17

Kerr effect, 423

laser bar, 263, 300
laser cavity, external, 142, 152
leaky-mode arrays, *see* arrays of antiguides; antiguided arrays; gain-guided arrays
lifetest results
  coherent arrays, 55
  incoherent arrays, 261, 285, 288–90, 320–5. 420
  single-element diodes, 288–90
  indium-containing quantum-well structures, 289, 290, 355
lifetests, *see* lifetest results
light-current characteristics
  amplifiers, 81, 90, 94, 98, 101, 105, 111, 117
  individually addressable edge-emitting arrays, 433, 435
  monolithic laser bar, 279, 281, 300
  monolithic two-dimensional array, 278, 304
  phase-locked antiguided array, 57
  vertical-cavity surface emitter, 381, 405
linear stability analysis, 230
linewidth enhancement factor α, 207, 228
locking range, 229, 230

M^2, 114
magneto-optic recording systems, 423
mass transport, 131
master oscillator power amplifiers (MOPAs), 1, 72
  discrete-element MOPA, 96
  monolithically integrated (M-MOPA), 74, 85
  monolithically integrated active grating (MAG-MOPAs), 85
  monolithically integrated flared-amplifier (MFA-MOPAs), 109
mean-time-to-failure (MTTF), 288, 318, 421
method of lines, 199
microchannel cooler, 263, 266–77
microlenses, 129
  anamorphic, 138
  cylindrical, 131
  diffractive, 135
  fiber, 131
  GRIN, 133
  mass transport, 131
  photolytic, 129
  photoresist, 129
microoptics, 128
  diffractive, 135
  refractive, 129
misfit, 339
mode-hop noise, 423
modeling of coherent diode-laser arrays
  analytic matching technique, 191, 197
  beam-propagation method, 193
  boundary conditions at dielectric interfaces, 185
  Bloch-function analysis, 28, 46, 189
  coupled-mode theory, 10, 188
  effective-index method, 185
  fast-Fourier-transform (FFT) methods, 194
  Fox–Li iteration, 193
  gain-sheet approximation, 194
  method of lines, 199
  paraxial approximation, 184
  Prony method, 196
  separation of variables, 186, 197, 214
  thermal effects, 208
  triangular mesh modeling techniques, 202
  two-dimensional analytic matching, 197
  vertical-cavity surface emitters, 213
modulation response, 61, 408
monolithic two-dimensional incoherent arrays, 271, 276, 277, 302, 358
multichannel optical recording, 415
  array size, 420
  astigmatism, 419
  Kerr effect, 423
  mode-hop noise, 423
  numerical aperture, 417, 418
multiple channels, 415
multiple stripe array, *see* gain-guided arrays

Nd:YAG, 257, 259, 283, 337, 362
neutron irradiation, 327
nonabsorbing-mirror (NAM) laser, 288, 314, 435
nonradiative recombination, 207
numerical aperture (NA), 417, 418

optical data storage, 414
  systems, 414, 415

parallel coupling, 3, 4, 62, 64
parallel data transmission, 414
paraxial approximation, 184

phase control, laser array, 161
phase equation, 229
phase-locked laser arrays, 1, 297, 400
  antiguided, 22, *see also* arrays of antiguides
  comparison to broad-area coherent devices, 1
  modeling of, 180
  positive-index guided, 7, *see also* arrays of positive-index guides
  Talbot-cavity, 149–61
  two-dimensional surface emitting, 62, 218, 219, 221, 400
  types of monolithic, 1
  vertical-cavity surface-emitting, 212, 218, 219, 221, 400
phase locking, 1, 229, 297
  stability of evanescent-wave type, 230
photorefractive effect, 143
π phase-shifters, 16, 400
plane-wave array analysis, *see* Bloch-function array analysis
polarization resolved near-field, vertical-cavity surface emitter, 380
polarization, vertical-cavity surface emitter, 378
power conversion efficiency, η_T, 309
  flared amplifiers, 110
  coherent arrays, 55
  incoherent arrays, 280
  monolithic two-dimensional incoherent arrays, 271
  single-element edge-emitting laser, 309, 434
  vertical-cavity surface-emitting laser, 381, 410
  wide-stripe edge-emitting laser, 309
Prony method, 196
propagation model for laser-array dynamics, 234

quantum-well (QW) amplifier structures, 74, 88, 110
quantum-well (QW) laser structures, 8, 26, 208, 256, 288, 296, 336, 344, 348, 371, 434
  critical thickness in strained-layer, 370
quantum well, 296
quasi-cw, 301, 324

rack and stack incoherent arrays, 277, 279, 302, 324, 357
radiance theorem, 125
radiance, 125
rare-earth-doped fiber amplifiers, 337
rastered multiple wavelength (RMW) laser array, 394, 395, 397
  thickness gradient in, 390, 393
reflection-modulated surface-emitting, coherent arrays, 212, 221
relative intensity noise (RIN), 424
relaxation oscillations, 229
reliability, *see* lifetest results
repetitive Q-switching, 237
rod illumination, 261
ROW array, 33, 39, 61, 64

saturable-absorber induced instabilities, 226, 237, 239, 243, 247, 249
self-pulsing, 230, 236, 239, 240, 249
separately addressed array, *see* individually addressed array
series coupling, 3, 4
sidemode suppression ratio, 378
  flared amplifier, 117
  vertical-cavity surface-emitting laser, 378, 408
single-pass amplifiers, 97
slab illumination, 259–61
solid-state lasers, 257, 337
  pumping of, 169, 173, 257–62
spatiotemporal instabilities, 241
spectral linewidth
  incoherent diode-laser arrays, 283–5
  monolithic surface-emitting amplifiers, 94, 119
  monolithic flared amplifiers, 117
  vertical-cavity surface-emitting lasers, 378, 379, 383, 385, 386, 408
stacked arrays, *see* rack-and-stack incoherent arrays
stacking density or pitch, incoherent arrays, 268, 270, 275
stimulated emission, diode lasers, 207
strained-layer lasers, 336
  visible, 362
  long-wavelength, 361
strained layer, 296, 338
  quaternary alloys, 362
streak camera measurements, 241, 242
Strehl ratio, 128
submounts, 427
supercritical Hopf bifurcation, 228, 230
superlattice reflector, 74, 83
superposition, laser beam, 140
suppression of self-pulsing, antiguided arrays, 239, 248
surface-emitting laser arrays
  for optical recording, 437
  grating-outcoupled, 83, 277, 357, 437
  horizontal-cavity angled facet, 62, 271, 276, 277, 302, 358
  hybrid incoherent, 277
  monolithic incoherent, 271, 276, 277, 302, 358

surface-emitting laser arrays (*cont.*)
  monolithic spatially coherent, 62, 212, 218, 219, 400
  monolithic temporally coherent, 83, 250, 357, 436
  reflection-modulated, coherent, 212
  vertical cavity, 212, 358, 368

Talbot cavity, 149
  fill factor considerations, 153
  linear, 158
  modal description of, 155
  two-dimensional, 160
Talbot effect, 147, 151
Talbot effect, fractional, 148
Talbot-type spatial filter, 47, 49, 155
tapered amplifier, *see* flared-amplifier MOPA
TE/TM modes, 185
temperature cycle tests, 326
tensile strain, 362
thermal effects, 208
thermal resistance, 263–5, 274, 427, 430
thermal rollover, 299, 307, 308
thermoelectric cooler, 263, 266
threshold current, 308
threshold-current density
  quantum-well structures, 296
  strained-layer structures, 296, 344, 350, 362
  antiguided-array structures, 49
transparency current density, 49, 309, 347
triangular mesh modeling techniques, 202
two-dimensional amplifier arrays, 82
two-dimensional incoherent arrays, 257–61
  CHIC cooler, 263, 266–70, 274
  high-average-power operation, 280, 285, 287
  hybrid, 277
  hydraulic power, 274, 275
  intensity uniformity, 259–61
  microchannel cooler, 263, 266–77
  monolithic, 271, 276, 277, 302, 358
  packaging, 263, 302
  performance, 279–84, 302, 324
  power requirements, 257
  rack and stack, 277, 279, 302, 324, 357
  spectral linewidth, 283–5
  stacking density or pitch, 268, 270, 275
  thermal power dissipation, 281
  thermal resistance, 263–6, 274
  wavelength uniformity, 259–61
T_0, *see* characteristic temperature

vertical cavity surface-emitting laser (VCSEL)
  bottom-emitting, 371, 381
  buried heterostructure, 410, 411
  coupled-cavity laser, 382
  distributed Bragg reflector (DBR), 370, 371, 372
  Fabry–Pérot mode, 371, 383, 388, 392, 396
  gain-guided, 373, 377, 378
  index-guided, 373, 410, 411
  long-wavelength, 411
  modulation response, 408
  near-field patterns, 377–80
  polarization, 378
  sidemode suppression, 378, 407
  spectral linewidth, 378, 379, 382, 385, 386, 408
  time-resolved spectra, 383–6
  top-emitting, 371, 381
  transverse modes, 377, 379, 410
  visible wavelength, 411
  wavelength chirp, 385, 408
vertical cavity surface emitting laser (VCSEL) arrays, 212, 358, 368
  bottom-emitting, 374, 409
  chirped layer, 395, 396
  crosstalk in, 388, 399, 407–9
  distributed Bragg reflector (DBR), 388, 392, 396, 398
  flip-chip bonding, 375, 405
  gain-guided, 374, 400
  graded thickness, 390, 393
  independently addressable, 374, 402
  index-guided, 373
  matrix-addressable, 375, 403
  modeling of phase-locked, 213
  multiple wavelength, 387, 398, 405
  optical filter for, 398
  rastered multiple wavelength, 394, 395, 397
  top-emitting, 403
  two-dimensional phase-locked, 212, 218, 219, 400
  wavelength chirp in, 408
  wavelength-division-multiplexing (WDM), 369, 388, 398

wall plug efficiency, *see* power-conversion efficiency
wavelength-division-multiplexing (WDM), 369, 388, 398
window laser, *see* nonabsorbing-mirror (NAM) laser

Y-guide laser arrays, 2, 17, 147, 239
YAG, 257–9, 261

Zernike phase contrast, 162